Cognitive Technologies

Managing Editors: D.M. Gabbay J. Siekmann

For further volumes:
http://www.springer.com/series/5216

Dov M. Gabbay

Reactive Kripke Semantics

Springer

Dov M. Gabbay
Department of Computer Science
Bar-Ilan University
Israel

King's College London
UK

University of Luxembourg
Luxembourg

Managing Editors

Dov M. Gabbay
Department of Computer Science
Bar-Ilan University
Israel

King's College London
UK

University of Luxembourg
Luxembourg

Jörg Siekmann
Deutsches Forschungszentrum
für Künstliche Intelligenz (DFKI)
Universität des Saarlandes
Saarbrücken
Germany

ISSN 1611-2482 Cognitive Technologies
ISBN 978-3-642-41388-9 ISBN 978-3-642-41389-6 (eBook)
DOI 10.1007/978-3-642-41389-6
Springer Heidelberg New York Dordrecht London

Library of Congress Control Number: 2013954970

Printed on acid-free paper

Springer is part of Springer Science+Business Media (www.springer.com)

Contents

Preface

The concept of reactivity has its roots in the ideas of my book on fibring logics [57]. When we fibre the two logics, say intuitionistic $\to$ and modal logic $\Box$, we can form formulas using both connectives, $\to$ and $\Box$. So we can write something like

$$\Box(A \to \Box B).$$

$\to$ should behave like intuitionistic implication and $\Box$ like a modality, say an **S4** modality. The area of fibring logics develops methodologies of how to get properties of the fibred modal (S4) intuitionistic logic from the properties of the individual logics involved. We can also fibre two modal logics, say an **S4** modality and a **K** modality. We can write

$$\Box_{\mathbf{S4}}\Box_{\mathbf{K}}A.$$

The first $\Box$ is an **S4** modality and the second is a **K** modality. We can also write

$$\Box\Box A$$

provided we agree that the first modality is **S4** and the second is **K**.

How about $\Box\Box\Box A$? Well, we need an agreement about the third modality. Once we have such an agreement, we know what $\Box\Box\Box A$ means.

Imagine for a moment that we write arbitrary formulas with one modality only, and that we have an agreement that if $\Box$ is nested in the formula to a depth k it must behave like a modality $\Box_k$. Say $\Box_1$ is a **K** modality. $\Box_2$ is an **S4** modality, $\Box_3$ is a **T** modality and then we repeat in a cycle, i.e. $\Box_{3m+n}$ is a $\Box_n$ modality, $0 \leq n \leq 2$. Thus $\Box(A \to \Box\Box B)$ is $\Box_{\mathbf{K}}(A \to \Box_{\mathbf{S4}}\Box_{\mathbf{T}}B)$. We can write axioms for this $\Box$ modality and show the axiom system to a third party without revealing our interpretation for $\Box$. We ask the question: what kind of semantics we should seek for $\Box$?

Our third party will probably proceed in a traditional manner and look for Kripke models (S, R), $R \subseteq S \times S$ and seek conditions on R. He does not know the complexity of the interpretation we had in mind and will probably fail to find such conditions. (Actually, there are logics for which there are no corresponding conditions on R.)

So what kind of semantical interpretation should we adopt for $\Box$? The idea we came up with is to think of $\Box A$ as telling us to which worlds to go in order to check the truth of A, rather than telling us the properties of R. So our model is a fixed general (S, R). A **K** modality says look at all accessible worlds, while a **T** or an **S4** modality tells us to look at the reflexive or respectively at the reflexive and transitive closures of R. So for example we have

$$\begin{aligned} t \vDash \Box_{\mathbf{K}} A &\text{ iff for all } y \text{ such that } tRy \\ &\quad \text{we have } y \vDash A. \\ t \vDash \Box_{\mathbf{T}} A &\text{ iff for all } y \text{ such that } y = t \text{ or } tRy \text{ we have } y \vDash A. \\ t \vDash \Box_{\mathbf{S4}} A &\text{ iff for all } y \text{ such that } y = t \text{ or for some } x_1, \ldots, x_m \\ &\quad m \geq 0 \text{ such that } tRx_1 \text{ and} \ldots x_{m-1}Ry \text{ we have } y \vDash A. \end{aligned}$$

So $\Box$ means where to look, and the "semantics" for $\Box$ is an algorithm for "where to look".

This work was done in 2002.

Once we think of $\Box$ in this way, as instructing us where to go and look, we can modify and enrich the interpretation by giving modified algorithms as to where to go. For example we can say

$$t \vDash \Box A \text{ iff for all } y \text{ such that } tRy \wedge \neg yRy \text{ we have } y \vDash A$$

We can also put obstacles along the way or give you money to travel and say "go as far as the travel money allows you to go".

To summarise, the basic model is $(S, R), R \subseteq S \times S$.

The traditional approach to semantics for $\Box$ is to associate conditions on R and maintain

$$t \vDash \Box A \text{ iff for all } y\ tRy \to y \vDash A$$

where R satisfies the conditions on R.

The new approach maintains the basic model (S, R) but as semantics for $\Box$ it associates an algorithm $\mathcal{A}$ involving R.

We have

$$t \vDash \Box A \text{ iff for all } y \text{ such that } \mathcal{A} \text{ can reach } y \text{ from } t \text{ we have } y \vDash A.$$

The idea of reactive arrows is a natural way of restricting the algorithm $\mathcal{A}$ (your travel).

Figure 0.1 is a simple illustration.

If you go from t to y then you cannot continue to b. However this is only because you passed through the connection $t \to y$. If you get to y via x, then you can go to b.

This idea was finalised in 2004 and turned out to be very applicable, as the various chapters in this book show. We are still applying the idea in many areas, and we have not exhausted its usefulness yet.

The reactive result depends on the direction of the algorithm.

Consider Figure 0.2

We have in this figure tRx and xRy and the bi-directional reactive double arrow

$$(t \to x) \twoheadleftarrow\!\!\twoheadrightarrow (x \to y).$$

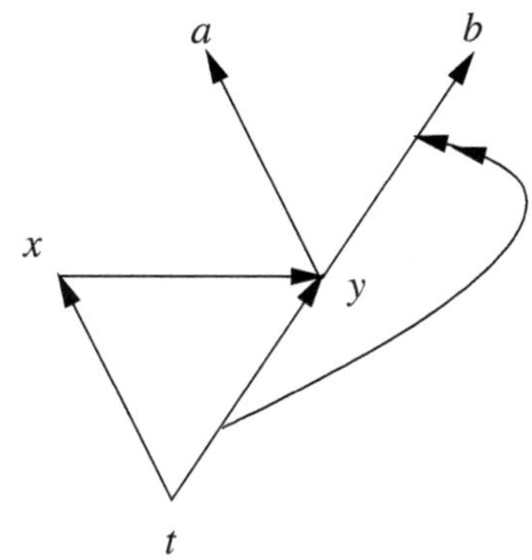

Fig. 0.1

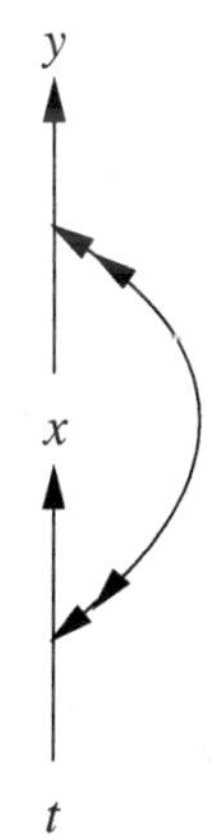

Fig. 0.2

So if we start from y and try to go backwards to t, we have to pass through x and the arrow $t \to x$ will be disconnected.

If we start from t and go to y though x, the connection $x \to y$ will be disabled.

Consider now the question $t \vDash \Box\Diamond\top$. Obviously we need to check whether y is connected to t, as y is the only point at distance 2 from t.

1. If we start at t and move towards y, we get $t \vDash \Box\Diamond\top$ iff $x \vDash \Diamond\top$ iff false, since $x \to y$ gets cancelled.
2. If we move from y backwards, $t \to x$ will get disconnected and therefore $t \vDash \Box\Diamond\top$ will be true.

The detailed theory of reactive Kripke semantics is developed in Chapters 1–7. Chapters 8–12 are applications in a variety of areas. The reader can see how diverse the application areas are!

Some chapters are based on my published or unpublished papers, as follows: Chapter 1 – [69]; Chapter 2 – [66]; Chapter 3 – [89]; Chapter 4 – [68]; Chapter 5 – [90]; Chapter 6 – [62]; Chapter 7 – [91]; Chapter 8 – [37]; Chapter 9 – [17]; Chapter 10 – [63]; Chapter 11 – [79]; Chapter 12 [71].

The research in Chapter 10 was done under the Russian–Israeli Scientific Research Cooperation, Project No. 206891: Combined Modal Logics.

1

A Theory of Hypermodal Logics: Mode Shifting in Modal Logic

1.1 Introduction

Hypermodal logics belong to a special kind of modal logics which appear in a variety of forms in many applications. These logics will be motivated and studied in later sections, but if I were to characterise them in one sentence I would say that these are modalities $\Box$ whose nature changes depending on where they appear in the wff. Thus, for example, when we write the formula $\Box(\Box A \to A)$, the outer modality can be a **T** modality while the inner modality can be a **K** modality.[1]

Semantically this means that evaluating $t \vDash \Box A$ in a Kripke model depends not only on t and on the meaning of $\Box$ (the kind of modality $\Box$ is), but also on how we got to t in the process of evaluation.[2] These distinctions will become clear later in the chapter. Proof-theoretically this kind of dependence means that the set of available proof rules changes as we progress through the proof.

Traces of these ideas have been around for a while. However, this is the first time that they are all addressed in one framework, the concept of *hypermodality* is coined and its formal properties studied.

Since our main example is a system involving **K** and **T** modalities, let us remind ourselves of some convenient semantic and proof-theoretical formulations of these systems.

Definition 1.1 (Semantics for strict implication).

1. *A Kripke propositional model has the form* $\mathbf{m} = (S, R, a, h)$, *where* S *is the set of possible worlds and* $R \subseteq S^2$ *is the accessibility relation.* $a \in S$ *is the actual world and* h *is the assignment, giving for each atomic* q *a subset* $h(q) \subseteq S$.

[1] The word 'hyper' was suggested to me by Michael Gabbay. It can be used to qualify the logical meaning of any symbol x and make it dependent on its occurrence in the wff. So, for instance, a *hypernominal* is a nominal whose meaning depends on its place in the wff. Similarly a *hyperSkolem-function*, *hyperquantifier*, etc.

[2] As we evaluate $s \vDash B$ inductively from the outside, the evaluation path leading to any $t \vDash \Box A$ indicates the position of this occurrence of $\Box A$ inside B.

D.M. Gabbay, *Reactive Kripke Semantics*, Cognitive Technologies,
DOI 10.1007/978-3-642-41389-6_1, © Springer-Verlag Berlin Heidelberg 2013

2. *Satisfaction for $t \in S$ is defined inductively as follows:*
   - $t \vDash q$ *iff* $t \in h(q)$ *for* q *atomic.*
   - $t \vDash A \wedge B$ *iff* $t \vDash A$ *and* $t \vDash B$
   - $t \nvDash \bot$
   - $t \vDash A \Rightarrow B$ *iff* $\forall s(tRs$ *and* $s \vDash A$ *imply* $s \vDash B)$.
   - $\mathbf{m} \vDash A$ *iff* $a \vDash A$.

   Here $\Rightarrow$ *is strict implication and* $\Box A$ *can be defined as* $(\top \Rightarrow A)$.
3. *We have completeness as follows, see [50].*
   - $\mathbf{K} \vDash A$ *iff in all models* $\mathbf{m} = (S, R, a, h)$ *we have* $\mathbf{m} \vDash A$.
   - $\mathbf{T} \vDash A$ *iff in all models* $\mathbf{m}$ *with* R *reflexive (i.e.* $(\forall x)(xRx)$ *holds) we have* $\mathbf{m} \vDash A$.
   - $\mathbf{S4} \vDash A$ *iff in all models in which* R *is reflexive and transitive we have* $\mathbf{m} \vDash A$.

Definition 1.2 (Proof theory for strict implication). *Consider for simplicity the language with* $\Rightarrow$ *only.*

1. *A data structure is a list of sets of sentences of the form* $(\Delta_1, \ldots, \Delta_n)$. *Our proof rules define the consequence relation* $(\Delta_1, \ldots, \Delta_n) \vdash B$. *The semantic meaning of* $\vdash$ *is as follows:*
   - $(\Delta_1, \ldots, \Delta_n) \vDash_{\mathbf{L}} B$ *iff for all Kripke models* $\mathbf{m}$ *of the logic* $\mathbf{L}$ *and all points* $t_1, \ldots, t_n$ *in* S *such that* $t_1Rt_2, t_2Rt_3, \ldots, t_{n-1}Rt_n$ *and such that* $t_i \vDash \Delta_i$, *we have that* $t_n \vDash B$.
2. *To see what kind of proof rules we can expect, consider the following.*
   Suppose $A \Rightarrow B \in \Delta_i$ *and* $A \in \Delta_j$. *This means semantically that* $t_i \vDash A \Rightarrow B$ *and* $t_j \vDash A$.
   If in the model t_iRt_j *holds, then also* $t_j \vDash B$. *If* $\mathbf{L} = \mathbf{K}$ *then* t_iRt_j *is assured to hold only if* $j = i + 1$. *If* $\mathbf{L} = \mathbf{T}$, *then* t_iRt_j *holds necessarily only if* $i \leq j \leq i + 1$. *If* $\mathbf{L} = \mathbf{S4}$ *then* t_iRt_j *is assured to hold only if* $i \leq j$.
   We thus get the following rule of $\Rightarrow E$:
   $(\Rightarrow E)$
   $$\frac{\Delta_1, \ldots, \Delta_i \cup \{A \Rightarrow B\}, \ldots, \Delta_j \cup \{A\}, \ldots, \Delta_n \vdash D}{\Delta_1, \ldots, \Delta_i \cup \{A \Rightarrow B\}, \ldots, \Delta_j \cup \{A, B\}, \ldots, \Delta_n \vdash D}$$
   Provided $j = i + 1$ *for* $\mathbf{K}$, $i \leq j \leq i + 1$ *for* $\mathbf{T}$ *and* $i \leq j$ *for* $\mathbf{S4}$.
3. *What would be the form of* $\Rightarrow I$, *that is the* $\Rightarrow$ *introduction rule? Suppose we want to show that* $\Delta_1, \ldots, \Delta_n \vDash A \Rightarrow B$. *This means that in* any *model* $\mathbf{m}$, *and* $t_1, \ldots, t_n$ *as before,* $t_n \vDash A \Rightarrow B$. *This means that for any* t_{n+1} *such that* t_nRt_{n+1} *and* $t_{n+1} \vDash A$ *we must have* $t_{n-1} \vDash B$. *But this simply means* $\Delta_1, \ldots, \Delta_n, \{A\} \vDash B$.
   We thus get the rule
   $(\Rightarrow I)$ $\quad (\Delta_1, \ldots, \Delta_n) \vDash A \Rightarrow B$ *iff* $(\Delta_1, \ldots, \Delta_n, \{A\}) \vDash B$.
   For completeness of these rules, see [55].
4. *Let us now look at the proof rules from the* Δ_j *point of view (i.e. as if we are 'living' in the world* t_j*). Suppose we have argued successfully and shown* A *is available to us in world* t_j*; (i.e. in* Δ_j*). We want to use an* $A \Rightarrow B \in \Delta_i, i \leq j$, *and perform modus ponens and get* $B \in \Delta_j$. *We first have to ask in which logic we are operating. If the logic is* $\mathbf{K}$, *then* $A \Rightarrow B \in \Delta_i$ *is usable to us iff* $j = i + 1$. *If*

the logic is **T**, *we can use* $A \Rightarrow B$ *also if it is in* Δ_j. *If the logic is* **S4**, *we can use any* $A \Rightarrow B$ *in any* Δ_i, *for any* $i \leq j$.
How do we show that $X \Rightarrow Y$ *holds at* Δ_j? *We start a new theory* Δ_{j+1}, *assume* X *is there and try to prove* Y. *Again if the logic is* **K**, *only* $A \Rightarrow B$ *from* Δ_j *can be used in* Δ_{j+1}. *If the logic is* **T**, *we can also use* $A \Rightarrow B$ *which are proved in* Δ_{j+1}. *If the logic is* **S4**, *we can use any* $A \Rightarrow B$ *from* $\Delta_1, \ldots, \Delta_{j+1}$.
In this context it is very easy to define the notion of mode of proof. *The* mode *of proof as applied to proofs in* Δ_{j+1} *tells us from which* Δ_i *we can import and use wffs* $A \Rightarrow B$.
5. *We can now talk about* changing modes. *We can have a rule of the following form*
   - *To show* $A \Rightarrow B$ *in* Δ_j, *when we are in mode* x, *assume* $\Delta_{j+1} = \{A\}$ *and try to prove* B *in a new mode* $y = \varepsilon(x)$, *where* ε *is the mode-changing function.*

The exact meaning of the above will be explained in the next example and beyond.

Let us consider some natural examples involving modes.

Example 1.3 (Modes of Proof with **K** *and* **T***).* This example illustrates how proof theory can change modes. The modal logics **K** and **T** are used. Consider modal logic formulated with strict implication $\Rightarrow$, conjunction $\wedge$, $\top$ and $\bot$.

Focus on a natural deduction proof theory for $\Rightarrow$ with a view to making $\Rightarrow$ either a **K** or **T** implication. Consider the following candidate formula to be proved:

$$\alpha = A \wedge (A \Rightarrow (B \Rightarrow C)) \Rightarrow (B \wedge (C \Rightarrow D) \Rightarrow D).$$

Consider the following proof of α in **T**:[3]

1. We show α, from the subproof in lines 1.1–1.4
   1.1. A, assumption
   1.2. $A \Rightarrow (B \Rightarrow C)$, assumption
   1.3. $B \Rightarrow C$, from 1.1 and 1.2. This step is allowed since the logic is **T**. It is not allowed if the logic is **K**.
   1.4. We show $B \wedge (C \Rightarrow D) \Rightarrow D$, from the subproof in lines 1.4.1–1.4.4.
      1.4.1. B, assumption
      1.4.2. $C \Rightarrow D$, assumption
      1.4.3. C, from 1.3 and 1.4.1. This step is available for **K** and for **T**.
      1.4.4. D, from 1.4.2 and 1.4.3. This step is correct if the logic is **T** but not if the logic is **K**.

Thus, α can be proved in the logic **T**.

Imagine now that we have $\Rightarrow$ as a hypermodality where the outermost $\Rightarrow$ is a **K** modality and the inner $\Rightarrow$ is a **T** modality and then the **T** and **K** modalities keep alternating with the nesting of $\Rightarrow$. We can rewrite α more explicitly as α'

[3] In the proof below think semantically. Let **m** be an arbitrary model. Then $a \vDash \alpha$ iff for all t, aRt and $t \vDash$ (1.1 and 1.2 and 1.3) imply $t \vDash$ 1.4. But $t \vDash$ 1.4 iff for any s, tRs, if $s \vDash$ (1.4.1 and 1.4.2) then $s \vDash$ 1.4.4.

Thus proof lines 1.1–1.4 take place in t and 1.4.1–1.4.4 take place in s.

$$\alpha' = A \wedge (A \Rightarrow_{\mathbf{T}} (B \Rightarrow_{\mathbf{K}} C)) \Rightarrow_{\mathbf{K}} (B \wedge (C \Rightarrow_{\mathbf{K}} D) \Rightarrow_{\mathbf{T}} D).$$

Let us now rewrite the proof above, annotating the modalities and proof rules available at each subproof. Our outer proof procedure is a **K** proof procedure. This makes the proof procedure of any immediate $X \Rightarrow Y$ subproof (i.e. of the form: to show $X \Rightarrow Y$: assume X show Y) a **T** procedure.[4]

1. Show α' using **K** rules. This follows from subproof 1.1–1.4 which is carried out using **T** rules:
   1.1. A, assumption
   1.2. $A \Rightarrow_{\mathbf{T}} (B \Rightarrow_{\mathbf{K}} C)$, assumption
   1.3. $B \Rightarrow_{\mathbf{K}} C$, from 1.1–1.2. We can execute modus ponens, since we are using **T** rules which include reflexivity.
   1.4. We show $B \wedge (C \Rightarrow_{\mathbf{K}} D) \rightarrow_{\mathbf{T}} D$ from subproof 1.4.1–1.4.4, using **K** rules (we are alternating the use of **K**, **T** rules as we go through the nesting).
      1.4.1. B, assumption
      1.4.2. $C \Rightarrow_{\mathbf{K}} D$, assumption
      1.4.3. C, from 1.3 and 1.4.1. This step is available using **K** rules.
      1.4.4. We are not allowed to get D from 1.4.2 and 1.4.3, since we are following **K** rules here!

So the proof fails.

The structure of the chapter is as follows: Section 2 will introduce hypermodalities from a variety of semantical points of view. Section 3 will semantically present a case study. Section 4 will discuss translations. Section 5 will axiomatise the case study, and Section 6 will conclude the chapter.

1.2 Hypermodalities

This section introduces hypermodality first intuitively and then more formally.

1.2.1 Intuitive presentation of hypermodality

The traditional semantics for propositional modal logic with $\Box$ (and $\neg, \vee, \wedge, \rightarrow$) offers Kripke models of the form (S, R, a, h), where S is the set of possible worlds, $R \subseteq S^2$ is the accessibility relation, $a \in S$ is the actual world and h is the *assignment*, associating with each atomic q a subset $h(q) \subseteq S$. (S, R, a) is called the *frame* of the model.

Completeness of a logical system **L** relative to a class of Kripke structures $\mathcal{K}$ is defined as follows:

[4] In semantical terms in a model (S, R, a), we are evaluating as follows: $a \vDash_{\mathbf{K}} \alpha'$ iff $\forall s(aRs \wedge s \vDash_{\mathbf{T}} A \wedge (A \Rightarrow_{\mathbf{T}} (B \Rightarrow_{\mathbf{K}} C))$ implies $s \vDash_{\mathbf{T}} B \wedge (C \Rightarrow_{\mathbf{K}} D) \Rightarrow_{\mathbf{T}} D)$.

- *Actual world completeness:*
  $\mathbf{L} \vdash A$ iff for all $(S, R, a, h) \in \mathcal{K}$ we have $a \vDash A$.

Note that we require satisfaction in the actual world only. This may turn out to be important later.

A class $\mathcal{K}$ of models is said to be a class of frames defined by a frame formula $\varphi(a, R)$ (in a possibly higher-order language with a and R) iff we have

- $(S, R, a, h) \in \mathcal{K}$ iff $(S, R, a) \vDash \varphi(a, R)$.

For example we may have $\varphi(a, R) \equiv aRa$. The logic of this class is not normal, (because $\Box A \to A$ is a theorem but $\Box(\Box A \to A)$ is not).

This semantics offers the following truth conditions for $\Box A$ at a world t.

- $t \vDash \Box A$ iff for all s such that tRs, $s \vDash A$.

Let us refer to the above as the *mode of evaluation* of $\Box$ at a possible world.

We now continue with our explanation of hypermodalities.

Consider the configuration in Figure 1.1 in which aRt, aRs and sRt all hold.

Fig. 1.1

Let us now evaluate at a the two wffs $A_0 = \Box\Box\Box q$ and $A_1 = \Box\Box q$.

- $a \vDash A_0$ iff $\forall xy(aRx \wedge xRy \to y \vDash \Box q)$
- $a \vDash A_1$ iff $\forall y(aRy \to y \vDash \Box q)$.

In the course of evaluating $a \vDash A_0$, we come to evaluate $t \vDash \Box q$, after having 'passed through' the points s and a.

In the course of evaluating $a \vDash A_1$, we come to evaluate $t \vDash \Box q$, after having 'passed through' the point a only.

The traditional evaluation of $t \vDash \Box q$ does not care about 'how we got to t' and how many points we 'passed through'.

When evaluating $t \vDash \Box A$ for a hypermodality $\Box$, however, we do care about the 'history' of previous evaluation points.

The *history* of points one passes through in the course of evaluation can be taken as a *historical mode parameter* $\alpha = (a, x_1, \ldots, x_n)$, where $aRx_1 \wedge x_1Rx_2 \wedge \ldots \wedge x_{n-1}Rx_n$ holds. We have

- $x_n \vDash_\alpha \Box A$ iff $\forall x_{n+1}(x_nRx_{n+1} \to x_{n+1} \vDash_{\alpha * (x_{n+1})} A)$, where $*$ is concatenation of sequences.

In traditional modal logic the evaluation at x_n does not depend on α, i.e. $\vDash_\alpha$ equals $\vDash$ for all α.

To give some intuition as to how the evaluation might depend on α, consider the connective *Soon A*. Consider a flow of time $\{1, 2, 3, 4, \ldots, \}$ and let us understand *Soon A* to mean that A will be true soon. The meaning of 'soon' depends on the moment of time. So assume that there exists a function f such that

- $m \vDash \mathit{Soon}\ A$ iff $n \vDash A$ for some n such that $m \leq n \leq f(m)$.

We can assume that f is a function such that $m < f(m)$ for all m. The interval $[m, f(m)]$ is the interval of time considered 'soon' from the point of view of world m. The above gives us a traditional temporal logic, with temporal accessibility relation R where

$$mRn \text{ iff } n \in [m, f(m)].$$

Let us examine the truth value of Soon^2A at time 1. We have:

- $1 \vDash \mathit{Soon}^2A$ iff $\exists n_1, n_2(1 \leq n_1 \leq f(1)$ and $n_1 \leq n_2 \leq f(n_1)$ and $n_2 \vDash A)$.

It may be the case that $n_2 < f(1)$, in which case $n_2 \vDash A$ is a witness for the truth of $1 \vDash \mathit{Soon}\ A$ as well for the truth of $1 \vDash \mathit{Soon}^2A$.

Our evaluation of $n_2 \vDash A$ in traditional temporal logic is independent of whether we come to consider it in the course of evaluating $1 \vDash \mathit{Soon}\ A$ or $1 \vDash \mathit{Soon}^2A$.

Let us now add a twist to our story. Imagine a time traveller who is trying to evaluate $t \vDash \mathit{Soon}\ A$ by actually travelling over the time points $t, t + 1, t + 2, t + 3, \ldots, t + f(t)$. The more he travels, the more impatient he becomes. Suppose in the course of evaluation of $1 \vDash \mathit{Soon}^3q$ he gets to the point of evaluating $n_2 \vDash \mathit{Soon}\ q$ and he is so impatient that he decides to look for q at the points n_2 and $n_2 + 1$ only. Thus we have two modes of evaluation for $n \models \mathit{Soon}\ q$, *impatient mode* and *patient mode*. We have

- $t \vDash_p \mathit{Soon}\ q$ iff $\exists n(t \leq n \leq f(t)$ and $n \vDash q)$.
- $t \vDash_i \mathit{Soon}\ q$ iff $t \vDash q$ or $t + 1 \vDash q$.

To summarise:

The mode of evaluation does not depend only on t but also on how we got to t and ow long it took.

Let us now go back to our background Kripke semantics; to models of the form (S, R, a, h). We are going to introduce evaluation modes into such semantics. It is convenient to regard the relation xRy as a classical formula $\Psi_{\mathbf{K}}(x, R, a, y)$ in the language of the relation R, the individual variables x, y and actual world constant a as follows

- $\Psi_{\mathbf{K}}(x, R, a, y) =_{\text{def}} xRy$

We have

- $t \vDash \Box A$ iff $\forall s(\Psi_{\mathbf{K}}(t, R, a, s) \rightarrow s \vDash A)$.

We can refer to $\Psi_{\mathbf{K}}$ as the mode of evaluation for $\Box$. It is fixed in the semantics and does not change. Intuitively it tells us, for a world x, how to evaluate $\Box A$ at x, namely where to look for worlds y where $y \models A$ must hold. The subscript **K** indicates that this formula is used in the case of **K** modality.

We can think of different formulas Ψ for the mode. Consider for example (see [50], **KB** is also known as **B**, the logic for R symmetric).

- $\Psi_{\mathbf{T}}(x, R, a, y) =_{\text{def}} xRy \vee x = y$
- $\Psi_{\mathbf{K4}}(x, R, a, y) =_{\text{def}} (\exists n \geq 1)xR^n y$.
  Where $xR^n y$ is defined by induction as:
  - $xR^0 y$ iff $x = y$
  - $xR^{n+1} y$ iff $\exists z(xRz \wedge zR^n y)$.
- $\Psi_{\mathbf{KB}}(x, R, a, y) =_{\text{def}} (xRy \vee yRx)$.

Clearly $\Psi_{\mathbf{K4}}(x, R, a, y)$ is not a first-order formula. It defines the transitive closure of R.

One can think of Ψ as changing the accessibility relation from R to $\lambda x \lambda y \Psi(x, y)$. Another way of looking at Ψ is that it gives us a new mode of how to use R in evaluating the truth value of $\Box A$. The latter view is more convenient to use because we will be shifting modes during the evaluation.

Let us write $\models_i$, to mean that the mode Ψ_i is used in the evaluation. Then $\models_{\mathbf{K}}$ for arbitrary frames (S, R, a) yields the logic **K**, $\models_{\mathbf{T}}$ yields the logic **T**, $\models_{\mathbf{KB}}$ yields the logic **KB** and $\models_{\mathbf{K4}}$ yields the logic **K4**.

Note that our starting point is a frame (S, R, a) with an arbitrary R. We define $\Psi_i(x, R, a, y)$ as a binary relation and use it to evaluate $\Box$. Thus in traditional terms the frame we are using is (S, Ψ_i, a) not (S, R, a). When we shift modalities, i.e. change from $t \models_i \Box A$ to $s \models_j \Box A$, it is like shifting from(S, Ψ_i, a) to (S, Ψ_j, a).

We need to be able to recognise the logic $\mathbf{L}_i$, defined by all frames of the form (S, Ψ_i, a), so that we can say that in mode i we let $\Box_i$ be this logic. This presents us with a technical problem. Let **L** be a familiar logic: can it be characterised by a Ψ as the logic of all frames of the form (S, Ψ, a)? We discuss this problem in Section 6.

We now give another example, an historical-mode-dependent modality.

Definition 1.4 (Historical hypermodality). *Let (S, R, a, h) be a Kripke model*

1. *A historical mode parameter α is any sequence of points of the form $\alpha = (t_0 = a, t_1, t_2, \ldots, t_n), n \geq 0$ such that for $i = 0, \ldots, n-1$, $t_i R t_{i+1}$ holds.*
2. *For each α, let $\Psi_\alpha(x, R, a, y)$ be a (possibly second-order) formula with two free variables x, y and constants R and a. Ψ_α is called a* mode formula.
3. *Satisfaction has the form $t \models_\alpha A$, and satisfies the following clauses*
   - *$t \models_\alpha q$ iff $t \in h(q)$, for q atomic.*
   - *$t \models_\alpha A \wedge B$ iff $t \models_\alpha A$ and $t \models_\alpha B$*
   - *$t \models_\alpha \neg A$ iff $t \not\models_\alpha A$.*
   - *$t \models_\alpha \Box A$ iff $\forall s(\Psi_\alpha(t, R, a, s) \to s \models_{\alpha*(s)} A)$.*

   We say A α-holds in the model if $a \models_\alpha A$.

Remark 1.5. The above definition is an extreme case, where different historical parameters α yield different modes Ψ_α. We can regard the family

$$\{\Psi_{(a,t_1,\ldots,t_k,t)}(t, R, a, y)\}$$

as a relation $\Psi(\alpha, y)$ between finite sequences α and points y, and consider the model (S, Ψ, a, h). We have

- $\alpha \vDash \Box A$ iff $\forall s(\Psi(\alpha, s) \to \alpha * (s) \vDash A)$.

Such models arise in connection with generalised quantifiers and automated deduction for modal logic. See [52, 54, 132]. The respective logic is decidable and axiomatisable and has the finite model property.

Example 1.6. Consider the modes

$$\Psi_1(x, y) = \Psi_{\mathbf{K}}(x, y) \equiv xRy$$
$$\Psi_0(x, y) = \Psi_{\mathbf{T}}(x, y) \equiv xRy \vee x = y$$

Let us shift alternately between the two modes, namely

- $t \vDash_i \Box A$ iff $\forall s(\Psi_i(t, s) \to s \vDash_{1-i} A)$

Let $\mathbf{K}[\Psi_{\mathbf{T}}, \Psi_{\mathbf{K}}] = \{A \mid a \vDash_0 A$, in any model $(S, R, a, h)\}$.

To see what our logic does, consider the formula $\Box(\Box A \to A)$. We have

- $a \vDash_0 \Box(\Box A \to A)$ iff $\forall x(aRx \vee a = x \to x \vDash_1 (\Box A \to A))$.
- $x \vDash_1 \Box A \to A$ iff $\forall y(xRy \to y \vDash_0 A) \to x \vDash_1 A$.

Example 1.7 (Quantifiers). If we apply the same ideas to quantifiers we get that an expression of the form $QxQyQzR(x, y, z)$ has the following optional meaning.

1. Generalised quantifiers, where Qu denotes a quantifier and $A(u, w)$ denotes the formula (matrix) it quantifies, and where the range of $QuA(u, \mathbf{w})$ depends on the free variables $\mathbf{w}$ in the matrix $A(u, \mathbf{w})$. Denoting the range by $V_{\mathbf{w}}$ we get

$$QxQyQzR(x, y, z) \equiv \forall x \in V_\varnothing \forall y \in V_x \forall z \in V_{x,y} R(x, y, z).$$

   This reading corresponds to the hypermodal logic of Remark 1.5.
2. If we want the meaning of Q to alternate between $\forall$ and $\exists$ we get

$$QxQyQzR(x, y, z) \equiv \forall x \exists y \forall z R(x, y, z).$$

This corresponds to Example 1.6 where $\Box$ alternates between a **T** and **K** modality.

This kind of Q quantifier is reminiscent of the natural language quantifier ‘any’, where its meaning (as $\forall$ or $\exists$) depends on its position in the sentence.

1.2.2 Formal presentation of hypermodality

This subsection gives a formal definition of hypermodality. It also develops some machinery for later sections.

We treat the simple case is where the number of modes is a finite set μ and there is a function ε for shifting modes. This case is given in the next definition.

Definition 1.8 (Mode shifting). *Let $\mu = \{\Psi_0, \dots, \Psi_k\}$ be a set of modes and let ε be a function assigning to each $0 \leq i \leq k$ a value $0 \leq \varepsilon(i) \leq k$.*

Let (S, R, a, h) be a Kripke model. We define the following (μ, ε) mode satisfaction in the model

- $t \vDash_i \Box A$ iff $\forall s(\Psi_i(t, s) \to s \vDash_{\varepsilon(i)} A)$.
- We say A is true in the model if $a \vDash_0 A$.

Definition 1.9.

1. *Let $\mathcal{K}$ be a class of models of the form (S, R, a, h). Let (μ, ε) be a mode system. We write $\mathcal{K} \vDash_{(\mu,\varepsilon)} A$ iff for every model (S, R, a, h) in $\mathcal{K}$ we have $a \vDash_0 A$.*
2. *Let $\mathbf{L}$ be a logic complete for a class $\mathcal{K}$ of Kripke models of the form (S, R, a, h). Let $\mathcal{K}[\mu, \varepsilon]$ be $\{A \mid \mathcal{K} \vDash_{(\mu,\varepsilon)} A\}$. We sometimes write $\mathbf{L}[\mu, \varepsilon]$ for $\mathcal{K}[\mu, \varepsilon]$, when the implicit dependence on $\mathcal{K}$ is clear.*

Obviously the nature of a hypermodal logic depends on (μ, ε) and its abstract properties and also on the class $\mathcal{K}$ of models chosen.

A general mode-shifting system (μ, ε) can be a sort of abstract non-deterministic automaton, shifting from mode to mode. When used to shift in any concrete model (S, R, a), it also interacts with the properties of R. In order to identify those abstract properties of this automaton which are independent of the particular (S, R, a), we can 'run' it on a 'free' model (S, R, a) and see what it does. To give an example of what we mean, consider the condition

- $t \vDash_m \Box A$ iff $\forall s(tRs \to s \vDash_n A)$.

This abstract condition does not imply that for some s, evaluation of a $\Box$ can occur both in m and in n modes, since in a general free model (S, R, a), tRs implies $s \neq t$. However, if the m mode is reflexive (as in the case of $\Psi_{\mathbf{T}}(t, s)$) then we know that such an s exists, even in the free (S, R, a) model.

We need some definitions.

In a free model of a binary relation of the form (S, R, a), we need to have infinitely many different non-related successors of $a, a_1, a_2, a_3, \dots$, each a_i of which has infinitely many all different successors $a_{i,1}, a_{i,2}, \dots$ and so on. Similarly for predecessors. Here is the definition:

Definition 1.10 (Free model of a binary relation).

1. *Let a be constant and let $\mathbf{a}_n, \mathbf{b}_n, n = 1, 2, \dots$ be formal pairwise different unary functions. Let $\mathbb{S}$ be the set of all Skolem terms constructed using these functions. Namely we have*

- *$a \in \mathbb{S}$*
- *If $x \in \mathbb{S}$ then $\mathbf{a}_n(x) \in \mathbb{S}$ and $\mathbf{b}_n(x) \in \mathbb{S}$*

Define a relation $\mathbb{R}$ on $\mathbb{S}$ by

- *$x\mathbb{R}y$ iff for some n either $y = \mathbf{a}_n(x)$ or $x = \mathbf{b}_n(y)$.*

2. *Let $\mathbb{R}^*$ be the reflexive and transitive closure of $\mathbb{R}$. Let for $t \in \mathbb{S}$, $\mathbb{S}_t = \{x \in \mathbb{S} | t\mathbb{R}^* x\}$. For ordinary modality $\Box$, we evaluate $t \vDash \Box A$ at points of $\mathbb{S}_t$.*

We now need some machinery for future use. Given a mode system (μ, ε), and a model (S, R, a, h), we begin evaluating any wff A at mode Ψ_0, i.e. $a \vDash_0 A$. Thus we can ask for a $t \in S$, what possible modes can be evaluated at t, if we try all evaluations $a \vDash_0 A$ for all A? The next definition deals with this.

Definition 1.11 (Worlds where evaluation is at mode m**).** *Let (S, R, a) be a frame, let (μ, ε) be a mode system, with $\mu = \{\Psi_0, \dots, \Psi_k\}$ and ε the mode shift function as defined in Definition 1.8.*

We first define by induction the (not necessarily disjoint!) sets $V_m \subseteq S, 0 \le m \le k$ as follows:

1. *$a \in V_0$ (0 is the start mode)*
2. *If $t \in V_m$ and $\Psi_m(t, s)$ holds and $m' \in \varepsilon(m)$ then $s \in V_{m'}$.*

For each $t \in S$ and mode m define the set S^m_t by induction on m. S^m_t is the set of all worlds that may be involved in an evaluation of the form $t \vDash_m A$, for some A.

3. *Let $\varepsilon^0(m) = \{m\}, \varepsilon^1(m) = \varepsilon(m), \varepsilon^{k+1}(m) = \bigcup_{y \in \varepsilon^k(m)} \varepsilon(y)$.*
4. *Let $S^m_{t,0} = \{t\}$ if $t \in V_m$, and $\varnothing$ otherwise.*
5. *If $x \in S^m_{t,k}$ we can inductively assume that $x \in V_n$, for some $n \in \varepsilon^k(m)$. Let y be any point such that $\Psi_n(x, y)$ holds and let $n' \in \varepsilon(n)$, (then $n' \in \varepsilon^{k+1}(m)$ and $y \in V_{n'}$ by our definitions above), then $y \in S^m_{t,k+1}$.*
6. *Let $S^m_t = \bigcup_k S^m_{t,k}$.*

The previous definition can be used to define properties of ε as it applies to μ.

Definition 1.12 (Coherence of a mode system).

1. *Let (μ, ε) be a mode system and let $(\mathbb{S}, \mathbb{R}, a)$ be the free model. Consider the sets V_m as defined on the free model. Let $\mu' = \{\Psi_{m_1}, \dots, \Psi_{m_n}\} \subseteq \mu$ be a maximal set of modes such that $\bigcap_j V_{m_j} \ne \varnothing$. We say (μ, ε) is coherent iff for any such maximal $\mu' = \{\Psi_{m_1}, \dots, \Psi_{m_n}\}$ we have that $\lambda y \bigwedge_j \Psi_{m_j}(x, y)$ is consistent as a property of x.*[5]
2. *In any model (S, R, a), let $V_{\mu'} = \bigcap_{m \in \mu'} V_m$. Note that any $t \in S$ is a member of a single $V_{\mu'}$. I.e. for $\mu' \ne \mu''$ we have $V_{\mu'} \cap V_{\mu''} = \varnothing$.*

Definition 1.13. *Let (μ, ε) be a coherent mode-shifting system. Let $I_\mu = \{\mu' | \mu'$ a maximal subset of models such that the intersection $V_{\mu'} = \bigcap_{m \in \mu'} V_m \ne \varnothing$ holds on the free model of Definition 1.10.*

Define a relation $<_\varepsilon$ on I_μ by

[5] By $\lambda z A(z)$ we mean the unary relation formed by the predicate $A(z)$. Some books use the notation $\hat{z}A(z)$ or $\{z | A(z)\}$ viewed as a one place relation.

- *$\mu' <_\varepsilon \mu''$ iff for some modes $\Psi_{m'} \in \mu'$, $\Psi_{m''} \in \mu''$ we have $m'' \in \varepsilon(m')$.*

1.3 Case Study: Shifting between K and T modalities

The language in this section contains one hypermodal symbol $\Box$.

This section will study in detail the logic $\mathbf{K}[\Psi_\mathbf{T}, \Psi_\mathbf{K}]$ of Example 1.6, (named in this section $\mathbf{H}_1^S$ for short).

We shall see that this logic is not complete for any Kripke or neighbourhood frames. We shall also axiomatise it. This will give us a feel for what hypermodal logics can do.

Definition 1.14 (Semantical definition of the hypermodal logic $\mathbf{H}_1^S$). *Consider the family of Kripke models of the form $\mathbf{m} = (S, R, a, h)$ where (S, R, a) is a tree.*[6]

We define two satisfaction relations $t \vDash_0 A$ and $t \vDash_1 A$ inductively as follows:

1. *$t \vDash_i q$ iff $t \in h(q)$, for q atomic and $i = 0, 1$.*
2. *$t \vDash_i A \wedge B$ iff $t \vDash_i A$ and $t \vDash_i B$,*
   $t \vDash_i \neg A$ iff $t \nvDash_i A$.
3. *$t \vDash_0 \Box A$ iff $t \vDash_1 A$ and $\forall s(tRs \to s \vDash_1 A)$.*
4. *$t \vDash_1 \Box A$ iff $\forall s(tRs \to s \vDash_0 A)$.*
5. *We say $\mathbf{m} \vDash A$ iff $a \vDash_0 A$.*
6. *Let $\mathbf{H}_1^S \vDash A$ mean that $\mathbf{m} \vDash A$ for all $\mathbf{m}$.*

Definition 1.15 (Translation of $\mathbf{H}_1^S$ into Standard K). *We define two translations τ_0 and τ_1 from $\mathbf{H}_1^S$ to $\mathbf{K}$.*

1. *$\tau_i(A) = A$, for A without modality, $i = 0, 1$.*
2. *$\tau_i(A \wedge B) = \tau_i(A) \wedge \tau_i(B)$*
   $\tau_i(\neg A) = \neg\tau_i(A)$
3. *$\tau_0(\Box A) = \tau_1(A) \wedge \Box\tau_1(A)$.*

---

[6] The modal logic **K** is complete for many classes of models. The widest class is where R is an arbitrary relation. A more strict class is where (S, R, a) is a tree. When we define modes on a class $\mathcal{K}$ of models we have to be careful which class to choose. We may get different hypermodal logics for different choices of classes of models, even though these classes characterise the same ordinary modal logic. Thus for $\mathbf{H}_1^S$ we start with trees as the class of models.

We now define our notion of a tree. Let W be the set of all finite sequences of the form $(-x_0, x_1, \ldots, x_m)$ where $x_0 \geq 0$ is a natural number and $x_i, i \geq 1$ are positive natural numbers. Let $\mathbf{p}$ be a function defined by $\mathbf{p}((-x_0, x_1, \ldots, x_n)) = (-x_0, x_1, \ldots, x_{n-1})$, for $n \geq 1$ and $\mathbf{p}((-x_0)) = (-x_0 - 1)$. For t, s *in* W, let tRs mean $t = \mathbf{p}(s)$. The full tree model is $(W, R, 0)$. We can now be specific about our tree (S, R, a). It must satisfy the properties

- $a \in S \subseteq W$
- $x \in S \wedge yRx \to y \in S$.

4. $\tau_1(\Box A) = \Box\tau_0(A)$.

Lemma 1.16. *Let* $\mathbf{m} = (S, R, a, h)$ *and let* $t \in S$ *be arbitrary and let* A *be any wff. Let* $\vDash_0, \vDash_1$ *be as before and let* $\vDash$ *be the traditional* **K** *satisfaction relation. Then for* $i = 0, 1$ *we have:*

$(*)$ $t \vDash_i A$ *iff* $t \vDash \tau_i(A)$.

Proof. By structural induction.

1. $(*)$ clearly holds for A without $\Box$.
2. The cases of $\wedge$ and $\neg$ are immediate.
3. $t \vDash_0 \Box A$ iff $t \vDash_1 A$ and $\forall s(tRs \rightarrow s \vDash_1 A)$ iff (by the induction hypothesis) $t \vDash \tau_1(A) \wedge \forall s(tRs \rightarrow s \vDash \tau_1(A))$ iff $t \vDash \tau_1(A) \wedge \Box\tau_1(A)$ iff $t \vDash \tau_0(\Box A)$.
4. $t \vDash_1 \Box A$ iff $\forall s(tRs \rightarrow s \vDash_0 A)$ iff (by the induction hypothesis) $\forall s(tRs \rightarrow s \vDash \tau_0(A))$ iff $t \vDash \Box\tau_0(A)$ iff $t \vDash \tau_1(\Box A)$.

The following series of Lemmas, Examples and Definitions serves to acquaint us with the logic $\mathbf{H}_1^S$ as well as providing tools for axiomatising $\mathbf{H}_1^S$.

Lemma 1.17. *Let* A *be any wff. Then*

1. $\tau_0(\Box^0 A) = \tau_0(A)$.
2. $\tau_0(\Box A) = \tau_1(A) \wedge \Box\tau_1(A)$
3. $\tau_0(\Box^2 A) = \Box\tau_0(A) \wedge \Box^2\tau_0(A)$
4. $\tau_0(\Box^3 A) = \Box\tau_1(A) \wedge \Box^2\tau_1(A) \wedge \Box^3\tau_1(A)$
5. $\tau_0(\Box^4 A) = \Box^2\tau_0(A) \wedge \Box^3\tau_0(A) \wedge \Box^4\tau_0(A)$
6. $\tau_0(\Box^5 A) = \Box^2\tau_1(A) \wedge \Box^3\tau_1(A) \wedge \Box^4\tau_1(A) \wedge \Box^5\tau_1(A)$
7. For $n \geq 2$ *and* $0 \leq i \leq 1$
$\tau_0(\Box^{2n+i} A) = \bigwedge_{m=n}^{2n+i} \Box^m\tau_i(A)$.

Proof.

1. $\Box^0 A = A$
2. By definition.
3. $$\begin{aligned}\tau_0(\Box^2 A) &= \tau_1(\Box A) \wedge \Box\tau_1(\Box A)\\ &= \Box\tau_0(A) \wedge \Box\Box\tau_0(A)\end{aligned}$$
4. $$\begin{aligned}\tau_0(\Box^3 A) &= \tau_1(\Box^2 A) \wedge \Box\tau_1(\Box^2 A)\\ &= \Box\tau_0(\Box A) \wedge \Box^2(\tau_0(\Box A))\\ &= \Box(\tau_1(A) \wedge \Box\tau_1(A)) \wedge \Box^2(\tau_1(A) \wedge \Box\tau_1(A))\\ &= \Box\tau_1(A) \wedge \Box^2\tau_1(A) \wedge \Box^2\tau_1(A) \wedge \Box^3\tau_1(A)\\ &= \Box\tau_1(A) \wedge \Box^2\tau_1(A) \wedge \Box^3\tau_1(A).\end{aligned}$$
5. – 6. check directly and for the induction case you can also use 7. below.
7. We use induction on $n \geq 2$.

renewcommand $*$ 11.5

$$\begin{aligned}\tau_0(\Box^2(\Box^{2n+i} A)) &= \Box(\tau_0(\Box^{2n+i} A) \wedge \Box^2\tau_0(\Box^{2n+i} A))\\ &= \Box \bigwedge_{m=n}^{2n+i} \Box^m\tau_i(A) \wedge \Box^2 \bigwedge_{m=n}^{2n+i} \Box^m\tau_i(A)\\ &= \bigwedge_{m=n+1}^{2(n+1)+i-1} \Box^m\tau_i(A) \wedge \bigwedge_{m=n+2}^{2(n+1)+i} \Box^m\tau_i(A)\\ &= \bigwedge_{m=n+1}^{2(n+1)+i} \Box^m\tau_i(A).\end{aligned}$$

Example 1.18. For atomic q

1. $\tau_0(\Box^2 q) = \Box q \wedge \Box\Box q$
2. $\tau_0(\Box^3 q) = \Box q \wedge \Box^2 q \wedge \Box^3 q$
3. $\tau_0(\Box^4 q) = \Box q' wedge' sqare^2 q \wedge \Box^3 q \wedge \Box^4 q$
4. For $n \geq 2, \tau_0(\Box^{2n+i} q) = \bigwedge_{m=n}^{2n+i} \Box^m q.$
5. $\tau_0(\neg q \wedge \Box^2 q) = \neg q \wedge \Box q \wedge \Box\Box q)$
6. $\tau_0(\neg q \wedge \Box^2 q \wedge \Diamond(\neg q \wedge A)) = \neg q \wedge \Box q \wedge \Box^2 q \wedge \tau_1(A).$
7. $\tau_0(\neg q \wedge \Box^2 q \wedge \Diamond(\neg q \wedge \Box B)) = \neg q \wedge \Box q \wedge \Box^2 q \wedge \Box\tau_0(B)$

Example 1.19. Let q^i_j be a double-indexed sequence of atoms. Let

1. $\beta^i_1(q^i_1) = \neg q^i_1 \wedge \Box^2 q^i_1.$
2. $\beta^i_2(q^i_1, q^i_2) = \neg q^i_2 \wedge \Box^2 q^i_2 \wedge \Diamond(\neg q^i_2 \wedge \Box\beta^i_1).$
3. $\beta^i_{n+1}(q^i_1, \ldots, q^i_{n+1}) = \neg q^i_{n+1} \wedge \Box^2 q^i_{n+1} \wedge \Diamond(\neg q^i_{n+1} \wedge \Box\beta^i_n)$
4. Note that:
$$\begin{aligned}\tau_0(\beta^i_1) &= \neg q^i_1 \wedge \Box^2 q^i_1 = \beta^i_1\\ \tau_0(\beta^i_{n+1}) &= \neg q^i_{n+1} \wedge \Box^2 q^i_{n+1} \wedge \Box\tau_0(\beta^i_n)\end{aligned}$$
Using induction we get:
$$\tau_0(\beta^i_{n+1}) = \bigwedge_{j=0}^{n} \Box^j(\neg q^i_{n+1-j} \wedge \Box^2 q^i_{n+1-j})$$
5. We show that if $t_{n+1}Rt_n \wedge \ldots \wedge t_2Rt_1$ and $t_{n+1} \vDash_0 \beta^i_{n+1}$ then $t_1 \vDash_0 \beta^i_1$.

The proof is by induction

Case $n = 1$.

$$\begin{aligned}&t_2 \vDash_0 \beta^i_2 \text{ iff } t_2 \vDash_0 \neg q^i_2 \wedge \Box^2 q^i_2 \wedge \Diamond(\neg q^i_2 \wedge \Box(\neg q^i_1 \wedge \Box^2 q^i_1))\\ &\text{iff } t_2 \vDash_0 \neg q^i_2 \wedge \Box^2 q^i_2 \text{ and } \forall t_1(t_2Rt_1 \to t_1 \vDash_0 \neg q^i_1 \wedge \Box^2 q^i_1).\end{aligned}$$

Hence if t_2Rt_1 and $t_2 \vDash_0 \beta^i_2$ then $t_1 \vDash_0 \beta^i_1$.

Case n.

$$\begin{aligned}&t_{n+1} \vDash_0 \beta^i_{n+1} \text{ iff } t_{n+1} \vDash_0 \neg q^i_{n+1} \wedge \Box^2 q^i_{n+1} \wedge \Diamond(\neg q^i_{n+1} \wedge \Box\beta^i_n)\\ &\text{iff } t_{n+1} \vDash_0 \neg q^i_{n+1} \wedge \Box^2 q^i_{n+1} \text{ and } \forall t_n(t_{n+1}Rt_n \to t_n \vDash \beta^i_n).\end{aligned}$$

Hence by induction for all $t_n, \ldots, t_1$ if $t_{n+1}Rt_n \wedge \ldots \wedge t_2Rt_1$ and $t_{n+1} \vDash_0 \beta^i_{n+1}$ we get $t_1 \vDash \beta^i_1$.

Lemma 1.20. *Let A be such that*

$$\mathbf{H}^S_1 \vDash A$$

then

$$\mathbf{H}^S_1 \vDash \Box^{2n} A.$$

Proof. We have $\mathbf{K} \vdash \tau_0(A)$ by assumption.

$$\tau_0(\Box^2 A) = \Box\tau_0(A) \wedge \Box^2\tau_0(A)$$
$$\tau_0(\Box^{2n} A) = \bigwedge_{m=n}^{2n} \Box^m\tau_0(A).$$

In both cases we have $\mathbf{K} \vdash \tau_0(\Box^{2n} A)$ and hence $\mathbf{H}_1^S \models A$.

Lemma 1.21. *Let q be an atom, then*

1. $\mathbf{H}_1^S \not\models \Box(\Box q \to q)$
2. $\mathbf{H}_1^S \models \Box q \to q$
3. $\mathbf{H}_1^S$ *is not closed under substitution of provably equivalent formulas.*

Proof.

1. $\tau_0(\Box(\Box q \to q)) = \tau_1(\Box q \to q) \wedge \Box\tau_1(\Box q \to q) = (\Box q \to q) \wedge \Box(\Box q \to q)$ and this is not a theorem of **K**.
2. $\tau_0(\Box q \to q) = q \wedge \Box q \to q$ which is a theorem of **K**.
3. We have $\mathbf{H}_1^S \models \top \leftrightarrow (\Box q \to q)$ but $\mathbf{H}_1^S \not\models \Box\top \leftrightarrow \Box(\Box q \to q)$.

Lemma 1.22. $\mathbf{H}_1^S$ *is not complete for any class of Kripke frames.*

Proof. Assume otherwise and get a contradiction. Let (S, R, a) be a frame where all theorems of our logic hold under any assignment h and let h_0 be an assignment such that $a \models_{h_0} \Diamond(\neg q \wedge \Box q)$. Such a frame and an assignment must exist in view of the previous lemmas.

Since every instance of $\Box q \to q$ must be valid at a under any assignment we must have that aRa holds. We also have since $a \models_{h_0} \Diamond(\neg q \wedge \Box q)$ that there exists an a_1 with aRa_1 such that $a_1 \models_{h_0} \neg q$ and $a_1 \models_{h_0} \Box q$ and therefore a_1Ra_1 does not hold.

Consider now $\Box^2(\Box q \to q)$. This is a theorem of the logic and must therefore hold at a under h_0.

But how can it hold?, We have $aRa \wedge aRa_1$ and $a_1 \models_{h_0} \Box q$ and $a_1 \models_{h_0} \neg q$.

A contradiction.

The next lemma shows that our logic is not complete for any class of neighbourhood frames. Although the case of Kripke frames follows from that, we think including both proofs is instructive for the case of other logics.

Definition 1.23. *Neighbourhood models have the form $(S, \mathbb{F}, a, h)$ where S is the set of possible worlds, $a \in S$ is the actual world and $\mathbb{F}$ is a family of filters, associating with each $t \in S$ a filter $\mathbb{F}_t$.*[7] *h is the assignment to atoms. The following is the truth condition for $\Box$.*

[7] A filter $\mathbb{F}_t$ is a family of subsets of S satisfying

1. $S \in \mathbb{F}_t$
2. $\varnothing \notin \mathbb{F}_t$
3. $X \in \mathbb{F}_t \wedge X \subseteq Y$ implies $Y \in \mathbb{F}_t$
4. $X \in \mathbb{F}_t$ and $Y \in \mathbb{F}_t$ implies $X \cap Y \in \mathbb{F}_t$.

- *$t \vDash_h \Box A$ iff $\|A\|_h \in \mathbb{F}_t$ where $\|A\|_h = \{x \mid x \vDash_h A\}$.*

We say A is true in the model iff $a \vDash_h A$.

Lemma 1.24. $\mathbf{H}_1^S$ *is not complete for any class of neighbourhood frames.*

Proof. Assume otherwise, then since $\neg\Box(\Box q \to q)$ is a consistent formula of the logic, there must exist a neighbourhood model $(S, \mathbb{F}, a, h_0)$ such that for any assignment h and for any A

$$a \vDash_h \Box A \to A$$
$$a \vDash_h \Box^2(\Box A \to A)$$

while for h_0 and q

$$a \nvDash_{h_0} \Box(\Box q \to q).$$

Let $Q_0 = \|q\|_{h_0}$.

The above conditions mean the following second-order conditions hold for arbitrary $X \subseteq S$

$$\forall X[X \in \mathbb{F}_a \to a \in X] \tag{1.1}$$

1.1. holds because $\Box A \to A$ mush hold at a under any assignment and any A.

Since $a \nvDash_{h_0} \Box(\Box q \to q)$, this means that $\|\Box q \to q\|_{h_0} \notin \mathbb{F}_a$. This implies that $\{y \mid y \nvDash_{h_0} \Box q$ or $y \vDash_{h_0} q\} \notin \mathbb{F}_a$, which further implies that

$$\bar{Q}_0 = \{y \mid Q_0 \notin \mathbb{F}_y \text{ or } y \in Q_0\} \notin \mathbb{F}_a.$$

We summarise the latter as condition (2):

$$\bar{Q}_0 = Q_0 \cup \{y \mid Q_0 \notin \mathbb{F}_y\} \notin \mathbb{F}_a. \tag{1.2}$$

The third condition is obtained from the fact that for any h and any $A, a \vDash_h \Box\Box(\Box A \to A)$. This gives us the following condition for any atom x and h

$$\{z \mid z \vDash_h \Box(\Box x \to x)\} \in \mathbb{F}_a.$$

but

$$z \vDash_h \Box(\Box x \to x) \text{ iff } \|\Box x \to x\|_h \in \mathbb{F}_z.$$

Thus we get

$$\{z \mid \{y \mid \|x\| \notin \mathbb{F}_y \text{ or } y \vDash_h x\} \in \mathbb{F}_z\} \in \mathbb{F}_a.$$

Since h is arbitrary we get condition (3).

$$\forall X[\{z \mid X \cup \{y \mid X \notin \mathbb{F}_y\} \in \mathbb{F}_z\} \in \mathbb{F}_a] \tag{1.3}$$

We can now proceed to get a contradiction:

Let $X = Q_0$ in 1.3. We therefore get that:

$$Y_0 = \{z \mid Q_0 \cup \{y \mid Q_0 \notin \mathbb{F}_y\} \in \mathbb{F}_z\} \in \mathbb{F}_a.$$

By 1.1. we get that $a \in Y_0$ and hence

$$Q_0 \cup \{y \mid Q_0 \notin \mathbb{F}_y\} \in \mathbb{F}_a$$

which contradicts 1.2.

Remark 1.25. Notice that we have not used any special properties of the filters $\mathbb{F}_t$. This means that any extension of **K** which proves all instances of $\Box A \to A, \Box^2(\Box A \to A)$ but not $\Box(\Box A \to A)$ cannot be complete for any neighbourhood frames. For example we can take **K4.3** [**T**, **K**] (i.e. assume R is linear, transitive and irreflexive).

1.4 Translations of Hypermodality

This section discusses the relationship of hypermodality and other kinds of modalities, namely many-dimensional modal logics and multimodal logics.

We can view the system $\mathbf{H}_1^S$ as a two-dimensional system. Instead of writing $t \vDash_i A$ we write $(t, i) \vDash A$ and consider our set of possible worlds as $S \times \{0, 1\}$. We then have a relation R_2 defined as follows:

$$(t, 0)R_1(s, i) \text{ iff } i = 1 \wedge (t = s \vee tRs)$$
$$(t, 1)R_2(s, i) \text{ iff } i = 0 \wedge tRs.$$

Figure 1.1 will become Figure 1.2, where $\to$ now shows R_2 accessibility, and $\twoheadrightarrow$ shows R_1 accessibility.

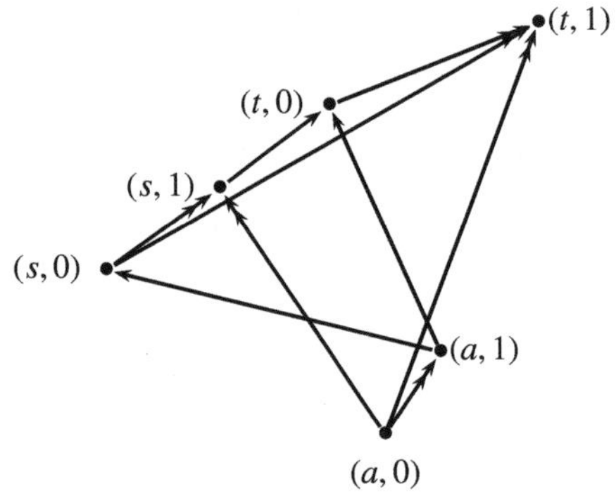

Fig. 1.2

The two-dimensional view requires a further restriction on the assignment h, namely that for all atomic q and $t \in S$ and $i, j \in \{0, 1\}$ we have:

$$(t, i) \in h(q) \text{ iff } (t, j) \in h(q)$$

Therefore this point of view does not yield completeness for frames!

We now turn to the connection of hypermodality and multimodal logics.

A multimodal logic with modalities $\Box_0, \Box_1$ can be characterised by Kripke models of the form (S, R_0, R_1, a, h) where $R_i \subseteq S^2, i = 0, 1$ and the truth table for $\Box_i, i = 0, 1$ is as follows:

- $t \vDash \Box_i A$ iff $\forall s(tR_i s \to s \vDash A)$.

If we want $\Box_0$ to be a **T** modality we can assume that R_0 is reflexive. We can view a hypermodal system with $\Box$ as a fragment of a multimodal logic, where $\Box$ can be $\Box_i$, depending on its position in the formula.

Let ν_i be two translations from the hypermodal language into the multimodal language with $\Box_0$ and $\Box_1$. We have

- $\nu_i(A) = A$, for A atomic
- $\nu_i(\neg A) = \neg\nu_i(A)$
- $\nu_i(A \wedge B) = \nu_i(A) \wedge \nu_i(B)$
- $\nu_0(\Box A) = \Box_0\nu_1(A)$
- $\nu_1(\Box A) = \Box_1\nu_0(A)$.

Thus

- $\nu_0(\Box q \to q) = \Box_0 q \to q$
- $\nu_0(\Box(\Box q \to q)) = \Box_0(\Box_1 q \to q)$
- $\nu_0(\Box\Box(\Box q \to q)) = \Box_0\Box_1(\Box_0 q \to q)$
- $\nu_0(\Box\Box q \to \Box q) = \Box_0\Box_1 q \to \Box_0 q$.

Our hypermodal logic for $\Box$ with modes Ψ_0, Ψ_1 based on the class of models $\{(S, R, a, h)\}$ is translated into the multimodal logic with $\Box_0, \Box_1$ based on the class of models $\{(S, \Psi_0, \Psi_1, a, h)\}$. Ψ_0, Ψ_1 are based on R. Thus our modal logic $\mathbf{H}_1^S$ when viewed as a multimodal logic is special; the relations used for $\Box_0$ and $\Box_1$ are both defined from a single underlying relation R.

We can define the notion of a multimodal frame (S, R_0, R_1, a, h) as *generated* from a frame (S, R, a, h) iff there exist generators $\Psi_i(x, R, a, y), i = 0, 1$ such that $R_i = \lambda x\lambda y\Psi_i, i = 0, 1$.

This means for our case that we can take R_1 as our R and assume that R_0 is the reflexive closure of R_1. So let (S, R_0, R_1, a, h) be a bimodal logic where $R_1 = R$ and R_0 is the reflexive closure of R. Consider (S, R, a, h) as a monomodal model of $\mathbf{H}_1^S$.

Lemma 1.26. *For the above translation we have:*

(*) $t \vDash_i A$ iff $t \vDash \nu_i(A)$.

Proof. This can be proved by structural induction on A.

The cases of atomic A and of classical connectives $\wedge$ and $\neg$ present no difficulties. Let us check the case of $\Box$:
$t \vDash_0 \Box A$ iff $t \vDash_1 A$ and $\forall s(tRs \to s \vDash_1 A)$ iff $t \vDash \nu_1(A)$ and $\forall s(tRs \to s \vDash \nu_1(A)$ iff $t \vDash \Box_0\nu_1(A)$ iff $t \vDash \nu_0(\Box A)$.

Similarly $t \vDash_1 \Box A$ iff $\forall s(tRs \to s \vDash_0 A)$ iff $\forall s(tRs \to s \vDash \nu_0(A))$ iff $t \vDash \Box_1\nu_0(A)$ iff $t \vDash \nu_1(\Box A)$.

(*) presents us with a conceptual query. We saw that $\mathbf{H}_1^S$ is not complete for any family of Kripke frames. Is the bimodal logic with $\Box_0, \Box_1$ complete for a family of Kripke frames?

The answer is yes, the family of frames of the form (S, R_0, R_1, a) where R_0 is the reflexive closure of R_1.[8]

Does this not imply that $\mathbf{H}_1^S$ is complete for frames? The answer is no. When we examine $\Box q \to q$ in a frame (S, R, a), since it is a theorem of $\mathbf{H}_1^S$ for all q, we conclude that R must be reflexive. If we examine $\nu_0(\Box q \to q) = \Box_0 q \to q$ in models (S, R_0, R_1, a), we conclude R_0 is reflexive, which is OK.[9]

Compare, for example, $\mathbf{H}_1^S \models \Box q \to q$ but $\mathbf{H}_1^S \not\models \Box\Box q \to \Box q$. The translation $\nu_0(\Box\Box q \to \Box q)$ is $\Box_0\Box_1 q \to \Box_0 q$ which does not hold in the bimodal logic.

The reader may be under the intuitive impression that somehow a hypermodal logic is a composition of several modalities $\Box_1, \ldots, \Box_n$ used according to position in the formula. Thus $\Box$ is sometimes $\Box_i$, sometimes $\Box_j$, etc. The reader may think that the study of hypermodality should be reduced to special cases of the study of multimodal logic. This is not the general case. The interaction between $\Box_1, \ldots, \Box_n$ may make $\Box$ a completely new modality.[10] Example 1.6 below illustrates this point.

1.5 Axiomatising $\mathbf{H}_1^S$

1.5.1 Methodological Discussion

Before giving a syntactical axiomatisation for our logic, we need to discuss methodology. There are different kinds of proof methods around. There are Hilbert type systems, Gentzen type systems, tableaux systems, natural deduction systems and labelled deductive systems. The difference between these methods, when applied to modal logics, manifests itself in how much of the geometry of the Kripke structure (possible worlds semantics) is brought into the syntax.

We defined modal logic satisfaction in terms of truth in the actual world. Thus the set of theorems of a modal logic **L** is the set

---

8 The proof is standard. We take as axioms and rules all **K** axioms for $\Box_1$ and **T** axioms for $\Box_0$, together with the additional interaction axioms $\Box_0 A \to \Box_1 A$ and $\Diamond_0 A \to A \vee \Diamond_1 A$.

To prove completeness, let S be the set of all complete consistent theories. Let $\Delta R_i \Theta$ be defined as 'for all $\Box_i A \in \Delta$ we have $A \in \Theta$'. For atomic q, let $\Delta \models q$ iff $\Delta \vdash q$ and show by induction that for all A, $\Delta \models A$ iff $\Delta \vdash A$. The axioms $\Box_0 A \to \Box_1 A$ and $\Diamond_0 A \to A \vee \Diamond_1 A$ ensure that R_0 is the reflexive closure of R_1.

9 Another point to watch for is substitutivity of equivalents. $\Box q \to q$ is equivalent to $\top$ in $\mathbf{H}_1^S$ but $\Box\top$ is not equivalent to $\Box(\Box q \to q)$.

However, as a two-dimensional formula, $\Box q \to q$ is $\Box_0 q \to q$ and indeed $\Box_1\top$ is equivalent to $\Box_1(\Box_0 q \to q)$.

The non-substitutivity of equivalents, in itself, does not imply non-completeness for a class of frames. Consider the class of all frames (S, R, a) such that R is transitive and aRa holds. Let the logic be the set of all wffs valid at the actual world a. This logic is finitely axiomatisable (with modus ponens only, without necessitation, since it is not normal) and it has no substitutivity of equivalents.

10 The theory of Abelian groups, for example, is a special case of the theory of groups but it is so special that it is completely different.

$$\{A | a \vDash A, \text{ for all models } (S, R, a, h)\}.$$

Axiomatising the logic means generating this set. In the case of $\mathbf{H}_1^S$ it means generating the set of all formulas A such that $a \vDash_0 A$, in any model (S, R, a, h); see Definition 1.14.

We know, however, that $a \vDash_0 A$ may depend, in its semantic evaluation, on other points in the model; for example it may depend on whether a point b such that $aRb, a \neq b$ exists such that $b \vDash_1 B$, for some subformula B.

The Hilbert type axiomatisation, because it generates only As such that $a \vDash_0 A$, cannot directly talk about such bs. We can indirectly write $a \vDash_0 \Diamond B$ but this means $\exists x[(x = a \vee aRx) \wedge x \vDash_1 B]$. We cannot force the geometrical condition $x \neq a$.

There are other ways around the problem. Suppose we had an atomic q such that $a \vDash_0 \neg q$ and $\forall x(aRx \wedge a \neq x \rightarrow x \vDash_1 q)$, then we could say what we want by writing $a \vDash_0 \Diamond(q \wedge B)$.

Can we say that such a q exists? That depends on the properties of our logic. Indeed, we can write $a \vDash_0 \neg q \wedge \Box\Box q$. This will give us such a q. Thus to axiomatise our logic as a Hilbert system, we need to say more about other points and we need to use such qs to enable us to say what we need. This approach gives us a rather complicated Hilbert axiomatisation of our logic.

The source of the complication lies in the fact that a Hilbert system allows us to code information of the form $b \vDash_i B, bRc$, etc. only though $\vDash_0$ validity in the actual world and the use of the connectives $\Diamond$ and $\Box$. It makes the axioms complicated for the case of our logic because of the multiple modes available. Traditional Hilbert axiomatisations of **K** happen to be able to manage because there is only one mode.

It is not that our two-mode modal logic is complicated, it is that Hilbert style axiomatisations are geometrically poor.

Let us now turn to Gentzen style axioms. This style allows for a richer geometry. We can look at sequents of the form

$$A_0, A_1, \ldots, A_n \vDash B.$$

It is up to us to interpret this sequent semantically. Let us, for the case of ordinary modal logic, take the following interpretation

- $A_0, A_1, \ldots, A_n \vDash B$ iff in any Kripke model (S, R, a, h) and any $t_0 = a, t_1, \ldots, t_n$ such that $t_0Rt_1, t_1Rt_2, \ldots, t_{n-1}Rt_n$ and such that $t_i \vDash A_i, i = 0, \ldots, n$ we have $t_n \vDash B$.

This kind of geometry gives us better expressive power reflecting more of the possible world semantics in the syntax. In the case of two modes 0 and 1 we can venture and propose to amend the definition to:

- $A_0, A_1, \ldots, A_n \vDash B$ iff for all models and all $t_0 = aRt_1, \ldots, t_{n-1}Rt_n$ we have that if $t_i \vDash_i A_i, i = 0, \ldots, n$ then $t_n \vDash_n B$, where $\vDash_m$ is taken to be $\vDash_0$ if m is even and $\vDash_1$ if m is odd.

It is likely that we can find a simpler axiomatisation using the above sequents because of their better expressive power.

In fact, we can put more of the geometry into the syntax. Authors have proposed double sequents for modal and substructural logics; we can allow sequents of the form

$$(A_0, A'_0), (A_1, A'_1), \ldots, (A_n, A'_n) \vDash (B, B')$$

with the following validity definition

- $(A_0, A'_0), \ldots, (A_n, A'_n) \vDash (B, B')$ is valid iff for every model and every $t_0 = aRt_1, \ldots, t_{n-1}Rt_n$ such that for all $0 \le i \le 1, t_i \vDash_0 A_i$ and $t_i \vDash_1 A'_i$ we have $t_n \vDash_0 B$ and $t_n \vDash_1 B'$.

It is obvious that the more semantical information we put into the syntax the easier it becomes to axiomatise. We therefore adopt the following approach. Rather than sneak in the information through the geometry of the sequents, double sequents, hyper-sequents, special connectives or whatever, we shall go all the way and use the theory of labelled deductive systems. This allows for part of the Kripke structure to be represented in the syntax as labels. See [55]. Thus we have letters standing for labels (which stand for possible worlds) and a relation ρ on labels (standing for the accessibility R) and our theory (sequents) have the form ($m_t, n = 0$ or 1)

$$\Delta = \{(m_t, t) : A_t, t_i \rho t_j\} \vDash (n, s) : B.$$

Validity is defined as follows:

- $\Delta \vDash (n, s) : B$ is valid iff for every Kripke model (S, R, a, h) and every function g assigning to labels t values $g(t) \in S$ such that if $t\rho t' \in \Delta$ then $g(t)Rg(t')$ holds, we have the following:
  If for all $(m_t, t) : A_t \in \Delta$ we have $t \vDash_{m_t} A_t$ then also $s \vDash_n B$ holds.

In this chapter we shall give two extreme axiomatisations, the Hilbert one and the labelled deductive systems one.

1.5.2 Hilbert axiomatisation

We shall give a Hilbert axiomatisation of $\mathbf{H}_1^S$. We need first to adopt a different point of view in axiomatising the traditional logic **K**. We saw that in $\mathbf{H}_1^S$, $\Box q \to q$ for q atomic is a theorem, so is $\Box^{2n}(\Box q \to q)$ but not $\Box(\Box q \to q)$.

Also the rule of substitution of logical equivalents does not hold. We have $\mathbf{H}_1^S \vDash \top \leftrightarrow (\Box q \to q)$ but $\mathbf{H}_1^S \nvDash \Box\top \leftrightarrow \Box(\Box q \to q)$.

We shall therefore formulate an axiomatisation of traditional **K** without necessitation and substitution.

Definition 1.27 (Another formulation of modal logic K). *Consider the following formulation of modal* **K**.
Axioms

1. *A if A is a substitution instance of a truth function tautology*
2. *$\Box A$ if A is as in (1)*
3. *$\Box(A \to B) \to (\Box A \to \Box B)$, A, B arbitrary*

4. $\Box(\Box(A \to B) \to (\Box A \to \Box B))$, *$A, B$ arbitrary.*

Rules

$$\frac{\vdash A, \vdash A \to B}{\vdash B}$$

$$\frac{\vdash A}{\vdash \Box\Box A}.$$

Note that we replaced necessitation

$$\frac{\vdash A}{\vdash \Box A}$$

by 2-necessitation

$$\frac{\vdash A}{\vdash \Box^2 A}.$$

To compensate for the loss of necessitation we had to also take as an axiom $\Box A$ *for any axiom* A *of* **K**.

Definition 1.28 (The modal logic $\mathbf{H}_1$). *Let* $\mathbf{H}_1$ *be the system obtained by extending* **K** *under the formulation of the previous definition 1.27 with the axioms*

5. $\Box A \to A$, *for* A *such that* $\mathbf{K} \vdash \tau_1(A) \wedge \Box\tau_1(A) \to \tau_0(A)$.
   Note that we will have shown that $\Box(\Box A \to A)$ *cannot be proved in* $\mathbf{H}_1$. *If it is added as an axiom we get the traditional logic* **T**.
6. *If* $\mathbf{K} \vdash \tau_0(A) \to \tau_0(B)$ *then* $\mathbf{H}_1 \vdash A \to B$.

The above is not a good axiomatisation. It makes use of **K** provability (which is technically OK, since **K** is decidable) and of the translations τ_0 and τ_1. It is illuminating, however, and affords a neat, almost trivial completeness proof.

We shall give a better Hilbert axiomatisation later.

We saw that $\mathbf{H}_1^S$ can be translated into a bimodal logic with $\Box_0, \Box_1$ and that the Hilbert axiomatisation of the bimodal logic is relatively easy. If we can define inside $\mathbf{H}_1^S$, using $\Box$ only, the two modalities $\Box_0$ and $\Box_1$, we can then give an axiomatisation of $\mathbf{H}_1^S$. The key to defining $\Box_0, \Box_1$ is irreflexivity. Let q be an atom and assume that $t \vDash_0 \neg q \wedge \Box\Box q$. This means that $t \vDash \neg q$ and $\forall x, y(tRx \wedge xRy \to x \vDash q$ and $y \vDash q)$.

Now imagine that $t \vDash_0 \Diamond(\neg q \wedge A)$ holds. This forces $t \vDash_1 A$. In our logic, a theory Δ is $\{A \mid t \vDash_0 A\}$ for some t. We can get the $\vDash_1$ part of Δ by looking at $\Theta = \{A \mid \Delta \vdash \Diamond(\neg q \wedge A)\}$ provided q is such that $\Delta \vdash \neg q \wedge \Box\Box q$. Together (Δ, Θ) constitute a possible world t because they contain in them both $\vDash_0$ and $\vDash_1$ satisfaction.

$$\Delta = \{A \mid t \vDash_0 A\}, \Theta = \{A \mid t \vDash_1 A\}.$$

So to give an effective axiomatisation we need irreflexivity rules involving $\neg q \wedge \Box\Box q$.

Definition 1.29 (IRR Hilbert System for $\mathbf{H}_1^S$). *The following axiomatisation defining the system* $\mathbf{H}_1^S$ *makes use of the well known Gabbay Irreflexivity Rule, (see [53]).*

Axioms*: (E, F are wffs without $\Box$).*

1. $A \wedge \Box A$, *where* A *is a substitution instance of a truth functional tautology.*
2. $\Box(A \to B) \to (\Box A \to \Box B)$.
3. $\Box(\Box(A \to B) \to (\Box A \to \Box B))$.
4. $\Diamond\top$.
5. $\neg E \wedge \Box^2 E \wedge Y \to \Diamond(\neg E \wedge Y)$, *where* $Y = A$ *or* $Y = \Box A$, *for* A *without* $\Box$.
6. $\neg E \wedge \Box^2 E \wedge \Diamond(\neg E \wedge A) \wedge \Diamond(\neg E \wedge B) \to \Diamond(\neg E \wedge A \wedge B)$.
7. $\neg E \wedge \Box^2 E \to \Diamond(\neg E \wedge A) \vee \Diamond(\neg E \wedge \neg A)$.
8. $\Box A \wedge \neg E \wedge \Box^2 E \to \Diamond(\neg E \wedge A)$.
9. $\Box X \wedge \neg E \wedge \Box^2 E \to \Diamond(\neg E \wedge \Box(\neg \wedge \Box^2 F \to \Diamond\neg F \wedge X))$.
10. $\neg E \wedge \Box^2 E \wedge \Diamond A \wedge \Diamond(\neg E \wedge \Box Y \wedge \Box\Box Y' \wedge \neg A \wedge \Box\Box X) \to \Diamond(A \wedge X \wedge Y \wedge \Box Y')$. *where* Y, Y' *are without* $\Box$.
11. $\neg E \wedge \Box^2 E \wedge \Diamond(C \wedge E \wedge Y) \to \Diamond(\neg E \wedge \Diamond(Y \wedge E \wedge \Diamond(C \wedge E)))$.

Rules

MP:
$$\frac{\vdash A; \vdash A \to B}{\vdash B}$$

IRR:
$$\frac{\vdash \neg q \wedge \Box^2 q \to A}{\vdash A}$$
where q *is an atom not in* A.

2-necessitation:
$$\frac{\vdash A}{\vdash \Box^2 A}$$

IRR^n:
$$\frac{\vdash \bigwedge_{m=1}^{n} \beta_m^m \to A}{\vdash A}$$
where β_m^m *are as in Example 1.19 and* q_j^i *are all not in* A.

Lemma 1.30. *The axioms and rules given are sound for the semantics.*

Proof. By direct verification.

Definition 1.31. *A theory* Δ *is said to be* 0*-consistent (resp.* 1*-consistent) iff for no finite number of* $A_i \in \Delta$ *do we have* $\vdash \neg \bigwedge A_i$ *(resp.* $\vdash \Box\neg \bigwedge A_i$*).*

Lemma 1.32. *Any i-consistent theory can be extended to a complete i-consistent theory.*

Proof. We need to address the case $i = 1$.

Assume Δ_n is 1-consistent and let $\delta_{n=1}$ be the $n + 1$th wff. Can $\Delta_n \cup \{\delta_{n+1}\}$ and $\Delta_n \cup \{\neg\delta_{n+1}\}$ both be 1-inconsistent?

If so, we have for some A_i, B_j

$$\vdash \Box(\bigwedge A_i \to \delta_{n+1})$$
$$\vdash \Box(\bigwedge B_j \to \neg\delta_{n+1}).$$

hence $\vdash \Box(\bigwedge A_i \wedge \bigwedge B_j \to \bot)$.

A contradiction because of axiom $\Diamond\top$.

Definition 1.33. *A theory Δ is said to be an IRR theory if it is 0-consistent and for some sequence of atoms $q_j^i, i, j = 1, 2, 3, \ldots$ and for each $\beta_n^n(q_1^n, \ldots, q_n^n)$ we have that $\beta_n^n \in \Delta$.*

We shall see later how any 0-consistent formula A can be extended to a complete IRR theory. At this stage, let us show that every IRR theory has a model.

Definition 1.34. *(Δ, Θ) is a 0-1 pair iff*

1. *Δ is 0-consistent and complete and Θ is 1-consistent and complete.*
2. *For any A, B:*
   $\Box B \in \Delta \to B \in \Theta$
   $\Box^2 \neg A \wedge \Diamond A \in \Delta \to A \in \Theta$
3. *$A \in \Theta \to \Diamond A \in \Delta$, for all A*
4. *For any Y without $\Box$, $Y \in \Delta \to Y \in \Theta$*
5. *For any Y' without $\Box$:*
   $\Box Y' \in \Delta \to \Box Y' \in \Theta$

Lemma 1.35. *Let Δ be an IRR theory and assume $\neg q_1^1 \wedge \Box^2 q_1^1 \in \Delta$. Let $\Theta = \{A \mid \Diamond(\neg q_1^1 \wedge A) \in \Delta\}$. Then Θ is a complete 1-consistent theory such that (Δ, Θ) is a 0-1 pair.*

Proof. Axiom 7. gives us completeness. Assume Θ is not 1-consistent, then $\vdash \Box(\neg q_1^1 \to \neg \bigwedge_i A_i), A_i \in \Theta$. hence $\Box(\neg q_1^1 \to \neg \bigwedge_i A_i) \in \Delta$. But from axiom 6 we get $\Diamond(\neg q_1^1 \wedge \bigwedge_i A_i) \in \Delta$, a contradiction.

The 0-1 pair property of (Δ, Θ) follows from axioms 5, 8.

Lemma 1.36. *Let (Δ, Θ) be a 0-1 pair, with Δ an IRR theory.*

Assume $\Diamond A \in \Theta$.

Let $\Delta'_{1,A} = \{B \mid \Box B \in \Theta\} \cup \{A\}$. Then $\Delta'_{1,A}$ is 0-consistent and can be completed to a 0-consistent IRR theory $\Delta_{1,A}$.

Proof. Otherwise

$$\vdash \bigwedge B_i \to \neg A, \Box B_i \in \Theta.$$

Hence $\vdash \Box^2(\bigwedge B_i \to \neg A)$.
Hence $\Box^2(\bigwedge B_i \to \neg A) \in \Delta$ and from Definition 1.34 (2).

$$\Box(\bigwedge B_i \to \neg A) \in \Theta.$$

Hence $\bigwedge \Box B_i \to \Box\neg A \in \Theta$ and so $\Box\neg A \in \Theta$, a contradiction.

We can complete $\Delta'_{1,A}$ to a 0-consistent complete theory $\Delta_{1,A}$.

Note that since Δ is an IRR theory, $\beta_{n+1}^{n+1} \in \Delta$, namely $\neg q_{n+1}^{n+1} \wedge \Box^2 q_{n+1}^{n+1} \wedge \Diamond(\neg q_{n+1}^{n+1} \wedge \Box\beta_n^{n+1})$ is in Δ and so $\beta_n^{n+1} \in \Delta_{1,A}$.

These considerations show that $\Delta_{1,A}$ is IRR. For let p_i^n be a renaming of $q_{i+1}^{n=1}, i = 1, \ldots, n$. Then $\beta_n^n(p_1^n, \ldots, p_n^n) = \beta_n^{n+1}$.

Lemma 1.37. *Let $(\Delta, \Theta), \Delta_{1,A}$ be as before. Let $\Theta_{1,A} = \{A \mid \neg q_1^2 \wedge \Box^2 q_1^2 \wedge \Diamond(\neg q_1^2 \wedge A) \in \Delta_{1,A}\}$.*

Then as proved before, $(\Delta_{1,A}, \Theta_{1,A})$ are a 0-1 pair. Furthermore, for any X, if $\Box X \in \Delta$, then $X \in \Theta_{1,A}$.

Proof. Assume $\Box X \in \Delta$, then from axiom 9 we get that $\Box(\neg q_1^2 \wedge \Box^2 q_1^2 \to \Diamond(\neg q_1^2 \wedge X)) \in \Theta$ and hence $X \in \Theta_{1,A}$.

Lemma 1.38. *Let (Δ, Θ) be as before and assume $\neg A \in \Theta$ and $\Diamond A \in \Delta$. Then $\Theta'_{0,A} = \{A\} \cup \{B \mid \Box B \in \Delta\} \cup \{C \mid \Box\Box C \in \Theta\} \cup \{Y \wedge \Box Y' \mid \Box Y \wedge \Box^2 Y' \in \Theta, Y, Y'$ without $\Box\}$ is 1-consistent and can be completed to a 1-consistent theory $\Theta_{0,A}$.*

Proof. Otherwise

$$\vdash \Box(\bigwedge_i B_i \to \neg(A \wedge Y \wedge \Box Y' \wedge \bigwedge_j C_j))$$

hence $\Box\neg(A \wedge Y \wedge \Box Y' \wedge \bigwedge_j C_j) \in \Delta$ but also $\Box\Box \bigwedge_j C_j \in \Theta$.

From axiom 10 we get that $\Diamond(A \wedge Y \wedge \Box Y' \wedge \bigwedge_j C_j) \in \Delta$, a contradiction. We can extend $\Theta'_{0,A}$ to a complete 1-consistent theory $\Theta_{0,A}$.

Lemma 1.39. *Let $(\Delta, \Theta), \Theta_{0,A}$ be as before. Then*

$$\Delta'_{0,A} = \{B \mid \Box B \in \Theta\} \cup \{\Diamond C \mid C \in \Theta_{0,A}\} \cup \{Y \mid Y \in \Theta_{0,A}, Y \text{ without } \Box\}$$

is 0-consistent.

Proof. Otherwise let E be such that $\neg E \wedge \Box^2 E \in \Delta, \neg E \in \Theta$ and $E \in \Theta_{0,A}$. Such E exists by construction. We have

$$\vdash \bigwedge_i \Box B_i \to (Y \wedge E \to \neg \bigwedge_j \Diamond(C_j \wedge E \wedge Y))$$

hence

$$\vdash \Box^2(\bigwedge_i \Box B_i \to (Y \wedge E \to \bigwedge_j \Diamond(C_j \wedge E \wedge Y)))$$

hence

$$\Box(\bigwedge_i \Box B_i \to (Y \wedge E \to \neg \bigwedge_j \Diamond(C_j \wedge E \wedge Y))) \in \Theta$$

hence

$$\Box(Y \wedge E \to \neg \bigwedge_j \Diamond(C_j \wedge E \wedge Y)) \in \Theta$$

but

$$\vdash \Box(\neg \bigwedge \Diamond D_j \to \neg\Diamond \bigwedge D_j)$$

hence

$$\Box(\Diamond \bigwedge(C_j \wedge E \wedge Y) \to \neg(E \wedge Y)) \in \Theta.$$

But by construction of $\Theta_{0,A}$ we have $\Diamond(\bigwedge_i C_i \wedge E \wedge Y) \in \Delta$. From axiom 11 we have $\Diamond(Y \wedge E \wedge \Diamond(\bigwedge C_i \wedge E \wedge Y)) \in \Theta$, a contradiction.

Definition 1.40. *Let (Δ, Θ) and (Δ', Θ') be two 0-1 pairs. We say $(\Delta, \Theta)R(\Delta', \Theta')$ iff the following holds:*

1. *If $\Box A \in \Theta$ then $A \in \Delta'$.*
2. *If $\Box A \in \Delta$ then $A \in \Theta'$.*

The previous series of lemmas proved the following:

Lemma 1.41. *Let (Δ, Θ) be an IRR 0-1 pair then*

1. *If $\Diamond A \in \Theta$ then there exists an IRR 0-1 pair (Δ', Θ') such that $(\Delta, \Theta)R(\Delta', \Theta')$ and $A \in \Delta$.*
2. *If $\Diamond A \in \Delta$ and $\neg A \in \Theta$ then there exists an IRR 0-1 pair (Δ', Θ') such that $(\Delta, \Theta)\, R(\Delta', \Theta')$ and $A \in \Theta'$.*

Proof. From previous lemmas.

Lemma 1.42. *Let A be a consistent wff. Then A has a model.*

Proof. Since A is consistent, so is $\Delta_0 = \{A\} \cup \{\beta_n^n \mid n = 1, 2, \ldots,\}$, for β_n^n using completely new atoms $\{q_j^i\}$. This follows from the IRR^n rule. We can complete Δ_0 to a complete IRR theory Δ and construct the IRR 0-1 pair (Δ, Θ). Let S be the set of all 0-1 IRR theories. Let R be as before and let $a = (\Delta, \Theta)$. Let h be defined by

$$(\Delta, \Theta) \in h(q) \text{ iff } q \in \Delta$$

(we know that $q \in \Delta$ iff $q \in \Theta$ for an IRR 0-1 pair!)

We claim for any pair $(\Gamma_0, \Gamma_1) \in S$ and any A

(*) $(\Gamma_0, \Gamma_1) \vDash_i A$ iff $A \in \Gamma_i, i = 0, 1$.

This is proved by structural induction on A.

1.5.3 LDS axiomatisation

We now develop the basic concepts of an LDS theory and a consequence relation for the same logic $\mathbf{H}_1^S$. The general theory of LDS is presented in [55]. See also Definition 1.2.

Definition 1.43 (Labelled theories).

1. *Consider a formal first-order language with atomic terms $D = \{t_1, t_2, \ldots\}$, a binary relation ρ, equality $=$ and a distinguished constant $\mathbf{d} \in D$. This is the language of the algebra of labels. D stands for a set of possible worlds, $\mathbf{d}$ stands for the actual world and ρ stands for the accessibility relation.*
2. *A literal has the form $x = y, x \neq y, x\rho y, \neg x\rho y$ where $x, y \in D$.*
3. *A declarative unit has the form $(m, t) : A$ where $m \in \{0, 1\}, t \in D$ and A is a wff of the modal language with $\Box$.*
4. *A (labelled) theory is a set of declarative units and literals.*

Definition 1.44 (Labelled rules). *This definition lists proof rules to be used to define a consequence relation for* $\mathbf{H}_1^S$.

1. *Axioms*
   $(m,t) : A$, *for A a substitution instance of a truth functional tautology.*
2. $$\frac{(m,t) : A; (m,t) : A \to B}{(m,t) : B}$$
3. a) $$\frac{(m,t) : \Box A; t\rho s}{(1-m,s) : A}$$
   b) $$\frac{(1-m,s) : A; t\rho s}{(m,t) : \Diamond A}$$
4. a) $$\frac{(0,t) : \Box A}{(1,t) : A}$$
   b) $$\frac{(1,t) : A}{(0,t) : \Diamond A}$$
5. $$\frac{(m,t) : \bot}{(n,s) : A}$$
6. $$\frac{(1,t) : \Diamond A}{\textit{create a } \text{new } s, t\rho s \textit{ with } (0,s) : A}$$
7. $$\frac{(0,t) : \Diamond A; (1,t) : \neg A}{\textit{create a new } s, t\rho s \textit{ with } (1,s) : A}$$
8. $$\frac{(m,t) : A}{(1-m,t) : A}$$
   for A not containing $\Box$
9. *Rules 6–7 are referred to as* creative, *the others as* non-creative.

Definition 1.45 (Consequence). *Let* Δ *be a theory and let* $(m,t) : A$ *be a declarative unit. We define the notion of* $\Delta \vdash_n (m,t) : A$. *n is the number of creation steps used.*

1. $\Delta \vdash_0 (m,t) : A$ *if* (m,t) *appears in* Δ *and* $(m,t) : A$ *can be reached after a finite number of applications of the non-creative rules of the previous definitions to the units in* Δ.
2. $\Delta \vdash_0 \Delta'$ *if* $\Delta \vdash_0 (x,s) : B$ *for all* $(x,s) : B$ *in* Δ'.
3. $\Delta \vdash_{n+1} (m,t) : A$ *if* Δ *is obtained from a* Δ', Δ'' *such that* $\Delta \vdash_0 \Delta'$ *and* Δ'' *is obtained from* Δ' *by the application of a creative rule (and the new declarative unit added) such that* $\Delta'' \vdash_n (m,t) : A$.
4. $\Delta \vdash (m,t) : A$ *if for some n,* $\Delta \vdash_n (m,t) : A$.
5. Δ *is consistent if for no* (m,t) *do we have* $\Delta \vdash (m,t) : \bot$.

Definition 1.46. *Let* Δ *be a theory and* (S,R,a,h) *a model. Let g be a function* $g : \Delta \mapsto S$ *such that for any t appearing in* Δ, $g(t) \in S$.

Assume g satisfies the following:

- *If* $\pm t\rho s \in \Delta$ *then* $\pm g(t)Rg(s)$ *holds.*

- *If* $t = s$ *(resp.* $t \neq s$*) in* Δ *then* $g(t) = g(s)$ *(resp.* $g(t) \neq g(s)$*) holds.*
- $g(\mathbf{d}) = a$.
- *If* $(m, t) : A \in \Delta$ *then* $g(t) \vDash_m A$ *holds.*

then we say that (S, R, a, h, g) is a model of Δ.

Theorem 1.47. *If* Δ *is consistent then it has a model.*

Proof. The usual Henkin/tableaux construction.

1.6 Conclusion: Hypermodality in Context

One can certainly look at the hypermodal framework as a way of significantly broadening the expressive power of modal logic in an interesting way. However, there is another, more general way of looking at hypermodality, as a manifestation of a more general proof-theoretic mechanism. This point of view will now be explained. It is best done by an example.

Consider **S4** strict implication, $A \Rightarrow B$. We can understand it as having a future temporal meaning. Read $A \Rightarrow B$ as an *insurance policy*, saying that whenever an accident happens, (i.e. A holds) then remedies are guaranteed (B holds). It is clear that the accident must happen after the policy was taken out. Thus if we list all data available according to the earlier-later temporal occurrence then modus ponens can be effected only in the form

$$A \Rightarrow B, A \vdash B$$

and not in the form

$$A, A \Rightarrow B \vdash ?B.$$

We can be even more careful and record that B does not come at the same time as A, but afterwards. (Compare with Definition 1.2, items 2 and 4.)

Now let us look at automobile insurance known as *Home Start*. This allows you to call the company any time if your car does not start and they will come and start/charge your battery, wherever you are. Arrival is guaranteed within the hour. There is a catch, however. You are allowed to use this service only a fixed number of times (I think it is four times). So let us see what this means in terms of modus ponens. We will have to write something like

$$(\text{mode } n)\ A \Rightarrow B, A \vdash B\ (\text{mode } n - 1)$$

where the (mode n) says how many times you can use $A \Rightarrow B$.

The proof theory will have to tell us how to change mode with every use of modus ponens.

This example is very realistic and can become very complex. I can now make my general point:

Logic is now evolving and becoming more and more complex in response to the needs of computer science and artificial intelligence. These disciplines are building devices which help/replace humans in their daily activities. To successfully sell such devices to the public, we need to properly model (in logic) various aspects of human activity and use these logical models to help build these devices. The urgency of this need is fuelling the development of new concepts in logic. The process of modelling is the same as the intuitive analysis traditionally done by philosophical logic, only more urgent and more computational. The notion of mode is one such concept arising in this context.

I believe it is going to be central in the modal logic of the future.

2

Introducing Reactive Kripke Semantics and Arc Accessibility

2.1 Motivation and Background

2.1.1 The reactive idea[1]

Traditional modal logic uses possible world semantics with accessibility relation R. When we evaluate a formula such as $B = \Box p \wedge \Box^2 q$ in a Kripke model $\mathbf{m} = (S, R, a, h)$ (S is the set of possible worlds, $a \in S$, $R \subseteq S^2$ and h is the assignment) the model $\mathbf{m}$ does not change in the course of evaluation of B. We say the model $\mathbf{m}$ is not *reactive*. It stays the same during the process of evaluation.

To make this point absolutely clear, consider the situation in Figure 2.1 below[2]

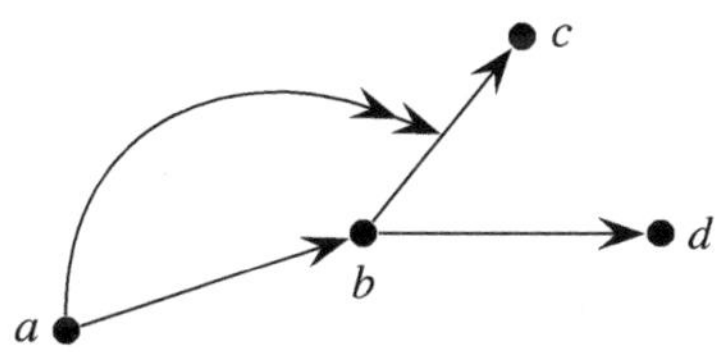

Fig. 2.1

To evaluate $a \vDash \Box^2 q$, we have to check $b \vDash \Box q$. We can also check another formula at b, say, $b \vDash \Box p$. In either case the worlds accessible to b are c and d.

We *do not* say that since $b \vDash \Box q$ started its evaluation at world a as $a \vDash \Box^2 q$ and continued to $b \vDash \Box q$, then the accessible worlds to b are now different. In other words

[1] The idea of reactivity was first presented at CombLog 04, July 28–30, 2004, and published in the proceedings [88]. See `www.cs.math.ist.utl.pt/comblog04/talks.html`

[2] Single arrows indicate point-to-point accessibility, double arrows indicate point-to-arc accessibility.

D.M. Gabbay, *Reactive Kripke Semantics*, Cognitive Technologies,
DOI 10.1007/978-3-642-41389-6_2, © Springer-Verlag Berlin Heidelberg 2013

the model does not react to our starting the evaluation of $a \vDash \Box^2 q$ by changing the accessible worlds at b (for example, see the double arrow in Figure 2.1, which indicates that we can disconnect the accessibility of node c to b) and therefore allowing us to see a different set of accessible worlds when we continue the evaluation of $b \vDash \Box q$.

The evaluation of $\Box$ at b *does not* depend, in traditional Kripke semantics, on how we 'got' to b.

This chapter addresses the case where the semantics does change (or react) under us as we evaluate a formula. This idea makes the evaluation of a wff at a world t dependent on the route leading to t. Thus we get a new kind of semantics, the reactive semantics. This semantics is stronger than ordinary Kripke semantics as there are axiomatic propositional logics with one modality $\Box$ which cannot be characterised by a class of traditional Kripke models but which are complete for a class of reactive Kripke models.

The idea of dynamic interaction in logic, classical or non-classical, is not new. Interactive logic has been done in many guises. The most well-known example is the game-theoretic semantics for the classical quantifiers, and hence for modalities, which has extensive applications in natural language analysis and in theoretical computer science. A related interactive system is dialogue logic with its vast applications to logic and to argumentation theory. The paradigm of these systems is different from what we are offering here. The interactive paradigm involves two players over a fixed object (a model or a database), who have their respective goals and rules which allow them to manipulate the object and each other in order to win the interaction. Our paradigm is different. As we discuss later in Section 2.2, we consider systems which react and change because of built-in faults and remedies which get activated while they are used. Applied in this context we can roughly say that we allow the object as well as the rules constraining the players to react to the moves of the players and try to influence them. As the players make their moves the object which they manipulate as well as the game rules governing them can change.[3]

Meanwhile let us give an example based on Figure 2.1, to give the reader some immediate comparison.

Example 2.1. Consider the Kripke model described in Figure 2.1. It has the nodes $S = \{a, b, c, d\}$ with $R = \{(a, b), (b, c), (b, d)\}$. Let the actual world be a.

Let the assignment h for the atom q be $h(q) = \{a, b, d\}$. Thus q is false at world c but holds at worlds a, b and d.

Let us evaluate $a \vDash^1_h \Box^2 q$ in a traditional manner (use superscript 1 for traditional)

$$a \vDash^1_h \Box^2 q \text{ iff } b \vDash^1_h \Box q \text{ iff both } c \vDash^1_h q \text{ and } d \vDash^1_h q \text{ iff false.}$$

3 If the reader insists on the interactive game-theoretical point of view then we can regard the reactive paradigm as adding an additional player to the game which is trying to modify and obstruct the two existing players. Whether this point of view is technically correct and/or useful is a matter for investigation towards the end of the chapter, after we define our reactive semantics. This additional player may not have any strategy beyond interfering with the game of the other two players. It may be regarded as a faulty environment which changes unpredictably. Michael Gabbay suggested that perhaps we can study a third player whose winning strategy is to prevent the other two players from having a winning strategy.

If we allow the model to be reactive (use superscript 2) we get

$$a \vDash_h^2 \Box^2 q \text{ iff } b \vDash_h^2 \Box q \text{ iff } d \vDash_h^2 q \text{ iff true.}$$

under $\vDash_h^2$, once we passed through node a, c was no longer accessible from b when we came to evaluate $b \vDash_h^2 \Box q$.[4]

The reader might wish to say that we have two models here, one with c accessible to b and one with c not accessible to b.

This view does not work in general. Suppose the reactive model switches bRc on and off each time we make a move. Then whether bRc holds or not depends on how many moves we make to get to b. To illustrate this point, let us modify the model and assume that aRa holds and evaluate $a \vDash \Box^3 q$.

$$a \vDash_h^2 \Box^3 q \text{ iff both } (*)\ a \vDash_h^2 \Box^2 q \text{ and } (**)\ b \vDash_h^2 \Box^2 q \text{ hold.}$$

Since we made only one move, the node c is not accessible to b.

Let us continue:

$$(*)\ a \vDash_h^2 \Box^2 q \text{ iff both } (*1)\ a \vDash_h^2 \Box q \text{ and } (*2)\ b \vDash_h \Box q \text{ hold.}$$

Since we now made two moves, we have that c is again accessible to b and hence $(*2)$ is evaluated at b with bRc active.

On the other hand, $(**)$ is evaluated at b with bRc not active.

We shall see later that this reactive semantics is not reducible to the traditional semantics.

Let us now check how a game-theoretical semantics would evaluate $\Box^2 q$ in this model. We have two players **A** and **B**. **A** claims that $\Box^2 q$ holds and **B** claims it does not hold. **B** moves by choosing an accessible point (e.g. b) and challenges **A** to claim that $\Box q$ holds at b. The game goes on and **A** must have a strategy for winning. It is clear that the model is fixed in this game.

We can add a third player **C** who makes changes to the model (and the game rules, if you want) in reaction to the moves made by **A** and **B** and we require a winning strategy for **A** against both **B** and **C**.

2.1.2 Examples motivating the reactive idea

Before we continue with more technical material, let us motivate our idea of reactive semantics and consider some case studies.

Airline example

We begin with a very simple and familiar example. Consider Figure 2.2.

Figure 2.2 gives the possible flight routes for the aeroplanes of TUA (Trans Universal Airlines). It is well known that many features of a flight depend on the route.

---

[4] Think of node a transmitting a signal along the double arrow and disconnecting the arrow from b to c.

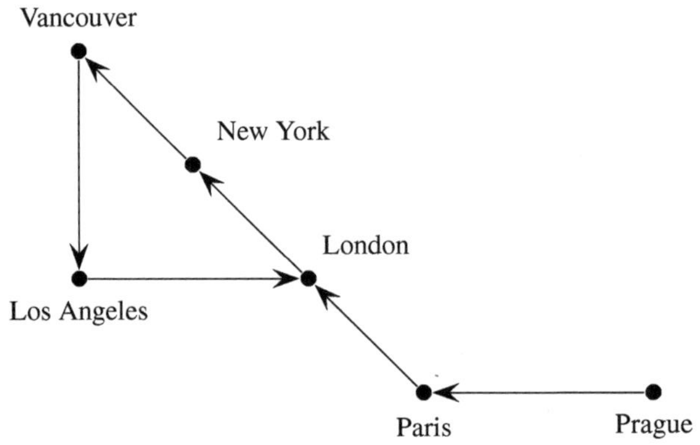

Fig. 2.2

These include the cost of tickets, as well as the right to take passengers at an airport. The right to take passengers at an airport depends on the flight route to that airport and on bilateral agreements between the airlines and governments. Thus, for example, flights to New York originating in London may take on passengers in London to disembark in New York. However, a flight starting at Paris going to New York through London may not be allowed to pick up passengers in London to go on to New York. It is all a matter of agreements and landing rights. It is quite possible, however, that on the route Prague–Paris–London–New York, the airline is allowed to take passengers in London to disembark in New York. We can describe the above situation in Figure 2.3.

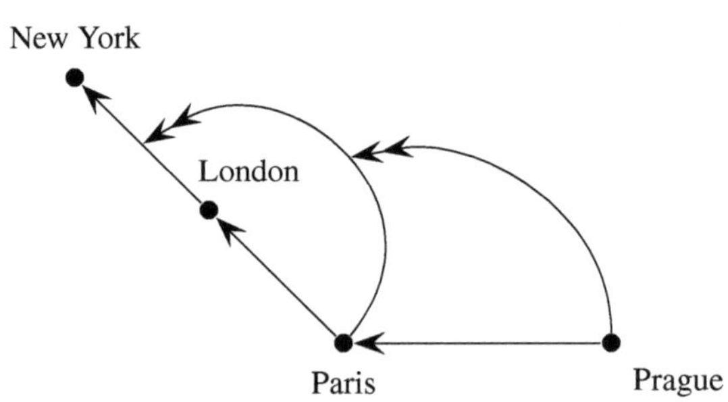

Fig. 2.3

The double-headed arrow from Paris to the arc London→New York indicates a cancellation of the 'passenger' connection from London to New York. The double-headed arrow from Prague to the double arrow arc emanating from Paris indicates a cancellation of the cancellation.

Figure 2.3 looks like a typical reactive Kripke model, where we have arcs leading into arcs.

Let us see more examples of this.

Inheritance networks example

This example offers a different point of view on arc semantics, coming from the non-monotonic theory of inheritance networks. Consider Figure 2.4.

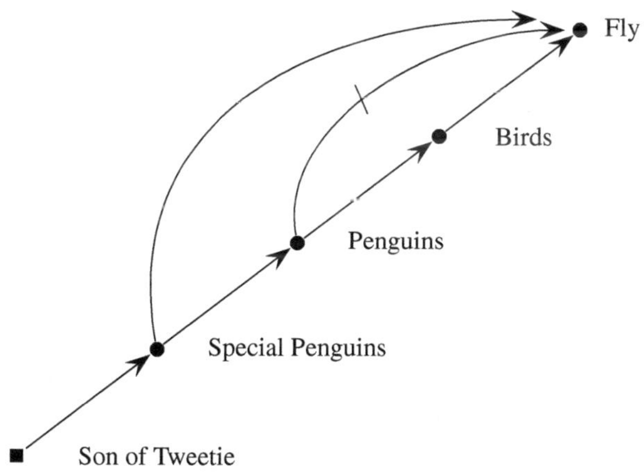

Fig. 2.4

In Figure 2.4, the circular nodes are predicates, such as Fly, Birds, etc. The arrows indicate inheritance, so for example, we have $\forall x(\text{Bird}(x) \to \text{Fly}(x))$. The arrows with a bar indicate blockage, for example $\forall x(\text{Penguin}(x) \to \neg\text{Fly}(x))$. The square nodes indicate instantiation, so Son of Tweetie is a special penguin.

Figure 2.4 is the kind of figure one finds in papers on inheritance networks. The figure indicates that Penguins are Birds, that Birds Fly but that Penguins do not Fly. However, Special Penguins do Fly and the Son of Tweetie, a rare bird, is a Special Penguin, and therefore does Fly. The arrow with the bar on it blocks the information from flowing from the Penguin node to the Fly node. The theory of inheritance networks spends a lot of effort on algorithms that allow us to choose between paths in a network so that we can come up with the desired intuitively correct answers. In the case of Figure 2.4 we want to get that Son of Tweetie does fly, since we have the most specific information about him. It is not important to us in this chapter to take account of how inheritance theory deals with this example. We want to look at the example from our point of view, using our notation, as in Figure 2.5.

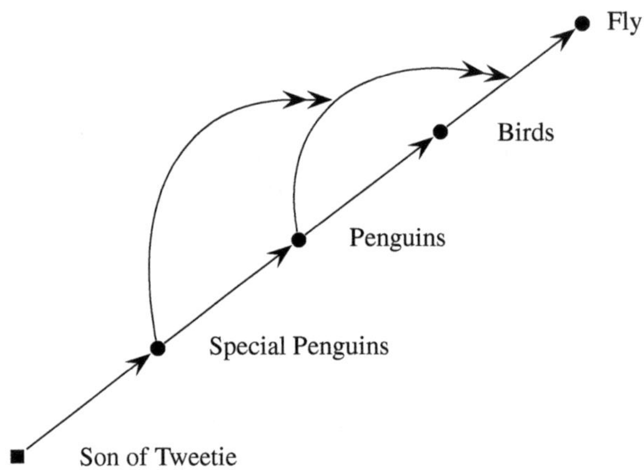

Fig. 2.5

In Figure 2.5 the double headed arrow $\twoheadrightarrow$ emanating from Penguins attacks the arrow from Birds to Fly, and the double arrow emanating from Special Penguins attacks the double arrow emanating from Penguins and attacking the arrow from Birds to Fly. This is not how inheritance theory would deal with this situation but we are not doing inheritance theory here. Our aim is to motivate our approach and what we need from the inheritance example is just the idea of the algorithmic flow of information during the dynamic evaluation process.[5]

We have already put forward the idea of reactive and dynamic evaluation in earlier papers and lectures (see [63]). A typical example we give is to consider $t \vDash \Diamond A$. In modal logic this means that there is a possible world s such that we have $s \vDash A$. We take a more dynamic view of it.

We ask: where is s? How long does it take to get to it? and how much does it cost to get there?

The reader should recall the way circumscription theory deals with the Tweetie example, see [121, section 4.1, especially page 324]. We write

- Birds $(x) \wedge \neg Ab_1(x) \rightarrow$ Fly(x)
- Penguins$(x) \rightarrow$ Birds(x)

[5] It is our intention to explore whether our idea of double headed arrows cancelling other arrows can simplify inheritance theory algorithms.

- Penguins(x) $\wedge \neg Ab_2(x) \to Ab_1(x)$
- Special Penguins(x) $\to$ Penguins(x)
- Special Penguins (Son of Tweetie)
- Special Penguins (x) $\to Ab_2(x)$.

"$Ab(x)$" stands for "x is abnormal". If the clause $C(x) \to B(x)$ represents the arc $C \to B$ then $C(x) \wedge \neg Ab(x) \to B(x)$ represents the situation in Figure 2.6

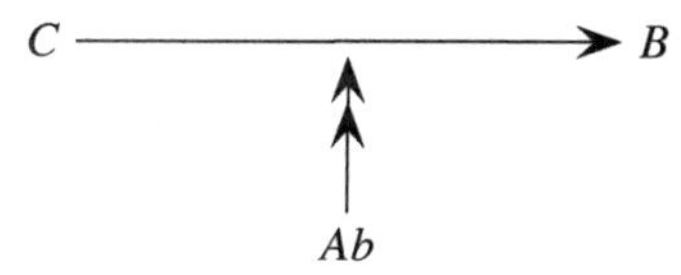

Fig. 2.6

A technical example

It is now time to give a technical example. Consider Figure 2.7 below. This figure displays a past flow of time. The node t is the present moment and a single-headed arrow from one node to another, say from s to t, means that t is in the immediate future of s. We use the modality $\Box$ to mean 'always in the immediate past'. Thus the accessibility relation R of Figure 2.7 is as follows:

- $tRs, tRb, bRs, sRa, tRt, bRb, sRs$ and aRa.

The double-headed arrows cancel the accessibility relation.

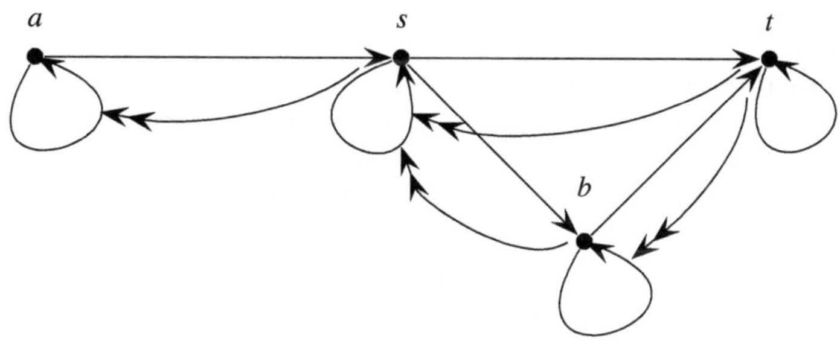

Fig. 2.7

Let us calculate $t \vDash \Box^3 q$ in Figure 2.7.

Initial Position: Starting point is t and all arrows are active.
Step 1: Send double arrow signal from t to all destinations inverting the active/inactive status of all destination arrows. Then go to all accessible worlds (in this case s and b) and evaluate $\Box^2 q$ there. If the result is positive 1, representing 'true' at all nodes, then send 'success' back to node t.
Step 2: Evaluate $\Box^2 q$ at nodes b and s.
Subcase 2.b. Evaluation at b: First we send a double arrow signal from b to all destinations reversing the activation status of these destinations. Thus the single arrow from s to s will be re-activated and we will evaluate $\Box q$ at s at the next step 3 (with s accessible to itself). b is not accessible to itself because its arrow has been deactivated by t at Step 1.
Subcase 2.s. Evaluation at s: First we send a double arrow signal to reverse the status arrow from a to a. Then we evaluate $\Box^2 q$ at s with s not accessible to itself, since the arrow from s to s was deactivated by t at Step 1.
Subcase 2.t. Evaluation at t: This can go on, but we shall not continue as we trust that the reader has got the idea by now.

Note that if we start at t and evaluate $B = \Box^3 q \wedge \Box^2 q$, we will get that $\Box q$ must be evaluated at s in two ways: once with s not accessible to itself (coming from b via $t \vDash \Box^3 q$) and once with s accessible to itself (coming from t via $t \vDash \Box^2 q$).

Let us now calculate $t \vDash \Diamond^3 q$ in Figure 2.7.

Initial Position: Starting point is t and all arrows are active.
Step 1: Send double arrow signal from t to all destinations inverting the active/inactive status of all destination arrows. Then go to one of the accessible worlds (in this case t, s or b) and evaluate $\Diamond^2 q$ there. If the result is positive 1 at this node, then send 'success' back to node t.
Step 2: Evaluate $\Diamond^2 q$ at nodes b, s or t.
Subcase 2.b. Evaluation at b: First we send a double arrow signal from b to all destinations reversing the activation status of these destinations. Thus the single arrow from s to s will be re-activated and we will evaluate $\Diamond q$ at s at the next step 3 (with s accessible to itself). b is not accessible to itself because its arrow has been deactivated by t at Step 1.
Subcase 2.s. Evaluation at s: First we send a double arrow signal to reverse the status arrow from a to a. Then we evaluate $\Diamond^2 q$ at s with s not accessible to itself, since the arrow from s to s was deactivated by t at step 1.

This can go on, but we shall not continue as we trust that the reader has got the idea by now. For the case of $\Diamond$ we make a non-deterministic choice. The model is not sensitive to whether we come to a point because we are evaluating a $\Box$ or a $\Diamond$. If we want this kind of sensitivity we can have arrows of the form $\twoheadrightarrow_{\Box}$ and $\twoheadrightarrow_{\Diamond}$.

Such distinctions may be desirable in dealing with quantifier games, where changes may be different for the cases of $\forall$ and $\exists$.

Tax example

Having explained the technical side of our reactive (changing) semantics, let us give some real examples.

House prices in London have gone up a great deal. An average upper middle class family is liable to pay inheritance tax on part of the value of their house (suppose that if the house is valued over £500,000, for example, then there is tax liability on £250,000). Some parents solve the problem by giving the house as a gift to their children. If at least one of the parents remains alive for seven years after the transaction, then current rules say that no tax is due. Consider therefore the following scenario:

1. current date is April 2004
2. parents gave house as a gift to children in 1996
3. parents continued to live in house as guests of the children.

(1)–(3) above imply that (4):

4. if parents both die in March 2004, then no tax is liable.

To continue the story, there were rumours that the tax authorities were going to change the rules in April 2004, declaring that if the parents remain living in the house after it was given as a gift, then the gift does not count as such and there is tax liability. The rumours also said that this law is going to apply *retrospectively*.

Thus we have that (5) holds:

5. If parents both die in March 2005, then tax is liable.

We assume that (4) still holds even after the new law as we cannot imagine that the UK tax inspector would be reopening closed files and demanding more tax.

The way to represent (4) and (5) is to use two-dimensional logic. We write $t \vDash_s A$ to mean at time t A is true given the point of view proposed or held at time s.

Thus $2005 \vDash_{2003} \neg(5)$ holds, because from the 2003 laws point of view (before legislation) no tax is liable ((5) says tax is liable)). But $2005 \vDash_{2004} (5)$ also holds, because according to 2004 legislation tax is liable.

So far we have no formal problem and no need for our new semantics, because we can write

- $t \vDash_t \Box A$ iff for all future s, $s \vDash_s A$.

In other words we evaluate sentences at time t according to the point of view held at the very same time t.

The problem arises when we want to formalise the following scenario. The parents die in 2003. The lawyer is dealing with the estate. We do not know when he is going to finish. When he submits the paperwork then the tax liability in 2003 is judged according to the time of submission. Now the second index s in $t \vDash_s A$ behaves like a reactive model as we are evaluating

$$2003 \vDash_{\text{time lawyer submits}} (4).$$

Salesman example

Consider the simple graph of Figure 2.8.

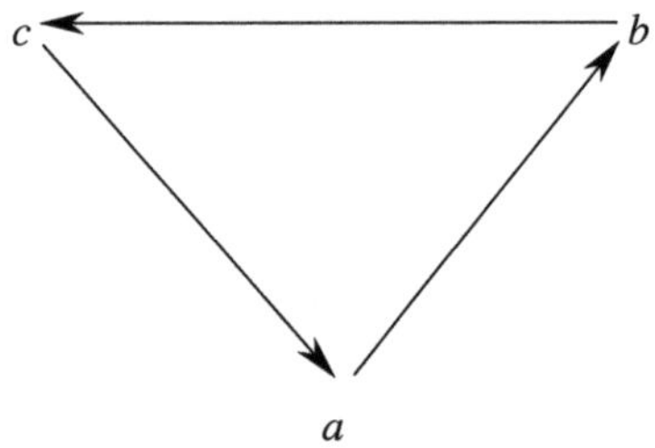

Fig. 2.8

A salesman wants to traverse this graph in such a way that he doesn't pass along the same edge twice. Such problems are very common in graph theory. The simplest way of implementing this restriction is to cancel an edge once it has been used. Figure 2.9 will do the job.

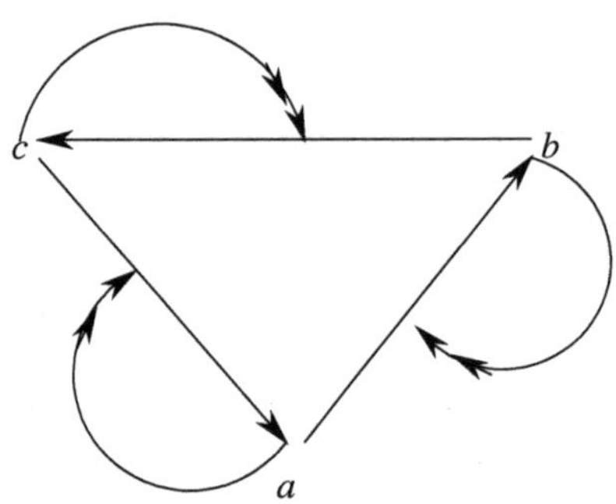

Fig. 2.9

One cannot always implement the salesman problem in this way. It depends on the graph we deal with. However, it is one more reason for considering arc accessibility.[6]

Resource example

Consider a road system as in Figure 2.10.

[6] For the salesman example we need accessibility from arcs to arcs. So in Figure 2.9, rather than have the double arrow $(b,(a,b))$, we need the double arrow $((a,b),(a,b))$. The arc cancels itself as we go through it.

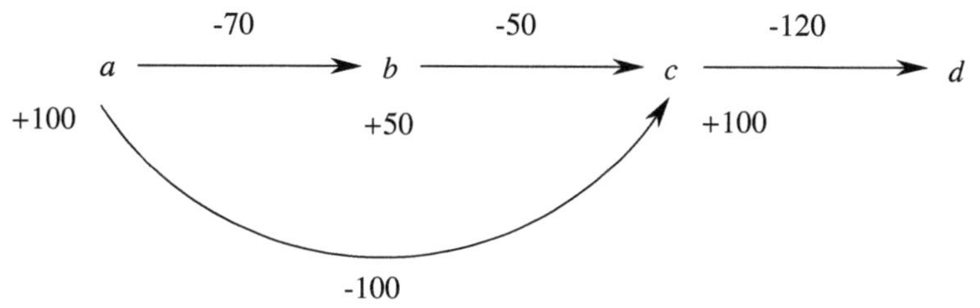

Fig. 2.10

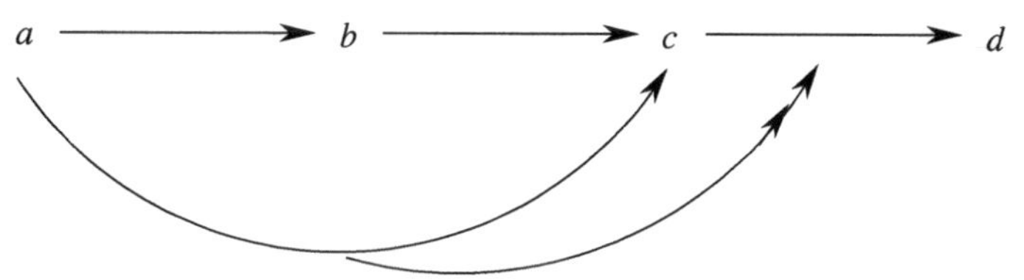

Fig. 2.11

Assume that whenever we drive along a road we need to pay a fee to pass. Payment must be in cash and it is very expensive. We can withdraw money from cash machines along the way to pay for our passage as follows:

Point *a*:	Cash withdrawal \$100 is available.
	Road *a* to *b*: costs \$ 70.
Point *b*:	Cash withdrawal \$50 available.
	Road *b* to *c*: costs \$50
Point *c*:	Cash withdrawal \$100 is available.
	Road *c* to *d*: costs \$120.

We start with no money at all at point *a*. If we drive from *a* to *c* directly, we can get \$100 from the cash machine at *a*, and this will get us to *c* with no money left and so we cannot get enough cash at *c* to pay for the passage from *c* to *d*. However, if we pass through *b*, we can withdraw money at *b* and later at *c* and we will have enough to pay for the passage from *c* to *d*.

Figure 2.10 is essentially an ordinary annotated graph describing the resource situation but the qualitative situation (where money considerations are hidden) can be described in Figure 2.11. The upshot of Figure 2.11 is that the arc from *a* to *c* sends a signal to cancel the arc from *c* to *d*. This is also the first case where we have double arrows going from arc to arc.

Flow products example

Ordinary products of Kripke frames are defined in a straightforward manner. Given two frames (S_1, R_1), and (S_2, R_2), we form the product space $S = S_1 \times S_2$, and define two modalities $\Box_1, \Box_2$ on pairs as follows:

- $(a, b) \vDash \Box_1 A$ iff $\forall x(aR_1x \to (x, b) \vDash A)$
- $(a, b) \vDash \Box_2 A$ iff $\forall y(bR_2y \to (a, y) \vDash A)$.

Figure 2.12 shows the configuration for the case of $S_1 = S_2 = N$, the set of natural numbers

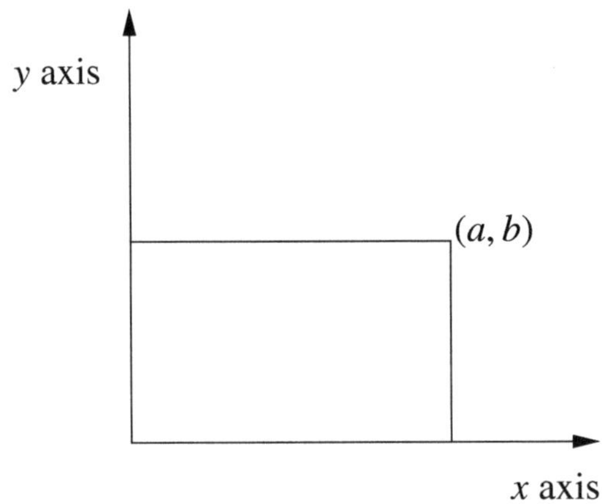

Fig. 2.12

$\Box_1$ shifts the x coordinate, leaving b fixed, and $\Box_2$ shifts the y coordinate, leaving a fixed.

We have among other things that $\Box_1\Box_2 = \Box_2\Box_1$. Products spaces are used whenever we deal with two independent modal or temporal aspects. This is a very active area of many-dimensional modal logics, and also related to classical predicate logic with a fixed number of variables. See our book [70].

The authors introduce in [63] the concept of flow products. Imagine the x axis is space (measured in kilometers) and the y axis is time measured in hours. Any shift in space will necessarily cause a shift in time. Assuming speed of 1 mile per hour, we get

- $(a, b) \vDash \Box_1 A$ iff $\forall u((a + u, b + u) \vDash A)$
- $(a, b) \vDash \Box_2 A$ iff $\forall u((a, b + u) \vDash A)$.

When we move in space, time shifts.

We can view this as a reactive model. Imagine for any point (a, b) the following double arrows exist

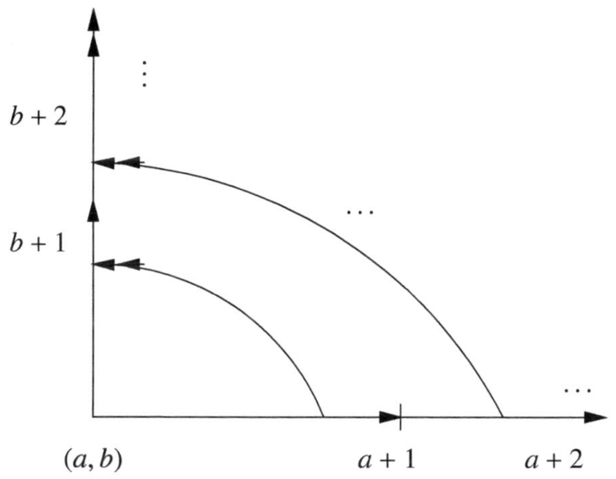

as we move from a to $a + 1$, the connection from b to $b + 1$ is switched off. Therefore we get $(a, b) \vDash \Box_1 A$ in the reactive model iff $\forall u((a + u, b) \vDash A$ in the new model after firing the reaction to moving through $a + 1, a + 2, \ldots, a + u$, which disconnects b from $b+1, b+2, \ldots, b+u)$. Hence $(a+u, b) \vDash \Box_2 B$ is now connected to $(a+u+1, b+u+1) \vDash B$, i.e. it is as if we are at the point $(a + u, b + u)$.

Of course a general flow product model will require a shift function $f(a, b, x)$ on the x-coordinate, satisfying

$$(a, b) \vDash \Box_1 A \text{ iff } \forall x(aR_1 x \to (x, f(a, b, x)) \vDash A.$$

$f(a, b, x)$ tells us how much the y coordinate shifts. f must satisfy some additivity properties, such as

- $aR_1 x \to bR_2 f(a, b, x)$
- $aR_1 x_1 \wedge x_1 R_1 x_2 \wedge aR_1 x_2 \to f(a, b, x_2) = f(x_1, f(a, b, x_1), x_2)$.

Flow products will be studied in Chapter 10.

2.1.3 Plan of this chapter

The next section, Section 2.2 describes the reactive paradigm in general and proceeds to give some specific examples from different application areas. These areas are studied in separate research papers. We also discuss several options for defining reactivity. The rest of the chapter gives some technical results in the area of modal logic. There is a lot to do here and the main bulk of the results is postponed to later chapters. In order to give some formal results in this chapter, we want to show that the class of switch reactive Kripke models can characterise more syntactic modal axiomatic systems than the class of ordinary Kripke models. This is done in Section 2.4. To prepare the ground for this result we first need to devote Section 2.3 to hyper-modalities, being a related but not an identical concept to the concept of reactive modalities. Section 2.4 concludes with two examples showing that the two concepts (hyper-modality and reactive modality) are different.

We now have two options of how to continue this chapter:

Option 1

Get more mathematical results for reactive modality. For example, characterise some modal axioms by conditions on models (on R) etc. Investigate complexity, introduce proof theory, etc, etc. See references.

Option 2

Continue with some general conceptual analysis of the concept of reactivity and comparison with relevant literature.

We follow the second option.

Section 2.5 introduces non-deterministic reactive Kripke models, Section 2.6 discusses connections with fibring logics and Section 2.7 discusses how to introduce dedicated reactivity operators.

Each of these topics has enough problems to investigate to merit a separate research paper.

We continue with four appendices which give further results and comparison with the literature.

2.2 The Reactive Paradigm in General

Consider a system S containing components and some internal connections and procedures governing the interactive behaviour of these components in response to external inputs. We can think of S, for example, as a graph (to be traversed or manipulated) or as an automaton which responds to inputs and changes states, or as a Kripke model in which a formula is evaluated, or as a bureaucratic system responding to an application form or even as a more familiar daily object such as a washing machine, a television set or a car.

We can represent such a system as a network containing possibly labelled vertices representing the components and arrows between the vertices representing possible "control flows" within the system of whatever the system's internal processes do.

Figure 2.13 is a typical situation we want to consider:

The arrows from a to b and from c to d indicate that internally whatever the system does in response to input or command, there is a possible path where component a passes the "flow" to component b and similarly, maybe in another part of the system, there is a connection between c and d.

So far we have nothing more than a possible network graphic representation of some system.

Now comes the reactive idea. Consider the possibility that the system develops faults due to overuse or stress. The double arrow from the arc (a,b) to the arc (c,d) (which we denote by $((a,b),(c,d))$) indicates how the fault may develop. When we pass from a to b we put internal pressure on the connection from c to d. If significant pressure builds up on the connection from c to d it may fail. For example everyone

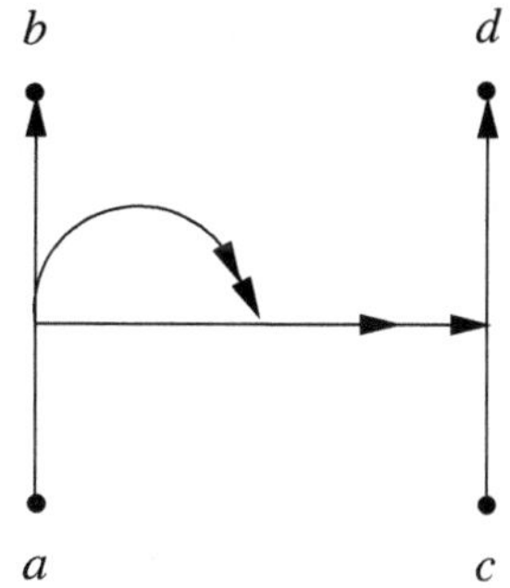

Fig. 2.13

who has driven a car a long distance during a very hot day knows that the engine may overheat. This is an example of a fault. In general these faults are predictable and are predetermined by the construction of the system. They are represented in Figure 2.13 by the double arrows.

In fact, Figure 2.13 also indicates a remedy to this pressure; this is a double arrow from the arc (a,b) to the double arrow $((a,b),(c,d))$, namely $((a,b),((a,b),(c,d)))$.

The double arrows represent known weaknesses or remedies in the system. We can look at the double arrows as indicating the ways in which the system reacts and adjusts itself under pressure. The simplest model is a once only bridge in Figure 2.14.

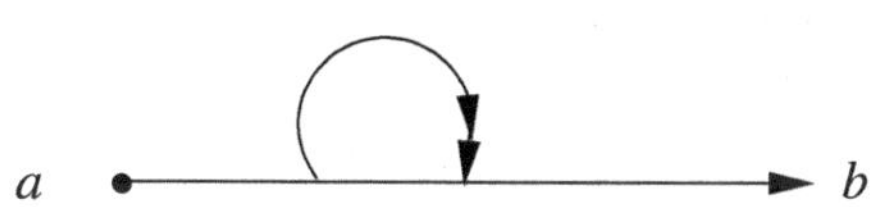

Fig. 2.14

When we cross the bridge from a to b, a signal (representing stress) is sent to the bridge to collapse and so after we cross the bridge the arc from a to b is cancelled.

The above is the basic intuition behind the idea of reactivity. Think of it as "fault" reactivity, or better as "fault-remedy" reactivity.

We now elaborate more about this concept to distinguish it from other somewhat similar concepts.

The "fault" reactivity should be distinguished from the idea of including dynamic metalevel operators in the input. To give an example of such metalevel operators, consider the possibility that the system may have several modes of operations, say *mode* 1 ,..., *mode n*.

The input may contain instructions of the form J_i (jump to mode i) which tells the system how to operate on the input. This is a metalevel instruction inducing a change in the system. It is different from "fault" reactivity, though the latter can be represented case by case in it. If we replace the double arrow to (c,d) by the metalevel instruction "disconnect(c,d)", we might be able to simulate the fault induced by an input by adding (interleaving) some "disconnect" instructions to the input to form a new input.

This option will be discussed later in the chapter in a case-by-case analysis. It is not reactivity but "fibring" of metalevel instructions giving the illusion of reactivity.

The notion of fault reactivity should also be distinguished from the notion of "sabotage", independently introduced by J. van Benthem [154] and further developed by P. Rohde [141]. The notion is similar; we envisage a saboteur disconnecting components within the system. There are both conceptual and technical differences between the two notions ("fault" reactivity and "sabotage" reactivity) and this will be analysed in detail in Appendix D. For the moment we note that faults and remedies are built into the system while sabotage is not and also sabotage is presented as a metalevel connective and fault is not. For more detail, see Appendix D.[7]

It turns out that as we apply this idea in various application areas where different non-reactive systems are used we can get new kinds of systems, the reactive systems, which can have better applicability in each respective area.

We give seven immediate examples:

1. Ordinary Kripke models become reactive Kripke models which can change during the process of evaluation, affording a wider class of models for modal logic.
2. Ordinary non-deterministic automata become reactive automata whose transition table changes every time they make a transition, see [37]. Although it is shown in [37] that every reactive automaton is equivalent to another ordinary non-reactive automaton, the reactive automata can perform tasks (recognise inputs) with many fewer states and represent substantial savings in computational costs!
3. An ordinary graph becomes a reactive graph where there are also arrows from edges to edges, see [62]. This area is very promising and is now under investigation.
4. Reactive grammars and rewrite systems [17]. For reactive rewrite systems we get new hierarchies of languages. Again this area is actively being investigated.
5. A proof system becomes a reactive proof system which changes as we advance in the proof process, see [57, 17] and Chapter 9 below.
6. Using reactive arrows we can solve contrary-to-duty paradoxes in Deontic logic, as well as model delegation in Talmudic logic , see [80, 81, 2].
7. In argumentation networks reactive arrows become higher-level attacks and this allows us to generalise the notions of various networks, see [80, 82, 15].

In each case a new class of models/graphs/automata/proof theory is introduced and some natural questions can be asked.

The following are some sample questions:

[7] Some sabotage may be non-deterministic, like water damage running through the system. This can be modelled as a non-deterministic reactive system. The possible options for water damage can be foreseen!

1. What is the expressive power/complexity of the new class and what can it do or not do?
2. What do traditional investigations yield when applied to this new class? (E.g. completeness theorems, cut elimination, correspondence theory, axioms, hierarchies of automata, theorems about cycles or classes of graphs, etc, etc.)
3. How does this concept relate to other dynamic/change concepts already existing in the literature?
4. Can the new models be reduced to known models via traditional interpretations? What is the edge/advantage we have by using the new reactive models?
5. Comparison and evaluation of the options for representing reactivity in the object level vs. metalevel.

Let us give some concrete examples, which will help us get a feel for the notion of reactivity.

Example 2.2 (Modal logic). Here we take the reactive Kripke models of the previous sections. The system is the evaluation process (for a given formula). The reactive parameter is a Kripke model with a point of evaluation t. The environment (or the faults in the system) is realised by the double arrows $\twoheadrightarrow$, which change the accessibility relation every time we make a move (i.e. every time we continue with the evaluation).

This example generalises slightly, giving it a familiar everyday meaning. Think of a network of nodes representing various components of a system. The arrows represent connections between components and a path through this graph following the arrows represents a way in which the system can be used. Imagine that the system is prone to faults and failures. So the use of one component may send a (double arrow) signal to other components and weaken or influence them.

We know this to be true for many real-life systems. So the reactivity of a network is just a measure of the ways it can fail and disappoint you!

Example 2.3 (Quantifier games). Recall the basic game semantics for the quantifiers. Given a classical model $\mathbf{m}$ and a formula, say $\Psi = \forall x \forall y \exists z(\varphi(x, y, z))$ with $\varphi(x, y, z)$ quantifier free, we play a game over the model between two players, **A** (claiming that Ψ holds) and **B** (claiming that Ψ does not hold). At step 0, **A** puts forward Ψ. At step 1 **B** challenges by choosing a in the domain of $\mathbf{m}$, and continues by choosing an element b in the domain of $\mathbf{m}$. It is the task of player **A** to supply a c such that $\varphi(a, b, c)$ holds. Player **A** has a winning strategy iff he has a function f such that for any a, b the element $c = f(a, b)$ is such that $\mathbf{m} \vDash \varphi(a, b, f(a, b))$.

In this case the moves are the choices of elements. The reactive environment player **E** has a tinkering function $\tau : (a, \mathbf{m}) \mapsto \mathbf{m}_a^\tau$ which changes a model $\mathbf{m}$ and an element a in the domain of $\mathbf{m}$ into a new model $\mathbf{m}_a$ with the same domain.

Thus in order to win, **A** has to have a winning function $\lambda\tau\lambda x\lambda y f_\tau(x, y)$ such that

$$\forall\tau\forall a\forall b\, \mathbf{m}_{a,b}^\tau \vDash \varphi(a, b, f_\tau(a, b)).$$

So as we can see, this definition covers the modal logic case, if viewed through its translation to classical logic.

For consider a reactive Kripke model **m** with actual world a. Consider the evaluation $a \vDash \Box^2\Diamond q$. This is translated into classical logic as

$$\Psi = \forall x \forall y \exists z (aRx \wedge xRy \rightarrow yRz \wedge Q(z)).$$

The function τ for changing the model is governed by the double arrows $\twoheadrightarrow$ and we clearly have a special case of the quantifier game's tinkering.

We get in this case the formula Ψ':

$$\Psi' = \forall x \forall y \exists z [aRx \wedge xR_a y \rightarrow yR_{a,x}z \wedge Q(z)]$$

where R_t denotes the new accessibility relation obtained from R after implementing the double arrows emanating from it.

Example 2.4 (Automata). A non-deterministic automaton $\mathcal{A} = (S, M, a, F, \Sigma)$ is characterised by a set of states S, an initial state $a \in S$, a set $F \subseteq S$ of final states and an alphabet Σ. M is a function giving for each state $t \in S$ and a letter $\sigma \in \Sigma$ a new set of states $M(t, \sigma) \subseteq S$, which are the states that the automaton $\mathcal{A}$ can non-deterministically move to.

Another way to view the automaton is as a multi-modal Kripke model of the form (S, R_σ, a, F), where R_σ for $\sigma \in \Sigma$ is defined by $xR_\sigma y$ iff $y \in M(x, \sigma)$.

A word of the form $(\sigma_1, \ldots, \sigma_n)$ is said to be recognisable by the automaton $\mathcal{A}$ iff there exists a sequence of states $x_1, \ldots, x_n$ such that $aR_\sigma x_1 \wedge \ldots \wedge x_{n-1}R_{\sigma_n}x_n$ holds and $x_n \in F$. If the atom q is assigned true on the set F, then $(\sigma_1, \ldots, \sigma_n)$ is recognisable by $\mathcal{A}$ iff $\mathcal{A} \vDash \Diamond_{\sigma_1} \ldots \Diamond_{\sigma_n} q$, where $\Diamond_\sigma$ corresponds to $R_\sigma, \sigma \in \Sigma$.

Thus naturally we can define a reactive automaton by adding double arrows $\twoheadrightarrow_\sigma$ for every $\sigma \in \Sigma$ in the places we want. Figure 2.15 is an example of such an automaton. Let $\Sigma = \{\sigma_1, \sigma_2\}, S = \{a, b, c\}$ and $F = \{c\}$.

We have

$$R_{\sigma_1} = \{(a, c), (b, c)\}$$
$$R_{\sigma_2} = \{(a, b), (b, c)\}$$

Figure 2.15 indicates the available double arrows. For example, the figure shows that as we move through node a in response to the letter σ_1 we disconnect (a, c) and (b, c) from the R_{σ_1} relation. If we move through node a in response to the letter σ_2 then we disconnect (a, b) from R_{σ_2}.

It is interesting to note that according to this figure, if we move out of state b in response to the letter σ_2 then we disconnect the connection (b, c) in the R_{σ_1} relation but not in the R_{σ_2} relation.

So there can be an interplay between the modalities here.

The reader should note that we can get a new hierarchy of automata here by taking the usual hierarchy and making it reactive. See Chapters 8 and 9 below.

The next remark, however, lists some possible options for creating reactivity.

Remark 2.5 (Options for the reactive property). We saw that the basic idea of a reactive action is to change the model every time a move is made. To explain our options on how to change the model consider Figure 2.16 below.

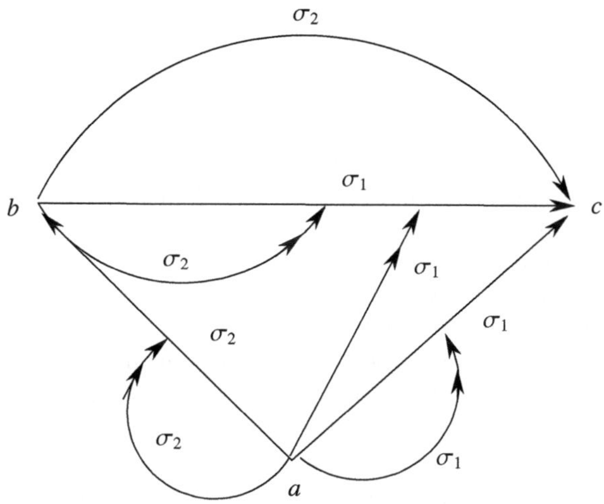

Fig. 2.15

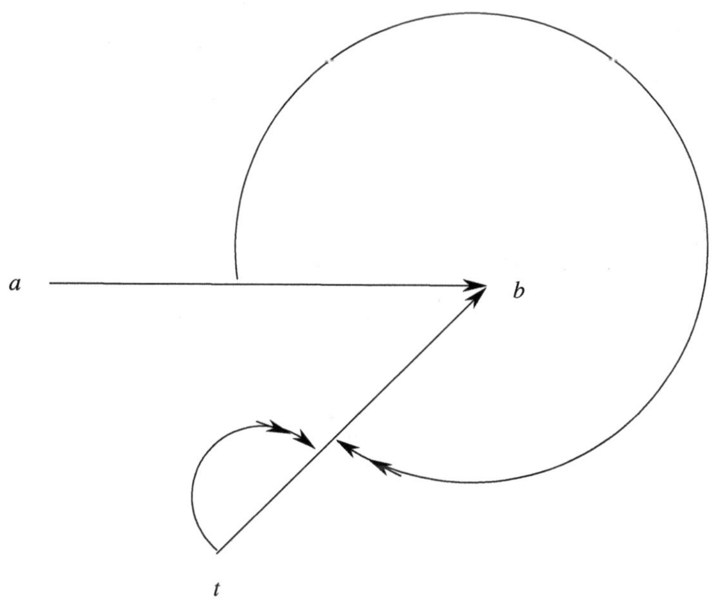

Fig. 2.16

From the point t arrows and double arrows emanate. Some may be active and some not. Also observe that we have a double arrow emanating from an arrow. Let us list what we have:

1. $t \to b$

2. $t \twoheadrightarrow (t \to b)$
3. $a \to b$
4. $(a \to b) \twoheadrightarrow (t \to b)$.

Our first list of options relates to what kind of arrows we allow. Do we want only arrows emanating from a point or do we also allow arrows emanating from arcs?

Items 1–3 above emanate from points and item 4 emanates from an arc.

The second list of options has to do with how we use the arrows to change the model as we pass through a point or an arc.

The first possibility is the switch-like use of the arrows. At any given moment some of the arrows are active (on) and some are dormant (off). When we pass through a point or an arc a signal is sent along the arrows emanating from the said point or arc which reverses the status of the target arrows at the destination. This is a simple switch action. In general, the reactivity can be more intelligent. The second possibility is to allow each node and arc from which arrows emanate to decide, depending on the state (on/off) of the target arrows, which of the target arrows to switch on and which to switch off. Care must be taken to ensure that this decision process is of the same complexity as the original non-reactive model. So for example in the case of the reactive finite state automaton of Example 2.4, the decision of each node which double arrows to activate should be done by another automaton.

Consider Figure 2.17 below

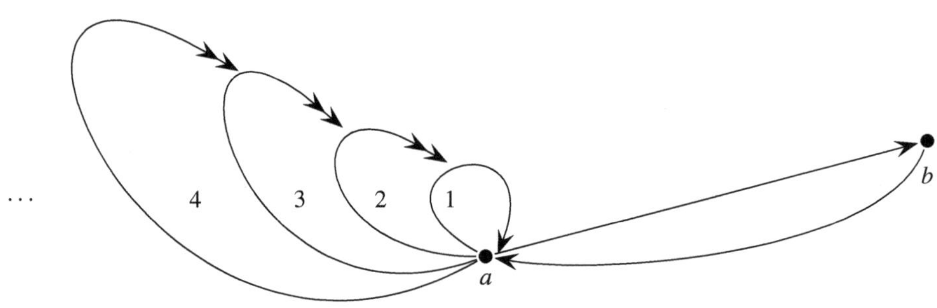

Fig. 2.17

Suppose we start with all arcs being switched on at the starting point a. As we move out of a to b, a switch behaviour will switch off all arcs $1, 2, 3, \ldots$. If we move through node a again, these arcs will be switched on again.

A more intelligent option might switch them on selectively one at a time, with each passage through the node a. This would allow us to use them as markers emulating aspects of a stack. Note that the idea is intuitively sound, and is not just a technicality. It makes sense to give a node some intelligence to decide how to react, based on the situation it 'sees'.

The above options were deterministic. The most general option was to attach an automaton at each exit point of arrows (of any kind) to decide what to switch on and what to switch off. It is also possible to make these decisions non-deterministic or probabilistic. Consider Figure 2.16. We can make all double arrows in this figure non-deterministic. So as we pass, for example, from node a to node b the double arrow to the arc $t \to b$ (namely $(a \to b) \twoheadrightarrow (t \to b)$), if active, may or may not (non-deterministically) send a signal.

Similarly we may attach a probability to such a connection, say 0.7, and so with probability 0.7 a signal will be sent. We shall elaborate more about this option in a subsequent paper.

Example 2.6 (Intelligent switches). Previous examples used on and off switches. The present example uses an intelligent switching system. In fact, we build on Example 2.4 (automata) and show in the present example how to simulate a stack automaton. We make sure the intelligent switching process is also done by a finite automaton. Consider the model described in Figure 2.18

This model has two relations R_{σ_1} and R_{σ_2}. It corresponds to an automaton with alphabet $\{q_1^+, q_1^-, q_2^+, q_2^-, t, a\}$ and stack letter α. The initial state is a and the terminal state is t. It is designed to recognise words of the form $\sigma_i^m \sigma_j^m, i \neq j, m \geq 1$.

The starting state is a. Upon seeing σ_i the automaton moves to state q_i^+ and writes α in the stack. It continues to write α as long as it sees σ_i. When it sees $\sigma_j, j \neq i$ it moves to q_i^- and starts deleting from the stack.

The double arrows are intelligent. The ones emanating from node a or a q^+ node activate the lowest non-active arrow or double arrow at α. The ones emanating from a q^- deactivate the highest active double arrow from α, providing α is active. If q^- sees that all arrows at α are not active it starts to activate arrow 2 and higher. So if arrow 1 is active then q^- deactivates the top arrow and if arrow 1 is not active then q^- activates the first non-active arrow above arrow 1. We need to assume that only a finite number of arrows and double arrows from α are active at any given time.

The following table describes the moves of the automaton of Figure 2.18. Note that only arrows at α may switch. The initial position is state a with all arrows at α not active. The terminal state is t.

So let us simulate an input computation.

Starting state
All double arrows and arrows are active except those at α which are not active.

Input step 1
We can assume without loss of generality that we get σ_1. So we move along R_{σ_1} from node a to node q_1^+. This move activates arrow 1 at α. The more σ_1s we see the more double arrows at α are activated in sequence. So if we see a total of m σ_1s, i.e. σ_1^m we get that arrows $1, \ldots, m$ are active. The first time we see σ_2 we move from q_1^+ to q_1^- along R_{σ_2}. There is no R_{σ_2} double arrow from q_1^+ to α.

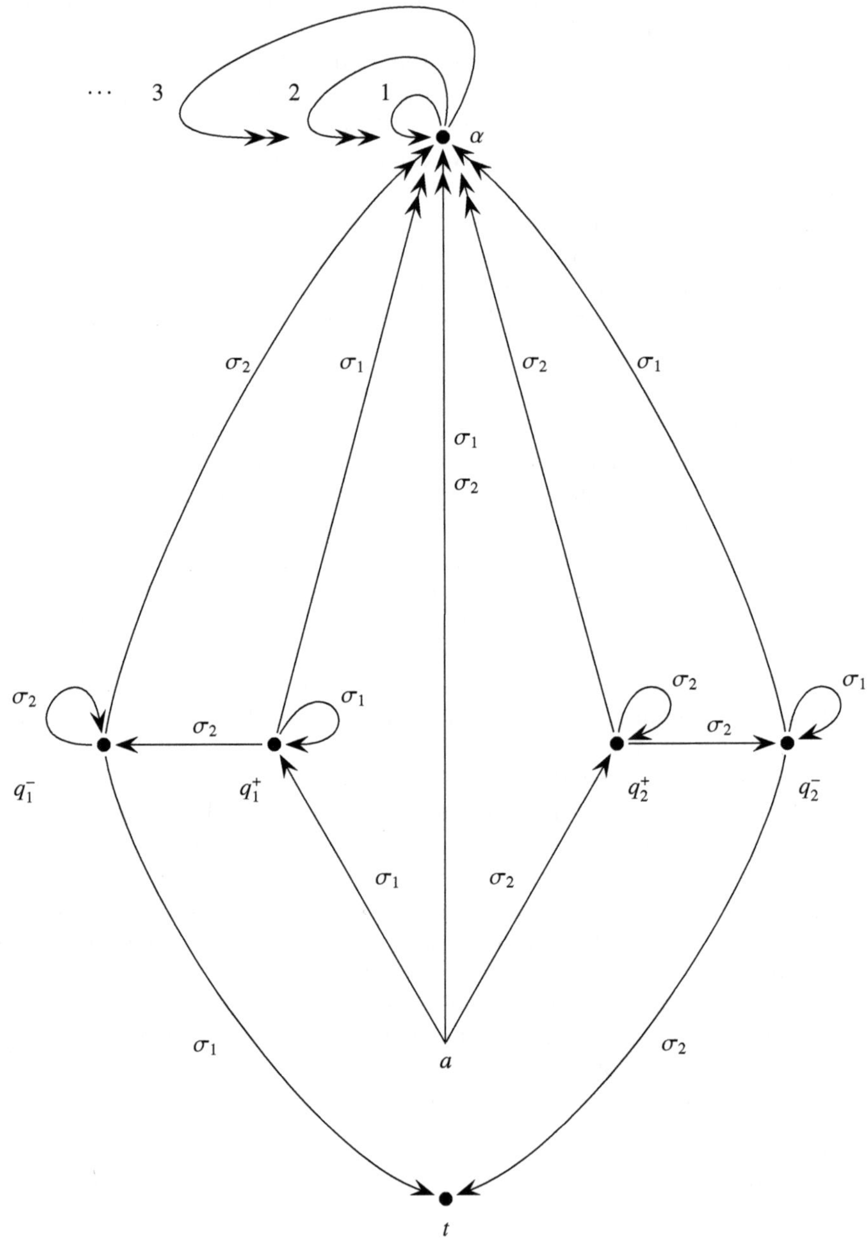

Fig. 2.18

If we continue to get σ_2 at q_1^- we cancel an arrow at α. The minute we get σ_1 again or the input finishes we move to a terminal state t. If the number of σ_2 is equal to σ_1, we end up stopping at t with no α arrow active. If the number of σ_2s is less than the number of σ_1s, we end up stopping with some α arrows active. If the number of σ_2s is larger than the number of σ_1s we are faced with a situation where the q_1^- automaton sees no arrow connections at α at all and needs to decide what to do. We can tell it to activate arrow 2, leaving arrow 1 not active and to continue to activate arrows $3, 4, \ldots$ as long as the input is σ_2. If the input is empty or q_1^- sees σ_1 again then it stops.

Example 2.7 (Options for interpreting necessity when double arrows emanate from arcs). Consider Figure 2.19 below:

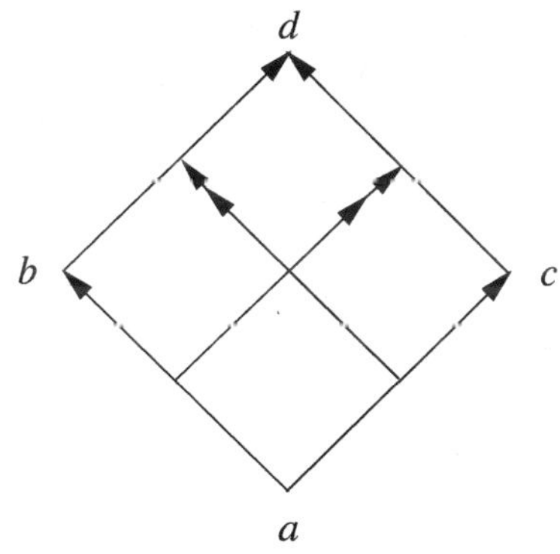

Fig. 2.19

In this figure we have two double arrows emanating from arcs. We ask: What semantic meaning can we give to $\Box$?

We want to evaluate

$$a \vDash \Box\Box\bot$$

How do we go about it? Do we go to every accessible point one by one and check whether $\Box\bot$ holds? Or do we go to all points *simultaneously*?

When double arrows emanate from points, there is no difference in the mode of operation, but when they emanate also from arcs, then there is a difference

Case 1 Separate evaluation of $\Box$
When we move from a to c, the arc (c, d) is still connected and so we have that $c \vDash \neg\Box\bot$. Similarly when we move from a to b, the arc (b, d) is still connected and so we have that $b \vDash \neg\Box\bot$.

Therefore, according to the separate evaluation of $\Box$, we get that $a \vDash \Box\neg\Box\bot$.

Case 2 Simultaneous evaluation of $\Box$
If we move to b and to c simultaneously and evaluate at b and at c both arcs (b, d) and (c, d) are disconnected and so we have $b \vDash \Box\bot$ and $c \vDash \Box\bot$ and hence $a \vDash \Box\Box\bot$.

It makes more sense to adopt the separate evaluation of $\Box$ because of the traditional connection of $\Box$ with $\Diamond$, namely $\Box = \neg\Diamond\neg$. We have:
$a \vDash \Diamond A$ iff for some accessible point s, when we move to s we have $s \vDash A$.

We also have $a \vDash \Box A$ iff for all accessible points s when we move to s we have $s \vDash A$.

To preserve the duality $\Diamond = \neg\Box\neg$ we must adopt the separate evaluation of $\Box$.

2.3 Connection with Hyper-modalities

In Chapter 1, we introduced hyper-modal logics. We showed that such modalities cannot always be characterised by a class of Kripke frames. However, there is hope that our new reactive semantics might provide frames for some of these modalities. This section and the next study the connection.

A hyper-modality $\Box$ is a modality which changes its nature depending on where in a formula it appears. So for example, in the formula $B = \Box^3 q \wedge \Box^2 q$, the inner modalities may not have the same meaning as the outer ones. To illustrate this point consider the arrangement of Figure 2.20

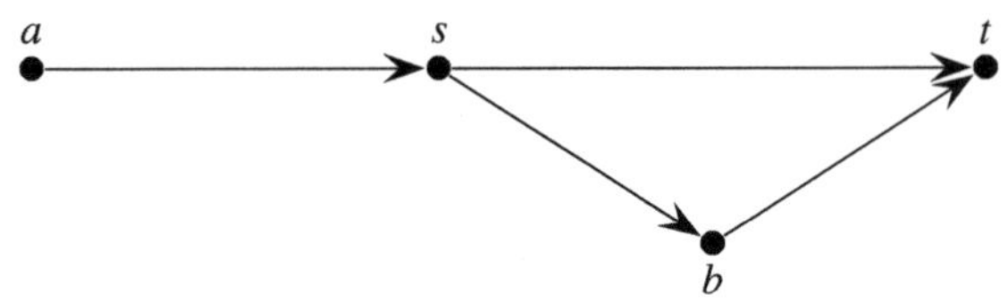

Fig. 2.20

t is *now*, s is in the past of t and so are a and b. We consider two past operators

- $H_{\mathbf{K}}A$ saying A was true at all the immediately past moments of time

and

- $H_{\mathbf{T}}A$ saying $A \wedge H_{\mathbf{K}}A$.

We let $\Box$ alternate between $H_{\mathbf{K}}$ and $H_{\mathbf{T}}$, starting with $H_{\mathbf{T}}$.

Thus $B = \Box^3 q \wedge \Box^2 q$ reads

$$H_{\mathbf{T}}H_{\mathbf{K}}H_{\mathbf{T}}q \wedge H_{\mathbf{T}}H_{\mathbf{K}}q.$$

Let us evaluate B at t. We write $t \vDash_{\mathbf{K}} A$ when we are evaluating A in the $\mathbf{K}$ mode and $t \vDash_{\mathbf{T}} A$ when we are evaluating A in the $\mathbf{T}$ mode.

Writing the above in full we have:

(*1) $t \vDash_{\mathbf{K}} \Box A$ iff for all immediately past points n we have $n \vDash_{\mathbf{T}} A$.
(*2) $t \vDash_{\mathbf{T}} \Box A$ iff $t \vDash_{\mathbf{K}} A$ and for all immediately past points n we have $n \vDash_{\mathbf{K}} A$.

Let us now evaluate $t \vDash (\Box^3 q \wedge \Box^2 q)$ in the flow of Figure 2.20. We have (remember we start with $\vDash_{\mathbf{T}}$):

- $t \vDash_{\mathbf{T}} \Box^3 q$ iff first $s \vDash_{\mathbf{K}} \Box^2 q$ and second $b \vDash_{\mathbf{K}} \Box^2 q$ and third $t \vDash_{\mathbf{K}} \Box^2 q$ iff first $a \vDash_{\mathbf{T}} \Box q$ and second $s \vDash_{\mathbf{T}} \Box q$ and third $b \vDash_{\mathbf{T}} \Box q$ and $s \vDash_{\mathbf{T}} \Box q$.
- $t \vDash_{\mathbf{T}} \Box^2 q$ iff first $s \vDash_{\mathbf{K}} \Box q$, and second $b \vDash_{\mathbf{K}} \Box q$ and third $t \vDash_{\mathbf{K}} \Box q$.

Since both $t \vDash_{\mathbf{T}} \Box^3 q$ and $t \vDash_{\mathbf{T}} \Box^2 q$ must hold, we see that we need to evaluate both $s \vDash_{\mathbf{T}} \Box q$ and $s \vDash_{\mathbf{K}} \Box q$.

This means that we cannot make the evaluation of $\Box q$ at s be dependent solely on the properties of the set $\{y \mid yRs\}$. We do need the dependency on the **T** and **K** modes.

Indeed, we axiomatise in Chapter 1 a modal logic with only the connective $\Box$ with the property that this logic can be characterised by the two **K** and **T** modes but it cannot be characterised by any class of frames. This shows that mode shifting is a genuinely stronger instrument of defining modal logics than imposing conditions on the accessibility relation R. We will show in this chapter that this logic can be characterised by a class of reactive models.

Let us now proceed to show the connection with the reactive semantics of this chapter.

First observe that the modes described above change the meaning of $\Box$. The modes do not change the semantics. In other words, the geometry of Figure 2.20 remains fixed. The model has not changed during the course of evaluation of $t \vDash B$. We want to show that we can achieve the same effect by changing the semantics as we evaluate.

Consider Figure 2.21.

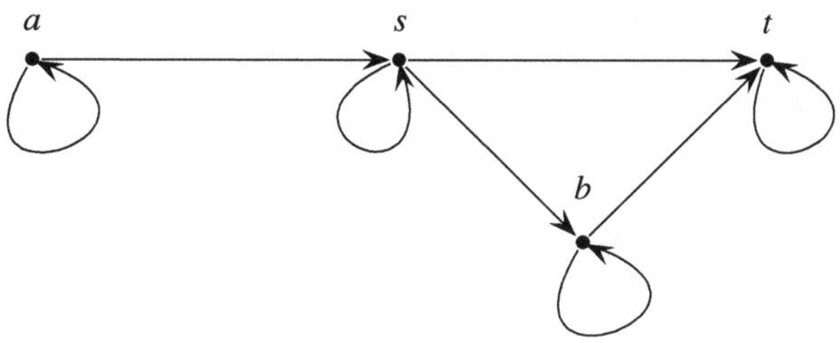

Fig. 2.21

Figure 2.21 describes the same flow of time as Figure 2.20 with the addition of the property that time is reflexive. Now suppose we have two modes of evaluation, $\vDash_{\mathbf{T}}$, where we evaluate $\Box$ in the reflexive mode (i.e. Figure 2.21), and $\vDash_{\mathbf{K}}$, where we evaluate in the irreflexive mode (i.e. in Figure 2.20).

Let us spell it out clearly:

(*3) $t \vDash_{\mathbf{K}} \Box A$ iff we first deactivate all reflexive arrows in the accessibility relation R and then ask for $s \vDash_{\mathbf{T}} A$ to hold at every s which is accessible to t.

(*4) $t \vDash_{\mathbf{T}} \Box A$ iff we first reactivate all cancelled reflexive arrows in R and then ask for $s \vDash_{\mathbf{K}} A$ to hold at every s accessible to t.

Clearly the evaluation of $t \vDash B$ will end up the same whether we view it as shifting the meaning of $\Box$ or shifting the underlying accessibility relation in the model.

Let us view the changing of the semantics as disconnecting or reconnecting arrows (accessibility) in the model. This we have already done in the technical example of Section 1.3. See Figure 2.7.

It is not difficult to see how a general arc model can be constructed in which $\Box$ alternates between a **T** and a **K** modality.

To do this properly, we first need a formal definition of the reactive semantics for modal logic.

2.4 Switch Reactive Kripke Models

We now give a definition of what we call switch reactive Kripke models, giving rise to what we call switch arc modal logics. In these models the reactive double arrows are just on/off switches, and they emanate from points.

Definition 2.8. *Let S be a set of possible worlds and $a \in S$ is the actual world. An arc-accessibility relation on S is defined as follows:*

1. *The set of all arcs $\mathcal{A}$ is defined by*
   1.1. *$S \subseteq \mathcal{A}$, these are 0 level arcs.*
   1.2. *If $\alpha \in \mathcal{A}$ is an n level arc and $s \in S$ then $\beta = (s \to \alpha)$ is an $n+1$ level arc.*
2. *A subset $\mathcal{R} \subseteq \mathcal{A}$ is an arc relation.*
3. *An arc-Kripke model has the form $(S, \mathcal{R}, \mathcal{R}^*, a, h)$, where $\mathcal{R}$ and $\mathcal{R}^*$ are arc-accessibility relations and $\mathcal{R} \subseteq \mathcal{R}^*$ and h is an assignment giving to each $t \in S$ and each atom q a value $h(t, q) \in \{0, 1\}$. The part of the model $(S, \mathcal{R}, \mathcal{R}^*, a)$ is called the model frame and $a \in S$ is the actual world of the frame.*
4. *Note that in this definition all arcs emanate from points.*

Definition 2.9. *Let $\mathbf{m} = (S, \mathcal{R}, \mathcal{R}^*, a, h)$ be a model.*

Define $a \vDash A$, by structural induction.

1. *$a \vDash q$ if $h(a, q) = 1$, for q atomic.*
2. *$a \vDash A \wedge B$ iff $a \vDash A$ and $a \vDash B$.*
3. *$a \vDash \neg A$ iff $a \nvDash A$.*

4. *$a \vDash \Box A$ iff for all s in S such that $(a, s) \in \mathcal{R}$ we have that $s \vDash A$ in the model $\mathbf{m}_s = (S, \mathcal{R}_a, \mathcal{R}^*, s, h)$, where $\mathcal{R}_a$ is obtained from $\mathcal{R}$ as follows:*

$$\mathcal{R}_a = \mathcal{R} - \{\alpha \mid (a \to \alpha) \in \mathcal{R} \wedge \alpha \in \mathcal{R}\} \cup \{\alpha \mid (a \to \alpha) \in \mathcal{R} \wedge \alpha \notin \mathcal{R} \wedge \alpha \in \mathcal{R}^*\}.$$

   This is a switch satisfaction clause for $\Box$.
5. *A model $\mathbf{m} = (S, \mathcal{R}, \mathcal{R}^*, a, h)$ is said to be of level $\leq n$, $n \geq 1$, if all its arcs in $\mathcal{R}^*$ are of level $\leq n$.*
   For example, a model of level ≤ 2 can contain only arcs of the form $t \to s$ or of the form $r \to (t \to s)$, where $r, t, s, \in S$.
6. *Let $\mathbf{K}^n_{\mathcal{A}}$, $n \geq 1$ be the set of wffs valid in the class of all reactive Kripke models of level $\leq n$. Note that $\mathbf{K}^1_{\mathcal{A}}$ is ordinary modal $\mathbf{K}$.*
   Let $\mathbf{K}_{\mathcal{A}}$ be the set of wffs valid in the class of all reactive Kripke models. This is the basic reactive analogue of modal $\mathbf{K}$.
7. *A modal logic $\mathbf{L}$ is said to be a reactive modal logic if for some class $\mathcal{K}$ of reactive models $\mathbf{L} = \{A \mid A$ is valid in all models of $\mathcal{K}\}$.*
8. *Note that to be more precise these models use only arcs emanating from points and the reactivity acts like a switch.*
9. *Note that the evaluation indicated in item 4 above of $a \vDash \Box A$ and $a \vDash \Box B$ for any two wffs A and B is done independently of one another and in parallel. We have $\Box(A \wedge B) \leftrightarrow \Box A \wedge \Box B$ and both A and B in this case get evaluated at the model with $\mathcal{R}_a$.*
   This gives us scope to introduce the connective "$\wedge\_$", where $X \wedge\_ Y$ reads X and then *Y. So $a \vDash \Box A \wedge\_ \Box B$ will evaluate $a \vDash \Box A$ first. In the course of the evaluation of $\Box A$ the relation $\mathcal{R}$ will become $\mathcal{R}_a$ and then we evaluate $a \vDash \Box B$ in the model $(S, \mathcal{R}_a, \mathcal{R}^*, a, h)$.*

Example 2.10. Consider the two-point model of Figure 2.22 with $S = \{a, b\}$:

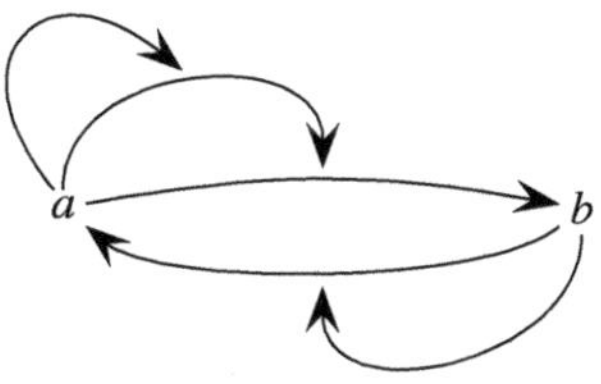

Fig. 2.22

Let $\mathcal{R} = \mathcal{R}^* = \{a \to b, b \to a, a \to (a \to b), a \to (a \to (a \to b)), b \to (b \to a)$.

Consider a model

$$\mathbf{m} = (S, \mathcal{R}, \mathcal{R}^*, a, h).$$

Here $\mathcal{R}_a = \{b \to a, b \to (b \to a), a \to (a \to (a \to b))\}$.

$$\mathcal{R}_{a,b} = \{b \to (b \to a), a \to (a \to (a \to b))\}$$
$$\mathcal{R}_{a,b,a} = \{b \to (b \to a), a \to (a \to (a \to b)), a \to (a \to b)\}.$$

Note that $a \to b$ is not restored until $\mathcal{R}_{a,b,a,b,a}$.

Definition 2.11. *Let (S, R) be a set S with a binary relation $R \subseteq S^2$. Let τ be a Horn clause theory in the language with R containing universal clauses of the form (universal closure) $(\bigwedge_{i=1}^{n} x_i R y_i \to xRy)$. Let $R^\tau \supseteq R$ be defined as the smallest extension of R such that $(S, R^\tau) \vDash \tau$. R^τ can be constructed by induction as the closure of R under all instances of τ as follows:*

1. *Let $R_0 = R$.*
2. *Let $R_{n+1} = \{(a, b) \mid$ for some clause L in τ of the form (universal closure) $(\bigwedge x_i R y_i \to xRy)$ and a substitution θ in the variables of L such that $\theta(y_i), \theta(x_i) \in S, \theta(x) = a, \theta(y) = b$, we have that $\bigwedge \theta(x_i) R_n \theta(y_i)$ hold.*
3. *Let $R_\tau = \bigcup_n R_n$.*

Definition 2.12.

1. *Let $\mathbb{N}$ be the set of natural numbers $\{0, 1, 2, \ldots\}$. Let $\mathbb{N}^*$ be the set of all finite sequences of natural numbers including the empty sequence $\varnothing$. Define $\alpha < \beta$, for $\alpha, \beta \in \mathbb{N}^*$ by*

   $$\alpha < \beta = (\textit{ definition}) \textit{ for some } m \in \mathbb{N}, \beta = \alpha * (m),$$

   where $*$ is concatenation.
   Let $<^*$ be the transitive closure of $<$.
2. *A tree T is a nonempty subset of $\mathbb{N}^*$ such that if $\beta \in T$ and $\alpha < \beta$ then $\alpha \in T$.*
3. *Let τ be a Horn clause theory on $<$.*
   Let $(T, <)$ be a tree and let $(T, <^\tau)$ be its τ-closure. Define $\mathcal{R}_\tau$ as follows:
   - $(\alpha \to \beta) \in \mathcal{R}_\tau$ *if* $(\alpha, \beta) \in <^\tau$.
   - $(\gamma \to (\alpha \to \beta)) \in \mathcal{R}_\tau$ *whenever* $\gamma <^\tau_* \alpha$ *and* $\gamma <^\tau_* \beta$ *hold where* $<^\tau_*$ *is the transitive closure of* $<^\tau$.
4. *Let $\mathcal{K}_\tau$ be the class of all models of the form $(T, <, \mathcal{R}_\tau, \delta, h)$, where T is a tree and $\delta \in T$.*
5. *Let H_τ be in the set of all wffs A such that A holds in all models of $\mathcal{K}_\tau$.*

Example 2.13. Let $\tau = \{\forall x(xRx)\}$. Then the models of $\mathcal{K}_\tau$ have the form $(T, <, \mathcal{R}_\tau, \delta, h)$, where $\mathcal{R}_\tau = \cup\{\gamma \to (\alpha \to \alpha) \mid \gamma \leqq^* \alpha\}$. This is so since $<^\tau$ is $< \cup\{(\alpha, \alpha) \mid \alpha \in T\}$.

It is easy to see that the meaning of $\Box$ alternates between **K** and **T** modalities because $\mathcal{R}$ switches the reflexive arcs on and off. Thus the logic H_τ of our example is the same as the logic H_1^S of Section 3 of Chapter 1.

We know from [69] that the following is a Hilbert axiomatisation of H_τ. We make use of our irreflexivity rule, see [56].

Axioms: (E, F are wffs without $\Box$).

1. $A \wedge \Box A$, where A is a substitution instance of a truth functional tautology.
2. $\Box(A \to B) \to (\Box A \to \Box B)$.
3. $\Box(\Box(A \to B) \to (\Box A \to \Box B))$.
4. $\Diamond\top$.
5. $\neg E \wedge \Box^2 E \wedge Y \to \Diamond(\neg E \wedge Y)$, where $Y = A$ or $Y = \Box A$, for A without $\Box$.
6. $\neg E \wedge \Box^2 E \wedge \Diamond(\neg E \wedge A) \wedge \Diamond(\neg E \wedge B) \to \Diamond(\neg E \wedge A \wedge B)$.
7. $\neg E \wedge \Box^2 E \to \Diamond(\neg E \wedge A) \vee \Diamond(\neg E \wedge \neg A)$.
8. $\Box A \wedge \neg E \wedge \Box^2 E \to \Diamond(\neg E \wedge A)$.
9. $\Box X \wedge \neg E \wedge \Box^2 E \to \Diamond(\neg E \wedge \Box(\neg E \wedge \Box^2 F \to \Diamond\neg F \wedge X))$.
10. $\neg E \wedge \Box^2 E \wedge \Diamond A \wedge \Diamond(\neg E \wedge \Box Y \wedge \Box\Box Y' \wedge \neg A \wedge \Box\Box X) \to \Diamond(A \wedge X \wedge Y \wedge \Box Y')$ where Y, Y' are without $\Box$.
11. $\neg E \wedge \Box^2 E \wedge \Diamond(C \wedge E \wedge Y) \to \Diamond(\neg E \wedge \Diamond(Y \wedge E \wedge \Diamond(C \wedge E)))$.

Rules

MP: $$\frac{\vdash A; \vdash A \to B}{\vdash B}$$

IRR: $$\frac{\vdash \neg q \wedge \Box^2 q \to A}{\vdash A}$$
where q is an atom not in A.

2-necessitation: $$\frac{\vdash A}{\vdash \Box^2 A}$$

IRR^n: $$\frac{\vdash \bigwedge_{m=1}^{n} \beta_m^m \to A}{\vdash A}$$
where β_m^m are as defined In Example 1.19 and q_j^i are all not in A.

Example 2.14. We now give an example of a class of reactive Kripke models characterising a logic which cannot be presented as a hyper-modal logic. Consider one-point models of the following form, see Figure 2.23.

Let
$$\begin{aligned} \mathrm{arc}_0 &= a \to a \\ &\vdots \\ \mathrm{arc}_{n+1} &= (a \to \mathrm{arc}_n) \end{aligned}$$

Consider models $\mathbf{m}_k$ of the form
$$\mathbf{m}_k = (\{a\}, \{\mathrm{arc}_0, \mathrm{arc}_k\}, \{\mathrm{arc}_n \mid n \leq k\}, a)$$

The model $\mathbf{m}_k$ is a one-node model with arcs as in Figure 2.23, where only the arcs $a \to a$ and $a \to (a \to \ldots (a \to a) \ldots)$ (k arrows) are active.

The following points are clear for the class of models $\{\mathbf{m}_k\}$.

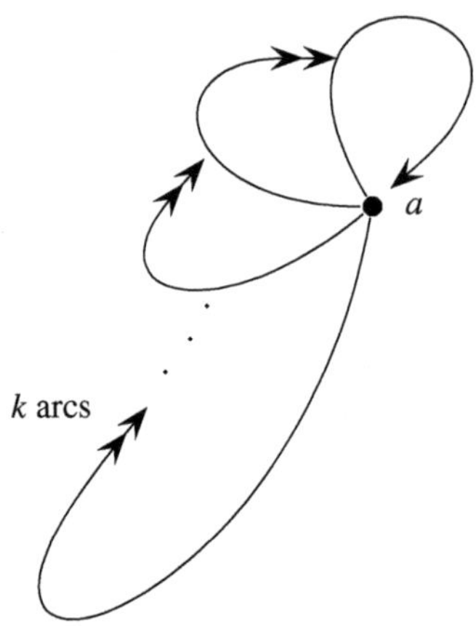

Fig. 2.23

1. $\Diamond^n\Box\bot$ is consistent for all n.
2. The following holds in the logic **L** of this class

$$\vDash \Diamond^n\Box\bot \to \bigwedge_{m\leq n}(\Diamond^m A \Leftrightarrow A)$$

   for all wffs A without $\Box$.

We claim the logic **L** cannot be presented as a hyper-modal logic.

For suppose there is a class of traditional Kripke models of the form (S, R, a) and a sequence $(\Psi_1, \ldots, \Psi_k)$ of conditions characterising modalities $\Box_1, \ldots, \Box_k$ such that the evaluation of $\Diamond$ cycles through this sequence (see Definition 1.8).

Then for a high enough n, the meaning of $\Diamond$ in $\Diamond^n\Box\bot$ starts repeating itself.

Let (S, R, a) be a model of $\Diamond^n\Box\bot$, for n large enough. By property 2, all $\Diamond^m, m \leq n$ satisfy $\vDash_a \Diamond^m A \Leftrightarrow A$, for A without $\Box$. Thus we must have $\Psi_i(a, R) = \{(a, a)\}$. Therefore, how can we also have $\vDash_a \Diamond^n\Box\bot$?

This example shows that there is a class of reactive models of finite-level defining a modal logic **L** which is not a hyper-modal logic.

Example 2.15. We now exhibit a hyper-modal logic which cannot be characterised by a class of finite level reactive models. Consider the situation in Figure 2.24

Let $\Psi_1 = \{(a, a)\}$ and let $\Psi_2 = \{(a, a), (a, b)\}$. Consider the sequence (Ψ_1, Ψ_1, Ψ_2). This means that $\Box$ is interpreted as seeing only the a node twice and then it can also see b once before repeating. Let **L** be the hyper-modal logic defined by this set-up.

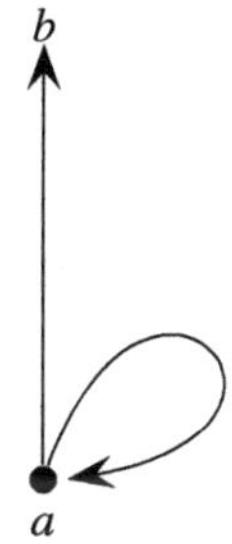

Fig. 2.24

To implement this logic in a reactive model we need to switch the arc $a \to b$ on and off using other arcs in the repeating sequence $(-, -, +)$.

Figure 2.25 can help visualise the situation:

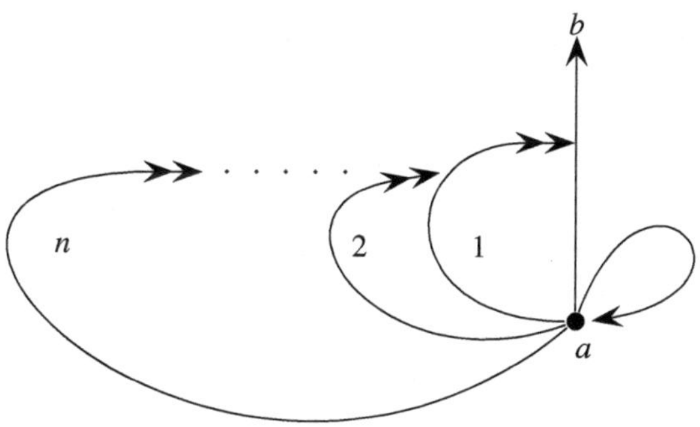

Fig. 2.25

We note that if the connection $a \to b$ must cycle through $(-, -, +)$ then connection $a \to (a \to b)$ (i.e. arc_1 in Figure 2.25) must cycle through $(-, +, +)$ and arc_2 cycle throughe $(+, -, +)$ and so on.

However in any finite-level model, the highest-level arcs cannot change status. Hence modalities of the form $\Box^n$ for high enough n cannot be implemented in any given model. We conjecture that if we allow models of unbounded level then all reasonable hyper-modalities can be implemented. In our case the sequence $(-, -, +)$ can be implemented by a single model of infinite level.

Remark 2.16. Let us give some thought to the problem of how to axiomatise the basic reactive modal logic $\mathbf{K}_{\mathcal{A}}$ (see Definition 2.9, item 6). Let Δ be a consistent complete theory of this logic. Then for some model $\mathbf{m} = (S, \mathcal{R}, \mathcal{R}^*, a, h)$ we have $\Delta = \{A \mid a \vDash A\}$. So Δ must contain within it enough information to reconstruct a canonical model with a suitable arc accessibility. How this can be is not immediately clear. In ordinary modal logic we can construct accessibility $\Theta R \Theta'$ by letting it be

$$\forall\alpha(\Box\alpha \in \Theta \to \alpha \in \Theta')$$

We can do the same here and this will give us the arrows but how do we define the double arrows?

For that matter, we still have to check whether $\mathcal{K}_{\mathcal{A}}$ is axiomatisable or is recursively enumerable. See Chapter 5.

2.5 Non-deterministic Reactive Kripke Models

Let us begin by considering again the basic situation of Figure 2.1. This is a simple reactive Kripke model. Let us make it non-deterministic. This would mean that as we move from node a to node b, the double arrow $a \twoheadrightarrow (b \to c)$ may or may not (non-deterministically) fire. So to evaluate $a \vDash \Box^2 q$, we need to evaluate $b \vDash \Box q$. If the double arrow does fire, then the connection $b \to c$ is cancelled and only $b \to d$ is active. If the double arrow does not fire, then $b \to c$ is also active when we evaluate $b \vDash \Box q$.

So in this model (the non-deterministic model of Figure 2.1), there are two possible non-deterministic valuations of $a \vDash \Box^2 q$: one with $b \to c$ active at the time of the evaluation of $b \vDash \Box q$ and one where it is not active. We say that a non-deterministic model $\mathbf{m}$ can satisfy a wff A if there is at least one non-deterministic valuation of A in the model in which A gets the value $\top$. We say that A holds in the model if it holds (gets $\top$) under all non-deterministic valuations in the model.

Example 2.17. Consider the model of Figure 2.26. Then as a non-deterministic model we have that $\Diamond^n \Box\bot$ can always be made false, by never activating the double arrow in the evaluation. But it can also be satisfied by activating the double arrow at the $n-1$th step and then the model becomes irreflexive at point b and $\Box\bot$ then holds at b. Thus we see that such models can satisfy both a wff A and its negation $\neg A$, because different non-deterministic options can come to bear. However, only one of A and $\neg A$ can hold in the model. We can see already that we need to be careful with this concept. Consider the axiom $\Box(A \wedge B) \to \Box A \wedge \Box B$. This is valid in any reactive Kripke model (provided the evaluations of $\Box A$ and of $\Box B$ are done in parallel, see Definition 2.9). It is not so in a non-deterministic model, for consider $\Diamond\Box(p \wedge \neg p) \wedge \Diamond\Box(q \wedge \neg q)$ where p and q are different atoms. When we evaluate $a \vDash \Diamond\Box(p \wedge \neg p)$ in the model of Figure 2.26, we may take the non-deterministic choice of not firing the double arrow and thus get value false. When we do it again

Table 2.1

state	stack at α	Input letter	Reaction
a	all arrows not active	σ_i	move to state q_i^+ and activate arrow 1 at α
a	some arrows active	any	do not care. Case will not arise
q_i^+	arrows $1, \ldots, m$ are active $m \geq 1$	σ_i	stay at q_i^+ and activate arrow $m + 1$
q_i^+	arrows $1, \ldots, m$ are active	$\sigma_j, j \neq i$	move to q_i^-
q_i^+	not the above	any	don't care. Case will not arise
q_i^-	arrows $1, \ldots, m$ are active $m \geq 1$	$\sigma_j, j \neq i$	stay at q_i^- and deactivate arrow m.
q_i^-	arrows $1, \ldots, m$ are active, $m \geq 1$	σ_i or no input	move to t and deactivate arrow m
q_i^-	all arrows at α are not active or arrows $2, \ldots, m$ are active $m \geq 2$ and arrow 1 not active	no input or input σ_i	move to t, activate arrow $m + 1$. If no arrow at α is active then activate arrow 2.
q_i^-	same as previous	$\sigma_j, j \neq i$	remain at q_i^-, activate arrow $m + 1$ or if no arrows at α are active then activate arrow 2.
q_i^-	Different from previous	any	don't care. Case will not arise.
t	any	no input. Terminal position	no reaction. Terminal position.

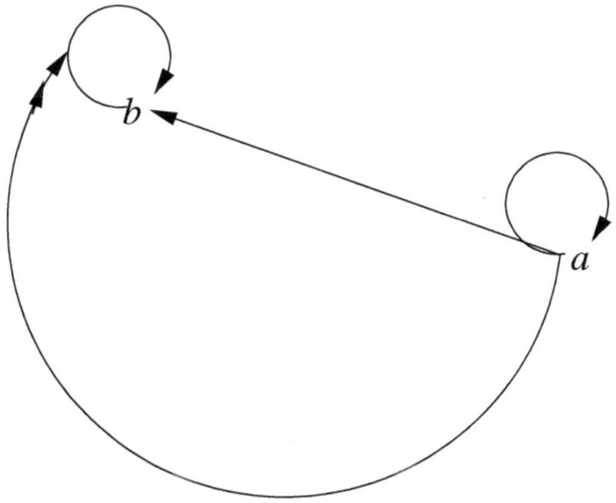

Fig. 2.26

for the second copy $\Diamond\Box(q \wedge \neg q)$, we may choose to fire and thus get value true. We are going to have to formulate our definitions carefully here.

The single non-deterministic model of Figure 2.26 is equivalent to the two ordinary Kripke models of Figure 2.27, $\mathbf{m}_1$ and $\mathbf{m}_2$

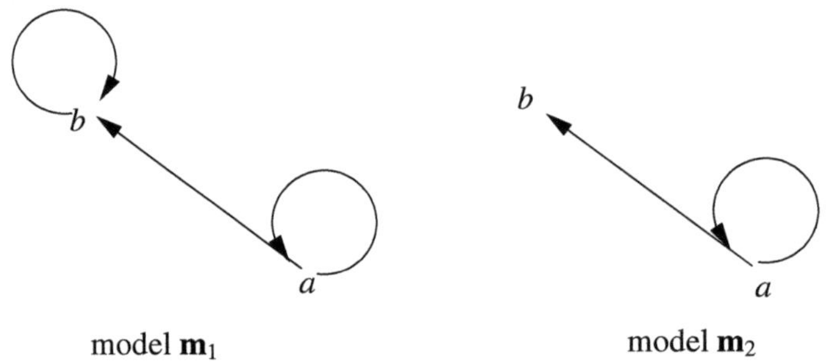

Fig. 2.27

During the evaluation at node a we can change our view of whether node b is reflexive or not.

Definition 2.18. *A non-deterministic choice of switches in a model* $\mathbf{m} = (S, \mathcal{R}, \mathcal{R}^*, a, h)$ *is a pair of sets* ρ_b^+, ρ_b^- *for each* $b \in S$ *satisfying the following:*

$$\rho_b^+ \subseteq \{\alpha \mid (b \to \alpha) \in \mathcal{R} \wedge \alpha \in \mathcal{R}^*\}$$
$$\rho_b^- \subseteq \{\alpha \mid (b \to \alpha) \in \mathcal{R} \wedge \alpha \notin \mathcal{R} \wedge \alpha \in \mathcal{R}^*\}$$

Definition 2.19 (Non-deterministic switch Kripke models).

1. *By a model* $\mathbf{m} = (S, \mathcal{R}, \mathcal{R}^*, a, h)$ *we mean a model as defined in Definition 2.8.*
2. *By a sequence of selections of non-deterministic choices of switches in a model* $\mathbf{m}$ *we mean a sequence of the form* $\rho_{a,k}^{\pm}, k = 0, 1, \ldots$ *where for each k,* $\rho_{a,k}^{\pm}$ *is a non-deterministic choice as defined in the previous definition.*
3. *Given a wff A and a node a in a model* $\mathbf{m}$*, we define the universal non-deterministic evaluation* $a \models\!\equiv_k A$ *in parallel to Definition 2.9 as follows (where k is a natural number representing a universal ticking clock, meaning that this is the kth time that a* $\Box$ *is evaluated at the node a).*
   a) $a \models_k q$ *if* $h(a, q) = 1$*, for q atomic and any* $k \geq 0$.
   b) $a \models_k A \wedge B$ *iff* $a \models_k A$ *and* $a \models_k B$
   c) $a \models_k \neg A$ *iff* $A \not\models_k A$
   d) $a \models_k \Box A$ *iff for all* $s \in S$ *such that* $(a, s) \in \mathcal{R}$*, we have that* $s \models_{k+1} A$ *in the model* $\mathbf{m}_s^k$*, where*
   $$\mathbf{m}_s^k = (S, \mathcal{R}_a^k, \mathcal{R}^*, s, h)$$
   and where
   $$\mathcal{R}_a^k = (\mathcal{R} - \rho_{a,k}^+) \cup \rho_{a,k}^-.$$

e) A formula A is said to be satisfiable in a model **m** *if for some non-deterministic choice sequence* $\rho^{\pm}_{t,k}, t \in S, k = 0, 1, 2, \ldots$. *We have that* $a \models_0 A$. *Note that it is possible to have that both* $a \models_0 A$ *and* $a \models_0 \neg A$.

f) A formula is said to hold in a model **m** *if its negation is not satisfiable in* **m**.

g) A modal logic **L** *is said to be non-deterministically reactive if for some class of reactive Kripke models we have*

$$\mathbf{L} = \{A \mid A \textit{ is non-deterministically valid in all models in the class}\}.$$

Example 2.20. This example explains the 'ticking clock' of the previous definition.

Consider the frame of Figure 2.26, and consider $a \vDash \Diamond\Box(p \wedge \neg p)$ and again $a \vDash \Diamond\Box(q \wedge \neg q)$.

In the first evaluation we activate the double arrow immediately and get true and in the second evaluation we do not activate it and get false. To avoid such a situation, we introduce a universal counting clock, counting the number of nested modalities. We now write

$$a \models_0 \Diamond\Box(P \wedge \neg p) \text{ iff } a \models_1 \Box(p \wedge \neg p) \text{ or } b \models_1 \Box(p \wedge \neg p)$$

We require always the same non-deterministic choice for any evaluation of the form $t \models_k \Box$, with the same t and k.

If we have this we get that $\Box(A \wedge B) \to \Box A \wedge \Box B$ holds

2.6 Connection with Fibring Logics

This section gives some methodological remarks and points out a connection with fibring logics, see [57].

2.6.1 Methodological considerations

Consider two modal logics $\mathbf{L}_1$ and $\mathbf{L}_2$. Assume $\mathbf{L}_1$ has a **K** modality, $\Box_1$ and let $\mathbf{L}_2$ have a **K4** modality, $\Box_2$. We know that $\mathbf{L}_1$ is complete for all Kripke frames of the form (S_1, R_1, a_1), with $a_1 \in S_1, R_1 \subseteq S_1^2$, and $\mathbf{L}_2$ is complete for all Kripke frames of the form (S_2, R_2, a_2) with $a_2 \in S_2, R_2 \subseteq S_2^2$ where R_2 is a transitive (irreflexive) relation. When we combine these two logics in a way which is generally called *fusion* (or *dovetailing* in [57]), we get completeness for the class of all Kripke models of the form (S, R_1, R_2, a) with R_2 transitive irreflexive and where the truth table for $\Box_i$ is defined as follows:

$$(*1) \qquad t \vDash \Box_i A \text{ iff } \forall s(tR_i s \to s \vDash A) \text{ for } i = 1, 2.$$

The above combination of $\mathbf{L}_1$ and $\mathbf{L}_2$ assumes that the languages $\mathbf{L}_1$ and $\mathbf{L}_2$ are disjoint, and to the extent that they share the classical connectives, they agree on these connectives.

Assume now that $\mathbf{L}_1$ and $\mathbf{L}_2$ contain an additional modality $\Box$, i.e. $\mathbf{L}_1 = \mathbf{L}_1(\Box, \Box_1)$ and $\mathbf{L}_2 = \mathbf{L}_2(\Box, \Box_2)$, where $\Box$ is a $\mathbf{T}$ modality. Now the models for $\mathbf{L}_i$ have the form (S_i, R, R_i, a_i) where $R \subseteq S_i^2$ and R is reflexive. Obviously $\mathbf{L}_1$ and $\mathbf{L}_2$ which contain the *same* symbol for modality $\Box$, must agree on the nature of this modality (namely that it is a $\mathbf{T}$ modality). We can therefore safely combine them (in the same way as before) and get a logic $\mathbf{L}_{1+2}$ with $\Box, \Box_2, \Box_2$. This logic is complete for models of the form (S, R, R_1, R_2, a) where $R \subseteq S^2$ is reflexive and $R_2 \subseteq S^2$ is irreflexive and transitive, and $R_1 \subseteq S^2$.

The question arises what can we do if $\mathbf{L}_1$ and $\mathbf{L}_2$ do not agree on the modality $\Box$? What if $\mathbf{L}_1$ has $\Box$ as a $\mathbf{K}$ modality while $\mathbf{L}_2$ has $\Box$ as a $\mathbf{T}$ modality? How do we do this? To put it bluntly, can we combine two systems for the same modality $\Box$, where one says $\Box$ is a $\mathbf{K}$ modality and the other says that $\Box$ is a $\mathbf{T}$ modality?

To make the problem more acute we can have modalities with contradictory conditions on R. For example, $\mathbf{L}_1$ can ask $\forall x \exists y x R y$ (the axiom $\Box\Diamond\top$) and $\mathbf{L}_2$ can ask $\exists x \forall y \neg x R y$ (the axiom $\Diamond\Box\bot$).

One way to combine two different systems for the same symbols is to timeshare the symbol, i.e. $\Box$ can sometimes be an $\mathbf{L}_1$ modality and sometimes an $\mathbf{L}_2$ modality. This is the way processes share the same resource; they timeshare according to a certain protocol μ.

So let us examine modal logics with a single modality $\Box$, where the meaning of $\Box$ is timeshared between, say a $\mathbf{K}$ and a $\mathbf{T}$ modality.

The questions we ask are the following:

(Q1) How do we define and execute the timeshare?
(Q2) What is the semantics corresponding to (Q1)?

Let us start by choosing something simple. Let the meaning of $\Box$ alternate between a $\mathbf{T}$ and a $\mathbf{K}$ modality. Thus a wff of the form

$$\alpha = \Box(\Box q \wedge \Box\Box\Box p)$$

will read as

$$\beta = \Box_{\mathbf{T}}(\Box_{\mathbf{K}} q \wedge \Box_{\mathbf{K}}\Box_{\mathbf{T}}\Box_{\mathbf{K}} p).$$

Thus the meaning of $\Box$ alternates according to its nesting depth in the formula.

So now we know how to read α, namely we read it as β, but what is the semantics for β? Being a mono-modal logic, we expect a Kripke semantics with a single accessibility relation R, i.e. models of the form (S, R, a). So how can R be and not be reflexive alternately? This is impossible!

To this end we need a new point of view on how we see our semantics. Let us consider the notion of a *mode of search* for a possible world. Namely let $\Psi(x, y)$ be (a possibly higher order) binary wff on the classical model (S, R, a). Let

(∗2) $\qquad t \vDash_\Psi \Box A$ iff for all s such that $\Psi(t, s)$ holds we have $s \vDash A$.

Ψ is the mode of search for worlds s. The following are examples:

- For $\mathbf{K}$ modality $\Psi_{\mathbf{K}}(x, y) \equiv xRy$

- For **T** modality $\Psi_{\mathbf{T}}(x,y) \equiv x = y \vee xRy$
- For **K4** modality $\Psi_{\mathbf{K4}}(x,y) \equiv$ for some $n \geq 0$ we have $xR^n y$.
  Note that $\Psi_{\mathbf{K4}}$ is not first-order.

The truth condition for a system with alternating **K** and **T** modality can be written as follows:

(∗3)
- $t \vDash_{\mathbf{K}} \Box A$ iff for all s, $\Psi_{\mathbf{K}}(t,s) \to s \vDash_{\mathbf{T}} A$.
- $t \vDash_{\mathbf{T}} \Box A$ iff for all s, $\Psi_{\mathbf{T}}(t,s) \to s \vDash_{\mathbf{K}} A$.

The logic with (S,R,a) where $\Box$ alternates in meaning as above can be seen as one way of fibring disagreeing modalities via a timesharing protocol.

Let us use the following terminology:

- use the term *C-fibring* for *Compromise Fibring* when we combine two logics which do not agree on the common part.
- Use the words *μ-Protocol* for any procedures μ for timesharing the common symbols.

We now continue and discuss another 'twist' in our way of thinking. Obviously when $\Box$ is interpreted as $\Box_{\mathbf{T}}$, we use $\Psi_{\mathbf{T}}$. How about modalities complete for classes of models whose relations cannot be nicely defined via some Ψ (even second order)? We certainly can form models of the form (S,R_1,R_2,a) and let the meaning of $\Box$ alternate between R_1 and R_2. However, this is not neat, because with one symbol $\Box$ we expect semantics with one symbol R or some generalisation of it.

We need some inspiration at this junction. So here it is:

Recall that *reactive models* are models whose accessibility relation changes as we go along. Figure 2.28 illustrates the basic mechanism:

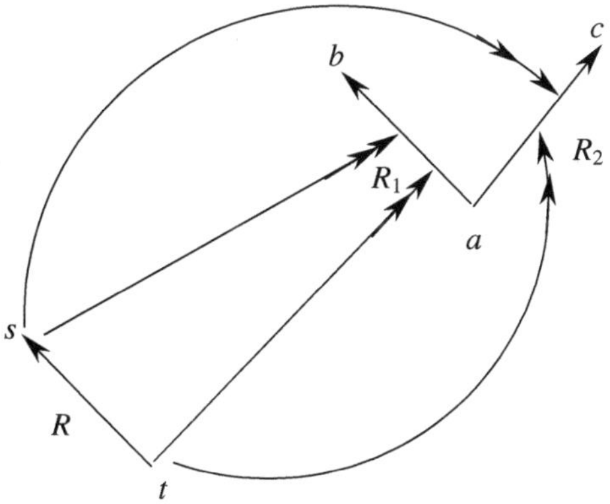

Fig. 2.28

We evaluate $t \vDash \Box q$ at moment t. We assume that at each moment, either all the R_1 or all the R_2 (but not both) connections are on. When we move from $t \vDash \Box q$ to

s, i.e. to $s \vDash q$, we send a signal from t ($\twoheadrightarrow$) to all other R_1 and R_2 connections. If a connection receives the signal and it is on, then it turns off. If it is off then it turns on. So when we come to s, we switch from R_1 to R_2 or from R_2 to R_1 depending on which one is active. In general, we deal with reactive models as in Section 2.2.

Now in order to be able to compromise fibre two logics with the same modality but different reactive Kripke semantics, we need to allow arrow $\twoheadrightarrow$ to cancel other arrows up to arbitrary finite iterations. This yields Definitions 2.8 and 2.9 of Section 2.3.

We have now completed our methodological motivation for reactive Kripke models. For more on compromise fibring, see [58].

2.6.2 Formal compromise fibring

It is now time to give a formal definition of C-fibring.

Definition 2.21.

1. *Let (S, R, a) be a Kripke frame. By a* finite path *we mean a sequence of elements of the form $\mathbf{t} = (a, a_1, \ldots, a_n)$ such that $aRa_1 \wedge a_1Ra_2 \wedge \ldots a_{n-1}Ra_n$ holds. We also include the empty sequence as a path.*
2. *Let $t \in S$, we define the model $(S_{[t]}, R_{[t]}, t)$ as follows.*
   a) *Let $S_{[t]}$ be the smallest subset $S' \subseteq S$ such that $t \in S'$ and if $x \in S'$ and xRy then $y \in S'$.*
   b) *Let $R_{[t]}$ be $R \upharpoonright S_{[t]}$.*

Definition 2.22. *Let $\mathcal{K}_1, \ldots, \mathcal{K}_n$ be n classes of Kripke frames. Let $\mathbf{L}_1, \ldots, \mathbf{L}_n$ be the logics they define as frames in a language with the syntactical modality $\Box$.*

We define a C-fibred general model for $\Box$ as follows:

1. *$(S, R, a, \mathbf{f})$ is a C-fibred model for $\mathbf{L}_1 + \mathbf{L}_2 + \ldots + \mathbf{L}_n$ iff $\mathbf{f}$ is a function giving for each finite path $\mathbf{t} = (a, a_1, \ldots, a_n)$ a value $\mathbf{f}(\mathbf{t}) = R^{\mathbf{f}}_{\mathbf{t}} \subseteq R_{[a_n]}$ such that for all $m \leq n$, the model $(S_{[a_n]}, R^{\mathbf{f}}_{\mathbf{t}}, a_n)$ is in $\mathcal{K}_m$.*
2. *We define satisfaction in $(S, R, a, \mathbf{f})$ as follows (under an assignment h to the atoms) for $\mathbf{t} = (a, a_1, \ldots, a_n)$:*
   - *$a_n \vDash_{\mathbf{t}} \Box A$ iff for all $s \in S_{[t]}$ such that $tR^{\mathbf{f}}_{\mathbf{t}}s$ we have that $s \vDash_{\mathbf{t}*(s)} A$, where $*$ is concatenation of sequences.*
3. *Note that the condition $R^{\mathbf{f}}_{\mathbf{t}} \subseteq R_{[a_n]}$ is not restrictive because we can always start with $R = S^2$.*
4. *Let $\mathbf{L}_1 + \ldots + \mathbf{L}_n$ be the logic defined by all models $(S, R, a, \mathbf{f})$.*

Remark 2.23. Note that ordinary fibring of two modalities $\Box_1$ and $\Box_2$ can be captured by this semantics. Let us pretend that $\Box_1 = \Box_2 = \Box$ and that $\mathbf{L}_1$ reads $\Box$ as $\Box_1$ for a class of models $\mathcal{K}_1$ and similarly $\mathbf{L}_2$ for $\mathcal{K}_2$. Consider all models with frames (S, R_1, R_2, a) where $(S, R_1, t) \in \mathcal{K}_1$ and $(S, R_2, t) \in \mathcal{K}_2$ for all $t \in S$. Let R be defined as $R_1 \cup R_2$. Let $\mathbf{f}(a, a_1, \ldots, a_n) = R_1$ if n is odd and R_2 if n is even. Then $\Box_1$ and $\Box_2$ can be translated in a *context-sensitive way* as seen by the following example.

Let $\alpha = \Box_1(\Box_1 q \wedge \Box_2\Box_1 p)$. Then α is translated into β.

$$\beta = \Box(\Box^2 q \wedge \Box\Box p).$$

We translate each $\Box_i$ from outside in. We translate each $\Box_i$ as either $\Box$ or $\Box\Box$ depending on whether it gives us an odd or even nesting of boxes as required by the index i. **This translation is not faithful. More work needs to be done on this idea.**

2.7 Dedicated reactivity connectives

Previous sections dealt with a modality $\Box$ that both sent the evaluation to all accessible worlds and at the same time activated all reactivity double arrows. Let us see how we can separate these two functionalities using two separate connectives. Consider the simple model in Figure 2.29

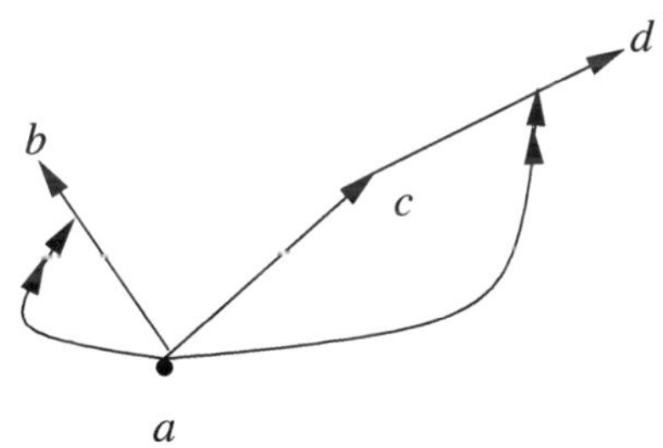

Fig. 2.29

When we evaluate $a \vDash \Box q$, we move to the evaluation of $b \vDash q$ and $c \vDash q$. The node a also sends a signal along the double arrows at a and as a result the arcs (a, b) and (c, d) are disconnected.

Let us separate these two operations. Let us look at the fault disconnection signal separately from the ordinary modality operation. Let us have an ordinary modality $\vec{\Box} A$ and a reactive modality $\overset{\twoheadrightarrow}{\Box} A$ operating as follows:

1. $a \vDash \vec{\Box} A$ iff for all y such that aRy we have $y \vDash A$. This is an ordinary modality moving along the single arrows.
2. $a \vDash \overset{\twoheadrightarrow}{\Box} A$ iff $a \vDash A$ in the model obtained by letting all reactive double arrows do their job!
   So this is a modality which changes the model (set of possible worlds and its accessibility relation). It is like a **Jump** operator of the form

   $$a \vDash_{\text{model 1}} \textbf{Jump}\, A \text{ iff } a \vDash_{\text{model 2}} A.$$

   The move from one model to the other is done by activating the double arrows emanating from a.

So, for the situation in Figure 2.29 we have that

$$a \vDash \overset{\twoheadrightarrow}{\Box} A \text{ in the model of Figure 2.29}$$
$$\text{iff}$$
$$a \vDash A \text{ in the model of Figure 2.30.}$$

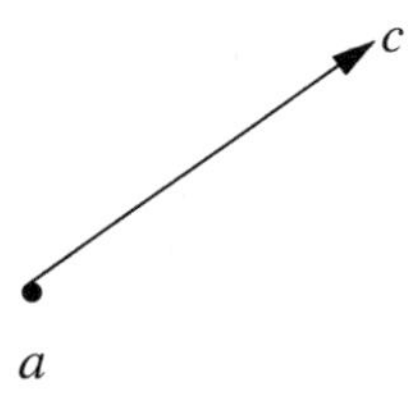

Fig. 2.30

Having isolated the reactive operation, we would like to check the interdefinability properties of $\Box$, $\overrightarrow{\Box}$ and $\overset{\twoheadrightarrow}{\Box}$.

We will see that matters are not so simple.

We would like to have maybe that

$$\Box = \overrightarrow{\Box}\overset{\twoheadrightarrow}{\Box}$$

or

$$\Box = \overset{\twoheadrightarrow}{\Box}\overrightarrow{\Box}$$

or some combination

$$\Box = \textit{combination}\,(\overrightarrow{\Box}, \overset{\twoheadrightarrow}{\Box}).$$

However this is not so simple because $\Box$ fires both $\overrightarrow{\Box}$ and $\overset{\twoheadrightarrow}{\Box}$ simultaneously. If we do $\overrightarrow{\Box}$ first, we can no longer disconnect the (c,d) arc. If we do $\overset{\twoheadrightarrow}{\Box}$ first, we no longer can get to b.

If we use $a \vDash \Box\top$, we do not get the effect of $\overset{\twoheadrightarrow}{\Box}$ because we move to the accessible worlds b and c.

Thus, no matter what we do, we have a problem interdefining these connectives.

In fact $\overset{\twoheadrightarrow}{\Box}$ is a metalevel connective, operating on the model (disconnecting accessibility arrows) and is different in nature from the evaluation of the ordinary connective $\overrightarrow{\Box}$. Once we have the concept of metalevel connectives we can talk naturally about the simultaneity of several actions. Let "||" indicate simultaneous operation of actions, being another metalevel connective.

Then we can write

$$\Box = \overrightarrow{\Box} \parallel \overset{\twoheadrightarrow}{\Box}$$

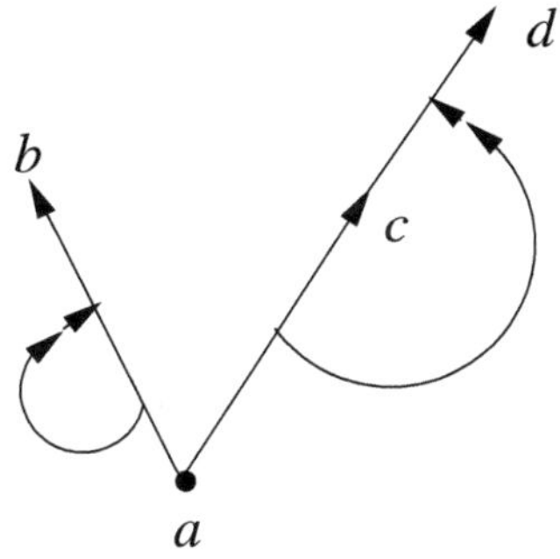

Fig. 2.31

The difference between the connectives is even more serious. Consider Figure 2.31.

Figure 2.31 is like Figure 2.29 except the "fault" double arrows emanate from the arcs and not from the point. Do we want $\overset{\twoheadrightarrow}{\Box}$ to fire all double arrows from arcs or only one of them? Note that when we write $a \vDash \Diamond A$, we either go to $c \vDash A$ in which case only (c,d) is disconnected or we go to $b \vDash A$ in which case only (a,b) is disconnected.

It looks as though we need, in addition to the modalities $\overset{\rightarrow}{\Box}$ and $\overset{\rightarrow}{\Diamond} = \neg \overset{\rightarrow}{\Box} \neg$ six additional fault connectives to be able to express all our options:

1. $a \vDash \overset{\twoheadrightarrow}{\Box}_{\mathrm{point}} A$ activates all double arrows emanating from the point a.
2. $a \vDash \overset{\twoheadrightarrow}{\Box}_{\mathrm{arc}} A$ activates all double arrows emanating from all arcs beginning at a.
3. $a \vDash \overset{\twoheadrightarrow}{\Diamond}_{\mathrm{point}} A$ activates at least one double arrow emanating from a.
4. $a \vDash \overset{\twoheadrightarrow \mathrm{one}}{\Diamond}_{\mathrm{arc}} A$ activates at least one double arrow emanating from one arc coming out of a.
5. $a \vDash \overset{\twoheadrightarrow \mathrm{all}}{\Diamond}_{\mathrm{arc}} A$ activates *all* double arrows emanating from at least one arc coming out of a.
6. $a \vDash (X\|Y)A$. Execute XA and YA in parallel, simultaneously.
   Thus we have
   - $\Box = \overset{\rightarrow}{\Box} \parallel \overset{\twoheadrightarrow}{\Box}_{\mathrm{point}} \parallel \overset{\twoheadrightarrow}{\Box}_{\mathrm{arc}}$
   - $\Diamond = \overset{\rightarrow}{\Diamond} \parallel \overset{\twoheadrightarrow}{\Box}_{\mathrm{point}} \parallel \overset{\twoheadrightarrow \mathrm{all}}{\Diamond}_{\mathrm{arc}}$

   Note that we can also use $\|$ as a connective and write $(\Box A)\|(\Box B)$, which means evaluate $\Box A$ and $\Box B$ in parallel at the same time so they will interfere with each other. Compare with $\Box A \wedge \Box B$ where $\Box A$ and $\Box B$ do not interfere.

These are metalevel deletion operators brought into the object level by the use of connectives. Their detailed logical study is a big story on its own and needs to be done within a wider context. Deletion appears in abduction in logic and in handling databases as well as deletion of elements from models. See our anti-formula and anti-element papers [60, 61].

Appendix: Further Topics

2.7.1 Traversing a graph

Definition 2.24. *By a graph G we mean a set S with a binary relation $R \subseteq S^2$. Let $a \in S$ be the starting point. We write $G = (S, R, a)$.*

Definition 2.25. *1. By a Horn closure condition in the language of R we mean a clause of the form*

$$C : \bigwedge_i x_i R y_i \wedge \bigwedge_j u_j \neq v_j \to xRy.$$

2. *A substitution θ from $Z = \{x_i, y_i, u_j, v_j, x, y\}$ into S is a function θ assigning a value $\theta(z) \in S$ to each $z \in Z$.*
3. *We say a relation $R^* \subseteq S^2$ satisfies the clause C if for all θ, if $\theta(x_i)R^*\theta(y_i)$ holds and $\theta(u_j) \neq \theta(v_j)$ holds then $\theta(x)R^*\theta(y)$ also holds. We write $G^* = (S, R^*, a) \models C$. Let τ be a set of clauses. We say $G \models \tau$ iff $G \models C$ for all $C \in \tau$.*

Lemma 2.26. *Let $G = (S, R, a)$ be a graph and τ a set of clauses. Then there exists the smaller $R^* \subseteq R$ such that $G^* = (S, R^*, a) \models \tau$.*

Definition 2.27. *Let $G = (S, R, a)$ be a graph and let $\mu = (\tau_1, \ldots, \tau_k)$ be a sequence of sets of clauses. Let $R^*_1, \ldots, R^*_k$ be the closures of R under τ_i resp. Define a μ-path H through S as follows.*

The first element of H is $a_0 = a$. The next element of H is a_1 such that $a_0 R^*_1 a_1$ holds. We say a_1 is a τ_1 choice.

Assume a_n is a τ_n choice, then a_{n+1} is such that $a_n R^*_{n+1} a_{n+1}$ where $R^*_{n+1} = R^*_{i+1}$ if a_n is a τ_i choice, $i < k$ and $R^*_{n+1} = R^*_1$ if a_n is a τ_k choice.

Definition 2.28 (A reactive graph). *Let $G = (S, R, a)$ be a graph. Write $R = \{(x, y) \mid xRy \text{ holds}\}$. Define W as follows:*

$$\begin{aligned} W_0 &= S \times S \quad \text{connections of level } 0 \\ W_{n+1} &= S \times W_n \text{ connections of level } n+1 \\ W &= \textstyle\bigcup_n W_n \end{aligned}$$

We consider any set of connections of level ≥ 1 as a switch.

A subset $\bar{R}$ of W can be used to describe S paths H as follows.

1. *The initial element of H is an $a_0 = a$ initial set $\bar{R}_1 = \bar{R}$.*
2. *Assume $\bar{R}_n$ is defined and $\bar{R}_n \subseteq \bar{R}$.*
3. *We define $\bar{R}_{n+1}$.*
   Let $\bar{R}_{n+1} = \bar{R}_n = \{\alpha \mid (a_n, \alpha) \in R_n\} \cup \{\beta \mid (a_n, \beta) \in (\bar{R}) - R_n\}$.
   We assume (a_n, a_{n+1}) is in $\bar{R}_n$.

Conjecture: Let H_μ be the set of all possible paths defined on $G = (S, R, a)$ using $\mu = (\tau_1, \ldots, \tau_k)$. Then there exists an $\bar{R} \subseteq W$ that yields the same paths.

Definition 2.29. *1. A multigraph has the form $G = (V, E, s, r)$ where V is a set of vertices, E a set of edges, and $s, r : E \to V$ are maps describing the source and range of edges.*

2. A family of sets of edges, $R_H = R_h = (E_h, s_h, r_h)$, is called compatible if: for

$$e \in E_h \cap E_{h'}$$

we have

$$s_h(e) = s_{h'}(e) \text{ and } r_h(e) = r_{h'}(e) \; .$$

So it makes sense to consider a set of edges that is the union of the elements R_H as

$$\bigcup R_H = (\bigcup_{h \in H} E_h, s, r)$$

where $s(e) = s_h(e)$ and $r(e) = r_h(e)$ for any $h \in H$. In a similar way we define the intersection of sets of edges and the predicate $\in$ on edges and sets of edges of this form.

Definition 2.30. *A reactive multigraph has the form $G = (V, R_H)$ where $R_H = \{R_h = (E_h, s_h, r_h) \mid h \in H\}$ is a compatible family of sets of edges based on V indexed on H and (V, R_h, a) is a multigraph for all h. The elements of H are sequences of edges and satisfies:*
$\varepsilon \in H$ the empty path, corresponding to the empty sequence of edges
$p = e_1 \dots e_n \in H$ iff $e_1 \dots e_{n-1} \in H$ and $r_{e_1 \dots e_{n-1}}(e_{n-1}) = s_{e_1 \dots e_{n-1}}(e_n)$.
H is called the set of reactive paths of G.

Definition 2.31 (Switch reactivity).

1. Let $G = (V, R)$ be a multigraph and $R = (E, s, r)$. Define the total set of possible switches W_R as follows:

$$\begin{aligned} W_0 &= E \\ &\vdots \\ W_{n+1} &= \{(e, \alpha) \mid e \in E, \alpha \in W_n\} \\ &\vdots \\ W_R &= \textstyle\bigcup_n W_n . \end{aligned}$$

2. A switch graph $\mathbf{G}$ has the form $(V, R, \mathbf{R}_\varepsilon, \mathbf{R})$ where (V, R) is a multigraph, with set of edges "possible in the graph", $\mathbf{R}_\varepsilon \subseteq \mathbf{R} \subseteq W_R$, where $\mathbf{R}_\varepsilon$ is the initial configuration of the switch graph.

3. Let $\mathbf{R}' \subseteq \mathbf{R}$ and let $e \in E$. Define $\mathbf{R}'_e$ as follows:

$$\mathbf{R}'_e = \mathbf{R}' - \{\alpha \mid (e, \alpha) \in \mathbf{R}'\} \cup \{\beta \mid (e, \beta) \in \mathbf{R} \wedge (e, \beta) \notin \mathbf{R}'\}.$$

We say α is deactivated *and β is* activated *by passing through the node x, where $x = s(e)$.*

4. *Let $H = (e_1 \dots e_n)$. We let $\mathbf{R}_H$ be defined as follows:*

$$\mathbf{R}_{e_1} = (\mathbf{R}_\varepsilon)_{e_1}$$
$$\vdots$$
$$\mathbf{R}_{e_1 \dots e_{m+1}} = (\mathbf{R}_{e_1 \dots e_m})_{e_{m+1}}$$

Here we see $e \in E$ as edge $(e, s_{|e}, r_{|e})$.

5. *A switch graph has level n of reactivity if $\mathbf{R} \subseteq \bigcup_{i \leq n} W_i$.*

A switch graph is a particular case of a reactive graph, with:

$$R_h = \mathbf{R}_h \cap R$$

Problem 2.32. Can every reactive multigraph be represented by a switch graph?

2.7.2 Reactive Proof Theory

We give an example of how to do reactive proof theory. See also Chapter 9. A full analysis will be done in [58]. The proper environment for developing reactive proof theory is the methodology of Labelled Deductive Systems, see [55]. See also [20] where reactive rules were independently introduced.

The usual natural deduction propositional system has elimination rules and introduction rules. In principle an elimination rule has the form

$\mathbb{ER}$	$\dfrac{A_1, \dots, A_n}{B}$	where $A_1, \dots, A_n, B$, are well formed formulas.

and an introduction rule has the form of a subproof:

$\mathbb{IR}$: To show $\varphi(A_1, \dots, A_n, B)$, start a subproof with $A_1, \dots, A_n$ as additional assumptions and conclude the subproof successfully with obtaining B, where $\varphi, A_1, \dots, A_n, B$ are well formed formulas. φ is built up from $A_1, \dots, A_n, B$.

Well known examples of such rules are the implicational rules:

$\Rightarrow \mathbb{E}$	$\dfrac{A, A \Rightarrow B}{B}$

$\Rightarrow \mathbb{I}$	To show $A \Rightarrow B$, assume A and prove B

Definition 2.33 (Proofs of level $\leq n$). *Let $\mathbf{S}$ be a set of proof rules.*

1. *A line (in a proof) is a sequence of the form*

$$\ell : A, J, \rho, \alpha$$

where ℓ is a line reference (line number), A is the formula of the line, J is the justification of the line, and ρ is the set of proof rules active at this line and α is

a set of line references accessible at the line. ρ changes only in a reactive proof system. In an ordinary proof system ρ is always the full set of rules of the logic, and α contains all previous line references.
The justification in $\ell : B, J, \rho, \alpha$ can be either the word "assumption" or the phrase "A is obtained using the elimination rule $\mathbb{ER} : \dfrac{A_1, \ldots, A_n}{A}$, where A_i are formulas obtained in lines $\ell_i, i = 1, \ldots, n$ and ℓ_i are accessible", or the phrase "$A = \varphi(A_1, \ldots, A_n, B)$ where A is a formula of an introduction rule $\mathbb{IR}$ and the justification is a subproof π whose assumptions are $A_1, \ldots, A_n$ together with whatever is accessible at ℓ".

2. *A (non-reactive) proof π of level 0 is a sequence of numbered lines containing formulas and justifications of the form*

   line number: wff, Justification

   such that the beginning of the sequence contains formulas justified as "assumptions" and any subsequent line has the form $\ell : B, \mathbb{R}$ where $\mathbb{R}$ is an elimination rule of the form $\dfrac{A_1, \ldots, A_n}{B}$ where $\ell_i : A_i$ Justification are previous lines in the sequence. We say that the sequence is a proof of the formula of the last line from the initial sequence of assumptions.
3. *A proof of level $\leq n + 1$ is defined as a sequence π of lines containing formulas and justifications where the initial elements of the sequence are all justified as "assumptions" and each subsequent formula in line ℓ is either justified from previous lines using an elimination rule, as described in (2) above, or is a formula of the form $\varphi(A_1, \ldots, A_n, B)$ appearing in an introduction rule, justified by a proof π_φ of level $\leq n$, of B from the assumptions $A_1, \ldots, A_n$ and whatever is accessible to ℓ.*
   If the last line of π contains the formula E and the initial sequence of the proof is the assumptions $D_1, \ldots, D_k$, then we say that π is a level $\leq n + 1$ proof of E from $D_1, \ldots, D_k$.

Definition 2.34 (Reactive proof system). *A reactive proof system has reactive rules of the form*

$$\begin{array}{|l|}\hline \mathbf{R}_n : (\mathbb{R}, r_i^+, r_j^-) \\ i = 1, \ldots, k^+, j = 1, \ldots, k^-, n = 1, \ldots, m, r_i^\pm \leq m. \\ \hline \end{array}$$

where $\mathbb{R}$ is a rule (elimination or introduction). The reading of $\mathbf{R}_n$ is that if we use rule $\mathbb{R}$ then the rules $\mathbf{R}_{r_i^+}, i = 1, \ldots, k^+$ should be activated and rules $\mathbb{R}_{r_j^-}, j = 1, \ldots, k^-$ should be deactivated.

The notion of a proof is modified as follows.

At the start of the proof (line 1) we add a third component ρ indicating which rules are active and a fourth component α indicating which previous lines are accessible for the purpose of justification. The general nature of the logic will have procedures for telling us given a line ℓ in the proof and its α, what will be the α of the next line. For example, if in line ℓ the justification is "assumption", then line ℓ is available to be used in justification of the next line or for example if line ℓ is justified *by an elimination rule of the form* $\dfrac{A_1,\ldots,A_n}{B}$, *where* A_i *are from accessible lines* ℓ_i, *resp. then lines* ℓ_i *are no longer accessible at the next line (line* $\ell+1$*). Lines which are assumptions do not change which rules are active and which are not. A line which uses a rule* $\mathbf{R}_n$ *as justification (either elimination or introduction) activates or deactivates the other rules as indicated in the rule itself and the result of the change is the rules* ρ *which are available for the next line of the proof. The changes in accessibility* α *can depend on the logic at hand. For example as we have mentioned, a resource logic may make assumptions not accessible once they are used in a justification.*

Example 2.35. Take the ordinary natural deduction rules for classical logic including

$$\boxed{\dfrac{A, A \Rightarrow B}{B}}$$

and

$$\boxed{\dfrac{\bot}{B}}$$

and

To show $A \Rightarrow B$ assume A and prove B

Assume that using modus ponens deactivates the negation rule.

Thus $A, A \Rightarrow \bot \nvdash C$ by modus ponens alone using level 0 proofs because once we use modus ponens to get $\bot$ we do not have the negation rule anymore. We could work our way around this in this case, using level 1 proofs as follows:

1. A, assumption
2. $A \Rightarrow \bot$, assumption
3. $\bot \Rightarrow C$
   using $\Rightarrow$ introduction rule,
   3.1. $\bot$, assumption
   3.2. C from 3.1 and the negation rule
   3.3. Exit with $\bot \Rightarrow C$ proved
4. $\bot$, from 1 and 2 using modus ponens. Note that the negation rule which was used in the subproof at 3.2 does not cancel the modus ponens in the main proof.
5. C from 3 and 4 using modus ponens again.

2.7.3 J. van Benthem's Sabotage Modal Logic

J. van Benthem introduced sabotage modal logic in [154]. The idea was put forward by van Benthem (2002) and is based on games played on graphs where one player is trying to traverse the graph while another tries to delete edges to make the first player fail.

This is another instance of reactive behaviour, conceptually conceived as sabotage. In the context of Kripke models (S, R, a, h), van Benthem introduces two modalities, the usual evaluation modality (denoted in our notation by $\vec{\Box}$) and a sabotage modality denoted by $\diamondminus$.

There are several versions for $\diamondminus$[8]

$a \vDash \diamondminus_{\text{point}} A$ iff $a \vDash A$ in a new model in which some point $s \in S, s \neq a$ is deleted.

We can use the convenient notation $a \vDash_s A$.

$a \vDash \diamondminus_{\text{arc}} A$ iff $a \vDash A$ in a new model in which some arc $(t, s) \in R$ is deleted. Again we can use the notation $a \vDash_{(t,s)} A$.

Philip Rohde studied properties of these logics, with regard to complexity.

Another recent variation recommended by Benedikt Löwe, is to have the sabotage modality localised by turning the system two-dimensional.

$a, t \vDash \diamondminus_{\text{arc}} A$ iff for some s such that tRs we have $a, s \vDash_{(t,s)} A$.[9]

We now compare sabotage modality with our "fault-remedy" modality to give the reader a view of what is going on.

It is obvious that Professor van Benthem and myself share is the idea of reactivity, of working against a system which changes under you whether by built-in weaknesses and remedies or by sabotage. The actual modal logics produced emerge from the original point of view taken.[10]

---

[8] van Benthem has modalities indexed also by actions, but this is irrelevant to the system's conceptual aspects.

[9] To clarify the notation, let

- $\sharp(z_1, \ldots, z_n) \equiv \bigwedge_{i \neq j} z_i \neq z_j$
- $\sharp((x_1, y_1), \ldots, (x_m, y_m)) = \bigwedge_{i \neq j}(x_i \neq x_j \vee y_i \neq y_j)$

Let $t \vDash_{z_1,\ldots,z_n} A$ mean that A holds at t, in the model where all the different points $z_1, \ldots, z_n$ have been deleted.

Similarly, $t \vDash_{(x_1,y_1),\ldots,(x_n,y_n)} A$ means that A holds at t in the model where all the different arcs $(x_1, y_1), \ldots, (x_n, y_n)$ have been deleted.

Thus for point sabotage we have

- $t \vDash_{z_1,\ldots,z_n} \diamondminus A$ iff $\exists s \exists y(tRs \wedge \sharp(t, s, z_1, \ldots, z_n, y) \wedge s \vDash_{z_1,\ldots,z_n,y} A)$.

For arc sabotage we have:

- $t \vDash_{(x_1,y_1),\ldots,(x_n,y_n)} \diamondminus A$ iff $\exists s \exists u \exists v(tRs \wedge uRv \wedge \sharp((x_1, y_1), \ldots, (x_n, y_n), (t, s)) \wedge \sharp((x_1, y_1), \ldots, (x_n, y_n), (u, v))$ and $s \vDash_{(x_1,y_1),\ldots,(x_n,y_n),(u,v)} A)$.

[10] van Benthem says in [154] (this paper was already written in 2002) as follows:

> "In particular, the logical model-checking angle suggests a study of evaluation of first-order logic on structures *which change under evaluation*. E.g. an object might become unavailable once drawn from a domain, or a fact might change when inspected (think of measurement in quantum mechanics)."

In comparison, I say in [56]

> "This paper addresses the case where the semantics does change (or react) under us as we evaluate a formula. This idea makes the evaluation of a wff at a world

I also adopt the point of view that deletion is a metalevel notion and that there should be a logical discipline and stylised machinery for bringing it into the object level. This should be seen in a wider context of dynamic evaluations and dynamic operators on which there is a lot of literature. See [153] for a recent application, and the references there.

The rest of this comparison is simply technical, comparing formal options between the fault-remedy modality and the sabotage modality.

1. Fault modality can have arrow-to-arrow deletions on many levels. This comes from the original point of view.
2. Switch procedures also come naturally in this context.
3. Fault modality is a simultaneous parallel combination of metalevel changes and evaluation. This makes it possibly technically different from a logic where the functionalities are separated.
   We can simulate sabotage by double arrows and non-determinism. The operator $\bar{\Diamond}_{\mathrm{arc}}$, deleting an arc can be simulated, for example, by looking at double arrows going from each node to all arcs and non-deterministically activating one double arrow every time we make a move.
4. Conceptually the fault-remedy concept is universal, while the notion of sabotage is agent specific. Consider a case of one agent sabotaging the work of several other agents. Additional special effort is required by the saboteur to make it look like a fault (accident).[11]

> t dependent on the route leading to t. Thus we get a new kind of semantics, the reactive semantics."

Later in the paper, I go on to say

> "we have put forward the reactive and dynamic idea of evaluation in earlier papers and lectures. A typical example we give is to consider $t \vDash \Diamond A$. In modal logic this means that there is a possible world s such that $s \vDash A$. We take a more dynamic view of it. We ask: where is s? How long does it take to get to it? and how much does it cost to get there?"

[11] Think of the numerous detective novels where the murderer tries to make the murder look like an accident, and how the detective can show it was not.

3

Introducing Reactive Modal Tableaux

3.1 Setting the scene

The purpose of this section is to explain the idea of reactive semantics and reactive Beth tableaux. The best way to do it is to begin with an ordinary (non-reactive) example of a modal logic, look at its axioms and semantics and give the corresponding ordinary (non-reactive) Beth tableaux for it. We then proceed to turn the example reactive and see what kind of modifications we need to maeke to the Beth tableaux we have developed. These modifications will introduce us to our first example of reactive Beth tableaux. For general references on tableaux and on modal tableaux, see [22, 41, 42, 48, 101].

This will be done in the first section. The second section defines the general notion of Reactive Beth tableaux and the third section discusses our options for reactivity in general, including reactive Beth tableaux, reactive automata, reactive sets and more.

Let us begin with modal logic. We denote a Kripke frame by $\mathbf{F} = (S, R, t)$ where $S \neq \varnothing$ is the set of possible worlds and $R \subseteq S \times S$ is the accessibility relation and $t \in S$ is the actual world.

Our language contains the set of atoms Q and the usual constants and connectives $\{\top, \bot, \sim \wedge, \vee, \Diamond, \rightarrow \text{ and } \Box\}$.

An assignment h in a frame $\mathbf{F} = (S, R, t)$ is a function giving to each $q \in Q$ a subset $h(q) \subseteq S$. A Kripke model has the form $\mathbf{m} = (S, R, t, h) = (\mathbf{F}, h)$. We evaluate a wff in a model as in Definition 3.1 below. We call it the ordinary evaluation (as opposed to the reactive evaluation, which we shall describe later).

Definition 3.1 (Ordinary evaluation in Kripke models). *We write*

$$s \vDash A \textit{ in } \mathbf{m},$$

to mean

$$A \textit{ holds at } s \textit{ in the model } \mathbf{m} = (S, R, t, h).$$

The following are the clauses (we omit **m**, *h since the model does not change).*

D.M. Gabbay, *Reactive Kripke Semantics*, Cognitive Technologies,
DOI 10.1007/978-3-642-41389-6_3, © Springer-Verlag Berlin Heidelberg 2013

1. • *$s \vDash q$ iff $s \in h(q)$, for q atomic.*
   • *$s \vDash A \wedge B$ iff $s \vDash A$ and $s \vDash B$.*
   • *$S \vDash\sim A$ iff $s \nvDash A$.*
   • *The cases of $A \vee B, A \to B$ are defined in the standard way from $\sim$ and $\wedge$.*
2. *$s \vDash \Box A$ iff for all x such that sRx we have $x \vDash A$.*
3. *We say that* $\mathbf{m} = (S, R, t, h)$ satisfies A *(we write $\mathbf{m} \vDash A$) iff $t \vDash A$.*
4. *We say that A* holds *in the frame (S, R, t) iff for all possible assignments h in the frame we have $(S, R, t, h) \vDash A$.*
5. *Let $\mathbb{K}$ be a class of frames and let $\mathbb{L}(\mathbb{K})$ be the set of all sentences which hold in every frame of $\mathbb{K}$. We say that $\mathbb{L}(\mathbb{K})$ is the modal logic defined by $\mathbb{K}$.*

There are several ways of characterising the logic $\mathbb{L}(\mathbb{K})$ of Definition 3.1.

1. We can generate $\mathbb{L}(\mathbb{K})$ by giving a Hilbert or Gentzen or natural deduction system $\mathbf{S}$ for the language, and define the notion of $A \vdash_{\mathbf{S}} B$ which means that B is provable in the system $\mathbf{S}$ from A. This is an algorithmic notion defined using $\mathbf{S}$.
   A completeness theorem for $\mathbf{S}$ would mean the following:
   • $A \vdash_{\mathbf{S}} B$ iff for every model $\mathbf{m}$ of $\mathbb{K}$ we have that $\mathbf{m} \vDash A$ implies $\mathbf{m} \vDash B$.
2. We can present a tableaux system for $\mathbb{L}(\mathbb{K})$ which tries to build a countermodel for any pair $A \vdash ?B$, showing, if successful, a model $\mathbf{m}$ with frame in $\mathbb{K}$ such that $\mathbf{m} \vDash A$ but $\mathbf{m} \nvDash B$.
   A tableaux system is an algorithm which constructs (if possible) a countermodel by following closely the options allowed by the semantics $\mathbb{K}$.
   The completeness theorem would mean the following:
   • The tableau construction for $A \vdash ?B$ is successful iff there exists a model $\mathbf{m}$ of $\mathbb{K}$ in which A holds and B does not hold.

It is not easy to find a proof system $\mathbf{S}$ or a tableaux system for arbitrary $\mathbb{K}$. It is easier to find a tableaux system for a logic defined by a single frame. So this is what we discuss now and we choose a single extremely simple frame to illustrate our reactive ideas.

3.1.1 First case study

From the point of view of modal logic, classical logic is the logic of one-point Kripke frame $\mathbf{F} = (S, R, t)$ with a reflexive accessibility relation.

We want to explain the idea of reactive Beth tableaux using this modal logic. So consider the modal logic with the one-point frame $\mathbf{F}_0$ of Figure 3.1.

In this logic we have

$$\vDash A \leftrightarrow \Diamond A.$$

It is easy to put forward an axiom system for $\mathbb{L}(\mathbf{F}_0)$. We take any axiom system for classical logic and add the axiom $A \leftrightarrow \Diamond A$.

The tableaux for this logic are clear. We use the tableaux and rules for classical logic together with two additional rules for $\Diamond$ which identify $\Diamond A$ with A. Since we are going to turn this logic reactive, and formulate reactive tableaux for it, let us write the familiar tableau rules explicitly in a way which makes them ready for becoming reactive.

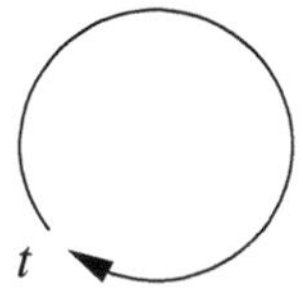

Fig. 3.1: Frame $\mathbf{F}_0$

Definition 3.2 (Tableaux for $\mathbf{F}_0$).

1. *A tableau τ is a two-compartment box with a label. The compartments contain formulas of the logic. The left compartment contains formulas intended to be true (hold) and the right compartment contains formulas intended to be false (not to hold). The label of the tableau contains information about the tableau including which possible world we are talking about, as well as additional information to be used when we turn the tableau reactive.*
Figure 3.2 shows an example of a tableau τ. The label is s and the formulas are A_i and $B_j, i = 1,\ldots,,n, j = 1,\ldots,k$.

	Left	Right
$\tau = s$:	$A_1,\ldots,A_n$	$B_1,\ldots,B_k$

Fig. 3.2: Tableaux τ

2. *A tableau is said to be closed if the same formula A appears in both the right and left box, or if $\bot$ appears in the left box or if $\top$ appears in the right box.*
A tableau is said to be a world (or a model) if all atoms of the language appear in it, and it is not closed.
3. *We consider a system of tableaux as a set $\boldsymbol{\tau}$ of tableaux. The system is closed if all tableaux in the system are closed.*

Definition 3.3 (Operations on a set τ of tableaux). *We start with a set $\boldsymbol{\tau}$ containing a single tableau as in Figure 3.2 where s is the actual world t. We perform successive operations on the elements of $\boldsymbol{\tau}$ to obtain successive sets of tableaux $\boldsymbol{\tau}_1, \boldsymbol{\tau}_2, \boldsymbol{\tau}_3 \ldots$. Each time we perform an operation on $\boldsymbol{\tau}_n$, we choose a tableaux $\tau \in \boldsymbol{\tau}_n$ and apply a rule on τ to obtain from it one or two simpler tableaux τ', τ'' and then we let $\boldsymbol{\tau}_{n+1} = (\boldsymbol{\tau}_n - \{\tau\}) \bigcup \{\tau', \tau''\}$. This way we keep on making the tableaux in $\boldsymbol{\tau}_n$ more and more simple until no more rules apply. Say we stop and cannot continue at stage $\boldsymbol{\tau}_m$. At this stage all the tableaux in $\boldsymbol{\tau}_m$ contain only atoms in their respective boxes (this will become apparent by looking at the rules below). We will get that if all tableaux in $\boldsymbol{\tau}_m$ are closed then there is no model which makes all formulas A_i true and all B_j false (see Figure 3.2).*

The operations are done by choosing a tableau $t \in \boldsymbol{\tau}$ for example:

$\tau = t_1$:	A_i	B_j

and choosing a formula A in the left or right box and operating on it using the tableaux rules.

The following are the rules for $\wedge$ and $\sim$ on the left and on the right. Our set of labels is $\{t\}$ with only one label. However, since we are going to turn the tableaux rules reactive we write the rules schematically as if more labels were available for use.

Left $\sim$ rule for label t_1
Replace

$$\tau = t_1 : \begin{array}{|l|l|} \hline A_i & B_j \\ \sim X & \\ \hline \end{array}$$

by

$$\tau_1 = t_2 : \begin{array}{|l|l|} \hline A_i & B_j \\ & X \\ \hline \end{array}$$

where $t_2 = t_1$.

Right $\sim$ rule for label t_1
Replace

$$\tau = t_1 : \begin{array}{|l|l|} \hline A_i & B_j \\ & \sim X \\ \hline \end{array}$$

by

$$\tau_1 = t_2 : \begin{array}{|l|l|} \hline A_i & B_j \\ X & \\ \hline \end{array}$$

where $t_2 = t_1$.

Left $\wedge$ rule for label t_1
Replace

$$\tau = t_1 : \begin{array}{|l|l|} \hline A_i & B_j \\ X \wedge Y & \\ \hline \end{array}$$

by the tableau

$$\tau_1 = t_2 : \begin{array}{|l|l|} \hline A_i & B_j \\ X, Y & \\ \hline \end{array}$$

where $t_2 = t_1$.

Right $\wedge$ rule for label t_1
Replace

$$\tau = t_1 : \begin{array}{|l|l|} \hline A_i & B_j \\ & X \wedge Y \\ \hline \end{array}$$

by the tableaux τ_1 and τ_2

$$\tau_1 = t_2 : \begin{array}{|l|l|} \hline A_i & B_j \\ & X \\ \hline \end{array}$$

$$\tau_2 = t_2 : \begin{array}{|c|c|}\hline A_i & B_j \\ & Y \\ \hline\end{array}$$

where $t_2 = t_1$.

Notice that the label t_1 *has not changed after the operations, we always have* $t_2 = t_1$.

We need not explicitly give the rules for $\vee$ *and* $\rightarrow$ *as* $\sim, \wedge$ *are sufficient for classical logic and anyway, the rules for* $\vee$ *and* $\rightarrow$ *are well known, and the reader can easily figure them out in our notation.*

For the logic with $\Diamond$*, we add the following rules:*

Left $\Diamond$ rule for label t_1
Replace

$$\tau = t_1 : \begin{array}{|c|c|}\hline A_i & B_j \\ \Diamond X & \\ \hline\end{array}$$

by

$$\tau_1 = t_2 : \begin{array}{|c|c|}\hline A_i & B_j \\ X & \\ \hline\end{array}$$

where $t_2 = t_1$.

Right $\Diamond$ rule for label t_1
Replace

$$\tau = t_1 : \begin{array}{|c|c|}\hline A_i & B_j \\ & \Diamond X \\ \hline\end{array}$$

by

$$\tau_1 = t_2 : \begin{array}{|c|c|}\hline A_i & B_j \\ & X \\ \hline\end{array}$$

where $t_2 = t_1$.

Theorem 3.4 (Completeness theorem). *Consider the logic for the frame* $\mathbb{F}_0$ *of Figure 3.1, and let A and B be two formulas. Then* $A \vDash B$ *iff the tableaux*

$$\tau_0 = \{t : \begin{array}{|c|c|}\hline A & B \\ \hline\end{array}\}$$

can be expanded into a closed system of tableaux by means of the reactive tableaux rules.

Proof. This is a well known process, the same as for classical logic. We use the additional axiom $\Diamond A \leftrightarrow A$ in the right and left rules for $\Diamond$ which basically eliminate $\Diamond$.

We now turn our system reactive. Consider the frame $\mathbb{F}_1$ of Figure 3.3

The set is $S = \{t\}$. The relation of accessibility is $\mathbb{R} = \{(t,t), ((t,t)(t,t))\}$. (t,t) is the arrow and $((t,t),(t,t))$ is the double arrow. The reactivity idea is expressed by the point of view that once we pass through the arrow, this activates the double arrow which disconnects the reflexive arrow. We are not giving a formal definition at this stage, instead we give an example. The formal definitions are Definitions 3.8 and 3.9.

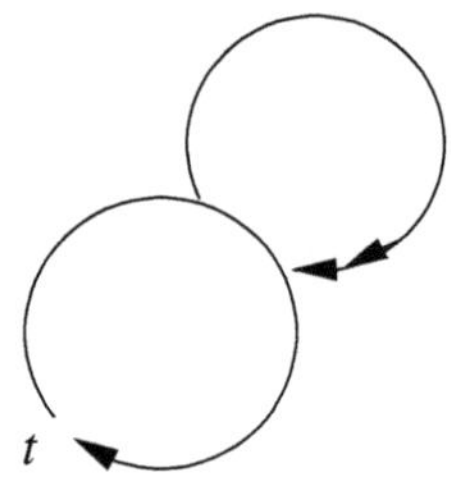

Fig. 3.3: Frame $\mathbb{F}1$

Example 3.5 (Reactive evaluation in $\mathbb{F}_1$). Let q be an atom and assume $t \vDash q$.

To evaluate $t \vDash ?\Diamond q$ we go to the accessible world t ($t\mathbb{R}t$ holds) and evaluate q. Since $t \vDash q$, we get that $t \vDash \Diamond q$.

Now let us evaluate $t \vDash ?\Diamond\Diamond q$. We go to the accessible world t and evaluate $t \vDash ?\Diamond q$. But now, having passed through the arrow (t, t) the double arrow $((t, t), (t, t))$ was activated, and disconnected the reflexive arrow (t, t). The model has changed. What we have now is a model with the frame $(\{t\}, \varnothing, t\})$. Hence now $t \vDash \Diamond q$ is false. Thus

$$t \nvDash \Diamond\Diamond q.$$

It is obvious that in a reactive model in general the value of $s \vDash ?\Diamond A$ depends not just on s but on the past history sequence of evaluation points $(s_1, \ldots, s_k, s)$ leading to s. So we must use the notation

$$(s_1, \ldots, s_k, s) \vDash ?\Diamond A.$$

The sequence of points leading to s will activate various reactivity double arrows and tell us what is the current accessibility situation at the point s.

Definition 3.6 (Reactive satisfaction in $\mathbb{F}_1$). *The following is the correct way of defining evaluation in our reactive model of Figure 3.3. We have paths t, tt, ttt, and we give evaluation clauses for each path. We need use only two labels, t and tt, because additional ts make no behavioural difference:*

- $t \vDash q$ *if* $t \in h(q), q$ *atomic*
- $tt \vDash q$ *if* $t \in h(q), q$ *atomic*
- *The evaluation at t, tt of $\wedge$ and $\sim$ is as in classical logic.*
- $t \vDash \Diamond A$ *iff* $tt \vDash A$
- $tt \nvDash \Diamond A$
- *A holds in the modal if $t \vDash A$.*

We now check what kind of logic we get from the satisfaction notion of Definition 3.6. Note that the notion of "logic" here is slightly different. These logics are not closed under arbitrary substitution.You can see this from the way we axiomatise our examples, see Definition 3.11.

Theorem 3.7. *The logic of the frame of Figure 3.3 as defined by the satisfaction notion of Definition 3.6 is not complete for any class of traditional Kripke frames*

Proof. Let us assume that the logic of the figure is complete for a class of Kripke frames. We get a contradiction. Since the formula $q \wedge \sim \Diamond\Diamond q$ is consistent in the logic of the figure, it must have a model. Let $\mathbb{K}$ be a class of traditional Kripke frames for which the logic is complete and let (S, R, t, h) be a model of $q \wedge \sim \Diamond\Diamond q$. The frame (S, R, t) must be a model of the logic under any assignment h to the frame.

Claim 1.
tRt must hold.

Otherwise, let p be atomic and let $t \vDash p$, but let $s \nvDash p$ for all $s \neq t$. Hence $t \vDash p \wedge \sim \Diamond p$.

This is impossible since in the reactive frame of Figure 3.3 we have $\vDash p \rightarrow \Diamond p$, for any atomic p, according to Definition 3.6.

Claim 2.
$q \rightarrow \Diamond\Diamond q$ holds under h.

This holds because tRt holds.

We now have a contradiction.

It is time to define the notion of switch reactive Kripke models. Section 3 will discuss other options for reactivity in Kripke models. The switch reactivity is the simplest.

Definition 3.8 (Switch reactive models).

1. *A switch reactive Kripke frame has the form $(S, \mathbb{R}, \mathbf{a}, t)$ where S is a non-empty set of possible worlds, $t \in S$ is the actual world and $\mathbb{R}$ is a reactive accessibility set of arrows on S defined as follows:*
   a) *$(t, s) \in S \times S$ is an arrow of level 1.*
   b) *If α is an arrow of level n and $(t, s) \in S \times S$ then $((t, s), \alpha)$ is an arrow of level $n + 1$.*[1]

   We assume that if $((t, s), \alpha) \in \mathbb{R}$ then $\alpha \in \mathbb{R}$.
   $\mathbf{a}$ is a $\{0, 1\}$ function on $\mathbb{R}$.
   If $\mathbf{a}(\alpha) = 1$, we say α is active.
   If $\mathbf{a}(\alpha) = 0$, we say α is not active.
   t is the actual world.

[1] An alternative way of defining $\mathbb{R}$ is as follows:
A k-level arrow is inductively defined as follows:

$$\mathbb{A}_0 = S \times S$$
$$\mathbb{A}_{k+1} = (S \times S) \times \mathbb{A}_k.$$

Let $\mathbb{A} = \bigcup \mathbb{A}_k$.
Let $\mathbb{R}$ be a subset of $\mathbb{A}$.

2. *Let* $(t, s) \in \mathbb{R}$*. Define* $\mathbf{a}_{(t,s)}$ *as follows*

$$\mathbf{a}_{(t,s)}(\beta) = 1 - \mathbf{a}(\beta), \text{ if } ((t, s), \beta) \in \mathbb{R}$$

$$\mathbf{a}_{(t,s)}(\beta) = \mathbf{a}(\beta) \text{ otherwise.}$$

Definition 3.9 (Satisfaction in switch reactive model). *Let* $\mathbf{m} = (S, \mathbb{R}, \mathbf{a}, t, h)$ *be a switch reactive model where* $(S, \mathbb{R}, \mathbf{a}, t)$ *is a switch reactive frame and* h *is an assignment into* S *giving for each atomic* q *a subset* $h(q) \subseteq S$*. We define the notion of* $\mathbf{m} \vDash A$.

1. $\mathbf{m} \vDash q$ *iff* $t \in h(q)$.
2. $\mathbf{m} \vDash A \wedge B$ *iff* $\mathbf{m} \vDash A$ *and* $\mathbf{m} \vDash B$
   $\mathbf{m} \vDash \sim A$ *iff* $\mathbf{m} \nvDash A$
   similarly for $\bot, \vee, \bot, \top$.
3. $\mathbf{m} \vDash \Diamond A$ *iff for some* $(t, s) \in \mathbb{R}$ *such that* $\mathbf{a}((t, s)) = 1$ *we have*

$$\mathbf{n} = (S, \mathbb{R}, \mathbf{a}_{(t,s)}, s, h) \vDash A$$

4. *Note that the real model is the active part of* $\mathbb{R}$*. When we evaluate* $\Diamond$ *we move to a new model with different active part.*
5. *Given a frame* $\mathbb{F} = (S, \mathbb{R}, \mathbf{a}, t)$*, the logic of the frame* $\mathbb{L}(\mathbb{F})$ *is the set of all wff* A *such that for all* h $(S, \mathbb{R}, \mathbf{a}, t, h) \vDash A$.

Corollary 3.10. *There exists a logic complete for a class of reactive frames but not complete for a class of ordinary frames.*

Definition 3.11 (Axiomatisiation of the logic of the reactive frame of Figure 3.3). *We offer the following system:*

1. *Take as axioms the following:*
   (*) All substitution instances of truth functional tautologies $A(q_1, \ldots, q_k)$.
2. *Let* $A(q_1, \ldots, q_k)$ *be a wff. A can be presented as a substitution result of the form*

$$A(q_1, \ldots, q_k) = B(q_1, \ldots, q_k, e_i/\Diamond B_i)$$

   $i = 1, \ldots, m$*, where* e_i *are new atoms and* B_i *are formulas using the atoms* $q_1, \ldots, q_k$ *and* $B(q_1, \ldots, q_k, e_1, \ldots, e_m)$ *is a formula without a modality.*
   This representation is unique.
   Let $B_A^\bot$ *be the formula* $B(q_1, \ldots, q_k, \bot, \ldots, \bot)$.
   We take the additional following axiom:
   (**) $\Diamond A$ *for any* A *such that* $B_A^\bot$ *is a truth functional tautology.*

Example 3.12. This example prepares the ground for giving a reactive tableaux algorithm for the logic of Figure 3.3. Although this logic is very simple, there is a technical problem with the reactive tableaux for it. This is typical of all reactive tableaux and arises from the very idea of reactivity. Consider Figure 3.3 and let p, q be two atoms such that $t \vDash p \wedge q$. Consider now $t \vDash ?\Diamond p \wedge \Diamond q$.

According to the definition of satisfaction, the evaluation is done in parallel.

$$t \vDash \Diamond p \wedge \Diamond q$$

iff

$$t \vDash \Diamond p \wedge t \vDash \Diamond q$$

iff

$$tt \vDash p \wedge tt \vDash q$$

When we give any tableaux algorithm the algorithm is sequential.

So if we apply a tableaux rule corresponding to $t \vDash \Diamond p$, we go to tt in the model and in the corresponding tableau and have $p, \Diamond q$ to evaluate and in tt, $\Diamond q$ is false!

Put differently, if we evaluate $\Diamond p$ first and pass through some arc to find a world in which p holds, the model might change. We should backtrack on the change when we evaluate $\Diamond q$.

So the tableaux need to be done in parallel.

Definition 3.13 (Tableaux system for the logic of the frame of Figure 3.3). *We modify the tableaux system of Definition 3.3 as follows:*

1. *We allow for two labels t and tt.*
2. *We require that all classical connectives in each box be handled first before approaching any $\Diamond$. For that purpose we regard any $\Diamond X$ formula as 'atomic' and reduce each tableaux to its 'atomic' components. The reason for this has to do with Example 3.12 and the need to handle all cases of $\Diamond$ in parallel.*
3. *The rules for $\sim$ and $\wedge$ remain the same as in Definition 3.3, where t_1 can now be either $t_1 = t$ or $t_1 = tt$. t_2 remains equal to t_1.*
4. *The rules for $\Diamond$ now split into two sets. For the case of $t_1 = t$, we take the rules of Definition 3.3 with $t_2 = tt$ (and not $t_2 = t_1$ as in Definition 3.3).*
   We have a condition here that all cases of $\Diamond X$, $\Diamond Y$ etc both on the right hand side or on the left hand side should be done in parallel. The classical connectives should be dealt with before $\Diamond$ is handled.
   For the case of $t_1 = tt$ we take the following rules:

   Left $\Diamond$ rule label tt
   Replace

   $tt :$

A_i	B_j
$\Diamond X$	

   by

   $tt :$

A_i	B_j
$\perp$	

   Right $\Diamond$ rule for label tt
   Replace

   $tt :$

A_i	B_j
	$\Diamond X$

by

tt :

A_i	B_j
	$\bot$

Theorem 3.14. *The tableaux system of Definition 3.13 is complete for the semantics of* $\mathbb{F}_2$ *and for the axiomatisation of Definition 3.11.*

Proof. Not too difficult.

Remark 3.15. The example we gave is rather simple. In Section 3.1.2 we give a slightly more complex example.

3.1.2 Second case study

We now consider another example. Consider frame $\mathbf{F}_2$ of Figure 3.4.

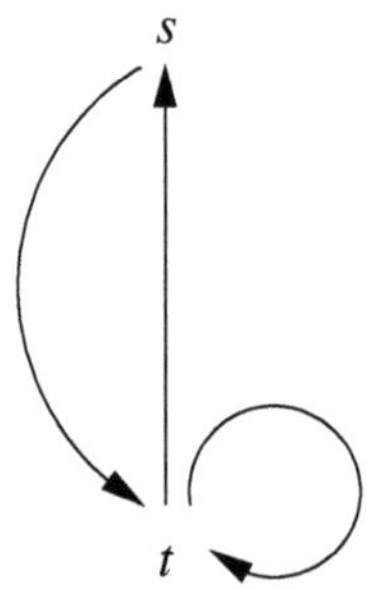

Fig. 3.4: Frame $\mathbf{F}_2$

We have $S = \{t, s\}$ and $R = \{(t,t), (t,s), (s,t)\}$.

As we have already said, there are several ways to characterise the logic of this frame.

1. Write a Hilbert or a Gentzen system for it and get a completeness theorem for it, namely:
   $A \vdash_{\text{system}} B$ iff for all assignments h to the frame $\mathbf{F}_2$ above we have $t \vDash_h A$ implies $t \vDash_h B$.
2. Provide a tableaux system for it and have the following completeness theorem
   $A \nvdash_{\text{system}} B$ iff there is successful tableaux procedure for $A = \top, B = \bot$.

The fact that we are dealing with the frame $\mathbf{F}_2$ of Figure 3.4 would be incorporated in the tableaux procedures.

We now present a reactive version of Figure 3.4. Consider Figure 3.5, displaying the frame $\mathbb{M}$.

In this Figure we added a double arrow from the arc (t,t) to itself. We can mathematically present the frame as $\mathbb{M} = (S, \mathbb{R}, t)$ where $S = \{t, s\}$ and $\mathbb{R} = \{(t,t), (t,s), (s,t), ((t,t),(t,t))\}$.

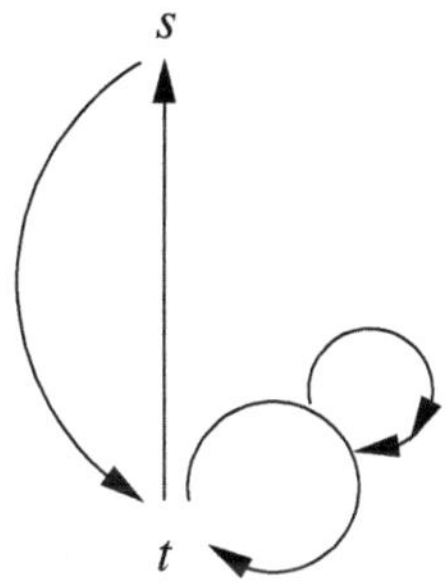

Fig. 3.5: Frame $\mathbb{M}$

The item $((t, t), (t, t))$ represents the double arrow. The evaluation of a formula at t under an assignment h to the atoms goes as in Definition 3.16 below. Here we use the full notation with the index.

Definition 3.16 (Reactive evaluation in the frame $\mathbb{M}$). *Let $x \in \{t, s\}$:*

1. • $x \vDash_h q$ *iff* $x \in h(q)$, *for* q *atomic.*
   • $x \vDash_h A \wedge B$ *iff* $x \vDash A$ *and* $x \vDash B$.
   • $x \vDash_h \sim A$ *iff* $x \nvDash A$.
2. $t \vDash_h \Box A$ *iff (a) and (b) hold.*
   a) $s \vDash_h A$, *in the frame* $\mathbb{M} = (S, \mathbb{R}, t)$.
   b) $t \vDash_h A$ *in the frame* $\mathbf{N} = (S, R_1, t)$ *where* $S = \{t, s\}$ *and* $R_1 = \{(t, s), (s, t)\}$. *See Figure 3.6.*

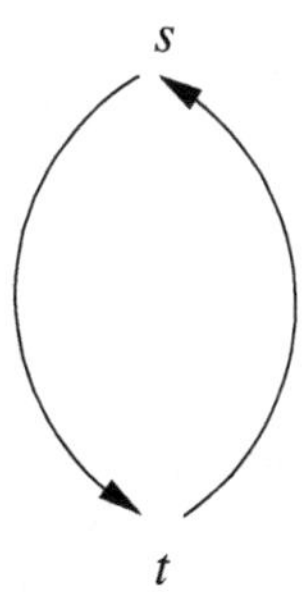

Fig. 3.6: Frame **N**

3. $s \vDash_h \Box A$ *in* $\mathbb{M}$ *or* $\mathbf{N}$ *iff* $t \vDash_h A$ *in* $\mathbb{M}$ *or* $\mathbf{N}$ *respectively.*

The way we view the change of model is as follows.

To evaluate $t \vDash \Box A$, we need to go to all accessible worlds from t. In our frame $\mathbb{M}$, these are the worlds t and s. To go to s we traverse the arc (t, s) and we get to s and at s A must hold. We thus get condition (a) above. We also go to t along the arc (t,t). As we traverse the arc (t,t), we activate the double arrow $((t,t),(t,t))$ which reactively disconnects the arc (t,t). (Think of it like a bridge collapsing because you passed a truck over it, or a use once only connection.) Now that we are at t, our frame has changed from $\mathbb{M}$ to $\mathbf{N}$. We evaluate A at t but the accessibility has changed, it is now the frame $\mathbf{N}$ not the frame $\mathbb{M}$.

Figure 3.7 shows the available paths.

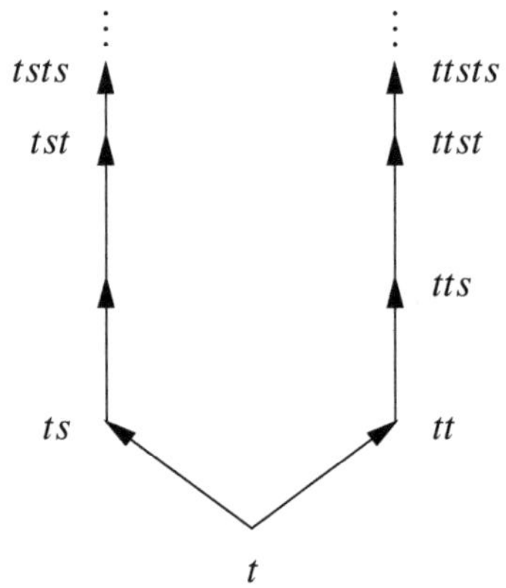

Fig. 3.7

The reactive evaluation in the fixed frame $\mathbb{M}$ does give rise to a set of theorems $\mathbb{L}(\mathbb{M})$. This set may be axiomatised as a Hilbert or Gentzen system or may be characterised by a tableaux system. Since any tableaux system closely follows the semantic evaluation procedures, this leads us to the idea of Reactive Tableaux.

The next section will abstractly define the notion of a reactive tableaux. In the rest of this section we need to show what kind of logic we get from the frame $\mathbb{M}$ and what kind of tableaux system corresponds to it.

First let us axiomatise the logics of frame $\mathbf{F}_2$ and $\mathbb{M}$ and show that the logic of $\mathbb{M}$ cannot be characterised by any class of ordinary frames.

Remark 3.17 (Methodological comments). We saw that reactive evaluation in the frame $\mathbb{M}$ moved to evaluation in the frame $\mathbf{N}$. This is typical. In general, since reactive evaluation in a reactive frame $\mathbb{N}$ changes the frame as we proceed in the evaluation, we end up with a complex interaction of evaluations in some traditional frames $\mathbf{N}_1,\dots,\mathbf{N}_k$.

This has implications for how we axiomatise and develop tableaux for the logic of $\mathbb{N}$. We need to understand the logics and tableaux of the traditional frames $\mathbf{N}_1,\dots,\mathbf{N}_k$.

We see that in principle we need a mutually recursive axiomatisation of several logics, all done together. This is already a new kind of game. We take several logics $\mathbb{L}_i$ and recursively define a new combination of them, where the exact way they are combined is dictated and guided by the reactivity involved.

This observation may necessitate different axiomatic formulations of $\mathbb{L}_i$, not necessarily the traditional ones, in order to facilitate their combination.

The next example shows how we need to reformulate the traditional modal **K**, in order to axiomatise the logic of $\mathbb{M}$.

We also see, by the way, that turning a logic reactive is a very special form of combining logics.

Definition 3.18 (Modal K without necessitation). *Consider the following formulation of modal logic* **K***, without the necessitation rule.*[2]

1. *a) All substitution instances of truth functional tautologies A are axioms.*
   b) If A is a substitution instance of a truth functional tautology then $\Box^m A, m \geq 1$ *is an axiom.*
2. *a) All substitution instances of* $(\Box(A \to B) \to (\Box A \to \Box B))$ *are axioms.*
   b) All substitution instances of $\Box^m(\Box(A \to B) \to (\Box A \to \Box B))$ *are axioms for any* $m \geq 1$.
3. *Note that the necessitation rule*
$$\frac{\vdash A}{\vdash \Box A}$$
   is derivable now for this formulation of **K**.

Definition 3.19 (The logic of the frame N). *The logic of the frame* **N** *of Figure 3.6 is easy to axiomatise. It is a normal logic. We can take the usual axioms of* **K** *with necessitation and add very simple axioms for* $\Diamond$*, as follows.*

1. *All substitution instances of truth functional tautologies.*
2. $\Box(A \to B) \to (\Box A \to \Box B)$
3. $\dfrac{\vdash A}{\vdash \Box A}$
4. $\dfrac{\vdash A \quad \vdash A \to B}{\vdash B}$
5. $\sim \Diamond A \leftrightarrow \Diamond \sim A$
6. $\Diamond\Diamond A \leftrightarrow A$.

Theorem 3.20. *The logic of Definition 3.19 is complete for the intended semantics of the frame* **N** *of Figure 3.6.*

Proof. Easy.

Definition 3.21 (The logic $\mathbb{L}(\mathbf{F}_2)$). *Consider the following modal axiom schemata, added to the axioms of modal logic* **K** *in the formulation without necessitation, as presented in Definition 3.18.*

[2] We need this complicated axiomatisation of **K** so that it becomes easy to axiomatise the non-normal logic of the frame $\mathbf{F}_2$. It also becomes easy to transform this axiomatisation into a reactive axiomatisation of the logic of the frame $\mathbb{M}$.

1. $A \to \Diamond A$
2. a) $\sim A \wedge \Diamond A \to \Box(A \wedge \Diamond \sim B \to \sim \Diamond B)$
   b) $\sim A \wedge \Diamond A \wedge \sim B \wedge \Diamond B \to \Diamond(A \wedge B)$
   c) $\sim A \wedge \Diamond(A \wedge B) \to \Box(A \to B)$
3. $A \to \Box\Diamond A$
4. *Note that since* **K** *was formulated without necessitation we do not have the axiom* $\Box(A \to \Diamond A)$.
   The logic is not normal and its theorems are what holds in the actual world only.
5. *We define the notion of* $\Delta \vdash_{\mathbb{L}(\mathbf{F}_2)} A$ *in this logic to be* $\Delta \cup$ **Axioms** $\vdash_{\mathbf{K}} A$, *where* **Axioms** *is the set of all substitution instances of the axioms (1)–(3) above.*
6. Δ *is a theory of the logic if* $\Delta \supseteq$ **Axioms**. *It is consistent if it is consistent in* **K**.

Lemma 3.22. *The logic* $\mathbb{L}(\mathbf{F}_2)$ *of Definition 3.21 is complete for the frame* $\mathbf{F}_2$.

Proof. We start by showing soundness. We check each axiom, they must all hold at t.

1. holds because tRt holds
2. a) Assume $t \vDash \sim A \wedge \Diamond A$. This means $s \vDash A$. We show that $t \vDash \Box(A \wedge \Diamond \sim B \to \sim \Diamond B)$. The only possible world accessible to t (out of $\{t, s\}$) in which A can hold is s. So if $s \vDash \Diamond \sim B$, this means $t \vDash \sim B$ and hence $s \vDash \sim \Diamond B$.
   b) Assume $t \vDash \sim A \wedge \Diamond A \wedge \sim B \wedge \Diamond B$. The only way this can hold at t is that $s \vDash A \wedge B$. So $t \vDash \Diamond(A \wedge B)$.
   c) Assume $t \vDash \sim A \wedge \Diamond(A \wedge B)$. This means $s \vDash B$. Hence $t \vDash \Box(A \to B)$ because the only accessible world to t in which A holds is s and $s \vDash B$.
3. Follows from the fact that in the frame R is symmetric.

To prove completeness we show every consistent theory has a model with frame $\mathbf{F}_2$. Let Δ be a complete and consistent theory of the logic of Definition 3.21. We can assume that for some wff α_0 we have $\sim \alpha_0 \wedge \Diamond\alpha_0 \in \Delta$. If no such α exists then for all α, $\sim (\sim \alpha \wedge \Diamond\alpha) \in \Delta$, i.e. $\alpha \vee \sim \Diamond\alpha \in \Delta$, which is $\Diamond\alpha \to \alpha \in \Delta$. Since $\alpha \to \Diamond\alpha$ is an axiom we get that $\alpha \leftrightarrow \Diamond\alpha \in \Delta$ for all α. This means Δ is essentially a consistent classical theory and can be given a one-point classical model.

So let us assume that for some $\alpha_0, \sim \alpha_0 \wedge \Diamond\alpha_0 \in \Delta$, and find a model for Δ.

Let $\Theta_0 = \{E \mid \sim E \wedge \Diamond E \in \Delta\} \cup \{D \mid \Box D \in \Delta\}$. We claim Θ_0 is consistent in **K**. Otherwise for some $D_1, \ldots, D_m, E_1, \ldots, E_k$ in Θ_0 we have

$$\mathbf{K} \vdash \bigwedge_j D_j \to \sim \bigwedge_i E_i,$$

where $\Box D_j \in \Delta, \sim E_i \wedge \Diamond E_i \in \Delta$, and so

$$\mathbf{K} \vdash \Box(\bigwedge_j D_j \to \sim \bigwedge_i E_i)$$

and by pushing $\Box$ through we get

$$\mathbf{K} \vdash \bigwedge_j \Box D_j \to \sim \Diamond \bigwedge_i E_i.$$

We now obtain a contradiction.

Since $\sim E_1 \wedge \Diamond E_1 \wedge \sim E_2 \wedge \Diamond E_2$ are in Δ, then so is $\Diamond(E_1 \wedge E_2)$. Therefore also $\sim (E_1 \wedge E_2) \wedge \Diamond(E_1 \wedge E_2)) \wedge \sim E_3 \wedge \Diamond(E_3) \in \Delta$ and hence $\Diamond(E_1 \wedge E_2 \wedge E_3) \in \Delta$. Continuing in this manner we get $\Diamond \bigwedge_i E_i \in \Delta$, a contradiction since Δ is consistent and $\mathbf{K} \vdash \bigwedge_j \Box D_j \to \sim \Diamond \bigwedge_i E_i$ and $\bigwedge_j \Box D_j \in \Delta$ and so $\sim \Diamond \bigwedge_i E_i \in \Delta$.

Therefore Θ_0 is consistent and can be extended in **K** to a complete theory Θ. We now show that if $A \in \Delta$ then $\Diamond A \in \Theta$. This holds because $A \to \Box\Diamond A \in \Delta$. We also have that if $\Box A \in \Delta$ then $A \in \Delta$. This follows the axiom $A \to \Diamond A$. We are now ready to construct a model for Δ.

Let $S = \{\Delta, \Theta\}$. We know $\Delta \neq \Theta$ because $\sim \alpha_0 \in \Delta$ and $\alpha_0 \in \Theta$. Let R be defined as $R = \{(\Delta, \Delta), (\Delta, \Theta), (\Theta, \Delta)\}$.

Let the actual world be Δ. Let $h(q) = \{X \in S | q \in X\}$.

We claim for any X, A we have

(*) $X \vDash A$ iff $A \in X$.

We need to check only the cases of $\Box A \in \Delta$ and $\Box A \in \Theta$.

Assume $\Box A \in \Delta$. Then $A \in \Delta$. Otherwise $\sim A \in \Delta$ and hence $\Diamond \sim A \in \Delta$, a contradiction.

Now assume $\sim \Box A \in \Delta$. Then if $\sim A \in \Delta$, we are finished. If $A \in \Delta$, we get $A \wedge \Diamond \sim A \in \Delta$ then by construction $\sim A \in \Theta$ and we are finished.

Now assume $\Box A \in \Theta$. Then $A \in \Delta$. Otherwise $\sim A \in \Delta$ and hence $\Box\Diamond \sim A \in \Delta$ and hence $\Diamond \sim A \in \Theta$ a contradiction.

Assume now that $\Diamond \sim A \in \Theta$. We need to show $\sim A \in \Delta$. If $A \in \Delta$ we get $\Box\Diamond A \in \Delta$ hence $\Diamond A \in \Theta$. But $\sim \alpha \wedge \Diamond\alpha \to \Box(\alpha \to (\Diamond \sim A \to \sim \Diamond A))$ holds and we get a contradiction.

This completes the proof of Lemma 3.22.

Lemma 3.23 (Axioms for $\mathbb{L}(\mathbb{M})$). *The following schemas hold in the reactive frame $\mathbb{M}$ under any assignment.*

a) The following axioms
- *(1*)* • $A \to \Box\Diamond A$
  - • $A \to \Diamond A$

  For any theorem A of the logic of the frame **N** *of Definition 3.19.*
- *(2*) Same as axiom (2) of Definition 3.21.*
- *(3*)* $A \wedge \Diamond \sim A \to \Box(\sim A \to \Diamond A)$.

b) $\sim A \wedge \Diamond A \to \Diamond(\sim A \wedge \Box A)$

c) If x is an atom then there exists a model for $\sim x \wedge \Diamond x$.

Proof. Axiom group (a) hold in the frame $\mathbb{M}$ for similar reasons as checked in the soundness part of Lemma 3.22.

We now check (b).

Assume $t \vDash \sim A \wedge \Diamond A$. This implies $t \vDash \sim A$ and $s \vDash A$. We now show $t \vDash \Diamond(\sim A \wedge \Box A)$. We use the fact that tRt holds. We go to t, to check whether $t \vDash \sim A \wedge \Box A$. Indeed, $t \vDash \sim A$. But now that we have passed through the connection tRt, because of reactivity, tRt no longer holds, it was disconnected. The only possible world accessible to t is s, and $s \vDash A$ hence $t \vDash \Box A$. See Definition 3.16, item 2(b).

c) Clearly holds

Lemma 3.24 (Reactive frames are stronger than ordinary frames). *The logic of the reactive frame $\mathbb{M}$ is not complete for any class of ordinary Kripke frames.*

Proof. Suppose there is a class $\mathbb{K}$ of ordinary frames for which $\mathbb{L}(\mathbb{M})$ is complete. This means that every instance of axioms (a), (b) of Lemma 3.23 holds at the actual world for any assignment h to any of the frames in $\mathbb{K}$. We also know that there exists a frame (S, R, t) and an assignment h such that (c) of Lemma 3.23 holds at t.

We shall use the above to derive a contradiction. We start with any frame (S, R, t) in which every instance of (a) of Lemma 3.23 holds under any h and see what are its properties. We prove (a), (b), (c) below:

(a) $tRs_1 \wedge tRs_2 \wedge t \neq s \wedge\ t \neq s_2 \Rightarrow s_1 = s_2$

Assume otherwise.

Let $\sim x$ hold at t and x hold at s_1 and s_2 and let u hold at s_1 and $\sim u$ hold at t and s_2. Figure 3.8 displays this situation.

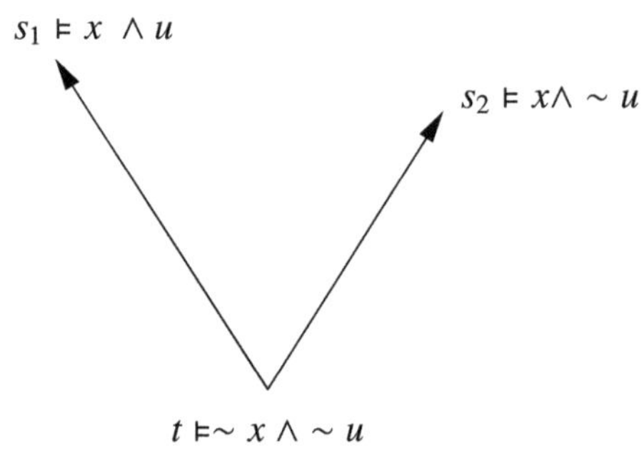

Fig. 3.8

Let $\alpha = x \vee u$.
Therefore by the axiom

$$\sim \alpha \wedge \Diamond(\alpha \wedge u) \rightarrow \Box(\alpha \rightarrow u),$$

and

$$t \vDash \sim \alpha \wedge \Diamond(\alpha \wedge u)$$

we have $t \vDash \Box(\alpha \rightarrow u)$ and so $s_2 \vDash u$, a contradiction.

(b) we claim that tRt holds. This follows from the axiom $A \rightarrow \Diamond A$, for otherwise we can let $t \vDash x$ and $\forall s(t \neq x \Rightarrow s \vDash \sim x)$ and get a contradiction.

(c) We claim that $tRs \rightarrow sRt$. This is because of the axiom $A \rightarrow \Box\Diamond A$.
For otherwise choose an atom x and let $t \vDash_h x$ and for all y, $sRy \Rightarrow y \vDash_h \sim x$. Since sRt does not hold, this is OK. Then we have $t \vDash_h x$ but $t \nvDash_h \Box\Diamond x$, since at s we have $s \vDash_h \Box \sim x$.

The above hold for any frame in $\mathbb{K}$.

4. We now get a contradiction. There exists a model (S, R, t, h) in $\mathbb{K}$ in which (c) of Lemma 3.23 holds. Namely $t \vDash_h \sim x \wedge \Diamond x$ holds in the model. Let s_1 be such that tRs_1 and $s_1 \vDash_h x$. So we must have also $t \vDash_h \Diamond(\sim x \wedge \Box x)$ from axiom (b) in Lemma 3.23. We show that this is impossible. For let s_2 be such that tRs_2 and $s_2 \vDash_h \sim x \wedge \Box x$. We have shown that in any frame of $\mathbb{K}$, $tRs_1 \wedge tRs_2 \wedge t \neq s_1 \wedge t \neq s_2 \Rightarrow s_1 = s_2$.
   We have according to our data that $t \neq s_1$ and $s_1 \neq s_2$ and tRs_1 and tRs_2. Therefore we must have $t = s_2$.
   But tRt holds and $t = s_2 \vDash_h \sim x \wedge \Box x$.
   A contradiction.

Example 3.25 (Tableaux for the logics of $\mathbf{F}_2$ *and of* $\mathbb{M}$*).* Let us make tableaux for the logics of $\mathbf{F}_2$ and of $\mathbb{M}$. We know that $\sim u \wedge \Diamond u \to \Diamond\Box u$ is refutable in the logic of $\mathbf{F}_2$ but is a theorem of $\mathbb{L}(\mathbb{M})$. So if we do tableaux for both we will see where the difference lies.

These two logics depend on frames with two possible worlds in them and only one reactive arrow. So our tableaux will take advantage of these special features and have the form

$$\begin{array}{r|c|c|} \beta: & A^2_1, \ldots, A^2_{m'} & B^2_1, \ldots, B^2_{n'} \\ \alpha: & A^1_1, \ldots, A^1_m & B^1_1, \ldots, B^1_n \end{array}$$

α, β are labels from $\{t, tt, s\}$, α can be t or tt and β is s.

A tableau is closed if

1. Some $A^k_i = B^k_j$
2. $\perp$ is on the left
3. $\top$ is on the right.

The following are the rules. They are specially tailored for this particular case.

For the formulas of the form $A = \sim X$ or $A = X \wedge Y$, the left rule and the right rule for a box in row α or β are the same as in classical logic. We replace the same way. The difference comes for the case of $\Diamond$.

Again we stress that we are required to do all classical connectives first for each label until we end up with only formulas which are either atomic or of the form $\Diamond X$. Only at this stage do we handle $\Diamond$.

In the reactive case of $\mathbb{M}$ all cases of $\Diamond X_i$ are done simultaneously in parallel.

This is a special feature of this logic because it has at most one reactive arrow. The general case is more complex. See Section 3.2.

$\Diamond$ rule for α

Let $\Diamond X_i$, $i \in I$ be on the left and $\Diamond Y_j$, $j \in J$ be on the right. Let I_1 be any subset of I. Replace

$$\tau = \begin{array}{r|c|c|} \beta: & A^2_i & B^2_j \\ \alpha: & A^1_i & B^1_j \\ & \Diamond X_i & \Diamond Y_j \end{array}$$

By some of the following, for all I_1

$\tau_{I_1} =$

$\beta:$	A_i^2 $X_I, i \in I - I_1$	B_j^2 $Y_j, j \in J$
$\alpha_1:$	A_i^1 $X_i, i \in I_1$	B_j^1 $Y_j, j \in J$

where $I_1 \neq \varnothing$.

$\tau_{\varnothing} =$

$\beta:$	A_i^2 $X_i, i \in I$	B_j^2 $Y_j, j \in J$
$\alpha_2:$	A_i^1	B_j^1

as follows.

Case $\mathbb{M}$

- If $\alpha = t$ then $\alpha_1 = tt$ and τ is replaced by all possible $\tau_{I_1}, I_1 \subseteq_{I_1 \neq \varnothing} I$.
- If $\alpha = tt$, then $\alpha_2 = tt$ and τ is replaced by $\tau_{\varnothing}$
- If $\alpha = t$ and $I = \varnothing$ then $\alpha_2 = t$.

Case $\mathbf{F}_2$
In this case $\alpha = \alpha_1 = \alpha_2 = t$ and τ is replaced by all possible τ_{I_1}, for all $I_1 \subseteq I$.

◊ rule for β
Replace

$\tau =$

$\beta:$	A_i^2 $\Diamond X_i, I \in I$	B_j^2 $\Diamond Y_j, j \in J$
$\alpha:$	A_i^1	B_j^1

by

$\tau =$

$\beta:$	A_i^2	B_j^2
$\alpha:$	$A_i^1, X_i, I \in I$	$B_j^1, Y_j, j \in J$

This is done for either of the logics and any α, β.

Example 3.26. Consider the tableau τ_0 below.

$\tau_0 =$

$s:$		
$t:$	$\sim u$ $\Diamond u$	 $\Diamond\Box u$

Let us operate on it using the rules of $\mathbf{F}_2$ and in parallel the rules of $\mathbb{M}$. The tableau should close for $\mathbb{M}$ and give a countermodel for $\mathbf{F}_2$.

The following sequence of replacements is what we get

$\tau_1 =$

$s:$		
$t:$	 $\Diamond\Box u$	u $\Diamond u$

Here we should be doing $\Diamond u$ and $\Diamond\Box u$ simultaneously, but $\Diamond u$ cannot go to α because u is false at α so we can do it first!

$\tau_2 =$

s :	u	
t :		u
	$\Diamond\Box u$	

We got from τ_1 to τ_2 using the $\Diamond$ rule for α. This rule gives two options, but one of them is a closed tableau in this case. So we took only one of the options. We have to use now the right $\Diamond$ rule for α, for the formula $\Diamond\Box u$. We get the tableau τ_3 with α_3.

$\tau_3 =$

s :	u	$\Box u$
α_3 :		$u \quad \Box u$

For logic $\mathbf{F}_2, \alpha_3 = t$. For the logic of $\mathbb{M}$, $\alpha_3 = tt$.

Using the right negation rule we get τ_4.

$\tau_4 =$

s :	u	
	$\Diamond \sim u$	
α_3 :	$\Diamond \sim u$	u

We continue by looking at the left $\Diamond$ rule for s, which yields:

$\tau_5 =$

s :	u	
α_3	$\Diamond \sim u$	u

In the logic $\mathbf{F}_2, \alpha_3 = t$ and so we replace τ_5 by two tableaux, one of them has u at the right hand side of the $\alpha_3 = t$ row. This is tableau τ_6 which gives us a countermodel. The other is τ_7 which is closed. For the case of $\mathbb{M}, \alpha_3 = tt$ and the only option is to replace τ_5 by τ_7 which is closed.

$\tau_6 =$

s :	u	
$\alpha_3 = t$:		u

$\tau_7 =$

s :	$u, \sim u$	
$\alpha_3 = tt$:		u

3.2 Reactive Beth Tableaux

We now discuss reactive Beth tableaux. First we need a general discussion of tableaux for modal logic and their relationship to modal labelled deductive systems.

Let $\mathbb{K}$ be a set of frames of any kind, ordinary frames or reactive frames. There are two main ways of presenting $\mathbb{K}$.

1. List the members of $\mathbb{K}$ or generate them inductively.
2. Use a meta-predicate Ψ in some language capable of talking about properties of frames and letting
$$\mathbb{K} = \{\mathbb{F} \mid \Psi(\mathbb{F}) \text{ holds}\}$$
The way we develop tableaux for $\mathbb{L}(\mathbb{K})$ depends on the way $\mathbb{K}$ is presented.

Suppose $\mathbb{K}$ is presented as a set of specific finite frames, namely $\mathbb{K} = \{\mathbb{F}_1, \mathbb{F}_2, \ldots\}$, where $\mathbb{F}_n$ are all finite. We can do tableaux for each $\mathbb{F}_i$ much in the same spirit as we did in Section 3.1 and prove a completeness theorem of the form:

$\mathbb{L}(\mathbb{K}) \vdash A$ iff for each $\mathbb{F}_i$, the A tableaux in $\mathbb{F}_i$ is closed.

The problem is more difficult when $\mathbb{K}$ is defined using a predicate Ψ. It is not simple even in the case of traditional Kripke frames with ordinary $R \subseteq S \times S$ and Ψ a condition on R. Take for example the logic $\mathbf{K} + \Diamond\Box A \vee \Diamond\Box \sim A$. This logic is complete for a class of frames defined by a second-order condition on R.
Let $E(x, y) = \exists a(xRa \wedge aRy)$. Then the condition is

$$\Psi(R) \equiv \forall x \forall T[(y \in T \to E(x, y)) \to \exists a(\forall z(aRz \to z \in T) \vee \forall z(aRz \to \sim z \in T))].$$

Another logic is provability logic, being

$$\mathbf{K} + \{\Diamond\Diamond A \to \Diamond A\} + \{\Diamond A \to \Diamond(A \wedge \Box \sim A)\}.$$

Provability logic is complete for the second-order condition saying that every ascending R-sequence of points has a last element.

To do tableaux for these logics we have to creatively make something tailor made. If we take the frames for these logics and turn them reactive then we need to know how to modify the tableaux systems for the logics and turn them reactive. This is not systematic. We therefore propose to solve the following problem:

Reactive Tableaux Problem
Given a tableaux system for some logic (never mind how we got it or what that logic is), how do we make it reactive? What are the features we change in it to make it reactive?

In the semantical case of Kripke models we know how to turn them reactive; we add the higher level arrows and allow the model to change as in Definition 3.8. We now ask how to do it in the syntactic case. Doing tableaux for reactive Kripke models gives us some clues. We make the tableaux dependent on the path of $\Diamond$ rules used, by using labels.

The story, however, is much more complex than that, as the examples below will show.

The following series of definitions intends to follow the way a reactive frame $\mathbb{F}_t$, with actual world t, changes in the course of evaluation along a path α into a new frame $\mathbb{F}_\alpha$. These concepts will help us formulate a recursive tableaux algorithm for the logic of the frame $\mathbb{F}_t$.

Definition 3.27 (How reactive frames change).

1. *Let* $\mathbb{F}_t = (S, \mathbb{R}, \mathbf{a}, t)$ *be a switch reactive frame.*
2. *Let* $\alpha = (t, t_1, \ldots, t_n), t_i \in S$ *be a sequence of worlds. Let* $\mathbf{e}(\alpha) = t_n$ *. We define* $\mathbb{F}_\alpha = (S, \mathbb{F}, \mathbf{a}_\alpha, \mathbf{e}(\alpha))$ *as follows:*
   a) *For* $\alpha = (t)$*, let* $\mathbb{F}_\alpha$ *be* $\mathbb{F}_t$ *as in (1).*

b) Assume $\mathbb{F}_\beta$ is defined for $\beta = (t, t_1, \ldots, t_n)$ and let $(t_n, t_{n+1}) \in \mathbb{R}$ be such that $\mathbf{a}_\beta(t_n, t_{n+1}) = 1$, i.e. $t_n \to t_{n+1}$ is active in $\mathbb{F}_\beta$. Then let $\mathbb{F}_{(\beta,t_{n+1})} = (S, \mathbb{R}, \mathbf{a}_{\beta,(t_n,t_{n+1})})$, (see Definition 3.8, item 2).

Definition 3.28 (Reactive tableaux). *A tableau has the form $(\tau, \mathbb{F}_\alpha)$, where $\mathbb{F}_\alpha$ is as in Definition 3.27 and τ is a function on S giving a pair of sets of formulas $\mathbf{L}^\tau(x)$ (left part of the tableau at x) and $\mathbf{R}^\tau(x)$, (right part of the tableau at x) for each $x \in S$. The formulas in $\mathbf{L}^\tau(x), \mathbf{R}^\tau(x)$ are either atomic or of the form $\Diamond A$.*

Definition 3.29 (Tableaux complexity). *let $(\tau, \mathbb{F}_\alpha)$ be a tableau. The complexity of $(\tau, \mathbb{F}_\alpha)$ is a pair of numbers (m, n) where m is the maximal number of nested $\Diamond$ in formulas of τ and n is the number of formulas with maximal number m. So $m \geq 0, n \geq 1$. Note that our tableaux rules will reduce complexity.*

Definition 3.30 (Tableaux feasibility). *Let $(\tau, \mathbb{F}_\alpha)$ be a tableau. We perform a feasibility check on $(\tau, \mathbb{F}_\alpha)$ as follows.*

Consider $\tau(\mathbf{e}(\alpha))$:

$\mathbf{e}(\alpha)$:		
	A_i	B_j
	$\Diamond X_r$	$\Diamond Y_w$

$\Diamond X_r, \Diamond Y_w$ are all the formulas of this form in the tableaux, $r = 1, \ldots, k_1$ and $w = 1, \ldots, k_2$.

First observe that if $k_1 \geq 1$ and there is no s such that $(\mathbf{e}(\alpha), s) \in \mathbb{R}$ is active then the tableau is not feasible. If there is such an s, then we can assume $k_1 \geq 1$, because we can always take $X_1 = \top$.

For each s such that the arc $(\mathbf{e}(\alpha), s) \in \mathbb{R}$ is active, and for each $\Diamond X_r$, consider the sets $(\mathbf{L}^{\tau,r,s}, \mathbf{R}^{\tau,r,s})$ where $\mathbf{L}^{\tau,r,s} = \mathbf{L}^\tau(s) \cup \{X_r\}$ and $\mathbf{R}^{\tau,r,s} = \mathbf{R}^\tau(s) \cup \{Y_w \mid w = 1, \ldots, k_2\}$. Consider this pair as a classical tableau and let $(\mathbf{L}_i^{\tau,r,s}, \mathbf{R}_i^{\tau,r,s})$ for $i = 1, \ldots, k_3(r)$ be all open options for this tableau. Consider the frame $\mathbb{F}_{\alpha,s}$ and let $\tau_i^{\alpha,s,r}$, $i = 1, \ldots, k_3(r)$ be the following tableaux for the frame $\mathbb{F}_{\alpha,s}$.

$$\tau_i^{\alpha,s,r}(x) = \begin{cases} \tau(x) \text{ for } x \neq \mathbf{e}(\alpha), s \\ \tau(x) - \Diamond X_r \text{ for } x = \mathbf{e}(\alpha), \mathbf{e}(\alpha) \neq s \\ (\mathbf{L}_i^{\tau,r,s}, \mathbf{R}_i^{\tau,r,s}) \text{ for } x = s, \mathbf{e}(\alpha) \neq s \\ (\mathbf{L}_i^{\tau,r,s} - \{\Diamond X_r\}, \mathbf{R}_i^{\tau,r,s}) \text{ for } x = s = \mathbf{e}(\alpha) \end{cases}$$

Notice that the tableau $\tau_i^{\alpha,s,r}(x)$ has lower complexity than the tableau τ^α, because we took out from it $\Diamond X_r$. We may have added more modal sentences but they are of lower complexity.

We now continue:

We define the predicate feasible$(\tau, \mathbb{F}_\alpha)$ *as follows:*

- feasible$(\tau, \mathbb{F}_\alpha)$ *iff*

$$\bigwedge_r \bigvee_s \bigvee_i \text{feasible}(\tau_i^{\alpha,s,r}, \mathbb{F}_{\alpha,s})$$

where $r = 1, \ldots, k_2$

$$s \in \{x \mid (\mathbf{e}(\alpha), x) \in \mathbb{R} \text{ is active}\}$$

and $i = 1, \ldots, k_3(r)$.

If there are no open options for any r *then* $(\tau, \mathbb{F}_\alpha)$ *is not feasible. If the tableau* $\tau(\mathbf{e}(\alpha))$ *contains no modalities then it is feasible if it is open.*

Definition 3.31 (Tableaux algorithm). *The following is a recursive algorithm for a tableaux system for a frame* $\mathbb{F}_t = (S, \mathbb{F}, a, t)$. *Let us ask whether* $A \vDash ?B$.

First consider the classical tableau with A on the left and B on the right. $c\tau : \boxed{A|B}$

Any formula $E(q_i)$ *of modal logic can be written uniquely as a substitution instance of a pure classical logic formula* $E'(q_i, p_j)$ *with additional atoms* p_j *where for the* p_j *we substitute unique modal formulas* $\Diamond E_j$. *Thus* $E = E'(q_i, P_j/\Diamond E_j)$. *Thus let us consider A and B in the tableau as classical formulas with 'atoms' which are either real atoms* q_i *or 'atoms' of the form* $\Diamond A_j$ *and* $\Diamond B_j$.

So we regard $c\tau$ *as*

$c\tau : \boxed{A'(q_i, \Diamond A_j) | B'(q_i, \Diamond B_j)}$

Develop this tableau classically. There are two possibilities.

1. *All its branches are closed, in which case* $A \to B$ *is an instance of a classical tautology.*
2. *Some branches are open. Let* $ct_1, \ldots, ct_{k_0}$ *be all the open tableaux.*

Let $c\tau_\delta$ *be a fixed one of these options. Write* $c\tau_\delta$ *in the notational form*

$c\tau_\delta$:

some atoms	*some other atoms*
$\Diamond X^\delta_r$	$\Diamond Y^\delta_w$

$r = 1, \ldots, k_1, w = 1, \ldots, k_2$

Then $\Diamond X^i_r$, $\Diamond Y^i_w$ *list the modal 'atoms' which are on the left and on the right (respectively) of the tableau* $c\tau_\delta$. *We know that if we can make* $\Diamond X^\delta_r$ *all true,* $r = 1, 2\ldots$ *and* $\Diamond Y^\delta_w$ *all false* $w = 1, 2, \ldots$ *then we get a countermodel.*

For a fixed δ, *we are in the situation of Definition 3.30, as follows. We are given* $\mathbb{F}_t$, *and we define a tableau function* τ *as in Definition 3.30*

$$\tau(t) = (\mathbf{L}^{c\tau_\delta}, \mathbf{R}^{c\tau_\delta})$$

where $\mathbf{L}^{c\tau_\delta}$ *is the left side of* $c\tau_\delta$ *and* $\mathbf{R}^{c\tau_\delta}$ *is the right hand side of* $c\tau_\delta$.

Let $\tau(x) = (\varnothing, \varnothing)$ *for* $x \neq t$. *We are now exactly in the situation of Definition 3.30, with* $\alpha = (t)$.

Therefore the tableau $(\tau, \mathbb{F}_t)$ *(which is really* $c\tau_\delta$ *evaluated at t) is feasible according to Definition 3.30 iff* $\bigwedge_r \bigvee_s \bigvee_i \text{feasible}((\tau_i^{t,s,r}, \mathbb{F}_{(t,s)}))$.

We now continue the tableaux feasibility construction for each $(\tau_i^{t,s,r}, \mathbb{F}_{(t,s)})$.

Note the complexity of the tableaux is reduced. We continue by induction until the tableaux we get have no $\Diamond$ *in them. This means they are classical tableaux and therefore are either closed or open. We can therefore know at this stage whether a countermodel exists or not.*

3.3 Conclusion

We discussed reactive Kripke semantics and showed that there are modal logics which cannot be characterised by classes of ordinary Kripke frames but which can be characterised by classes of reactive frames.

We offered a tableaux system for finite-frame reactive Kripke models. We made essential use of the fact that the frame is finite.

Completeness theorems for ordinary arbitrary frame reactive modal K can be found in Chapters 5 and 6.

Tableaux systems for ordinary reactive intuitionistic logic can be found in Chapter 4.

4

Reactive Intuitionistic Tableaux

4.1 Introduction

In Chapter 2 we introduced the idea of reactivity and studied reactive Kripke models for modal logics. In many published papers we studied other reactive systems such as reactive argumentation frames, reactive automata, reactive grammars, reactive preferential logics, reactive contrary to duties, reactive inheritance networks, and many more.

The purpose of this chapter is to introduce reactive intuitionistic frames (Kripke frames and Beth frames) and study their expressive power and properties.

Consider the network of Figure 4.1.

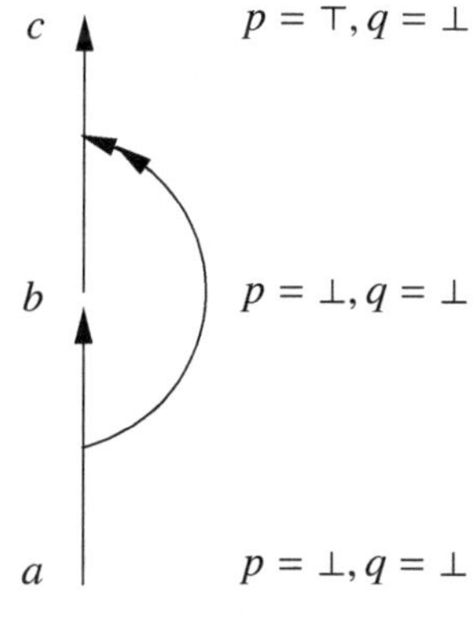

Fig. 4.1: E1

Let us first ignore the double arrow in Figure 4.1. Without the double arrow, Figure 4.1 is a three-point Kripke model for intuitionistic logic. So for example if we want to evaluate $a \nvDash p \to q$, we must check whether there exists a higher point x than a (including a itself) with $x \vDash p$ but $x \nvDash q$.

D.M. Gabbay, *Reactive Kripke Semantics*, Cognitive Technologies,
DOI 10.1007/978-3-642-41389-6_4, © Springer-Verlag Berlin Heidelberg 2013

The above definition is set-theoretical. The notion $x \vDash A$ is defined inductively, and the graph of Figure 4.1 without the double arrow is just a graph suggesting a Kripke model with a reflexive and transitive binary relation.

To be more precise, Figure 4.1 (without the double arrow) suggests a set $S = \{a, b, c\}$, a relation $R = \{(a, b), (b, c)\}$ and the reflexive and transitive closure of R being $R^* = \{(a, a), (a, b), (a, c), (b, b), (b, c), (c, c)\}$. If we use an explicit formula for R^*, we get: xR^*y iff $x = y$ or xRy or for some $k \geq 1$ and some $t_1, \ldots, t_k$, we have $xRt_1 \wedge t_1Rt_2 \wedge \ldots \wedge t_{k-1}Rt_k \wedge t_kRy$.

The assignment h to the atoms is also indicated in the Figure.

So the Kripke model is (S, R^*, a, h), with $h(p) = \{c\}$ and $h(q) = \varnothing$.

So to check whether $a \nvDash p \to q$ we simply ask set-theoretically whether $\exists x(aR^*x$ and $x \vDash p$ and $x \nvDash q)$.

To introduce the reactive approach we envisage ourselves walking along the arrows of the graph from point a onwards and at each point x that we pass, we evaluate $x \vDash p$ and $x \vDash q$ and compare. This is an actual walk and search along the graph.

The search forms a decision procedure for the model. If there is an x such that aR^*x and $x \vDash p$ and $x \nvDash q$ then we will walk into it sooner or later.

Now given this 'walk along the graph' point of view, the reactive double arrow makes sense. What it does is the following: As we cross from a to b, the double arrow gets activated and disconnects the path from b to c. So we do not get to the point c where $c \vDash p$ and $c \nvDash q$. Without getting to c, we will report that $a \vDash p \to q$ holds, because we cannot get to the counterexample, etc. So in the reactive model of Figure 4.1 with the double arrow, we have $a \vDash p \to q$.

We now sum up. We introduced two ideas here.

1. Evaluation in Kripke models is done by a 'walk along the arrows and check and report' policy.
2. Double arrows along the way can disconnect connections and control where we can go.[1]

Consider now Figure 4.2. In this Figure, when we walk along $a \to b \to d$ we cannot continue from d to e, because $d \to e$ gets disconnected. However, when we walk along $a \to c \to d$, we can continue to e because there is no double arrow along the path.

Section 4.2 gives the formal definitions involved and introduces the reactive models. We also show that we get a richer semantics than ordinary Kripke models.

The idea of reactivity is a general one and can apply to Beth models as well. Beth models are like Kripke models except the inductive truth definition is different. We need the notion of a Belt anti-chain of points. Given (S, R^*, a, h), and $t \in S$ then a set $T \subseteq S$ is a Belt anti-chain for t if all points of T are R^* incomparable and every maximal R^* chain beginning at t must meet the Belt T.

---

[1] There are more complex options for reactivity:
a) Double arrows can switch arrows on and off.
b) Double arrows can emanate from other double arrows.
c) We can have an inductively iterated version of the above.
In this chapter we are keeping the reactivity simple.

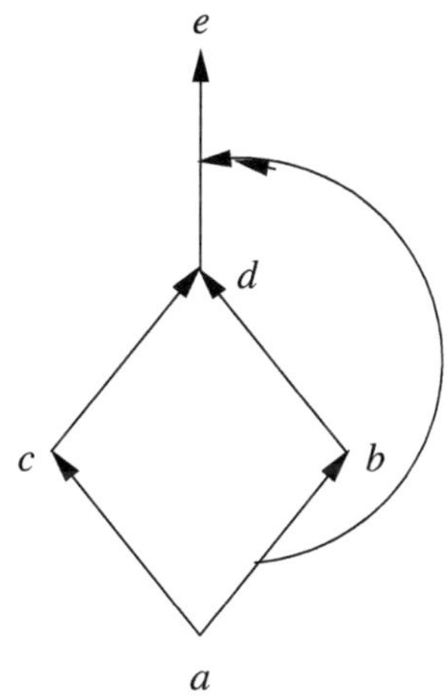

Fig. 4.2: E2

We have $t \vDash A$ iff there exists an antichain Belt T for t such that for all $x \in T, x \vDash A$.

Turning a model reactive is even easier if we give the correct definition of a reactive path. A reactive path beginning at t is a trace of a walk along the arrows from t onwards, where all double arrows are taken into account. So hopefully we can define reactive Beth models as well.

In Figure 4.2 there are two maximal reactive paths $a \to b \to d$ and $a \to c \to d \to e$.

4.2 Reactive Kripke Frames for Intuitionistic Logic

This section introduces reactive Kripke frames for intuitionistic logic and shows that intuitionistic propositional logic is complete for such frames. We also show that there are intermediate logics which are complete for a class of reactive Kripke frames but are not complete for any class of ordinary Kripke frames.

Thus reactive Kripke frames is a richer and stronger semantics than ordinary frames.

The above also means that we can study a richer class of intermediate logics, e.g. intermediate logics generated by finite reactive frames. We shall see in Remark 4.23 what kind of Heyting like algebras one gets from finite reactive frames.

To appreciate the opportunities opening for us through the notion of reactive Kripke frames for intuitionistic logic, consider a famous beautiful theorem of L. Maksimova [125].

- There are only seven intermediate logics which have interpolation.

Is this still true if we take into account logics generated by reactive frames? The notion of 'logic' may not be the same!

A later section will provide tableaux for logics defined by finite frames.

Definition 4.1 (Ordinary Kripke models for intuitionistic propositional logic). *A Kripke model has the form* $\mathbf{m} = (S, R, R^*, a, h)$ *where* S *is a non-empty set of worlds and* $R \subseteq S \times S$ *is a binary relation on* S. R^* *is the reflexive and transitive closure of* R,[2] $a \in S$ *is the actual world and* h *is an assignment, giving for each atomic* q *a subset* $h(q) \subseteq S$.

The system $\mathbf{F} = (S, R, R^*, a)$ *is called a Kripke frame.*

The following holds

$$t \in h(q) \wedge tR^*s \Rightarrow s \in h(q). \qquad (*)$$

The satisfaction relation is defined as follows, for $t \in S$ *and a propositional formula* A.

1. $t \vDash q$ *iff* $t \in h(q)$, *for* q *atomic.*
2. $t \vDash A \wedge B$ *iff* $t \vDash A$ *and* $t \vDash B$.
   $t \vDash A \vee B$ *iff* $t \vDash A$ *or* $t \vDash B$.
3. $t \vDash A \rightarrow B$ *iff for all* s *such that* tR^*s, *if* $s \vDash A$ *then* $s \vDash B$
4. $t \vDash \neg A$ *iff for all* s *such that* tR^*s *we have* $s \nvDash A$.
5. *We say* A *holds in the model iff* $a \vDash A$.
6. *Note we have not used* R *at all, only* R^*, *and that models have actual worlds and satisfaction in a model is defined as satisfaction in the actual world.*
   This presentation is for later comparison with the reactive case.

Theorem 4.2. *Intuitionistic propositional logic is complete for the semantics of Definition 4.1.*

Proof. Well-known result.

Definition 4.3 (Pre-reactive Kripke models). *A pre-reactive Kripke model has the form* (S, S^*, R, a, h) *where* S *is a non-empty set,* $a \in S, R \subseteq S \times S$ *is a binary relation (not necessarily reflexive or transitive) and* S^* *is the set of all* R *increasing sequences* β *of elements from* S *of the form* $\beta = (a_0, a_1, \ldots, a_n)$ *such that* $a_0 = a$ *and for* $i = 0, \ldots, n-1$ *we have* $a_i R a_{i+1}$. *We denote* a_n *by* $|\beta|$. h *is an assignment giving for each atomic* q *a subset* $h(q) \subseteq S$ *such that*

$$t \in h(q) \wedge tRs \Rightarrow s \in h(q).$$

We define satisfaction for β, *a sequence in* S^*, *as follows. (We need the notion of:* β' *is an extension of* β *iff* β *is an initial sequence of* β', *i.e.* $\beta = (a, t_1, \ldots, t_k)$ *and* $\beta' = (a, t_1, \ldots, t_{k+n})$, $n \geq 0$.*)*

---

[2] We have xR^*y iff $x = y$ or xRy or $\exists t_1, \ldots, t_m(xRt_1 \wedge t_1Rt_2 \wedge \ldots t_{m-1}Rt_m \wedge t_mRy)$, for some $m \geq 1$.

1. $\beta \vDash q$ *iff* $|\beta| \in h(q)$, *for* q *atomic.*
2. $\beta \vDash A \wedge B$ *iff* $\beta \vDash A$ *and* $\beta \vDash B$.
$\beta \vDash A \vee B$ *iff* $\beta \vDash A$ *or* $\beta \vDash B$.
3. $\beta \vDash A \to B$ *iff for every extension* β' *of* β, *we have that if* $\beta' \vDash A$ *then* $\beta' \vDash B$.
4. $\beta \vDash \neg A$ *iff for all extension* β' *of* β *we have* $\beta' \nvDash A$.
5. A *holds in the model if* $a \vDash A$.

Lemma 4.4. *Let* $\mathbf{m} = (S, S^*, R, a, h)$ *be a model. Let* R^* *be the reflexive and transitive closure of* R. *Consider the model* $\mathbf{n} = (S, R, R^*, a, h)$. *Then we have for every* $\beta \in S^*$

$$\beta \vDash A \text{ in } \mathbf{m} \text{ iff } |\beta| \vDash A \text{ in } \mathbf{n}.$$

Proof. By induction on A. The crucial point is $A \to B$.

1. Assume $\beta \vDash A \to B$. Then for all β' extending β we have that $\beta' \vDash A$ implies $\beta' \vDash B$.
   By the induction hypothesis we have if $|\beta'| \vDash A$ then $|\beta'| \vDash B$. We now show that $|\beta| \vDash A \to B$.
   Let s be such that $|\beta| R^* s$, then $|\beta| = s$ or $|\beta| R s$ or there exist $s_0, s_1, \ldots, s_m$ such that $s_0 = |\beta|$ and $s_m = s$ and for $i = 0, \ldots, m-1$ we have $s_i R s_{i+1}$. Hence $\beta' = \beta * (s_1, \ldots, s_m)$ is an extension of β with $|\beta'| = s_m = s$. Therefore if $s \vDash A$ then $|\beta'| \vDash A$, hence $\beta' \vDash A$ by the induction hypothesis, hence $\beta' \vDash B$, hence $|\beta'| \vDash B$, i.e. $s \vDash B$.
2. Now assume $|\beta| \vDash A \to B$. Let β' extend β. Hence $|\beta| R^* |\beta'|$. So if $\beta' \vDash A$ then $|\beta'| \vDash A$ hence $|\beta'| \vDash B$ hence $\beta' \vDash B$.

The proof for $\neg A$ is similar.

Remark 4.5. The second type of model is easier to turn reactive. In this new type of model, we view the evaluation of $A \to B$ at a node t as 'going along the relation R and at whatever point t' we reach, if $t' \vDash A$ then $t' \vDash B$.' So in this definition we actually have to traverse the arcs of the model.

Note that R need not be reflexive or transitive. We get these properties from the evaluation process. So consider Figures 4.3 and 4.4.

Fig. 4.3: A1

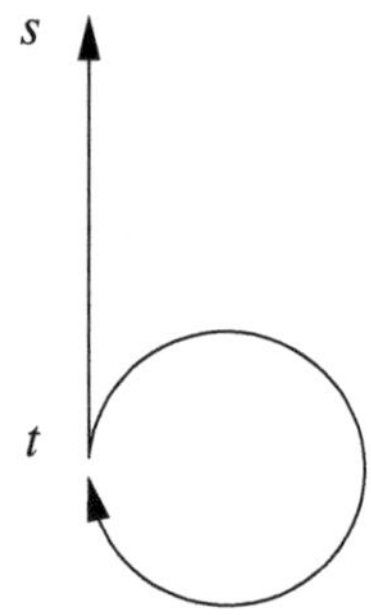

Fig. 4.4: A2

Figure 4.3 gives $S_1 = \{t, s\}, R_1 = \{(t, s)\}$. Figure 4.4 gives $S_2 = \{t, s\}, R_2 = \{(t, t), (t, s)\}$. We have $S_1^* = \{(t), (t, s)\}$ and $S_2^* = \{(t), (t, \underbrace{\ldots}_{m \text{ times}}, t, s) | m = 0, 1, 2, \ldots\} \cup \{(t, t \ldots)\}$. S_1^* corresponds to the ordinary Kripke model with two linear points $1 < 2$ (as in Figure 4.3) and S_2^* corresponds to the ordinary Kripke model with the infinite comb of Figure 4.5 such that the assignments to the points in $\{0, 1, \ldots\}$ are all identical (representing the point t) and also the assignment to the points in $\{w_0, w_1, w_2, \ldots, w_n \ldots\}$ are all identical, (representing the point s).

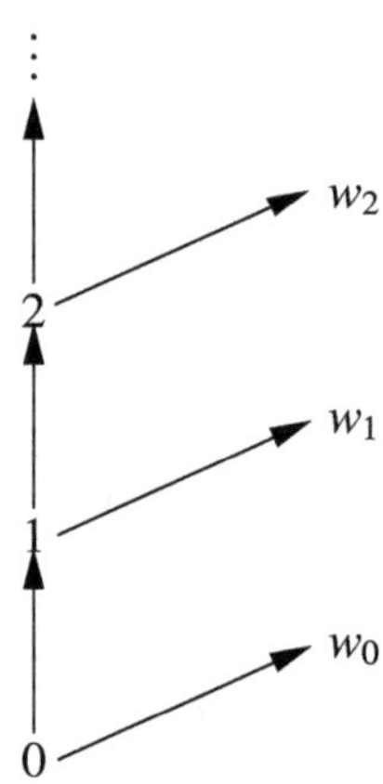

Fig. 4.5: A3

Definition 4.6 (Reactive intuitionistic Kripke frame). *A reactive intuitionistic Kripke frame has the form $(S, \mathbb{R}, a)$, where $a \in S$ and $\mathbb{R}$ is a set of pairs of the form*

1. *$(x, y) \in S \times S$ called arrows*
2. *$((x, y), (u, v)) \in S^2 \times S^2$ called reactive double arrows.*

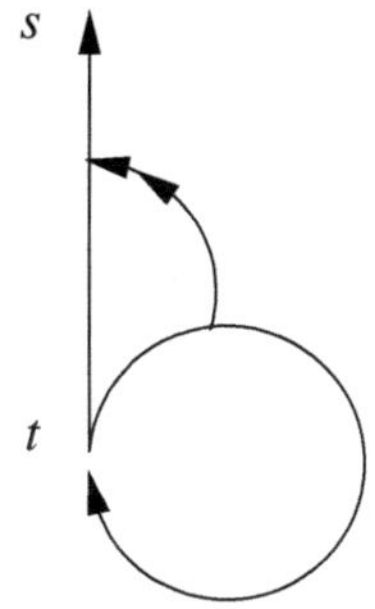

Fig. 4.6: A5

Example 4.7. A reactive frame, see Figure 4.6.
We have $\mathbb{R} = \{(t, s), (t, t), ((t, t), (t, s))\}$ and $a = t$.

Definition 4.8 (How reactivity operates).

1. *Let* $\mathbf{m} = (S, \mathbb{R}, a)$ *be a frame. Let* $(t, s) \in \mathbb{R}$ *be an arrow. Define* $\mathbb{R}_{(t,s)}$ *to be* $\mathbb{R}_{(t,s)} = \mathbb{R} - \{(x, y) \mid ((t, s), (x, y)) \in \mathbb{R}\}$. $\mathbb{R}_{(t,s)}$ *is the result of traversing the arc* $t \to s$ *and cancelling all connections as indicated by double arrows emanating from* $t \to s$.
2. *Let* $\beta = (a, t_1, \ldots, t_k)$. *We now define* $\mathbb{R}_\beta$ *by induction on k. For* $k = 1$, *we let* $\mathbb{R}_\beta = \mathbb{R}_{(a,t_1)}$, *provided* $(a, t_1) \in \mathbb{R}$.
Assume $\mathbb{R}_\beta$ *has been defined for* $\beta = (a, t_1, \ldots, t_k)$ *and assume* $(t_k, t_{k+1}) \in \mathbb{R}_\beta$. *Define* $\mathbb{R}_{\beta'}$ *for* $\beta' = (a, t_1, \ldots, t_{k+1})$ *to be* $\mathbb{R}_{\beta'} = (\mathbb{R}_\beta)_{(t_k,t_{k+1})}$.
3. *Let* $\mathbb{R}_\beta$ *be a reactive relation as defined in (2) where* $\beta = (a, t_1, \ldots, t_k)$.
Let
$$\begin{aligned} \beta_0 &= \beta \\ \beta_1 &= (a, t_1, \ldots, t_{k+1}) \\ &\vdots \\ \beta_n &= (a, t_1, \ldots, t_{k+n}). \end{aligned}$$
We say β_n *for* $n \geq 0$ *is a legitimate extension of* β *iff* $n = 0$ *or* $n \geq 1$ *and the following holds.*
   - $(t_k, t_{k+1}) \in \mathbb{R}_\beta$.
   - $\mathbb{R}_{\beta_1}$ *is obtained from* $\mathbb{R}_\beta$ *as in (1) above.*

     $\vdots$
   - $(t_{k+n-1}, t_{k+n}) \in \mathbb{R}_{\beta_{n-1}}$.
   - $\mathbb{R}_{\beta_n}$ *is obtained from* $\mathbb{R}_{\beta_{n-1}}$ *as in (1) above.*
4. *We say* β_1 *is an immediate legitimate extension of* β.

Lemma 4.9. *Let* $(S, \mathbb{R}_{(a)}, a)$ *be a reactive Kripke frame. Let* β *be a legitimate extension of* (a), *and* $\beta \neq (a)$. *Write* $\beta = (a, t_1, \ldots, t_k)$. *Then we have* $aRt_1, t_1Rt_2, \ldots, t_{k-1}Rt_k$, *where* $R = \{(x, y) | (x, y) \in S \times S \cap \mathbb{R}\}$.

Proof. By induction on k. For $k = 1$ we do have aRt_1.

We see from the construction of any $\mathbb{R}_\beta$ that we have $\mathbb{R}_\beta \subseteq \mathbb{R}$.

Hence if $\beta' = \beta * (t_{k+1})$ with $|\beta| = t_k$ and $(t_k, t_{k+1}) \in \mathbb{R}_\beta$, then we have $t_k R t_{k+1}$.

Definition 4.10 (Satisfaction in a reactive model).

1. *Let $(S, \mathbb{R}, a)$ be a reactive frame. Let $R = \{(x, y)|x, y \in S$ and $(x, y) \in \mathbb{R}\}$. Let h be an assignment such that*
   $t \in h(q)$ and tRs implies $s \in h(q)$.
   Let β be a legitimate extension of (a). Let $\mathbb{R}_\beta$ be the corresponding relation. Let $\mathbf{m}_\beta = (S, \mathbb{R}_\beta, |\beta|, h)$. Note that if $\gamma = (|\beta|, t_1, \ldots, t_m)$ is a legitimate extension of $|\beta|$ in $\mathbf{m}_\beta$, then $\beta^1 = \beta * (t_1, \ldots, t_m)$ is a legitimate extension of (a) in $(S, \mathbb{R}, a)$.
2. *We define satisfaction as follows:*
   - *$\beta \vDash q$ iff $|\beta| \in h(q)$, for q atomic*
   - *$\beta \vDash A \wedge B$ iff $\beta \vDash A$ and $\beta \vDash B$*
   - *$\beta \vDash A \vee B$ iff $\beta \vDash A$ or $\beta \vDash B$*
   - *$\beta \vDash A \to B$ in $\mathbf{m}_\beta$ iff for every legitimate extension β' of β, if $\beta' \vDash A$ in $\mathbf{m}_{\beta'}$ then $\beta' \vDash B$ in $\mathbf{m}_{\beta'}$.*
   - *$\beta \vDash \neg A$ in $\mathbf{m}_\beta$ iff for all legitimate extensions β' of β we have $\beta' \nvDash A$ in $\mathbf{m}_{\beta'}$.*
3. *We say $(S, \mathbb{R}, a, h) \vDash A$ iff $(a) \vDash A$, (in words: A holds in the model).*

Lemma 4.11. *Let $(S, \mathbb{R}, a, h)$ be a model and assume that $\beta \vDash A$ in $\mathbf{m}_\beta$ and that β' is a legitimate extension of β, then $\beta' \vDash A$ in $\mathbf{m}_{\beta'}$.*

Proof. By induction on A.

1. For A atomic, this follows from a previous Lemma 4.9.
2. The cases of $\wedge$ and $\vee$ are immediate.
3. Assume $\beta \vDash A \to B$, then for any legitimate extension β' of β, if $\beta' \vDash A$ in $\mathbf{m}_{\beta'}$ then $\beta' \vDash B$ in $\mathbf{m}_{\beta'}$.
   But now since any legitimate extension β'' of β' is also a legitimate extension of β we get that $\beta' \vDash A \to B$ in $\mathbf{m}_{\beta'}$.
4. The case of $\neg$ is similar.

Example 4.12 (Satisfaction in the frame of Figure 4.6). Let q be atomic. Let $h(q)$ be $\{s\}$. Then

1. $(t) \nvDash q$
2. $(t) \nvDash \neg q$
3. $(t) \nvDash \neg q \to q$.

(1) and (2) are clear. To show (3), note that $\beta = (t, t)$ is a legitimate extension of (t) and $\mathbb{R}_{(t,t)}$ is

$$\{(t, t), ((t, t), (t, s)\}.$$

In $(S, \mathbb{R}_{(t,t)}, (t, t)$ we have $(t, t) \vDash \neg q$ but $(t, t) \nvDash q$.

Lemma 4.13. *The logic defined by reactive satisfaction is intuitionistic logic.*[3]

Proof.

1. Since ordinary pre-reactive models are reactive models, (by Definition 4.3 and Lemma 4.4) the logic is not stronger than intuitionistic logic.
2. From Lemma 4.11 we see the logic is not weaker either.

Theorem 4.14. *The logic of the frame Figure 4.6 is not complete for any class of ordinary intuitionistic Kripke frames.*

Proof. The proof has four parts, (A)–(D).

(A) The following formulas hold in the frame of Figure 4.6, under any h,
   1. $x \vee (x \to (q \vee \neg q)), x, q$ atomic
   2. $(x \to y) \vee (y \to x), x, y$ atomic.

   We check (1).
   If $x = \bot$ at (t) then to falsify $x \to (q \vee \neg q)$ we need to go to (t, s) where x can hold. We cannot go to (t, t) because $(t, t) \nvDash x$. At (t, s) clearly $q \vee \neg q$ holds.
   We check (2).
   Assume $x \to y$ is false at (t). Then we have either
   $(t, \underbrace{\ldots}_{m \text{ times}}, t) \vDash x$ and $(t, \underbrace{\ldots}_{m \text{ times}}, t) \nvDash y$ or
   $(t, \underbrace{\ldots}_{n \text{ times}}, t, s) \vDash x$ and $(t, \underbrace{\ldots}_{n \text{ times}}, t, s) \nvDash y, m, n \geq 0$.
   In the first case, we have $(t) \vDash x$ and hence $(t) \vDash y \to x$.
   In the second case we have $(t, s) \vDash x, (t, s) \nvDash y$. Hence $(t) \vDash \neg y$ and so $(t) \vDash y \to x$.

(B) We now show that any ordinary frame which satisfies (1) and (2) under any h must be either a single point or the frame of Figure 4.3.
   This is well known because otherwise either (1) or (2) can be falsified. (1) is falsified by a three-point chain and (2) by a two-point anti-chain. So we can have neither.

(C) We now show that in Figure 4.3 or in a single point (3) must hold.
   3. $x \vee \neg x \vee (\neg x \to x)$.

   To falsify $x \vee \neg x$ we need Figure 4.3 with $t \nvDash x$ and $s \vDash x$ but from the latter it follows that $t \vDash \neg\neg x$ holds and hence $t \vDash \neg x \to x$.

(D) Our proof is concluded because Example 4.12 shows that (3) can be falsified in the frame of Figure 4.6.

Remark 4.15. It is helpful to have another view of Figure 4.6. The frame has two paths, as in Figure 4.7.

We can view Figure 4.7 as an ordinary three-point Kripke model with the understanding that the assignment at t and $(t, t, \ldots)$ is the same, i.e. for every q, $t \in h(q)$ iff $(t, t) \in h(q)$.

---

[3] Let $\mathbb{K}$ be a class of reactive frames. Let A be a wff which holds in any frame in the class under any assignment. Let **Logic**($\mathbb{K}$) be the set of all such wffs A. Then **Logic**($\mathbb{K}$) is a 'logic'. All its wffs are classical theorems but this 'logic' may *not be closed under substitution.*

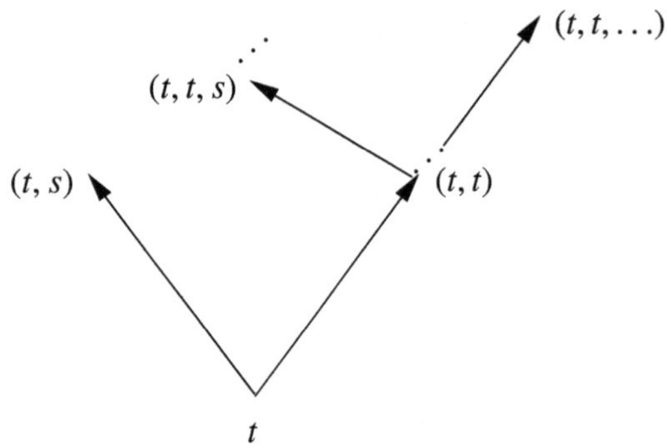

Fig. 4.7: A6

This is common to reactive models, that they can be 'unfolded' as models of paths with restrictions on the assignments. [62] studies such models. We examine this notion in the next section.

4.3 Folding reactive frames

We saw in the last section that the reactive frame of Figure 4.6 can be unfolded into the ordinary frame of Figure 4.7, with the added understanding that the points (t) and (t, t) must give the same values to the atoms. This unfolding process can be done in a systematic manner, and it seems to have significance for developing Beth tableaux for reactive intuitionistic logics. So in this section we study it in detail. We are going to unfold and then fold again.

Let $(S, \mathbb{R}, a)$ be a reactive frame. Let β be a legitimate path of the form $\beta = (a, t_1, \ldots, t_k)$. We saw that we can calculate $\mathbb{R}_\beta \subseteq \mathbb{R}$. Let $\mathbf{F}_\beta = (S, \mathbb{R}_\beta, |\beta|)$.

If S is finite and β ranges over all legitimate paths, we get only a finite number of different frames $\mathbf{F}_\beta$. Let us take advantage of this.

Definition 4.16 (Path equivalence relation). *Let* $\mathbf{m} = (S, \mathbb{R}, a, h)$ *be a reactive model. Define an equivalence relation on the paths of the model as follows:*

- $\beta \equiv \gamma$ *iff* $|\beta| = |\gamma|$ *and* $\mathbb{R}_\beta = \mathbb{R}_\gamma$.

Let Ω *be the set of equivalence classes,* $\{\beta| \equiv \mid \beta$ *a legitimate path extending* $(a)\}$.

Then Ω *is finite.*

Define ρ *(respectively* ρ_1*) on* Ω *as follows:*

- $\beta/\equiv \rho\gamma/\equiv$ *(respectively:* $\beta/\equiv \rho_1\gamma/\equiv$*) iff for some* $\beta_1 \equiv \beta$ *and* $\gamma_1 \equiv \gamma$ *we have* γ_1 *is a legitimate extension (respectively: an immediate legitimate extension) of* β_1.

Lemma 4.17. *ρ is reflexive and transitive, and is the reflexive and transitive closure of ρ_1.*

Proof. Reflexivity is not a problem. We show transitivity.

Claim: If β_1 is a legitimate extension of α_1 and $\beta_1 \equiv \beta_2$ and γ_2 is a legitimate extension of β_2, then there exists a $\gamma_1 \equiv \gamma_2$ such that γ_1 is a legitimate extension of α_1.

We now prove the claim:

1. Since β_1 is a legitimate extension of α_1, we have that $\beta_1 = \alpha_1$ or for some $k \geq 1$, we have $\beta_1 = \alpha_1 * (t_1, \dots, t_k)$ where $|\alpha_1| R t_1 \wedge t_1 R t_2 \wedge \dots t_{k-1} R t_k$ and $|\beta_1| = t_k$.
2. Since $\beta_1 \equiv \beta_2$ we get that $\mathbb{R}_{\beta_1} = \mathbb{R}_{\beta_2}$.
3. Since γ_2 is a legitimate extension of β_2, we have $\gamma_2 = \beta_2$ or for some $m \geq 1$ we also have $(s_1, \dots, s_m)$ such that $\gamma_2 = \beta_2 * (s_1, \dots, s_m)$ and $|\beta_2| R s_1 \wedge s_1 R s_2 \wedge \dots s_{m-1} R s_m$ and $|\gamma_2| = s_m$.
   Consider the path
   $$\gamma_1 = \alpha_1 * (t_1, \dots, t_k) * (s_1, \dots, s_m).$$
   where if $\beta_1 = \alpha_1$ then $(t_1, \dots, t_k)$ does not appear and if $\gamma_2 = \beta_2$ then $(s_1, \dots, s_m)$ does not appear in γ_1.
   It is clear that $|\gamma_1| = |\gamma_2|$.
   We want to show that
   $$\mathbb{R}_{\gamma_1} = \mathbb{R}_{\gamma_2}.$$
   Observe that $\mathbb{R}_{\beta_1} = \mathbb{R}_{\beta_2}$. Since γ_1 is an extension of β_1 along the sequence of nodes $(s_1, \dots, s_m)$ and γ_2 is the extension of β_2 along the sequence $(s_1, \dots, s_m)$ (same sequence) and they both start at $|\beta_1| = |\beta_2|$ with $\mathbb{R}_{\beta_1} = \mathbb{R}_{\beta_2}$, then they end up at the same relation, namely $\mathbb{R}_{\gamma_1} = \mathbb{R}_{\gamma_2}$. Hence $\gamma_1 \equiv \gamma_2$.
   We now finish the proof of Lemma 4.17:
4. Since $\gamma_1 \equiv \gamma_2$ and γ_1 extends α_1, we get $\alpha_1/\equiv \rho\, \gamma_2/\equiv$.
5. Clearly ρ is the reflexive and transitive closure of ρ_1.

Lemma 4.18. *Let* $\mathbf{m} = (S, \mathbb{R}, a, h)$ *and let* $\equiv, \Omega, \rho$ *be as in Definition 4.16. Consider* $\mu = (\Omega, \rho, a/\equiv, \mathbf{h})$*, as an ordinary Kripke model, where* $\mathbf{h}$ *is defined by*

$$\alpha/\equiv \in \mathbf{h}(q) \text{ iff } |\alpha| \in h(q).$$

Then for any A we have:

$$\alpha \vDash A \text{ in } \mathbf{m} \text{ iff } \alpha/\equiv\ \vDash A \text{ in } \mu.$$

Proof. By induction on A.

1. For q atomic this holds by the definition of $\equiv$.
2. The key case is that of $\rightarrow$.

Assume $\alpha \nvDash A \to B$, then for some β which is a legitimate extension of α we have $\beta \vDash A$ and $\beta \nvDash B$. But we also have in this case that $\alpha/\equiv \rho\, \beta/\equiv$ and by the induction hypothesis, $\beta/\equiv\ \vDash A$ and $\beta/\equiv\ \nvDash B$.

Now assume $\alpha/\equiv\ \nvDash A \to B$. Then for some $\gamma/\equiv$ we have $\alpha/\equiv \rho\, \gamma/\equiv$ and $\gamma/\equiv\ \vDash A$ and $\gamma/\equiv \nvDash B$. Therefore for some $\alpha_1 \equiv \alpha$ and $\gamma_1 \equiv \gamma$ we have $\gamma_1 = \alpha_1 * (t_1, \dots, t_k)$ and γ_1 is a legitimate extension of α_1. Hence since $\gamma_1 \equiv \gamma$, we get $\gamma_1/\equiv\ \vDash A$ and $\gamma_1/\equiv\ \nvDash B$. By the induction hypothesis we have $\gamma_1 \vDash A$ and $\gamma_1 \nvDash B$.

Now look at the two models $\mathbf{m}_\alpha = (S, \mathbb{R}_\alpha, |\alpha|)$ and $\mathbf{m}_{\alpha_1} = (S, \mathbb{R}_{\alpha_1}, |\alpha_1|)$. Since $\alpha_1 \equiv \alpha$, these two models are the same. So having $\gamma_1 = \alpha_1 * (t_1, \dots, t_k)$ with $\gamma_1 \vDash A$ and $\gamma_1 \nvDash B$ in $\mathbf{m}_{\gamma_1}$ implies that for $\delta = \alpha * (t_1, \dots, t_k)$ we also have $\delta \vDash A$ and $\delta \nvDash B$ in $\mathbf{m}_\alpha$. Hence $\alpha \nvDash A \to B$ in $\mathbf{m}$.

Corollary 4.19. *The proof of Lemma 4.17 actually gives us a stronger result. Suppose we have a situation as described in Figure 4.8 below, then there exist sequences of points from S of the form*

$$\begin{aligned}
(a, t_1^y, \dots, t_{k(y)}^y) &= \alpha(y)\\
(a, t_1^z, \dots, t_{k(z)}^z) &= \alpha(z)\\
(t_1^u, \dots, t_{k(u)}^u) &= \alpha(u)\\
(t_1^w, \dots, t_{k(w)}^w) &= \alpha(w)\\
(s_1^w, \dots, s_{m(w)}^w) &= \beta(w)\\
(t_1^v, \dots, t_{k(v)}^v) &= \alpha(v)
\end{aligned}$$

such that

$$\begin{gathered}
\alpha(y), \alpha(z), \alpha(y) * \alpha(u),\\
\alpha(z) * \alpha(v), \alpha(y) * \alpha(w),\\
\alpha(z) * \beta(w)
\end{gathered}$$

are all legitimate extensions of (a) and we have

$$\begin{aligned}
\alpha(y) &\equiv y\\
\alpha(z) &\equiv z\\
\alpha(y) * \alpha(u) &\equiv u\\
\alpha(z) * \alpha(v) &\equiv v\\
\alpha(y) * \alpha(w) &\equiv w\\
\alpha(z) * \beta(w) &\equiv w.
\end{aligned}$$

In other words, Figure 4.8 from μ can be realised by actual continuous legitimate paths in $\mathbf{m}$.

In fact, we can get a general result.

Define in $\mu = (\Omega, \rho, a/\equiv, \mathbf{h})$ the following immediate successor relation ρ_{11}:

- *$x/\equiv \rho_{11} y/\equiv$ iff $x/\equiv \rho\ y/\equiv$ and $x \not\equiv y$ and for all $z/\equiv$, if $s/\equiv \rho\ z/\equiv$ and $z/\equiv \rho\ y/\equiv$, then either $z \equiv x$ or $z \equiv y$.*

Since $x/\equiv\ \rho\ y/\equiv$ means that for some $(t_1, \dots, t_k, s_1, \dots, s_m)$ we have $\alpha = (a, t_1, \dots, t_k) \equiv x$ and $\beta = (a, t_1, \dots, t_k, s_1, \dots, s_m) \equiv y$. The condition $\alpha/\equiv \rho_{11}\ \beta/\equiv$

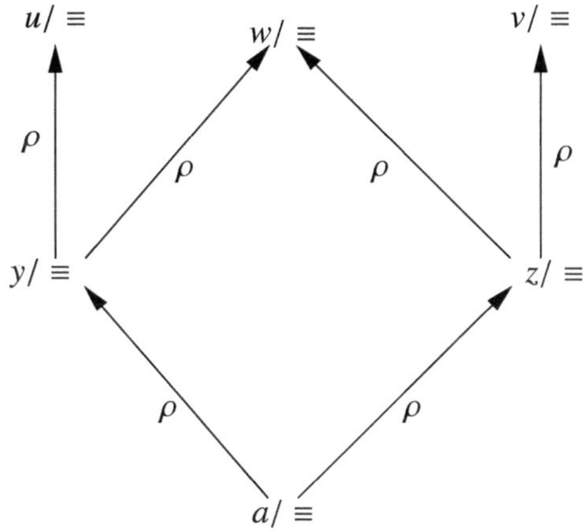

Fig. 4.8

means that each $\beta_j = \alpha * (s_1, \ldots, s_j)$ *for* $1 \leq j \leq m$ *is either* $\equiv$ *equivalent to* α *or to* β. *Since* $\alpha \not\equiv \beta$, *each* s_i *is either* t_k *or* s_m.

Note that ρ_1 *and* ρ_{11} *are different relations.* ρ_1 *is defined on legitimate sequences of* Ω *while* ρ_{11} *is defined on* $\Omega/\equiv$. *We may have that* $x\rho_1 y$ *holds but* $x/\equiv \rho_{11} y/\equiv$ *does not hold. However, if we have* $x/\equiv \rho_{11}\ y/\equiv$ *then for some* $x' \equiv x, y' \equiv y$ *we have* $x'\rho_1 y'$ *holds.*

The general result we get is the following:

Let $\mu_0 = (\Omega_0, \rho_{11} \upharpoonright \Omega_0, a/\equiv)$ *be any finite submodel of* Ω *with* ρ_{11} *restricted to* μ_0. *Then the model* μ_0 *can be realised by actual sequences from* **m**.

Remark 4.20 (Folding reactive frames). We started with a reactive model $\mathbf{m} = (S, \mathbb{R}, a, h)$ and converted it to a special model $\mu = (\Omega, \rho, a/\equiv, \mathbf{h})$. This model is special and we want to highlight some of its properties.

First note that if **m** is finite then we can effectively derive μ from **m** in tractable time. This observation is important because we shall use μ to develop Beth tableaux for **m**.

We now examine μ more closely.

The elements of Ω are equivalence classes of legitimate sequences β of **m**. We have

- $\alpha \equiv \beta$ iff $|\alpha| = |\beta|$ and $\mathbb{R}_\alpha = \mathbb{R}_\beta$.

Let $\mathbf{m} = (S, \mathbb{R}, a, h)$. Let $\mathbb{T}$ be the set of all legitimate sequences of **m**. We define two equivalence relations on $\mathbb{T}$.

- $\alpha \approx \beta$ iff $|\alpha| = |\beta|$
- $\alpha \approx_{\mathbb{R}} \beta$ iff $\mathbb{R}_\alpha = \mathbb{R}_\beta$

The relation $\equiv$ is the intersection of the above two relations. So if $\mathbb{E}_1, \ldots, \mathbb{E}_k$ are all the $\approx$ equivalence subsets of $\mathbb{T}$ and $\mathbb{D}_1, \ldots, \mathbb{D}_m$ are all the $\approx_{\mathbb{R}}$ equivalence subsets of

$\mathbb{T}$, then the equivalence subsets of $\equiv$ are all the different combinations of the form $\mathbb{T}_{i,j} = \mathbb{E}_i \cap \mathbb{D}_j$ $i = 1, \dots, k, j = 1, \dots, m$.

Actually, if $S = \{s_1, \dots, s_k\}$ then we can have

$$\mathbb{E}_i = \{\alpha \mid |\alpha| = s_i\}.$$

We have

$$\Omega = \{\mathbb{T}_{i,j} \mid \begin{matrix} i = 1, \dots, k; \\ j = 1, \dots, m \end{matrix}\}$$

Let us now look at the class $\mathbb{E}_i$, for fixed i. Let

$$\alpha = (a, t_1, \dots, t_r, e) \text{ and } \beta = (a, s_1, \dots, s_r, e)$$

be two elements of this class. We have $|\alpha| = |\beta| = e \in S$. Let us ask: Are α, β also in the same class $\mathbb{D}_j$ for some j? Well maybe they are, depending on whether $\mathbb{R}_\alpha = \mathbb{R}_\beta$. We can also regard $\approx$ as an equivalence relation on $\Omega = \{\mathbb{T}_{i,j}\}$:

- $\mathbb{T}_{i,j} \approx \mathbb{T}_{m,n}$ iff $i = m$.

This means that for any $\alpha \in \mathbb{T}_{i,j}, \beta \in \mathbb{T}_{i,n}$ we have $|\alpha| = |\beta|$.

But $|\alpha| = |\beta|$ means $\mathbf{h}(\alpha) = \mathbf{h}(\beta)$, since $\mathbf{h}(\alpha) = h(|\alpha|)$ and $\mathbf{h}(\beta) = h(|\beta|)$.

We can now view the model μ with $\approx$ as a model of the form

$$\mu = (\Omega, \rho, \approx, a/\equiv, \mathbf{h})$$

with an equivalence relation $\approx$ on Ω and the commitment:

$$x \approx y \to \mathbf{h}(x) = \mathbf{h}(y). \qquad (*)$$

This model satisfies special properties, because we know that $\mu = (\Omega, \rho, a/\equiv, \approx)$ arises from a model $\mathbf{m} = (S, \mathbb{R}, a)$ through a special construction.

We would like to list these special properties so that we can prove that any model μ with these properties comes from some reactive model $\mathbf{m}$.

We believe that this is possible to do.

Conjecture 4.21. It is possible to give conditions on ρ and $\approx$ of a folded Kripke model which ensure it is equivalent to a reactive Kripke model (without $\approx$).

Example 4.22 (Figure 4.6). We saw that the reactive Kripke frame of Figure 4.6 can be presented as the folded Kripke frame of Figure 4.7. We can also fold differently, as follows:

$$\Omega = \{t, s, (t,t)\}$$
$$\rho = \{(t,t), (t,s), ((t,t),s), (t,(t,t)), ((t,t),(t,t))\}$$

and we have $t \approx (t,t)$.

Remark 4.23. Let us see what is the status of folded Kripke frames in terms of Heyting algebras. An ordinary Kripke frame $\mathbf{m} = (\Omega, \rho)$ gives rise to a free Heyting algebra $\mathbf{H_m}$. When we add an equivalence relation $\approx$ to form $\mu = (\Omega, \rho, \approx)$ we are adding some equalities among the free generators of $\mathbf{H_m}$. These equalities generate a congruence relation $\approxeq$ on $\mathbf{H_m}$. If we let $\mathbf{H}_\mu = \mathbf{H_m}/\approxeq$ then we get the algebra corresponding to μ. It is not a free algebra.

4.4 Reactive Tableaux

We begin by explaining the intuitive idea of tableaux for reactive logics. Consider the tableau of Figure 4.9

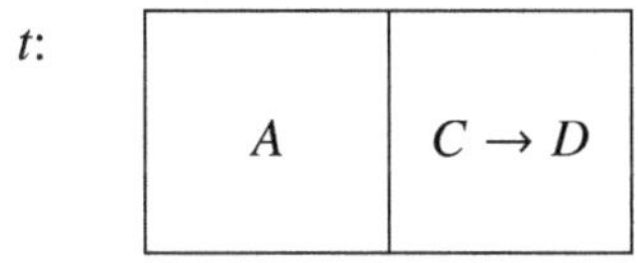

Fig. 4.9: tab1

The label of the tableau is t. This is usually the name of the possible world we are dealing with. A is on the left and so we want to make $t \vDash A$ and $C \to D$ is on the right hand side of the tableau, so we want to make $t \nvDash C \to D$.

To do the latter we need an accessible world s. such that tR^*s and $s \vDash C$ and $s \nvDash D$.

This means that we move into the following tableau in Figure 4.10.

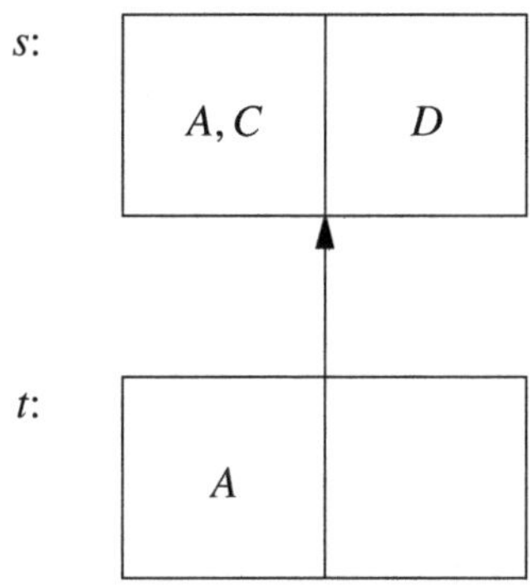

Fig. 4.10: tab2

A carries on into s and, in s, we put C on the left and D on the right.[4]

This is the usual tableau process for intuitionistic logic. If we have $\wedge$ and $\vee$ in the language, we might get different alternatives (tableau splitting). Let us assume our language contains only $\to$ so that we can concentrate on the differences between

[4] A carries on into s because in an intuitionistic Kripke model we have that for any A:

$$t \vDash A \text{ and } tRs \to s \vDash A.$$

Note that if for some reason we want that t and s agree on the values of atoms q, then any atom in the right hand side of t will also carry on into the right hand side of s.

ordinary tableaux and reactive tableaux, without the complexity generated by the presence of $\wedge$ and $\vee$.

In the case of the pre-reactive semantics of Definition 4.3 the tableau will have labels α, β which are paths. Here $\lambda|\alpha| = t$ and $|\beta| = s$. This is not an essential difference. The difference is essential in the reactive case because we must require that β is a legitimate extension of α. To do that we must record what $\mathbb{R}_\beta$ is.

So to simplify even further and allow us to present the essential ideas of the reactive tableau, let us assume our logic has a fixed finite reactive frame, $(S, \mathbb{R}, a)$.

In this case we get the following simplifications:

(S1.) Since the frame is finite and for any legitimate sequence β, $\mathbb{R}_\beta$ is smaller than $\mathbb{R}$, there is only a finite number of frames $(S, \mathbb{R}_i)$, $i = 1, \ldots, n$, that are in play,

(S2.) We can move to the finite folded Kripke frame $\mu = (\Omega, \rho, \approx, a)$ and do our tableaux on μ. This is significantly simpler because μ is like an ordinary Kripke frame with the additional simple condition (simple from the tableaux point of view) imposed by $\approx$.

The next definition gives the notion of Beth tableaux for the implicational fragment. Note three facts:

Fact 1

Every wff can be put in the form

$$E = [A_1 \to (\ldots \to (A_n \to q)\ldots)]$$

where q is atomic and each A_i is of the same form as E.

Fact 2

We need only two tableaux rules:

- To make E false (E on the right) at world t, find a world s such that $t\rho s$ and put all A_i on the left and q on the right.
- To make E true on the left at t when q is on the right at t, we must move one of the A_i to the right of t to make A_i false.

Fact 3

To accommodate $\approx$ we make some adjustments to the usual intuitionistic tableaux rules.

Definition 4.24 (Beth tableaux for folded Kripke frames). *Let $\mu = (\Omega, \rho, \approx)$ be a folded frame.*

1. *A tableau for μ is a pair of functions (τ^+, τ^-) from Ω into the set of wffs satisfying the following conditions:*

a) If $t\rho s$ holds in μ then $\tau^+(t) \subseteq \tau^+(s)$

b) If $t \approx s$ holds then $\tau^\pm(t)$ and $\tau^\pm(s)$ contain the same atoms (respectively)

2. *Let* **T** *be a family of tableaux τ as in (1). Let $\tau \in$* **T***. We define an operation which will split τ into several alternatives and we will replace τ in* **T** *by these alternatives to obtain a new family* **T′***.*

 Right hand operation

 Choose $t \in \Omega$ and choose E in $\tau^-(t)$. Get the following alternatives $\tau_s^\pm$ to replace τ by $\{\tau_s | s \in \Omega$ and $tRs\}$ where

$$\tau_s^-(x) = \begin{cases} \tau^-(x), \text{ for } x \not\approx s \\ t^-(x) \cup \{q\} \text{ for } x \approx s, x \neq t \end{cases}$$
$$\tau_s^-(t) = \tau^-(t) - \{E\} \cup \{q\}$$
$$\tau_s^+(x) = \begin{cases} \tau^+(x) \text{ if } x \not\approx s \\ \tau^+(x) \cup \{A_i | A_i \text{ atomic}\} \text{ if } x \approx s \text{ and } x \neq s \\ \tau^+(t) \cup \{A_j\} \text{ if } x = t \end{cases}$$

 Left hand operation

 Choose $t \in \Omega$ and choose E in $\tau^+(t)$ such that $q \in \tau^-(t)$.
 Recall that $E = [A_1 \to (\ldots \to (A_n \to q))]$.
 For each A_i form the following tableau τ_i.

$$\tau_i^+(x) = \tau^+(x), x \in \Omega$$
$$\tau^-(x) = \begin{cases} \tau^-(x) \cup \{A_i\}, \text{ if } A_i \text{ is atomic and } x \approx t. \\ \tau^-(x) \text{ if } x \not\approx t \\ t^-(t) \cup \{A_i\} \text{ if } x = t. \end{cases}$$

3. *A tableau τ is closed if for some $t \in \Omega$ we have $\tau^+(t) \cap \tau^-(t) \neq \varnothing$.*
4. **T** *is closed iff all of its tableaux are closed.*

Theorem 4.25 (Completeness). *The above procedure is complete for the logic defined by the frame μ.*

Proof. Modify a proof for the case of an ordinary intuitionistic frame.

Definition 4.26 (Beth tableau for reactive Kripke frame). *Let* $\mathbf{m} = (S, \mathbb{R}, a)$ *be a reactive frame. Let $\mu = (\Omega, \rho, \approx)$ be the folded frame effectively obtained from* **m***. Let A be a wff for which we seek a counter model. Execute the algorithm of Definition 4.24. If an assignment* **h** *is found which falsifies A in μ, then the corresponding assignment h will falsify A in* **m***. If the tableaux is closed and no such assignment exists, then no such assignment exists in* **m***. Thus the logic of* **m** *can be characterised using tableaux. Given a class $\mathbb{K}$ of finite reactive frames then the logic of $\mathbb{K}$ can be characterised as the intersection of the logics of its members. See Footnote 3.*

5

Completeness Theorems for Reactive Modal Logics

5.1 Overview

This chapter gives completeness theorems for some basic reactive Kripke models and semantics. This section will:

1. Refocus on reactivity following earlier chapters.
2. Discuss and compare the kind of Kripke semantics we get with reactivity with a view to placing the completeness theorem of this chapter in the context of other options.
3. Explain the challenges in obtaining completeness theorems.

5.1.1 Fibring and reactivity

Our starting point is an ordinary Kripke model for modal logic. This has the form $\mathbf{m} = (S, R, a, h)$ where S is a nonempty set of worlds, $a \in S$ is the initial (actual) world and $R \subseteq S \times S$ is the accessibility relation. The model also has an assignment h, giving for each atomic q a subset $h(q) \subseteq S$.

We shall focus on R.

Let us look at how the modality $\Diamond$ is evaluated at a point $t \in S$. This is done by clause $(*1)$

$(*1)$ $t \vDash \Diamond A$ iff $\exists s(tRs$ and $s \vDash A)$.

Let us write $(*1)$ slightly differently, stressing which model we are using

$(*2)$ $t \vDash \Diamond A$ in the model $\mathbf{m} = (S, R, a, h)$ iff for some $s \in S$ such that tRs we have $s \vDash A$ in $\mathbf{m}$.

Clause $(*2)$ can be changed to a fibring clause. Suppose we associate with every $s \in S$, a new Kripke model $\mathbf{m}_s = (S_s, R_s, s, h)$ and we modify $(*2)$ to be $(*3)$ for a fibred modality $\Diamond_F$ (see [57]).

$(*3)$ $t \vDash \Diamond_F A$ in $\mathbf{m}$ iff for some $s \in S$ such that tRs, we have $s \vDash A$ in $\mathbf{m}_s$.

D.M. Gabbay, *Reactive Kripke Semantics*, Cognitive Technologies,
DOI 10.1007/978-3-642-41389-6_5,

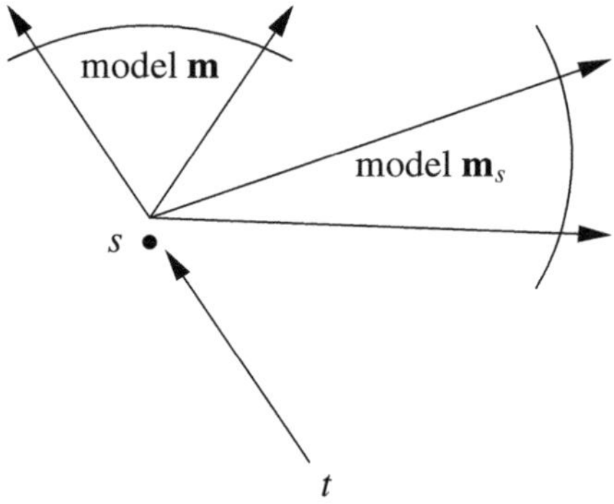

Fig. 5.1

Figure 5.1 describes the situation.

We switch evaluation at s from the model $\mathbf{m}$ to the model $\mathbf{m}_s$.

In a way this is a kind of multimodal logic with models of the form $\mathbf{n} = (S, R_s, s \in S, h)$ and the jump operators $J_s, s \in S$ and the modality $\Diamond$. Evaluation is two-dimensional as follows:

(∗4) $t \vDash_s \Diamond A$ iff for some $u, tR_s u$ and $u \vDash_s A$
$t \vDash_s J_u A$ iff $t \vDash_u A$

According to (∗4) the modality of (∗3) can be written as

(∗5) $t \vDash_s \Diamond_F A$ iff for some $u, tR_s u$ and $u \vDash_s J_u A$.

Note that public announcement logic (see references [156, 12]), is an example of this kind of operator. It has modalities of the form $\Diamond_A$, where A is a wff such that $t \vDash A$ holds. The truth condition is:

(∗6) $t \vDash \Diamond_A B$ in (S, R) iff $t \vDash B$ in (S_A, R), where $S_A = \{x \mid x \vDash A\}$.

The question to ask now is how does $\mathbf{m}_s$ relate to $\mathbf{m}$?

In reactive logic, we focus on R_s being a variation of R, obtained by switching on and off accessibility connections in R. If we denote $(x, y) \in R$ by $x \to y$ then the basic reactive unit is the double arrow which we have met previously. Using a family $\mathbb{R}$ of double arrows, we can move from one R_s to another $R_{s'}$.

We explain how reactivity can work in the next subsection.

5.1.2 Reactive Kripke models

Reactive Kripke models, discussed in Chapter 2, were introduced in 2004 [67] and an expanded version published in [66].

There are various options for kinds of reactivity

Example 5.1 (Reactive Kripke models). Consider the Kripke model of Figure 5.2. For the moment ignore the double-headed arrow in the diagram. If R denotes the accessibility relation then we have in this model aRc, aRb, cRd, bRd, dRe. Let us try to evaluate $a \vDash \Diamond\Diamond\Box\bot$.

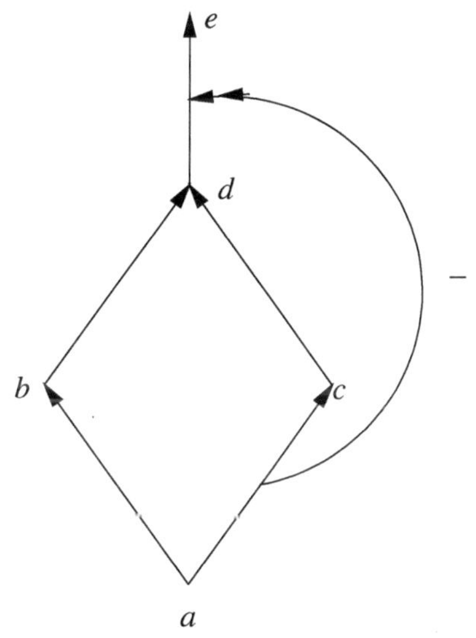

Fig. 5.2

For this to hold at a we need a path aRx, xRy and $y \vDash \Box\bot$. Such a path is not available in this model.

Now let us look at the double arrow from the arc (a,c) to the arc (d,e). This double arrow is decorated with the minus sign. We read its meaning as saying disconnect the accessibility dRe if you pass through the arc (a,c). Thus if we get to d through the path $aRb \wedge bRd$ then dRe holds. However, if we get to d through the path $aRc \wedge cRd$, then by the time we get to d, dRe does not hold. So in the reactive model of Figure 5.2 where the double arrow is active, we have that $a \vDash \Diamond\Diamond\Box\bot$ holds because of the path $aRc \wedge cRd$.

Thus in a reactive Kripke model the double arrows change the models.

Figure 5.3 shows what we see at each point depending on the paths. In fact the Kripke model of Figure 5.3 is an ordinary Kripke model where the nodes are paths. We must remember to give the same assignment to any atom x at points abd and acd. That is why we encircled them in the Figure.

We have all kinds of possibilities here for the double arrow. See Figures 5.4 and 5.5.

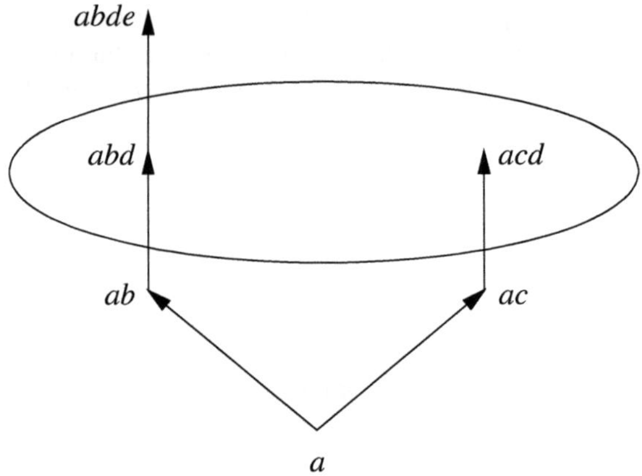

$\Delta = \{\Diamond\Diamond x \to \Box\Box x \mid x \text{ atomic}\}$
$\Diamond\Diamond\Box\bot, \Diamond\Diamond\Diamond\Box\bot$

Fig. 5.3

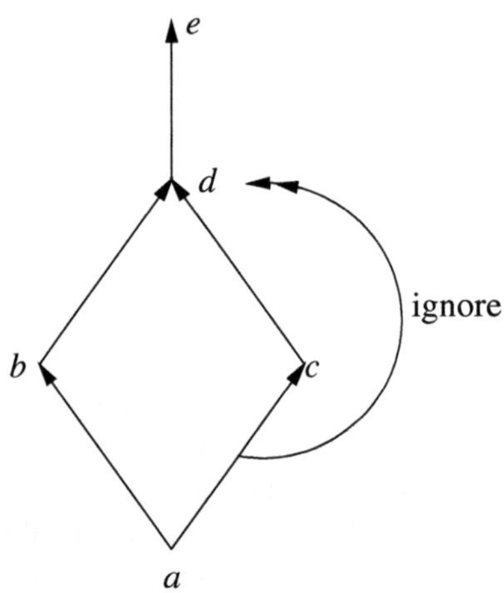

Ignore d

$\models \Diamond\Box\bot$
$\models \Diamond\Diamond\Diamond\Box\bot$

Various options for making a frame reactive using $\twoheadrightarrow$.

Fig. 5.4

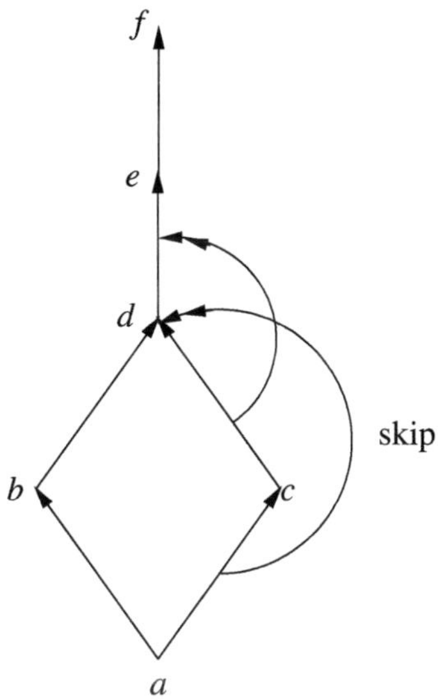

Skip d but do go through the arc (c,d) to get to e.
So at c we have $c \vDash \Diamond e$ even though we may have $d \vDash \neg e$ (we use the world name also as a nominal true exactly at that world).

Fig. 5.5

1. Double arrow to an arc can switch it off (or on), i.e. disconnect (or connect) it. If decorated with '−' then it switches the arc to *off* if it is *on* and leaves it as *off* if it is *off*. If decorated with '+' then it switches the arc *on* if it is *off* and leaves it *on* if it is *on*. If not decorated at all then it switches it *on* if it is *off* and *off* if *on*. This is called *switch reactivity*.
2. Double arrow to a point can delete it or can tell us to ignore it or tell us to skip it.

So for example in Figure 5.4, if point d is ignored then at point c the point d is ignored and so

$$a \vDash \Diamond\Box\bot$$

holds because of c. But we still have

$$a \vDash \Diamond\Diamond\Diamond\Box\bot$$

because we can still pass through d to get to e.

If we skip d then at c we see e directly. If we delete d then both d and e are deleted from the model, because e is now unreachable from point a.

Theorem 5.2. *There exist monomodal logics which are complete for a class of reactive Kripke frames but are not complete for any class of ordinary Kripke or neighbourhood frames, see Chapters 2 and 3.*

Example 5.3 (Sample application: deontic logic). See Chapter 11 and [64, 80, 81]. We model contrary-to-duty obligations using reactive Kripke models. See Figure 5.6.

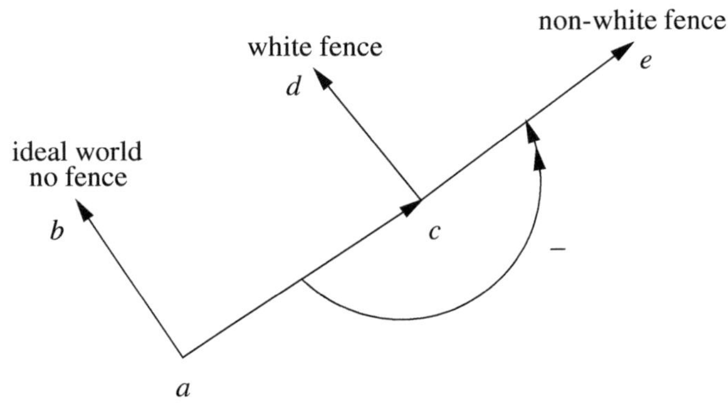

- There should be no fence
- If there is a fence it should be white

Fig. 5.6

There should be no fence (i.e. go to b). But if there is a fence (i.e. you go to c and because of that only worlds with a fence are accessible), then it must be white (do not go to e, go to d).

Example 5.4 (Sample application: reactive grammars). See Chapter 9.

We make a context-free grammar reactive by allowing the system to switch rules on and off as it progresses.

Figure 5.7 is such an example.

Reactive context-free grammar rules. (A rule contains a rewrite together with instructions on which rules to activate and which to deactivate.)

1. $Y \to \alpha X\beta$
   activate 1 and activate 2.
2. $X \to \gamma Z\delta$
   activate 2 and 3 and deactivate 1.
3. $Z \to aY$
   no change.
4. $Y \to \varnothing$
   deactivate 4, otherwise no change.

Fig. 5.7

Theorem 5.5. *A reactive context-free grammar can generate* $\{a^n b^n c^n \mid n = 1, 2, 3, \ldots\}$, *which ordinary CFGs cannot.*

Example 5.6 (Sample application: reactive automata). See Chapter 8.

An automaton is a state transition system responding to input. We can therefore change the automaton table as we go along. This will give us a reactive automaton. See Figure 5.8.

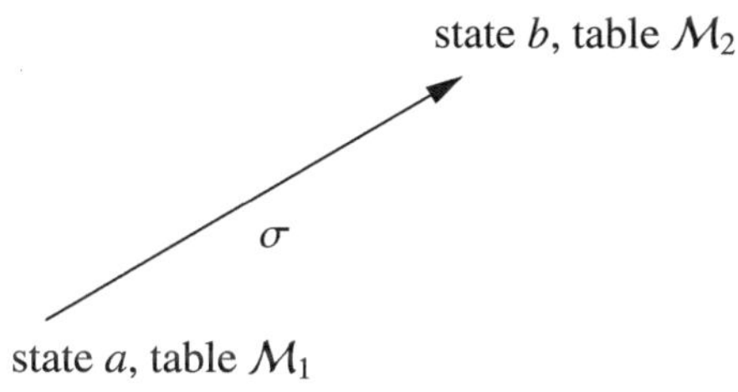

Fig. 5.8

Every reactive automaton is equivalent to an ordinary automaton recognising the same set of words. So we do not gain in expressive power by going reactive (as we do in the case of reactive context-free grammars) but we do gain simplicity.

Theorem 5.7. *Let* $\mathcal{A}$ *be an automaton (deterministic or non-deterministic) with* $k = \prod_{i=1}^{n} k_i$ *states. Then there exists an equivalent reactive automaton* $\mathcal{A}^*$ *(deterministic or non-deterministic, resp.) with* $k^* = \sum_{i=1}^{n} k_i$ *states.*

Example 5.8 (Sample application: networks). See [14], and Chapter 6. Figure 5.9 is an example of a network. It can be an argumentation network, neural network, transportation network, predator-prey biological network among many more.

Here the double arrows have their own meaning in each network and their availability greatly enhances the expressive power of such networks.

Example 5.9 (Sample application: reactive proof theory). Figure 5.10 is an example of a reactive proof rule.

Rule $\mathbf{R}_n$ says that if $A_1^n, \ldots, A_{k(n)}^n$ are available, then we can infer B_n but we need to activate Rules D_i^n and deactivate Rules E_j^n.

For example, Rule $\mathbf{R}_n$ can deactivate itself after use, as in linear logic. For reactive tableaux, see [89, 68].

Theorem 5.10. *See [62]. Every reactive Kripke model is equivalent to an ordinary Kripke model* $(S, R, \equiv)$ *with an additional equivalence relation* $\equiv$ *satisfying*

$$\neg\exists xyz(xRy \wedge xRz \wedge y \equiv z)$$

and vice versa.

Reactive networks

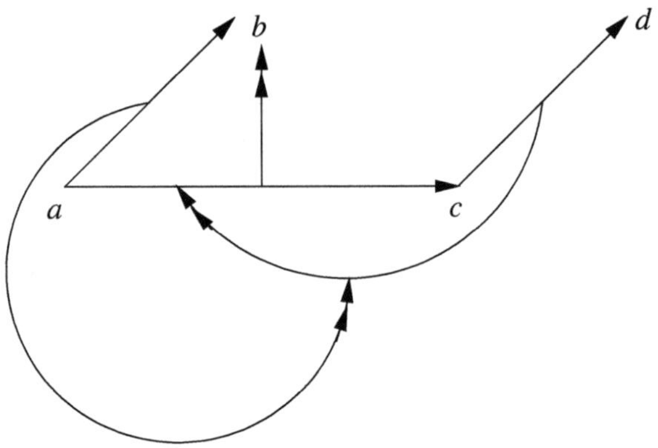

argumentation nets
transportation nets
neural nets
etc.

Fig. 5.9

Reactive proof theory
Rule $\mathbf{R}_n$

- $\dfrac{A_1^n \ldots A_{k(n)}^n}{B_n}$
- activate rules $D_1^n, \ldots, D_{m(n)}^n$
- deactivate rules $E_1^n, \ldots, E_{l(n)}^n$

Example: Goal-directed proof theory deactivates a rule once used!

Fig. 5.10

Figures 5.11, 5.12, 5.13 and 5.14 illustrate the need for the condition of the theorem. If $x \equiv y$ in Figure 5.11, we need different paths leading to x and y so that we can fold the model and make $x = y = z$, and then one path (leading to x) will see x_i as accessible and the other path (leading to y) will see y_i as accessible.

We get Figure 5.12.

The situation in Figure 5.13 does not allow us to have different paths to $z = x = y$ if we fold x and y together. All we can get is Figure 5.14.

5.1.3 Ternary relations

This section shows that reactive models can simulate a ternary relation and thus relevance logics and conditionals $A \Rightarrow B$ can be translated into a reactive modality

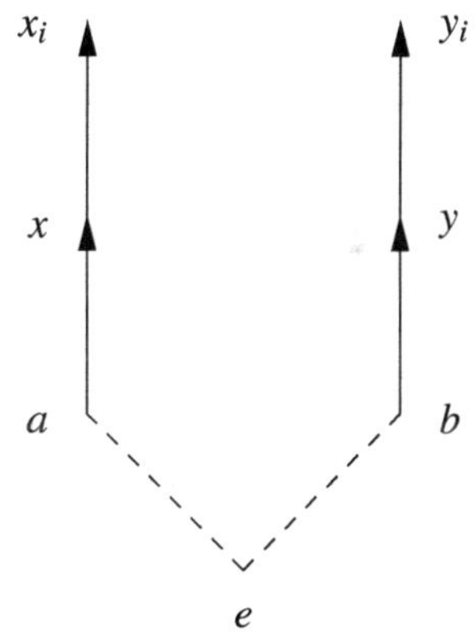

Fig. 5.11

To fold we must have $a \not\equiv b$

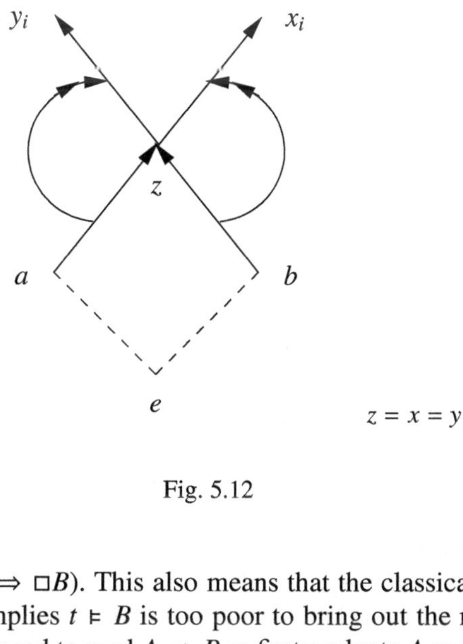

Fig. 5.12

$\Box. A \Rightarrow B = \text{def.}\Box(A \Rightarrow \Box B)$. This also means that the classical modal reading of $t \vDash A \rightarrow B$ as $t \vDash A$ implies $t \vDash B$ is too poor to bring out the richness of reactive Kripke models and we need to read $A \Rightarrow B$ as first evaluate A and *then* evaluate B.

So $t \vDash A \Rightarrow B$ iff whenever we evaluate $t \vDash A$ in the model and the model changes as a result of the reactivity and we get a new model, then $t \vDash B$ holds in the new model.

Example 5.11 (Reduction of three-place relation to reactive binary relation). This example shows how relevance logics, whose semantics requires a three-place relation $\rho(x, y, z)$, can be simulated using a reactive binary relation $\mathbb{R}$.

We have for relevance $A \Rightarrow B$ the following truth condition:

We cannot fold

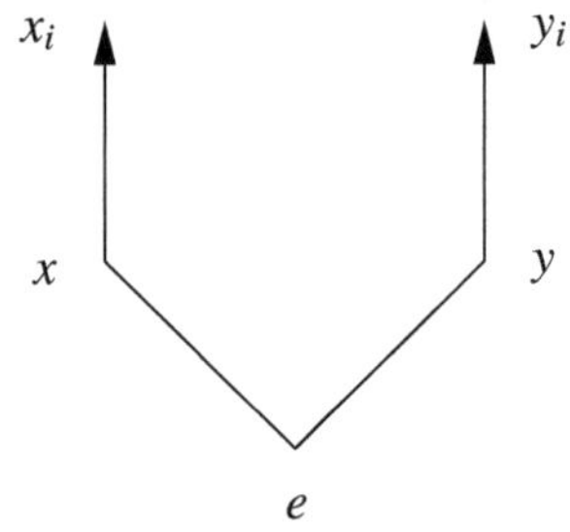

because we need different paths.

Fig. 5.13

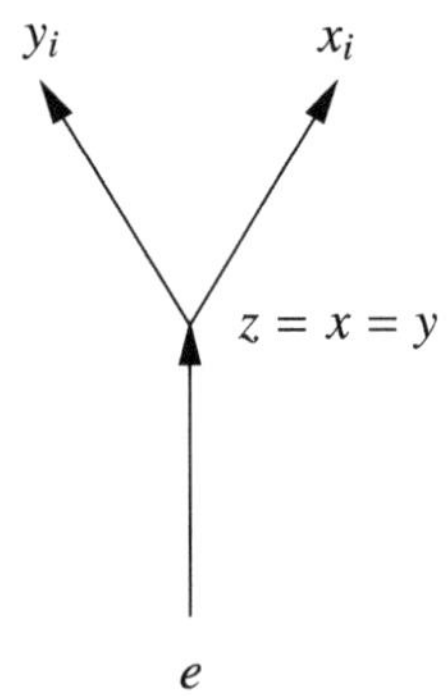

Fig. 5.14

$$x \vDash A \Rightarrow B \text{ iff } \forall yz(\rho(x, y, z))\text{and } y \vDash A \text{ imply } z \vDash B)$$

The idea is very simple. Consider $\rho(x, y, z)$ as a binary relation between the pair (x, y) and the point z. This we can represent in Figure 5.15.

In Figure 5.15, we know that (x, y) is related to z because as we go from x to y, we activate the connection $y \to z$. We also know that (x, y) is not related to u because as we go from x to y we cancel the connection $y \to u$.

Assume $\rho(x, y, z_i)$ holds but $\rho(x, y, v_j)$ does not. Let $\twoheadrightarrow +$ be an activating arrow from the arc $x \to y$ to all the z_i and $\twoheadrightarrow -$ be a deactivating arrow from the arc $x \to y$ to all the v_j.

The frame is written as $(S, \mathbb{R}, \mathbf{a})$, where $\mathbb{R}$ is now a relation containing arrows $x \to y$ (written $(x, y) \in \mathbb{R}$) , and double arrows $(x \to y) \twoheadrightarrow^{\pm} (u, v))$,written as $+((x, y), (u, v)) \in \mathbb{R}$ or $-((x, y), (u, v)) \in \mathbb{R}$.

$\mathbf{a}$ is a $\{0, 1\}$ function on arcs giving the active arcs and $\mathbf{a}_{(x,y)}$ is updated from $\mathbf{a}$ and $\mathbb{R}$ according to the $\pm((x, y), (u, r)) \in \mathbb{R}$.

$\mathbf{a}_{(t,s)}$ is modified from $\mathbf{a}$ as follows:

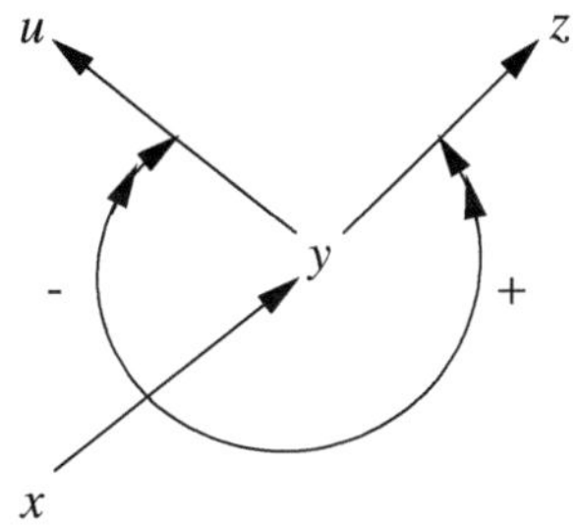

Fig. 5.15

- $\mathbf{a}_{(t,s)}(\beta) = 1-\mathbf{a}(\beta)$, if $[+((t, s), \beta) \in \mathbb{R}$ and $\mathbf{a}(\beta) = 0]$ or $[-((t, s), \beta) \in \mathbb{R}$ and $\mathbf{a}(\beta) = 1]$.
- $\mathbf{a}_{(t,s)}(\beta) = \mathbf{a}(\beta)$ otherwise.

Thus when we pass from x to y, we activate only those z_i such that $\rho(x, y, z_i)$ holds.

Thus if $A \Rightarrow B$ denotes the binary relevance or conditional connective whose semantics involves the ternary relation $\rho(x, y, z)$ and $\Box$ represents a reactive modality, then we translate

$$A \Rightarrow B = \text{def.}\Box(A \rightarrow \Box B)$$

We expect that for every frame (S, ρ) using $\Rightarrow$ there exists a frame $(S, \mathbb{R}, \mathbf{a})$ using $\Box$, such that for every A, B and $t \in S$ and h we have

$$t \vDash_h A \Rightarrow B \text{ in } (S, \rho, h) \text{ iff } t \vDash \Box(A \Rightarrow \Box B) \text{ in } (S, \mathbb{R}, \mathbf{a}, h).$$

The translation of $\rho(x, y, z)$ suggested in Figure 5.15 is not good enough. Consider the case of two points x, y and assume we have exactly $\rho(x, x, x)$ and $\rho(x, y, y)$

Figure 5.16 shows our proposed implementation.

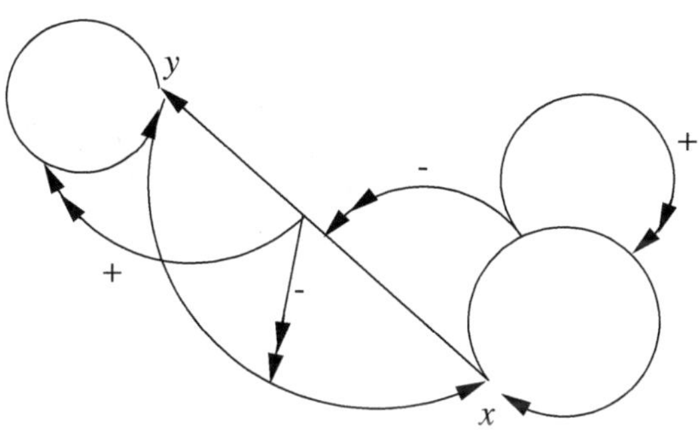

Fig. 5.16

From x we go to x. We cancel all arrows from x to points z such that $\neg\rho(x, x, z)$ and connect all arrows to z' such that $\rho(x, x, z')$ holds. So after going from x to x we are now at x again and we have no connection to y.

So let us evaluate at x the formula of relevance implication $A \Rightarrow (B \Rightarrow C)$.

$$x \vDash A \Rightarrow (B \Rightarrow C) \text{ iff } \forall zz'(\rho(x, z, z') \land z \vDash A \text{ implies } z' \vDash B \Rightarrow C).$$

In our case we need to check the cases of $\rho(x, x, x)$ and $\rho(x, y, y)$. Figure 5.16 shows that in both cases the relation works OK, but now we have to evaluate $x \vDash B \Rightarrow C$ and $y \vDash B \Rightarrow C$ (assuming $x \vDash A$ and $y \vDash A$ hold).

But we cannot continue correctly because the diagram has changed. What we need is to go back to the initial position!

To implement $\rho(x, y, z)$, we need to do the following:

1. As we move from x to y, we connect y only to $\{z|\rho(x, y, z)\}$. Call this action as 'making sure that $\lambda z\rho(x, y, z)$ is correct for (x, y)'.
2. If we continue now from y to z we do not want to activate the action described in (1) for (y, z), i.e. we do not want at this stage to 'make sure that $\lambda w\rho(y, z, w)$ is correct for (y, z)', we just want to get to z. So in (1) while we are dealing with $\lambda z\rho(x, y, z)$, we deactivate all actions for $\lambda w\rho(y, z, w)$. This requires double arrows of order 2.
3. As we actually pass through $y \to z$, we reactivate all actions for $\lambda w\rho(y, z, w)$, and let them propagate.

We therefore propose to translate $\rho(x, y, z)$ as in Figure 5.17.

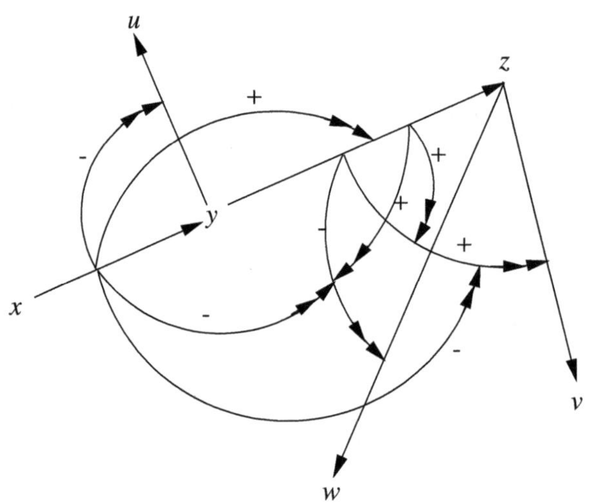

Fig. 5.17

In this figure, $\rho(x, y, z)$ and $\neg\rho(x, y, u)$ and $\rho(y, z, v)$ and $\neg\rho(y, z, w)$ hold.

Given a relation $\rho \subseteq S \times S \times S$, we simulate (S, ρ) by a frame $(S, \mathbb{R}, \mathbf{a})$ built as follows.

Let $\mathbf{a}(x, y) = 1$ for all $(x, y) \in S \times S$. Define $\mathbb{R}$ as follows:

(♯1.) $S \times S \subseteq \mathbb{R}$.
(♯2.) $+((x, y), (y, z)) \in \mathbb{R}$ if $\rho(x, y, z)$ holds, for all x, y, z.
(♯3.) $-((x, y), (y, z)) \in \mathbb{R}$ if $\sim \rho(x, y, z)$ holds, for all x, y, z.
(♯4.) if $\rho(x, y, z)$ holds and $\pm\rho(y, z, w)$ holds, then by (2) above $\pm((y, z), (z, w)) \in \mathbb{R}$ (respectively).
We add also $-((x, y), ((y, z), (z, w))) \in \mathbb{R}$ and also $+((y, z), ((y, z), (z, w))) \in \mathbb{R}$.

Now consider the relevance truth condition in a model (S, ρ, t) with a ternary ρ for the relevant $A \Rightarrow B$. It is

(*) $x \vDash A \Rightarrow B$ iff $\forall yz(\rho(x, y, z) \wedge y \vDash A$ implies $z \vDash B)$.

Consider now the evaluation of $\Box(A \rightarrow \Box B)$ at x in a reactive model $(S, \mathbb{R}, \mathbf{a}, x, h)$

(**) $x \vDash \Box(A \rightarrow \Box B)$ iff for all y such that $(x, y) \in \mathbb{R}$ and $\mathbf{a}(x, y) = 1$ (i.e. for all arcs $x \rightarrow y$ which are active), we have $y \vDash A \rightarrow \Box B$ holding in the model $(S, \mathbb{R}, \mathbf{a}_{(x,y)}, y, h)$ where $\mathbf{a}_{(x,y)}$ is the updated activeness function.

We continue:

(***) $y \vDash A \rightarrow \Box B$ iff whenever $y \vDash A$ then for all z such that $(y, z) \in \mathbb{R}$ and is $\mathbf{a}_{(x,y)}$ active we have $z \vDash B$ in the model $(S, \mathbb{R}, \mathbf{a}_{(x,y,z)}, z, h)$.

For both definitions (*) and (**) + (***) to be the same we need $\rho(x, y, z)$ to hold iff $\mathbf{a}_{(x,y)}(y, z) = 1$.

Indeed, this is what Figure 5.17 and items (♯1)–(♯4) show.

So we have the following:

1. Given a reactive frame $(S, \mathbb{R}, \mathbf{a})$ the relation $\rho(x, y, z)$ defined by $\mathbf{a}_{(x,y)}(y, z) = 1$ gives us a relevance frame (S, ρ). We have for $A \Rightarrow B = \text{def.}\Box(A \rightarrow \Box B)$ that $x \vDash A \Rightarrow B$ iff (**) + (***) hold iff (*) holds for ρ.
2. We have also seen that given (S, ρ), if we define $(S, \mathbb{R}, \mathbf{a})$ as in (♯1)–(♯4), we can simulate ρ.

Thus the translation of $A \Rightarrow B$ as $\Box(A \rightarrow \Box B)$ is faithful.

The interesting aspect of this example is that axioms on $\Rightarrow$ become genuine axioms on the reactivity of a Kripke frame.

5.2 Introducing the logic KR, Reactive modal K

We have two options for defining the reactive counterpart of modal **K**.

Option 1

Use only the modality which triggers the double arrows. Let us denote this by $\oslash A$ and its possibility by

$$\mathbb{P}A = \neg \oslash \neg A.$$

We have studied completeness theorems for such systems in [62].

Option 2

Study modal **K** with both $\oslash$ and $\mathbb{P}$ and the ordinary $\Box$ and $\Diamond$. We call this system **KR**. It has more expressive power and more applications, see Chapters 3 and 11.

This section deals with option 2. The reactivity we use is switch reactivity.

5.2.1 Introducing and discussing the reactive semantics

Definition 5.12 (Reactive K model).

1. *Let S be a non-empty set. A reactive relation on S has the form*

$$R = R_1 \cup R_2$$

where $R_1 \subseteq S^2$ and $R_2 \subseteq S^2 \times S^2$ such that the following holds

$$((x, y), (w, z)) \in R_2 \to (x, y) \in R_1 \wedge (w, z) \in R_1.$$

(This means $R_2 \subseteq R_1 \times R_1$.)
We can represent the elements of R_1 as '$x \to y$' and the elements of R_2 as '$(x \to y) \twoheadrightarrow (w \to z)$'. The elements of R_1 are called arrows and those of R_2 are called double arrows.
2. *A function $\mathbf{f} : R_1 \mapsto \{0, 1\}$ is called an activity function. When $\mathbf{f}(x, y) = 1$ we say $x \to y$ is 'on' or 'active' and we write $x \overset{+}{\to} y$. When $\mathbf{f}(x, y) = 0$, we say $x \to y$ is 'off' or 'not active' and we write $x \overset{-}{\to} y$.*
3. *An assignment h is a function giving each atomic q a subset $h(q) \subseteq S$.*
4. *A model has the form $\mathbf{m} = (S, R_1, R_2, \mathbf{f}, a, h)$ where $a \in S$ is the initial world.*

Definition 5.13 (Satisfaction). *Our language contains the classical connectives $\neg, \wedge, \vee, \to, \top, \bot$ and the two modalities $\Box$ and $\oslash$. We define satisfaction in a model as follows.*

1. *Let $(S, R_1, R_2, \mathbf{f})$ be given. Let $x \overset{+}{\to} y$ be in R_1. This means that $(x, y) \in R_1$ and $\mathbf{f}(x, y) = 1$. We define $\mathbf{f}[x, y]$ as follows:*

$$\mathbf{f}[x, y](u, v) = \begin{cases} \mathbf{f}(u, v) \text{ if } ((x, y), (u, v)) \notin R_2 \\ 1 - \mathbf{f}(u, v)) \text{ if } ((x, y), (u, v)) \in R_2 \end{cases}$$

2. *Given a model $\mathbf{m} = (S, R_1, R_2, \mathbf{f}, a, h)$ we define the notion of $\mathbf{m} \vDash A$ by induction as follows:*

- $\mathbf{m} \vDash q$ *iff* $a \in h(q)$, *for q atomic*
- $\mathbf{m} \vDash A \wedge B$ *iff* $\mathbf{m} \vDash A$ *and* $\mathbf{m} \vDash B$
- $\mathbf{m} \vDash \neg A$ *iff* $\mathbf{m} \nvDash A$
- *Similarly for the other classical connectives*
- $\mathbf{m} \vDash \Box A$ *iff for all y such that* $(a, y) \in R_1$ *we have* $\mathbf{n} = (S, R_1, R_2, \mathbf{f}, y, h) \vDash A$
- $\mathbf{m} \vDash \oslash A$ *iff for all y such that* $(a, y) \in R_1$ *and* $\mathbf{f}(a, y) = 1$ *we have* $\mathbf{n} = (S, R_1, R_2, \mathbf{f}[a, y], y, h) \vDash A$.

3. *The logic* **KR** *(reactive* **K***) is defined as the set of all wffs A such that for all models* $\mathbf{m}$ *we have* $\mathbf{m} \vDash A$.
4. *Note that we have frames of the form* $(S, R_1, R_2, \mathbf{f})$. *The logic is normal, as a and h can be chosen arbitrarily.*

Remark 5.14. Note that the function **f** introduced in the previous definition 5.12 actually defines another relation R_0 which is a subset of R_1. When **f** changes to $\mathbf{f}_{[x,y]}$, as in item 1 of Definition 5.13, then equivalently R_0 changes to $R_0[x, y]$. We use the notation with the function **f** to stress the fact that all reactivity changes arise from arrow connections from R_1.

We now address the task of axiomatising the logic **KR**. To be able to do that effectively, let us better familiarise ourselves with **KR** and its unique features.

Consider Figure 5.18.

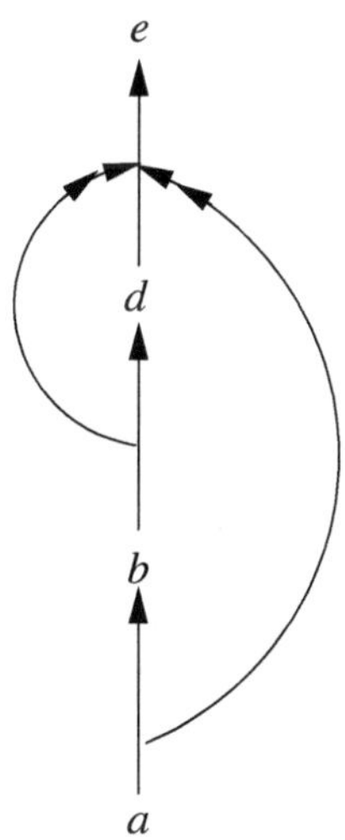

Fig. 5.18: JuneF1

Let our starting point be a. As we traverse the arc $a \to b$, there are two possibilities.

1. We ignore all double arrows and what they do.

2. We take account of what the double arrows say and what they do.

Let possibility 1 be formalised by $\Diamond$ and let possibility 2 be formalised by $\mathbb{P}$. So we can write: $a \vDash \Diamond q$ to mean traverse from a to the accessible world b and ignore all double arrows and we can write $a \vDash \mathbb{P}q$ to mean traverse from a to the accessible world b and take account of the double arrows.

Consider now

$$a \vDash \mathbb{P}\Diamond\Diamond\top$$

Here we move from a to b and activate the double arrow which disconnects the arc $d \to e$. We check:

$$b \vDash \Diamond\Diamond\top$$

and then

$$d \vDash \Diamond\top$$

Now the arc $d \to e$ is disconnected by a double arrow but as we are evaluating $\Diamond$ at node d and $\Diamond$ ignores all double arrows and their effects, as far as $\Diamond$ is concerned $d \to e$ is connected and hence $d \vDash \Diamond\top$ holds and therefore $a \vDash \mathbb{P}\Diamond\Diamond\top$ also holds.

The situation is different with

$$a \vDash \mathbb{P}\Diamond\mathbb{P}\top$$

Evaluating here takes us to $d \vDash \mathbb{P}\top$, but $\mathbb{P}$ is affected and takes account of double arrows and so $d \vDash \mathbb{P}\top$ does not hold and hence

$$a \vDash \mathbb{P}\Diamond\mathbb{P}\top$$

does not hold.

Consider now

$$a \vDash \mathbb{P}\mathbb{P}\mathbb{P}\top.$$

We can go from a to e taking account of double arrows. Moving from a to b switches the connection $d \to e$ off but continuing from b to d switches it on again and therefore $d \vDash \mathbb{P}\top$ holds.

Note that we cannot evaluate any $x \vDash \mathbb{P}q$, for arbitrary point x, without knowing how we got to x and which double arrows we activated along the path.

So asking does $d \vDash ?\mathbb{P}\top$ hold cannot guarantee an answer. We need to ask something like

$$a \xrightarrow{\Diamond} b \xrightarrow{\mathbb{P}} d \vDash ?\mathbb{P}\top$$

or

$$a \xrightarrow{\mathbb{P}} b \xrightarrow{\Diamond} d \vDash ?\mathbb{P}\top$$

or in general

$$(x_0 \xrightarrow{M_1} x_1 \xrightarrow{M_2} x_2 \to \ldots \xrightarrow{M_n} x_n) \vDash ?A$$

where $M_i \in \{\Diamond, \mathbb{P}\}$.

Note that we need a notion of *legitimate sequences*. This notion will be formally defined in Subsection 5.2.2 below. Meanwhile, we understand it intuitively. The sequence

$$a \xrightarrow{\mathbb{P}} b \xrightarrow{\Diamond} d \xrightarrow{\mathbb{P}} e$$

is not legitimate because by the time we get to d, $d \to e$ is disconnected. So to get to e from d we need to use $\Diamond$.

We thus get a bimodal logic with two modalities $\Diamond$ and $\mathbb{P}$ with evaluation of formulas depending on paths $(x_0, \ldots, x_n)$ but not as ordinary paths. We also need to know how we traverse them. So our paths have the form $\mathbf{t} = (x_0, M_1x_1, M_2x_2, \ldots, M_nx_n)$, where $\mathbf{t}$ is a legitimate sequence.

Our task in this section is to find axioms and prove completeness for **KR**. We might have thought that it would be useful to consider **KR** as a bimodal logic with two accessibility relations $R_{\mathbb{P}}$ and $R_{\Diamond}$, and see what properties they have. The problem is that the accessible worlds are the same for both relations, the difference being the way we traverse them. So really our most reasonable option is to consider a *transformed Kripke model* where the points/worlds are legitimate paths of the form

$$\mathbf{t} = (a, M_1x_1, \ldots, M_nx_n)$$

such that $aR_1x_1R_1x_2, \ldots, x_{n-1}R_1x_n$ holds.

Thus we start with

$$\mathbf{m} = (S, R_1, R_2, a, h)$$

for example with (S, R_1, R_2, a) is as in Figure 5.18, and then transform it into a model

$$\mathbf{m}^* = (S^*, R_{\Diamond}, R_{\mathbb{P}}, (a), h)$$

where S^* is the set of all legitimate sequences of the form $\mathbf{t} = (a, M_1x_1, \ldots, M_nx_n)$ where $aR_1x_1 \wedge x_1R_1x_2\wedge, \ldots, \wedge x_{n-1}R_1x_n$.

We define $R_{\Diamond}$ and $R_{\mathbb{P}}$ by

$$\mathbf{t}\, R_{\Diamond}\, \mathbf{t} * (My), \text{ when } x_nR_1y, \text{ and } M \in \{\Diamond, \mathbb{P}\}$$

$$\mathbf{t}\, R_{\mathbb{P}}\, \mathbf{t} * ((\mathbb{P}y), \text{ when } x_nR_1y \text{ and } \mathbf{t} * (\mathbb{P}y) \text{ is legitimate.}$$

where $*$ is concatenation of sequences.

Let $\mathbf{t} \vDash q$ iff $x_n \vDash q$, for q atomic.

Note that R_1 and R_2 are replaced by $R_{\Diamond}$ and $R_{\mathbb{P}}$. R_1 is used directly in the definition of the sequences $\mathbf{t}$ which go into S^* and R_2 is used in the notion of legitimate sequences.

So $\mathbf{m}$ is replaced by its transformed model $\mathbf{m}^*$, see Figure 5.19.

If we do this we can get a completeness theorem after some hard work. We are facing three formal tasks:

Task 1: Given a bimodal semantic model $\mathbf{m}^*$ with two binary relation R' and R'', and with an equivalence relation $\equiv$, such as the one in Figure 5.19, can we identify

whether it comes from a **KR** model **m** such as the one in Figure 5.18? In other words, can we identify when a bimodal model $\mathbf{n}^*$ with an equivalence relation $\equiv$ is actually a transformed version of another **KR** model **n**?
Task 2: Axiomatise the logic **K**.
Task 3: We need to systematically find corresponding axioms for various conditions on the **KR** relation R_1. This Task we postpone to a future paper.

Obviously the transformed models will play a part, but we note that the transformed models may not be in the same class as the original model. The lucky aspect for the case of **KR** is that such models $\mathbf{m}^*$ are also **KR** models because **KR** has no special requirements on the worlds or the relations.

The unlucky or difficult aspect of using the transformed models as tools is that we need to uniquely identify paths, and to do that we need nominals or second order propositional quantifiers or in short we need more expressive power at our service and such power is not available in **KR**. So our strategy is to use technical devices which eliminate the need for nominals.

So much for **KR**. If we look at Figure 5.18 we observe that the model is linear. If we were to insist on semantics which requires that (S, R_1) be linear then looking at the model $\mathbf{m}^*$ of all legitimate paths gives us a non-linear model. This is not good. We get a model which is not in the correct linear semantics. So how do we axiomatise the linear case? The answer is that I don't know yet. For the purpose of many applications such as modelling contrary-to-duties, **KR** is sufficient, see [79]. We postpone the general theory of reactive modal semantics until dedicated future papers.

We now give you an idea of what kind of axioms we need for **KR**.

We first note that both $\Box$ and $\oslash$ (the duals of $\Diamond$ and $\mathbb{P}$) are **K** modalities, satisfying the interaction axiom (for A not containing any $\oslash$).

$$\mathbb{P}A \to \Diamond A$$

To explain what additional axioms we need, consider again Figure 5.18 which describes S and compare with Figure 5.19 which describes S^*.

Figure 5.19 is a tree and so each node gives the sequence leading up to it. The indices on some nodes are there to help us talk about them. The actual sequences have no indices. So $(a\mathbb{P}b\Diamond d_3\Diamond e_3)$ is really the legitimate sequence $(a\mathbb{P}b\Diamond d\Diamond d)$.

Since the sequence $(a\mathbb{P}b\Diamond d\mathbb{P}e)$ is not legitimate, the point d_3 is $\Diamond$-connected only to e_3 and is not $\mathbb{P}$-connected to e.

We now show you how to get an axiom. The axioms which we get allow us to retrieve the double arrow from the graph. Since

1. $(a\Diamond b\Diamond d\mathbb{P}e)$ is legitimate

and

---

[1] This figure describes a bimodal logic model with a set of possible worlds $S*$ and with two accessibility relations, as indicated, and an additional equivalence relation $\equiv$ on worlds. The equivalence relation can be read from the fact that in the figure we named several nodes by the same letter. When node x is equivalent to node y, we are committed to give the same truth values to atoms at these nodes.

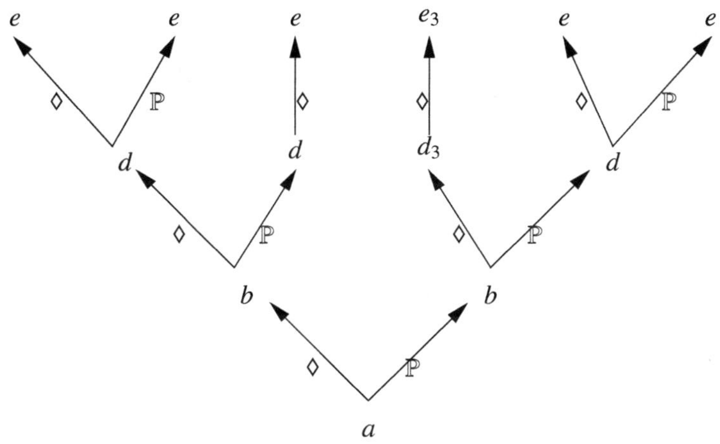

Fig. 5.19: JuneF2[1]

2. $(a\mathbb{P}b\Diamond d\mathbb{P}e)$ is not legitimate

we know that there must be a double arrow

$$(a \to b) \twoheadrightarrow (d \to e).$$

Similarly, we identify a double arrow from $(b \to d) \twoheadrightarrow (d \to e)$ since

3. $(a\Diamond b\Diamond d\mathbb{P}e)$ is legitimate

and

4. $(a\Diamond b\mathbb{P}d\mathbb{P}e)$ is not legitimate.

From this we know that if we use both double arrows, the move $d \xrightarrow{\mathbb{P}} e$ becomes possible. So

5. $(a\mathbb{P}b\mathbb{P}d\mathbb{P}e)$ is legitimate.

The axiom we get in this case is

Axiom: $(1) \wedge (2) \wedge (3) \wedge (4) \to (5)$.

This axiom is not written in the modal language of **KR**. To express it in **KR** we need extra technical manipulation. If we have nominals, i.e. we have names for worlds, say $\bar{a}$ names a, $\bar{b}$ names b, $\bar{d}$ names d and $\bar{e}$ names e, i.e. we have

$$y \vDash \bar{x} \text{ iff } y = x,$$

then we can express the following in the logic **KR** + nominals:

(*): "$(aM_1x_1M_2x_2\dots M_nx_n)$ is legitimate"

by writing $a \vDash \alpha_*$ where

$$\alpha_* = M_1(\bar{x}_1 \wedge M_2(\bar{x}_2 \wedge \ldots \wedge (M_{n-1}(\bar{x}_{n-1} \wedge M_n\bar{x}_n))) \ldots)$$

Thus our axiom for Figure 5.18 is

$$\alpha_1 \wedge \alpha_2 \wedge \alpha_3 \wedge \alpha_4 \to \alpha_5.$$

In general if $\mathbf{t} * (\mathbb{P}y)$ is legitimate (not legitimate) and $\mathbf{t}_1$ and $\mathbf{t}_2$ are obtained by changing a single $\Diamond$ into $\mathbb{P}$ at each place n_1 and n_2 respectively and $\mathbf{t}_1 * (\mathbb{P}y)$ and $\mathbf{t}_2 * (\mathbb{P}y)$ becomes not legitimate (resp. becomes legitimate) then $\mathbf{t}_{1,2} * (\mathbb{P}y)$ is unchanged, where $\mathbf{t}_{1,2}$ is obtained from $\mathbf{t}$ by changing both instances of $\Diamond$ (at places n_1 and n_2) into $\mathbb{P}$.

This gives us the general form of our additional axioms. However, as we said, to express the axioms we need nominals in the language. The challenge is to express the axioms in **KR** without the use of nominals, by using some other technical device. This is indeed what we are going to do now!

5.2.2 Manipulating finite bi-modal Kripke models

This subsection prepares the ground for our **KR** completeness proof. Our strategy is to construct first a finite bi-modal tree Kripke model for the modalities $\Diamond$ and $\mathbb{P}$ and then convert this model to a finite **KR** model. To achieve this, we need to learn how to manipulate finite bi-modal Kripke models. This is the task of the current subsection.

Definition 5.15.

1. *Let $\Diamond$ and $\mathbb{P}$ be two modalities. A finite Kripke model for these modalities has the form $\mathbf{m} = (S, \rho_1, \rho_2, a, h)$, where S is the finite set of possible worlds, $a \in S$ is the initial root (actual) world and ρ_1 and ρ_2 are binary relations on S. We require that $\rho_1 \cap \rho_2 = \varnothing$. h is an assignment giving each atomic q of the language a subset $h(q) \subseteq S$.*
2. ***m** is said to be a tree model if (S, ρ, a) is a tree with root a, where $\rho = \rho_1 \cup \rho_2$. This means that ρ is irreflexive and every node $x \in S$ has a unique path (leading up to it) of the form $(a, x_1, \ldots, x_n)$ such that $a\rho x_1 \wedge x_1\rho x_2 \wedge \ldots \wedge x_{n-1}\rho x_n$ holds and $x_n = x$.*
   Since $\rho = \rho_1 \cup \rho_2$ we also denote paths by
   $$a \xrightarrow{M_1} x_1 \xrightarrow{M_2} x_2 \longrightarrow \ldots \xrightarrow{M_n} x_n$$
   where $M_i \in \{\Diamond, \mathbb{P}\}$ and $x \xrightarrow{M} y$ means
   $$x\rho_1 y \text{ if } M = \Diamond \text{ and}$$
   $$x\rho_2 y \text{ if } M = \mathbb{P}.$$
   We call the above annotated paths.
   We also write
   $$\mathbf{t} = (a, M_1x_1, \ldots, M_nx_n)$$
   for annotated paths.

3. *Let* $\mathbf{m}$ *be a tree model. We define the notion of the level of a node* $x \in S$ *as follows: Let* n *be the length of the maximal path in* (S, ρ, a)*, i.e. there exists* $(a, x_1, \ldots, x_{n-1})$ *such that* $a\rho x_1 \wedge \ldots \wedge x_{n-2}\rho x_{n-1}$ *and for every path* $(a, y_1, \ldots, y_{m-1})$ *in* (S, ρ, a) *we have* $m \leq n$.
Then we define
   a) *The point* a *is of level* n.
   b) *If* x *is of level* $n-k$ *and* $x\rho y$ *then* y *is of level* $n-k-1$.
   c) *The level of the model* $\mathbf{m}$ *is the level of* a.
   Note that the level of a node x *is* $n-k$ *where* k *is its "distance" from the root* a.

Definition 5.16 (Path nominals for bi-modal models). *Let* $\mathbf{m} = (S, \rho_1, \rho_2, a, h)$ *be a finite tree bi-modal Kripke model. Let* δ *be a function associating with each* $x \in S$ *a formula* $\delta(x)$ *of the bi-modal language with* $\Diamond, \mathbb{P}$. *We say* δ *is a system of path nominals for* $\mathbf{m}$ *iff the following holds:*

1. $x \vdash \delta(x)$ *in* $\mathbf{m}$.
2. *Let* $x\rho_1 y$ *and* $x\rho_1 z$ *and* $y \neq z$ *hold. Then* $\delta(y) \wedge \delta(z)$ *has no finite tree model (i.e.* $S(y), S(z)$ *are semantically inconsistent).*
3. *If* x *is of level* k *then* $\delta(x)$ *has at most* $n-k$ *nested modalities.*

Remark 5.17. 1. Note that if a finite model $\mathbf{m}$ has a system of path nominals as in Definition 5.16, then the following hold for any $t \in S$
   a) $$t \vDash \mathbb{P}(A) \leftrightarrow \bigvee_{\substack{s \\ t\rho_2 s}} \mathbb{P}(\delta(s) \wedge A)$$
   $$t \vDash \neg\mathbb{P}(A) \leftrightarrow \bigwedge_{\substack{s \\ t\rho_2 s}} \mathbb{P}(\delta(s) \wedge \neg A)$$
   b) $$t \vDash \Diamond A \leftrightarrow \bigvee_{\substack{s \\ t\rho_1 s}} \Diamond(\delta(s) \wedge A)$$
   $$t \vDash \neg\Diamond A \leftrightarrow \bigwedge_{\substack{s \\ t\rho_1 s}} \Diamond(\delta(s) \wedge \neg A).$$
   c) If $M \in \{\mathbb{P}, \Diamond\}$ then we have
   $$t \vDash M(\delta(z) \wedge A) \wedge M(\delta(z) \wedge B) \leftrightarrow M(\delta(z) \wedge A \wedge B).$$

2. Note also that the restriction (3) on δ of Definition 5.16 is not really restrictive because we have the proposition (*):
   (*) If t is of level k then for any wff A there exists a formula A' with at most $n-k$ nested modalities such that $t \vDash A \leftrightarrow A'$.
   This can be proved by induction on $n-k$, where n is the level of the model $\mathbf{m}$.

Remark 5.18. Note that the concept of path nominals is intended to identify paths and not necessarily toidentify worlds. Consider Figure 5.20, describing a model with $q = \top$ everywhere and one modality $\Diamond$. We can use the system of path nominals δ as follows:

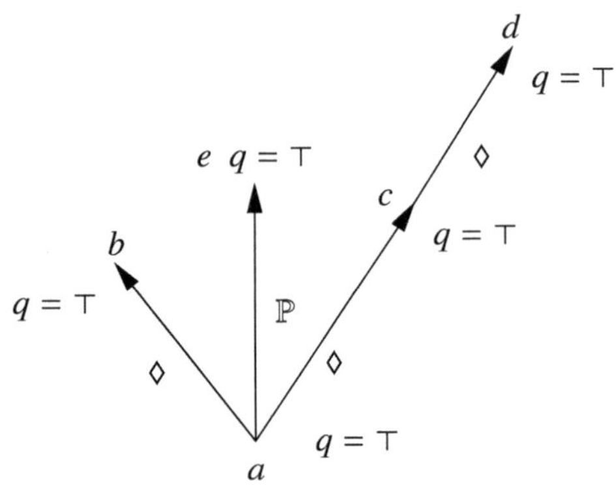

Fig. 5.20

$$\begin{aligned}\delta(e) &= \neg\Diamond\top \\ \delta(b) &= \neg\Diamond\top \\ \delta(d) &= \neg\Diamond\top \\ \delta(c) &= \Diamond\neg\Diamond\top \\ \delta(a) &= \Diamond\neg\Diamond\top \wedge \Diamond\Diamond\neg\Diamond\top\end{aligned}$$

We see that $\delta(b) = \delta(d)$, so these are not the traditional nominals. However $\delta(b) \wedge \delta(c)$ is not consistent. So the paths

$$\begin{aligned}&a \to c \to d \\ &a \to e \\ &a \to b\end{aligned}$$

can be identified uniquely.

Definition 5.19. *Let* $\mathbf{m} = (S, \rho_1, \rho_2, a, h)$ *be a finite tree model and let* δ *be a system of path nominals for* **m**. *Let* $M_i \in \{\Diamond, \mathbb{P}\}$ *and let* $(a, x_1, \ldots, x_n)$ *be a path.*

1. *Let* $\mathbf{t} = (a, M_1x_1, \ldots, M_nx_n)$ *be an annotated path in* **m**. *The notation means that we have* $a\rho_{M_1}x_1 \wedge \ldots \wedge x_{n-1}\rho_{M_n}x_n$ *where*

$$\rho_M = \rho_1 \text{ if } M = \Diamond$$

   and

$$\rho_M = \rho_2 \text{ if } M = \mathbb{P}.$$

   Thus the annotated path **t** *is more specific about how we get to* x_n.

2. *Consider the wff* $\Psi^\delta_{\mathbf{t}}(X)$ *defined as*

$$\Psi^\delta_{\mathbf{t}}(X) = M_1(\delta(x_1) \wedge M_2(\delta(x_2) \wedge \ldots \wedge M_n(\delta(x_n) \wedge X))\ldots).$$

3. *Define the modality* M' *as follows:*

$$M' = \Diamond \text{ if } M = \mathbb{P}$$
$$M' = \mathbb{P} \text{ if } M = \Diamond.$$

4. *Let* $\mathbf{t} = (a, M_1x_1, \ldots, M_nx_n)$*. Define* $\mathbf{t}'_{[j]} 1 < j \leq n$ *to be*

$$\mathbf{t}'_{[j]} = (a, M_1x_1, \ldots, M_{j-1}x_{j-1}, M'_jx_j, \ldots, M_nx_n)$$

Given $\mathbf{t}$ *consider* $\Psi^\delta_{\mathbf{t}'_{[j]}}(X)$*. We call this formula a one-point change of* $\Psi^\delta_{\mathbf{t}}(X)$*.*

5. *Let* $\mathbf{t} = (a, M_1x_1, \ldots, M_nx_n)$*. Assume* $x_n\rho_1 y$*. Then the formula* $\Psi^\delta_{\mathbf{t}}(\mathbb{P}\delta(y))$ *is called a* $\mathbb{P}$*-test at the point* x_n*.*

Remark 5.20. To explain the concepts of Definition 5.19, we consider the following two figures, Figure 5.21 and 5.22.

The first figure describes a **KR** model for $\Diamond$ and $\mathbb{P}$. The second figure describes the unfolding of this **KR** model into a bi-modal tree model for the two modalities $\Diamond$ and $\mathbb{P}$.

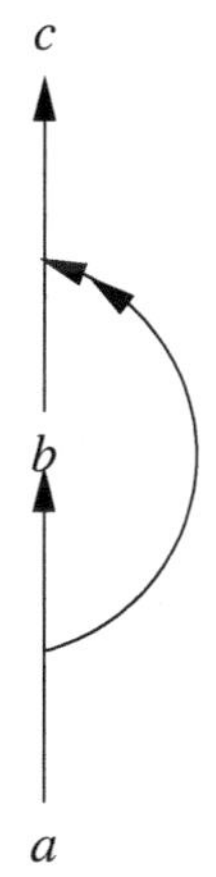

Fig. 5.21

A path nominal system can be defined on the model of Figure 5.22 as follows:

$$\begin{aligned}
\delta(c_1) &= \delta(c_2) = \delta(c_3) = \neg\Diamond\top \wedge \neg\mathbb{P}\top \\
\delta(b_1) &= \Diamond\delta(c_1) \wedge \mathbb{P}\delta(c_2) \\
\delta(b_2) &= \Diamond\delta(c_3) \\
\delta(a) &= \Diamond\delta(b_1) \wedge \mathbb{P}\delta(b_2).
\end{aligned}$$

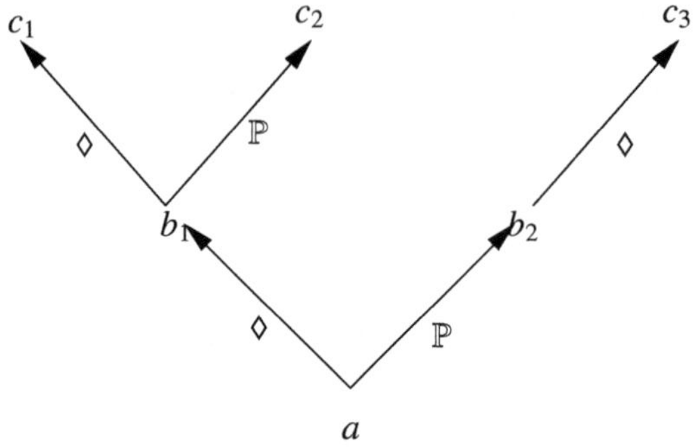

Fig. 5.22

If we are given the model of Figure 5.22 together with δ, how can we retrieve the model of Figure 5.21 from the model of Figure 5.22?

As part of the conversion/retrieval process, we need to identify the effects of the double arrow

$$(a \to b) \twoheadrightarrow (b \to c)$$

which resides in Figure 5.21 and has manifestations in Figure 5.22, namely that $b_2 \vDash \neg\mathbb{P}\delta(c_3)$. To achieve such an identification we use the $\mathbb{P}$-test formulas.

We formally want to convert the model of Figure 5.22 into the reactive bi-modal model of Figure 5.23 by adding a set R_2 of double arrows and an activity function $\mathbf{f}$. We let $R_1 = \rho_\Diamond \cup \rho_\mathbb{P}$ and let the $\mathbb{P}$ and $\Diamond$ modalities traverse all connections of R_1.

Obviously, adjustments are needed!

The end result we hope is a model equivalent to that of Figure 5.21.

1. Consider the formula
$$\Psi = \neg\mathbb{P}(\mathbb{P}\delta(c_3) \wedge \delta(b_2))$$
and consider Ψ'
$$\Psi' = \Diamond(\mathbb{P}\delta(c_3) \wedge \delta(b_2))$$
We have $a \vDash \Psi$ and $a \vDash \Psi'$, namely the value changed when we changed $\mathbb{P}$ to $\Diamond$! This invites $(a \to b_2) \twoheadrightarrow (b_2 \to c_3) \in R_2$.
2. Similarly, consider
$$\begin{aligned} \varphi &= \Diamond(\delta(b_1) \wedge \mathbb{P}\delta(c_2)) \\ \varphi' &= \neg\mathbb{P}(\delta(b_1) \wedge \mathbb{P}\delta(c_2)) \end{aligned}$$
We have $a \vDash \varphi$ and $a \vDash \varphi'$.
The correct way of dealing with this is to declare $a \to b_1$ as off. $\Diamond$ will ignore it and $\mathbb{P}$ will not be able to traverse it.
In this particular case we can also add the double arrows
$$(a \to b_1) \twoheadrightarrow (b_1 \to c_2)$$
as well as
$$(a \to b_1) \twoheadrightarrow (b_1 \to c_1)$$

3. Also consider

$$b_2 \vDash \Diamond\delta(c_3)$$

but

$$b_2 \vDash \neg\mathbb{P}\delta(c_3)$$

This is already taken care of by (1) above. Had it been the case that b_2 had no $\mathbb{P}$ predecessors, we would have declared $b_2 \to c_3$ as off! But this is not needed here.

The $\mathbb{P}$-tests show that if we change $\mathbb{P}$ to $\Diamond$ or $\Diamond$ to $\mathbb{P}$ the truth values switch. This calls for double arrows as discussed in Figure 5.23.

Now consider the model in Figure 5.23 as a **KR** model with the relation $R_1 = \rho_{\mathbb{P}} \cup \rho_{\Diamond}$ and R_2 as drawn. The assignment is the same as in Figure 5.21, i.e. points c_1, c_2, c_3 of Figure 5.23 get the same assignment as point c of Figure 5.21 and similarly points b_1, b_2 get the same as b, respectively.

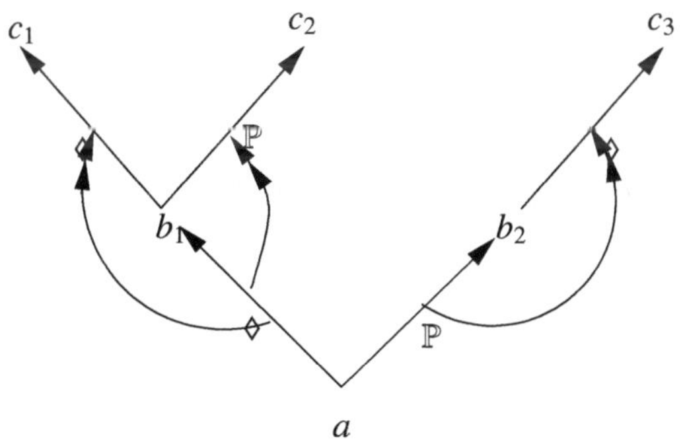

Fig. 5.23

So, when we evaluate $a \vDash \mathbb{P}A$ in Figure 5.21, we go to point b and activate the double arrow. When we evaluate $a \vDash \mathbb{P}A$ in Figure 5.23, we either go to b_1 or go to b_2 and activate the respective double arrow in each case.

When we go to b_2, the double arrow deactivates $b_2 \to c_3$ and so $\neg\mathbb{P}\delta(c_3)$ holds but $\Diamond\delta(c_3)$ still holds because $\Diamond$ ignores the effects of double arrows.

Similarly at b_1 we get the desired results. Thus the models of Figures 5.21 and 5.23 are equivalent.

Remark 5.21. Note in Definition 5.19 when we define $\Psi^{\delta}_{\mathbf{t}}(X)$ for a path

$$\mathbf{t} = (a, M_1x_1, \ldots, M_nx_n),$$

the formula $\Psi^{\delta}_{\mathbf{t}}(X)$ does not mention the path $\mathbf{t}$. This will be significant later on, when we define double arrows using the $\Psi^{\delta}_{\mathbf{t}}$ formulas.

To understand what is happening, consider Figure 5.20.

Let

$$t_1 = (a, \mathbb{P}e)$$
$$t_2 = (a, \Diamond b)$$

we have $\Psi^{\delta}_{t_1} = \Psi^{\delta}_{t_2}$.

Furthermore, if we consider

$$t_3 = (a, \Diamond c)$$

we have

$$a \vDash \Psi^{\delta}_{t_3}(\Diamond\top)$$

but also

$$a \vDash \Psi^{\delta}_{t_1}(\neg\Diamond\top)$$

Definition 5.22. *1. Let* $\mathbf{m} = (S, \rho_1, \rho_2, a, h)$ *be a tree model. Define a relation* $\equiv$ *on* S *by letting*

- $x \equiv y$ *iff* $\exists z(z\rho_i x \wedge z\rho_i y)$ *and* $i \in \{1, 2\}$ *and if* k *is the level of* x *and* y *then for all* A *of at most* $n - k$ *nested modalities we have* $x \vDash A$ *iff* $y \vDash A$.

2. let $\mathbf{m}' = (S' \subseteq S, \rho'_1, \rho'_2, a, h)$ *be defined as follows (we call* $\mathbf{m}'$ *the "reduced version" of* $\mathbf{m}$*):*

a) Let $a \in S'$.

b) Assume x *is in* S', *let* S^i_x *be a set of points from* S *such that*

- $x\rho_i z$ *for* $z \in S^i_x$
- *If* $z, z' \in S^i_x$ *and* $z \neq z'$ *then* $z \not\equiv z'$.
- *If* $x\rho_i z$ *then for some* $z' \in S^i_x$ *we have* $z \equiv z'$.

Also let $S^1_x \cup S^2_x \subseteq S'$.

Let S' *be the smallest subset of* S *closed under (a) and (b). Let* $\rho'_i = \rho_i \upharpoonright S'$.

3. A tree model $\mathbf{m}$ *is said to be reduced if* $\mathbf{m} = \mathbf{m}'$.

4. Note that under 2(b) above, we may choose different points, but all possible models are isomorphic. This is why we speak of "the reduced model".

Lemma 5.23. *Let* $\mathbf{m} = (S, \rho_1, \rho_2, a, h)$ *be a tree model and let* $\mathbf{m}' = (S', \rho'_1, \rho'_2, a, h)$ *be its reduced version. Assume* a *is of level* n. *Then for any* $x \in S'$ *of level* k *and any* A *with at most* $n - k$ *nested modalities we have* $x \vDash A$ *in* $\mathbf{m}$ *iff* $a \vDash A$ *in* $\mathbf{m}'$.

Proof. The proof is by induction on $n - k$ and presents no difficulties.

Lemma 5.24. *Let* $\mathbf{m}$ *be a reduced tree model. Assume our language is based on a finite set* $\mathbb{Q}$ *of atoms. Define* $\delta(x)$, $x \in S$ *by* $\delta(x) = \{A \mid x \vDash A$ *and* A *is based on* $\mathbb{Q}$ *and contains at most* $n - k$ *nested modalities, where* k *is the level of* $x\}$.

Then $\delta(x)$ *is a system of path nominals.*

Proof. This follows from the facts that the model is reduced and $\delta(x)$ is a complete theory for the language of formulas of up level $n - k$.

Lemma 5.25. *Let* $\mathbf{m} = (S, \rho_1, \rho_2, a, h)$ *be a reduced tree model for a finite language based on* $\mathbb{Q}$*, and let* $\delta(x)$ *be the path nominal system for* $\mathbf{m}$ *as defined in Lemma 5.24. Let* $t \in S$*. Then*

$$\delta(t) = \varepsilon_t \wedge \bigwedge_{\pi(t)} \Psi_{\pi(t)} \qquad (*)$$

where ε_t *is a conjunctive form, i.e.* $\varepsilon = \bigwedge_{q \in \mathbb{Q}} \pm q$ *and where* $\pi(t)$ *ranges over all maximal paths of the form* $(t, M_1 x_1, \ldots, M_n x_n)$ *and* Ψ_π *is* $M_1(\delta(x_1) \wedge M_2(\delta(x_3) \wedge \ldots \wedge M_n \delta(x_n)) \ldots)$*.*

Note that Ψ_π *is* $\Psi^\delta_\pi(\top)$ *as defined in Definition 5.19.*

Proof. By induction on $n - k$, where k is the level of t.

1. The case of $n - k = 0$ is clear.
2. The case of $n - k = 1$ is also clear since any theory of a node x of level $n - 1$ has the form $\delta(x) = \varepsilon \wedge \bigwedge_i \pm \mathbb{P}\varepsilon_i \wedge \bigwedge_j \pm \Diamond \varepsilon_j$.
3. Let t be such that $x_1, \ldots, x_m$ are its ρ_1 successors and $y_1, \ldots, y_r$ are its ρ_2 successors. We have $t\rho_1 x_1, \ldots, t\rho_1 x_m$, and $t\rho_2 y_1, \ldots, t\rho_2 y_r$.
   First we observe that the theory of node t is equivalent to

$$A_t = \varepsilon_t \wedge \bigwedge_i \Diamond \delta(x_i) \wedge \bigwedge_j \mathbb{P}\delta(y_j).$$

Next observe that by the induction hypothesis each $\delta(z), z \in \{x_i, y_j\}$ has the form

$$\begin{aligned}\delta(z) &= \varepsilon_z \wedge \textstyle\bigwedge_{\eta(z)} \Psi_{\eta(z)} \\ &= \delta(z) \wedge \textstyle\bigwedge_{\eta(z)} \Psi_{\eta(z)}\end{aligned}$$

where $\eta(z)$ ranges over all maximal paths starting at z.

Therefore A_t has the form

$$A_t = \varepsilon_t \wedge \bigwedge_{\substack{z \\ t\rho_1 z}} [\Diamond(\delta(z) \wedge \bigwedge_{\eta(z)} \Psi_{\eta(z)})] \wedge \bigwedge_{\substack{z \\ t\rho_2 z}} [\mathbb{P}(\delta(z) \wedge \bigwedge_{\eta(z)} \Psi_{\eta(z)})] \qquad (*1)$$

We want to rewrite equation (*) of Lemma 5.25 in the form (*1) above.

To achieve this let us first write each $\pi = (t, M_1 z, M_2 z_2, \ldots)$ as $\pi = (t, M_1 z, \eta)$.

Hence

$$\Psi_\pi = M_1(\delta(z) \wedge \Psi_\eta) \qquad (\sharp)$$

Second, let us use ($\sharp$) and rewrite (*) into two conjuncts:

$$(*) = \varepsilon_t \wedge \bigwedge_{\pi(t)} \Psi_{\pi(t)} = \varepsilon_t \wedge \bigwedge_{\substack{z, \eta(z) \\ t\rho_1 z}} \Diamond(\delta(z) \wedge \Psi_{\eta(z)}) \wedge \bigwedge_{\substack{z, \eta(z) \\ t\rho_2 z}} \mathbb{P}(\delta(z) \wedge \Psi_{\eta(z)})$$

Now since δ is a system of path nominals (see item 1(c) of Remark 5.17): we can continue and rewrite

$$= \varepsilon_t \wedge \bigwedge_{\substack{z \\ t\rho_1 z}} \Diamond(\delta(z) \wedge \bigwedge_{\eta(z)} \Psi_{\eta(z)}) \wedge \bigwedge_{\substack{z \\ t\rho_2 z}} \mathbb{P}(\delta(z) \wedge \bigwedge_{\eta(z)} \Psi_{\eta(z)}) = (*1)$$

Remark 5.26. Note that the representation of the path nominals in Lemma 5.25 uses formulas of the positive (in $\mathbb{P}$, $\Diamond$) form

$$\Psi^{\delta}_{\pi} = M_1(\delta(x_1) \wedge (\cdots M_n\delta(x_n))\ldots).$$

We do not use $\neg M$ in Ψ_π. So we do not use formulas of the form

$$M_1(\delta(x_1) \wedge (\ldots(M_n(\delta(x_n) \wedge \neg M\delta(y))))\ldots)$$

We now want to give definitions and theorems which convert a finite bi-modal model $\mathbf{m}$ into an equivalent finite reactive model $\mathbf{m}^\sharp$ (as in Definition 5.12). We need to develop some definitions of path nominals for such models. So we will define the notions of legitimate paths and path nominals for the models of Definition 5.12.

Definition 5.27 (Legitimate paths). *Let* $\mathbf{m} = (S, R_1, R_2, \mathbf{f}, a, h)$ *be a* **KR** *model as in Definition 5.12. We define by induction the notion of a legitimate path* $\mathbf{t} = (a, M_1x_1, \ldots, M_nx_n)$*, where* $aR_1x_1 \wedge \ldots x_{n-1}R_1x_n$ *hold and* $M_i \in \{\Diamond, \mathbb{P}\}, i = 1, \ldots, n$*. We also define the functions* $\mathbf{f_t}$, $\mathbf{t}$ *a legitimate path.*

1. *(a) is a legitimate path and* $\mathbf{f}_{(a)} = \mathbf{f}$.
2. *Assume that* $\mathbf{t} = (a, M_1x_1, \ldots, M_nx_n)$ *is a legitimate path and assume that* $\mathbf{f}_{(a,x_1,\ldots,x_n)}$ *has been defined. Let y be such that* x_nR_1y *holds. Then* $(a, M_1x_1, \ldots, M_nx_n, \Diamond y)$ *is a legitimate path and let* $\mathbf{f}_{\mathbf{t}*(\Diamond y)} = \mathbf{f_t}$.
3. *Assume as in (2) that* $\mathbf{t} = (a, M_1x_1, \ldots, M_nx_n)$ *is a legitimate path and that* $\mathbf{f_t}$ *has been defined. Let y be such that* x_nR_1y *and that* $\mathbf{f_t}(x_n, y) = 1$*. Then* $\mathbf{s} = (a, M_1x_1, \ldots, M_nx_n, \mathbb{P}y)$ *is a legitimate path, and let* $\mathbf{f_s} = \mathbf{f_t}[x_m, y]$.
4. *A legitimate path is maximal for M if it has the form* $(a, M_1x_1, \ldots, M_nx_n, \neg My)$*. This means that* x_n *is an endpoint, as far as M is concerned.*
5. *Let* S^π *be the set of all legitimate paths of S.*

Definition 5.28 (Path nominals for reactive models). *Let* $\mathbf{m} = (S, R_1, R_2, \mathbf{f}, a, h)$ *be a finite* **KR** *model. Let* δ *be a function defined on* S^π *giving for each legitimate path* $\mathbf{t} = (a, M_1x_1, \ldots, M_nx_n)$ *a wff* $\delta_\mathbf{t}$ *of the logic such that the following hold:*

1. *Let y, z be such that* $x_nR_1y \wedge x_nR_1z \wedge y \neq z$*. Let* $\mathbf{s}_1 = \mathbf{t}*(My)$ *and* $s_2 = \mathbf{t}*(Mz)$, $M \in \{\Diamond, \mathbb{P}\}$*. Then* $\delta_{\mathbf{s}_1} \wedge \delta_{\mathbf{s}_2}$ *has no reactive model, i.e. it is semantically not consistent.*
2. $\mathbf{t} \vDash \delta_\mathbf{t}$ *in the model* $\mathbf{m_t} = (S, R_1, R_2, \mathbf{f_t}, \mathbf{t}, h)$.
3. *Note that from (1) and (2) above it follows that for* $\mathbf{t}_i = (a, M_1x_1, \ldots, M_ix_i), i = 1, \ldots, n$ *we have*

   $a \vDash M_1(\delta_{\mathbf{t}_1} \wedge M_2(\delta_{\mathbf{t}_2} \wedge (\ldots \wedge M_{(n-1)}(\delta_{\mathbf{t}_{n-1}} \wedge M_n(\delta_{\mathbf{t}_n} \wedge A))))\ldots)$
   in $\mathbf{m} = (S, R_1, R_2, \mathbf{f}, a, h)$ *iff* $\mathbf{t} \vDash A$ *in* $\mathbf{m_t} = (S, R_1, R_2, \mathbf{f_t}, \mathbf{t}, h)$.

In other words, $\delta_{\mathbf{t}_i}$ *act as nominals for the path* $\mathbf{t}$.

4. *It will be convenient for us in the future to use another function $\delta^*_{\mathbf{t}}$ associating with each* **t** *a distinct atomic formula $q_{\mathbf{t}}$. We can abuse notation and write '$\delta^*_{\mathbf{t}}$' for '$q_{\mathbf{t}}$'.*
The atoms $q_{\mathbf{t}}$ (or $\delta^*_{\mathbf{t}}$ if we abuse notation) are intended to act as path nominals.

Definition 5.29. *Let* $\mathbf{m} = (S, \rho_1, \rho_2, a, h)$ *be a reduced finite tree model based on a finite set of atoms. By Lemmas 5.24 and 5.25*

$$\delta(t) = \varepsilon_t \wedge \bigwedge_{\pi(t)} \Psi_{\pi(t)}$$

is a path nominal system for **m**. *Let* $R_1 = \rho_1 \cup \rho_2$. *We shall now define a set* R_2 *of double arrows for* **m**, *as well as an activity function* **f**.

Let $t \in S$ *and let*

$$\begin{aligned}\pi &= (t, \mathbb{P}x_1, M_2x_2, \ldots, M_rx_r)\\ \pi' &= (t, \Diamond x'_1, M_2x'_2, \ldots, M_rx'_r)\end{aligned}$$

for $r \geq 1$.

Assume for $i = 1, \ldots, r$

$$\delta(x_i) = \delta(x'_i)$$

We distinguish several cases.

(a) **Case 1**. *For some* y *such that* $x_r\rho_2 y$ *(and therefore* $t \vDash \Psi^\delta_\pi(\mathbb{P}\delta(y))$*) we have* $t \vDash \Psi^\delta_{\pi'}(\neg\mathbb{P}\delta(y))$.
In this case introduce the double arrow $(t \to x_1) \twoheadrightarrow (x_r \to y)$ *into* R_2.
(b) **Case 2**. $u\rho_1 v$ *holds.*
In this case let $\mathbf{f}(u \to v) = 0$.
So all ρ_1 *arrows are off.*
(c) **Note following Case 2**. *Note that since Case 1 introduces double arrows only to* ρ_2 *connections, the function* **f** *does not change its 'off' value on all* ρ_1 *connections!*
Consider now the model $\mathbf{m}^\sharp = (S, R_1, R_2, \mathbf{f}, a, h)$. *Call this model the reactive conversion of* **m**. *In this model we read* $\mathbb{P}$ *and* $\Diamond$ *as* $\oslash$ *and* $\Box$ *respectively in Definition 5.13.*

Remark 5.30. 1. **Part 1. The question.** Consider model **m** of Definition 5.29 and its reactive conversion $\mathbf{m}^\sharp$. Take a $t \in S$ of level k and a wff A of at most $n - k$ nested modalities.
Question. Do we have (♯1) below?
(♯1) $t \vDash A$ in **m** iff $t \vDash A$ in $\mathbf{m}^\sharp$.
To be able to prove this we need some properties (axioms) of $\Diamond, \mathbb{P}$ which will manifest themselves as properties of R_2. With the right axioms (♯1) might hold. Let us examine (♯1) more closely. It is clear immediately that (♯1) is not properly formulated. The model $\mathbf{m}^\sharp$ is a reactive model and so we cannot ask the simple question for $t \in S$, does $t \vDash A$ in $\mathbf{m}^\sharp$, we need to know the legitimate path leading to t. So let

$$\mathbf{t} = (a, M_1x_1, \ldots, M_rx_r)$$

where $x_r = t$. We ask does (♯2) hold?

(♯2) $\mathbf{t} \vDash A$ in $\mathbf{m}$ iff $\mathbf{t} \vDash A$ in $\mathbf{m}^\sharp$.

Note that the question is not simple. Consider Figure 5.24 describing a model $\mathbf{m}$ (we are not showing h):

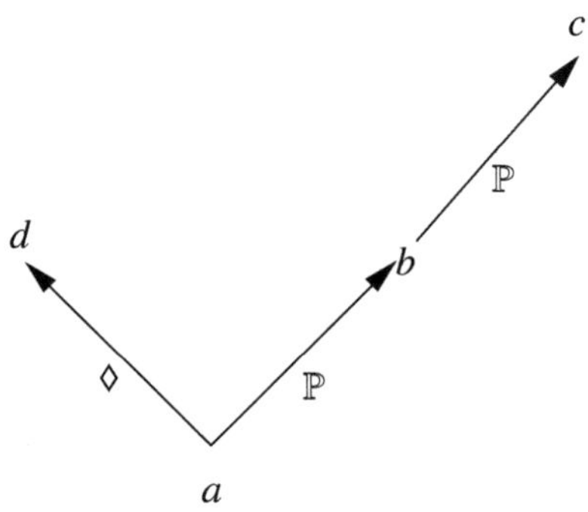

Fig. 5.24

The model $\mathbf{m}$ has two maximal paths

$$a \xrightarrow{\mathbb{P}} b \xrightarrow{\mathbb{P}} c$$

and

$$a \xrightarrow{\Diamond} d\,.$$

When we convert it to the model $\mathbf{m}^\sharp$ (we need to know the assignment h to convert and define R_2 and $\mathbf{f}$, but let us ignore that) we may get new paths, such as

$$a \xrightarrow{\Diamond} b \xrightarrow{\mathbb{P}} c$$
$$a \xrightarrow{\Diamond} b \xrightarrow{\Diamond} c$$
$$a \xrightarrow{\mathbb{P}} b \xrightarrow{\Diamond} c$$

and

$$a \xrightarrow{\mathbb{P}} d.$$

An examination of Case 2 of Definition 5.29 indicates that $a \xrightarrow{\mathbb{P}} d$ is not allowed but the other paths are allowed.

2. **Part 2. The answer.** What we need to show then are two items:

item (a) The double arrows in $\mathbf{m}^\sharp$ behave correctly.

item (b) The new paths obtained by allowing $\Diamond$ to traverse $\mathbb{P}$ connections for example, from t to s, (i.e. $t \xrightarrow{\mathbb{P}} s$), do not change the value of $\delta(t)$ at points t. Only the old paths (those already in $\mathbf{m}$) count for the value of $\delta(t)$.

Let us elaborate more on item (a) and item (b).
We start with item (a):
Let $\pi = (a, M_1 x_1, \ldots, \mathbb{P}x_j, \ldots, M_r x_r)$ be a fixed path which has $\mathbb{P}$ at the j position as indicated.
Let $\pi'_j = (a, M_1 x_1, \ldots, \Diamond x_j, \ldots, M_r x_r)$.
Let $x_r \rho_2 y$ hold.
Assume in the model $\mathbf{m}$ we have $a \vDash \Psi^{\delta}_{\pi}(\mathbb{P}\delta(y))$.
Let P^{π} be the set of all $j \in \{1, \ldots, r\}$ such that π has $\mathbb{P}$ in the j position.
Let $I, J \subseteq P^{\pi}$ be two disjoint sets such that $I \cup J = P^{\pi}$ and

a) $j \in J \to a \vDash \Psi^{\delta}_{\pi'_j}(\neg\mathbb{P}\delta(y))$

b) $j \in I \Rightarrow a \vDash \Psi^{\delta}_{\pi'_j}(\mathbb{P}\delta(y))$.

Then the number of elements in J is even.
This calls for the following axiom:

$$\Psi^{r}_{\pi}(\mathbb{P}\delta(y)) \wedge \bigwedge_{j\in J} \Psi^{\delta}_{\pi'_j}(\neg\mathbb{P}\delta(y)) \wedge \bigwedge_{j\in I} \Psi^{\delta}_{\pi'_j}(\mathbb{P}\delta(y)) \to \bot \qquad (\mathbb{A})$$

for any $\pi = (a, M_1 x_1, \ldots, M_r x_r)$ and where $x_r \rho_2 y$ holds and P^{π} and I, J are defined as above, and J has an odd number of elements.
The reason that J must be even is that if we traverse the path from a to t and trigger all double arrows as defined in Case 1 of Definition 5.29 then when we get to x_r we still have $x_r \vDash \mathbb{P}\delta_y)$ (since $x_r \rho_2 y$ holds). So the double arrows must cancel each other, so J must be even.
In fact, the situation is much more complex than that, as we shall now analyse.

3. **Part 3. The axiom.** Let $t \in S$. Let $\pi = (a, M_1 x_1, \ldots, M_r x_r)$ be the unique path in $\mathbf{m}$ leading up to x_r. Let $\pi^* = (a, M^*_1 x_1, \ldots, M^*_r x_r)$ be a legitimate path in $\mathbf{m}^\sharp$ leading to x_r such that π^* is obtained from π by changing some $\mathbb{P}$ into $\Diamond$.

   Assume that $x_r \xrightarrow{\mathbb{P}} y$ in $\mathbf{m}$. We want to make sure that $\mathbf{f}_{\pi^*}(x_r \to y) = 1$ in $\mathbf{m}^\sharp$. Figure 5.25 shows what we need to worry about

   The path π in $\mathbf{m}$ contains $x_1 \xrightarrow{\mathbb{P}} x_2$. This connection sends a double arrow to $x_r \xrightarrow{\mathbb{P}} y$ and also to $x_i \xrightarrow{\mathbb{P}} x_{i+1}$. If π^* goes through $x_1 \xrightarrow{\mathbb{P}} x_2$ in $\mathbf{m}^\sharp$ by using $\Diamond$, (i.e. $M^*_2 = \Diamond$), then the double arrows

$$(x_1 \to x_2) \twoheadrightarrow (x_i \to x_{i+1})$$

   and

$$(x_1 \to x_2) \twoheadrightarrow (x_r \to y)$$

   are not activated and no signal is sent to the arc $x_r \xrightarrow{\mathbb{P}} y$. So there are several possibilities of what can happen in $\mathbf{m}^\sharp$.

   a) $x_i \to x_{i+1}$ gets 'hit' by an odd number of active double arrows which were activated by π^* on its way to x_i, and is not active in $\mathbf{m}^\sharp$ and therefore the only way that π^* gets through to x_r is because it uses $x_i \xrightarrow{\Diamond} x_{i+1}$, (i.e. $M^*_{i+1} = \Diamond$).

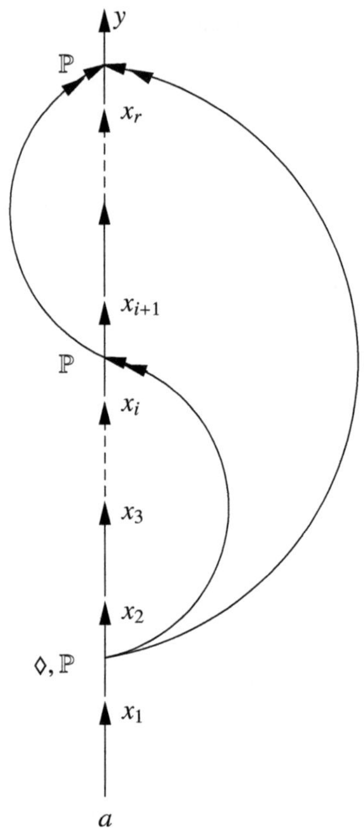

Fig. 5.25

b) $x_i \xrightarrow{\mathbb{P}} x_{i+1}$ is active in $\mathbf{m}^\sharp$ (receives an even number of double arrows) and π gets through $x_i \to x_{i+1}$ either as $\Diamond$ or as $\mathbb{P}$.

c) Depending on how π^* passes through $x_i \to x_{i+1}$ we will know whether the double arrow

$$(x_i \to x_{i+1}) \twoheadrightarrow (x_r \to y)$$

is activated in $\mathbf{m}^\sharp$ or not and so whether $x_r \xrightarrow{\mathbb{P}} y$ is 'hit' by it.

We need to show that the net effects in $\mathbf{m}^\sharp$ of all double arrows received by $x_r \to y$ from any active and $\mathbb{P}$-traversed $x_i \to x_{i+1}$ in π^* is that their number is even and therefore $x_r \to y$ is active in $\mathbf{m}^*$. We need to write an axiom to ensure that outcome.

We need some notation:

a) Given a path $\pi = (a, M_1x_1, \ldots, M_rx_r, \mathbb{P}y)$ let

$$\pi(i, j) = (x_i, M_{i+1}x_{i+1}, \ldots, M_jx_j)$$

$$\Psi^{\delta}_{\pi(i,j)}(\mathbb{P}\delta(x_{j+1})) = (\mathbb{P}(\delta(x_{i+1}) \wedge M_{i+2}(\delta(x_{i+2}) \wedge \ldots M_j(\delta(x_j) \wedge \mathbb{P}\delta(x_{j+1})))) \ldots)$$

$$\Psi^{*\delta}_{\pi(i,j)}(\neg\mathbb{P}\delta(x_{j+1})) = (\Diamond(\delta(x_{i+1}) \wedge M_{i+2}(\delta(x_{i+2}) \wedge \ldots \wedge M_j(\delta(x_j) \wedge \neg\mathbb{P}\delta(x_{j+1})))) \ldots)$$

Let E^{π} be all (i, j) such that $1 \leq i < k, 1 < j \leq k$ and there exists a double arrow $(x_i \to x_{i+1}) \twoheadrightarrow (x_j \to x_{j+1})$. This means that in **m** we have $x_i \vDash \alpha_{i,j(\pi,\pi^*)}$ where

$$\alpha_{i,j(\pi,\pi^*)} = \Psi^{\delta}_{\pi(i,j)}(\mathbb{P}\delta(x_{j+1})) \wedge \Psi^{*\delta}_{\pi(i,j)}(\neg\mathbb{P}\delta(x_{j+1}))$$

Let $\pi_i = (a, M_1x_1, \ldots, M_ix_i)$. Then we have $x_i \vDash \alpha_{i,j(\pi,\pi')}$ iff $a \vDash \Psi^{\delta}_{\pi_i}(\alpha_{i,j(\pi,\pi')})$. Let $\pi^* = (a, M^*_1x_1 \ldots M^*_rx_r, \mathbb{P}y)$.

Let $D^{\pi^*\pi}$ be the set of all i such that $x_i \xrightarrow{\mathbb{P}} x_{i+1}$ is in both π and in π^*. Define the predicate Active $(x_s \to x_{s+1}, \pi, \pi^*)$ by

- Active$((x_1 \to x_2), \pi, \pi^*)$
- Active$((x_{s+1} \to x_{s+2}), \pi, \pi^*)$ iff the set $\{i \mid 1 \leq i \leq s$ and Active$((x_i \to x_{i+1}), \pi, \pi^*)$ and $(i, s) \in E^{\pi}$ and $i \in D^{\pi,\pi^*}\}$ has an even number of elements.

To write an axiom $\mathbb{RA}$ expressing what is needed ($\mathbb{RA}$ for reactive axiom) we need to translate the predicate Active$(x_s \to x_{s+1}, \pi, \pi^*)$ into logic. This we do as follows: "**Active**" translates "Active"

- **Active**$((x_1 \to x_2), \pi, \pi^*) = \top$
- **Active**$((x_{s+1} \to x_{s+2}), \pi, \pi^*) =$
$$\bigwedge_{\substack{1 \leq i \leq s \\ i \in D^{\pi,\pi^*} \\ (i,s) \in E^{\pi}}} [\mathbf{Active}((x_i \to x_{i+1}), \pi, \pi^*) \wedge \Psi^{\delta}_{\pi_i}(\alpha_{i,s}(\pi, \pi^*))].$$
If the set $\{i \mid 1 \leq i \leq s$, such that $\delta(a) \vdash$ **Active**$((x_i \to x_{i+1}), \pi, \pi^*)$ in **m** and $(i, s) \in E^{\pi}$ and $i \in D^{\pi,\pi^*}\}$ has an even number of elements,
and
- **Active**$((x_s \to x_{s+1}), \pi, \pi^*) = \perp$ otherwise.
Note that in classical logic 'the set of true propositions $\{a_1, \ldots, a_k\}$ has an even number of elements' can be written as
$$[(\ldots(\neg a_1 \to \neg a_2) \to \neg a_3) \to \ldots) \to \neg a_n k)]$$

The axiom $\mathbb{RA}$ is

$$\mathbb{RA} = \bigwedge_{\pi^*,\pi} [\Psi^{\delta}_{\pi}(\mathbb{P}(\delta(y)) \to \mathbf{Active}((x_r \to y), \pi, \pi^*))]$$

Note that there is a restriction on π^* that it is only allowed to change some $\mathbb{P}x_i$ in π into a $\Diamond x_i$ in π^*.

Note that the axiom $\mathbb{RA}$ was defined using satisfaction in the model **m** through its system δ of path nominals. To write a syntactic axiom for the completeness proof we need to use variables for the syntactical description of **m** and define δ

syntactically. Let $\mu_{\mathbf{m}}, \delta_{\mathbf{m}}$ be the wffs doing this job. Then $\mu \to \mathbb{RA}(\mu)$ is an axiom for each μ. This is done in Section 5.3.
This will give us an infinite set of axioms for each syntactical finite tree with two successor functions.

4. **Part 4. Completeness.** To explain item (b), consider Figure 5.26

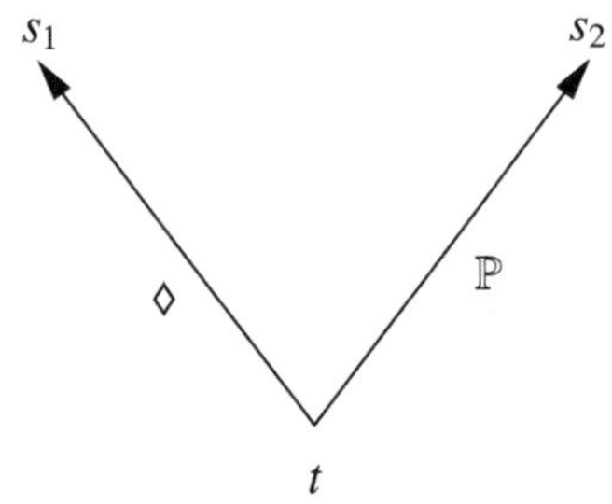

Fig. 5.26

Figure 5.26 describes the situation in model $\mathbf{m}$. In this model we have $t \vDash \mathbb{P}\delta(s_2)$ and $t \vDash \Diamond\delta(s_1)$. If in $\mathbf{m}^\sharp$ we allow $\Diamond$ to traverse also $t \xrightarrow{\mathbb{P}} s_2$, we may have a problem. Maybe we have that $t \vDash \neg\Diamond\delta(s_2)$ holds in $\mathbf{m}$ but if we allow $\Diamond$ to get to s_2 we will we get a contradiction? The answer is that we need not worry. In $\mathbf{m}^\sharp$, when we go to s_2 from t we trigger some double arrows and so we can prove that in $\mathbf{m}^\sharp$ $s_2 \nvDash \delta(s_2)$ (i.e. the model changes). We also show that if there are no double arrows emanating from $t \xrightarrow{\mathbb{P}} s_2$, then $t \vDash \neg\Diamond\delta(s_2)$ cannot be the case.
As for traversing $t \xrightarrow{\mathbb{P}} s_1$ in $\mathbf{m}^\sharp$, this has been forbidden by Case 2 of Definition 5.29.

We are now ready to state and prove our main lemma.

Lemma 5.31. *Let* $\mathbf{m}$ *and* $\mathbf{m}^\sharp$ *be two models as in Remark 5.30 and assume that* $\mathbf{m}$ *satisfies axiom* $\mathbb{RA}$ *of this remark.*

Then the following condition ($\sharp$) *holds for any path* $\pi = (a, M_1x_1, \ldots, M_rx_r)$ *in* $\mathbf{m}$ *and path* π^* *in* $\mathbf{m}^\sharp$

$$\pi^* = (a, M_1^*x_1, \ldots, M_r^*x_r)$$

where π *and* π^* *are related by the condition :* $M_i^* = \mathbb{P} \Rightarrow M_i = \mathbb{P}$.

($\sharp$) $\quad x_r \vDash \delta(x_r)$ *in* $\mathbf{m}$ *iff* $x_r \vDash \delta(x_r)$ *in* $\mathbf{m}^\sharp_{\pi^*} = (S, R_1, R_2, \mathbf{f}_{\pi^*}, a, h)$

where $\mathbf{f}_{\pi^*} = \mathbf{f}[a, x_1][x_1, x_2] \ldots [x_{r-1}, x_r]$, *as defined in Definition 5.12.*

Proof. By induction on $n - r$, where n is the level of a.

Case $(n-r) = 0$
In this case $\delta(x_r)$ is a classical model (i.e. ε is a conjunctive normal form of atoms from $\mathbb{Q}$) and so ($\sharp$) holds.

Case $(n-r) = 1$
In this case $\delta(x_r)$ has the form

$$\varepsilon \wedge \bigwedge_i \Diamond(\varepsilon_i \wedge \mathbb{P}\varepsilon_i) \wedge \bigwedge_j \Diamond(\eta_j \wedge \neg\mathbb{P}\eta_j)$$

where ε_i and η_j are classical models.

The same holds in $\mathbf{m}^\sharp$ because $\mathbb{P}A \to \Diamond A$ holds for A not containing $\mathbb{P}$ and in $\mathbf{m}^\sharp$ all arrows of the form $u \xrightarrow{\Diamond} v$ are off.

See Case 2 of Definition 5.29 and the Note there following case 2.

Case $(n-r) > 1$
The cases to check are those of $\Diamond$ and $\mathbb{P}$.
Subcase of $\Diamond$
Let $x_r\rho_1 y$ hold in $\mathbf{m}$. Then $x_r \vDash \Diamond\delta(y)$ in $\mathbf{m}$. Hence $y \vDash \delta(y)$ in $\mathbf{m}$ and by the induction hypothesis, $y \vDash \delta(y)$ in $\mathbf{m}^\sharp_{(\pi^*,\Diamond y)}$.

Since in $\mathbf{m}^\sharp$ we have $x_r R_1 y$, we also get that $x_r \vDash \Diamond\delta(y)$ in $\mathbf{m}^\sharp_{\pi^*}$.

Assume now that $x_r\rho_2 z$ holds in $\mathbf{m}$ and that $x_r \vDash \neg\Diamond\delta(z)$ in $\mathbf{m}$ and of course $x_r \vDash \mathbb{P}\delta(z)$ in $\mathbf{m}$. Look at x_r from the point of view of $\mathbf{m}^\sharp_{\pi^*}$; we see that $x_r R_1 z$ holds and therefore we have to ensure that we do *not* have that $\delta(z)$ holds at z in $\mathbf{m}^\sharp_{(\pi^*,\Diamond z)}$.

Note that the models $\mathbf{m}^\sharp_{(\pi^*,\Diamond z)}$ and $\mathbf{m}^\sharp_{(\pi^*,\mathbb{P}z)}$ may not be the same because in the model $\mathbf{m}^\sharp_{(\pi^*,\mathbb{P}z)}$ the activity function is $\mathbf{f}_{\pi^*}[x_r, z]$, while in the model $\mathbf{m}^\sharp_{(\pi^*\Diamond z)}$ the activity function is $\mathbf{f}_{\pi^*}$. We know by the induction hypothesis that in the model $\mathbf{m}^\sharp_{(\pi^*,\mathbb{P}z)}$ we have that $z \vDash \delta(z)$.

The question is does $z \vDash \delta(z)$ hold in $\mathbf{m}^\sharp_{(\pi^*,\Diamond z)}$? We want it *not* to hold.

This we now proceed to prove. We consider two possibilities for the arc $x_r \xrightarrow{\mathbb{P}} z$.
Possibility 1
There is at least one double arrow emanating from $x_r \xrightarrow{\mathbb{P}} z$ into the model $\mathbf{m}^\sharp_{(\pi^*,\mathbb{P}z)}$. This means that for some $(M_1w_1, \ldots, M_sw_s)$ we have $w_s\rho_1 y$ and $\Psi^\delta_{(x_r,\mathbb{P}z,M_1w_1,\ldots,M_sw_s)}(\mathbb{P}\delta(y))$ holds at x_r while $\Psi^\delta_{(x_r,\Diamond z,M_1w_1,\ldots,M_sw_s)}(\neg\mathbb{P}\delta(y))$ also holds at x_r.

We can assume that there is no double arrow of the form $(x_r \to z) \to (w_i \to w_{i+1})$ for $i < s$. This ensures that s is minimal and we don't have to worry about other double arrow changes along the way from x_r to w_s!

This means by Definition 5.29 that $(x_r \to z) \twoheadrightarrow (w_s \to y) \in R_2$.

We now consider $\delta(x_r)$. According to Lemma 5.25, we have that

$$\delta(x_r) \vdash \Diamond\Psi^\delta_{(M_1w_1,\ldots,M_sw_s)}(\neg\mathbb{P}\delta(y))$$

because

$$x_r \vDash \Psi^{\delta}_{(x_r,\Diamond z,M_1w_1,\ldots,M_sw_s)}(\neg\mathbb{P}\delta(y))$$

in $\mathbf{m}$. Thus $\delta(z) \vdash \Psi^{\delta}_{(M_1w_1,\ldots,M_sw_s)}(\neg\mathbb{P}\delta(y))$.

On the other hand in the model $\mathbf{m}^{\sharp}_{(\pi,\Diamond z)}$ we have that $z \vDash \Psi^{\delta}_{(M_1w_1,\ldots,M_sw_s)}(\mathbb{P}\delta(y))$ because the double arrow from $x_r \to z$ to $w_s \to y$ is ignored by $\Diamond$.

Therefore in the model $\mathbf{m}^{\sharp}_{\pi^*}$ we do not have $x_r \vDash \Diamond\delta(z)$.

Possibility 2

No double arrows emanate from $x_r \to z$. In this case for all $(M_1w_1,\ldots,M_sw_s)$ $x_r \vDash \Psi_{(x_r,\mathbb{P}z,M_1w_1,\ldots,M_sw_s)}(\mathbb{P}\delta(y))$ in $\mathbf{m}$ iff in $\mathbf{m}$ also $x_r \vDash \Psi_{(x_r,\Diamond z,M_1w_1,\ldots,M_sw_s)}(\mathbb{P}\delta(y))$.

But this means that at x_r, $\Diamond\delta(z) = \mathbb{P}\delta(z)$ in $\mathbf{m}$ so $x_r \vDash \Diamond\delta(z)$ (see Remark 5.26) contrary to our assumption.

Subcase of $\mathbb{P}$

We now examine the case of $\mathbb{P}$. Assume $x_r\rho_1 z$ and assume $x_r \vDash \neg\mathbb{P}\delta(z)$ in $\mathbf{m}$. This means that for all y such that $x_r\rho_2 y$ we have $y \vDash \neg\delta(z)$ in $\mathbf{m}$. Hence by the induction hypothesis, $y \vDash \neg\delta(z)$ in $\mathbf{m}^{\sharp}_{(\pi^*,\mathbb{P}y)}$. Hence we have that in $\mathbf{m}^{\sharp}_{\pi^*}$, $x_r \vDash \neg\mathbb{P}\delta(z)$. The reason for the last step is that any y such that $x_r\rho_1 y$ is not accessible for $\mathbb{P}$, i.e. such an arc $x_r \xrightarrow{\Diamond} y$ is off in $\mathbf{m}^{\sharp}$.

Now assume $x_r\rho_2 z$ in $\mathbf{m}$. Then $x_r \vDash \mathbb{P}\delta(z)$ holds in $\mathbf{m}$. Hence $z \vDash \delta(z)$ holds in $\mathbf{m}$. By the induction hypothesis, $z \vDash \delta(z)$ holds in $\mathbf{m}^{\sharp}_{(\pi^*,\mathbb{P}z)}$. If the connection $x_r \xrightarrow{\mathbb{P}} z$ is on in $\mathbf{m}^{\sharp}_{\pi^*}$ (i.e. $\mathbf{f}_{\pi^*}(x_r \xrightarrow{\mathbb{P}} z) = 1$), then we will get that in $\mathbf{m}^{\sharp}_{\pi^*}$, $x_r \vDash \mathbb{P}\delta(z)$.

How can it be the case that $\mathbf{f}_{\pi^*}(x_r \to z) = 0$? This can happen if along the path $\pi = (a, M_1x_1,\ldots,M_rx_r)$ an odd number of double arrows was triggered towards the arc $x_r \to z$. To see what this means we go to Case 1 of Definition 5.29. But then let us also consider Remark 5.30 and axiom $\mathbb{RA}$. We will be in violation of axiom $\mathbb{RA}$ if an odd number of double arrows hits $x_r \to z$. The axiom says that $x_r \to z$ must be active. This completes the induction case and the Lemma is proved.

We now have almost all that we need for the completeness proof. We basically take axiom $\mathbb{RA}$ and construct a Henkin type reduced model $\mathbf{m}$, convert it to $\mathbf{m}^{\sharp}$ and use Lemma 5.31.

This we shall do in the next Section 5.3.

5.3 Completeness theorem for KR

This section provides an axiom system and a completeness theorem for the logic **KR** as defined semantically in Definition 5.12. Our strategy is as follows:
Given a formula A, we want to construct a model for it. A has a finite number of atoms and no more than n nested modalities, and so we expect to construct a finite model of level n with an assignment to the atoms of A. We need to find the right axioms which will enable us to carry out the construction of the correct model.

We begin by syntactically defining all finite models. This is done in Definition 5.32. We also define syntactically a natural system of path nominals for each model and using it we can talk syntactically about paths π and their associated paths π^*.

Armed with this syntactical capability, we can write axiom 3 of Definition 5.35. The formulation of this axiom was discussed at length on page 151 in part 3 of Remark 5.30.

We mention that this axiomatisation is brute force (see discussion below in Remark 5.38), we cannot get around this without possibly adding more connectives.

The reader should note that there is a difference of culture here between the pure mathematical logician and the applied logician. The pure logician would tend to axiomatise with the minimum number of connectives and the minimum number of variables used. The Polish school of logic, for example, is well known for its minimal axiomatisations of classical logic and its variants, an ongoing activity which was slowed down only by the Second World War. See as a sample [152].

The applied logician would be more inclined to add connectives and constants as long as they have a natural meaning in the application area. We shall address this possibility in the concluding section.

5.3.1 Methodological presentation of the axioms for KR

Definition 5.32 (Syntactical models). *Let $\mathbb{Q} = \{q_1, \ldots, q_k\}$ be a set of atoms and let S be an infinite set of names of possible worlds. We shall define inductively the family of all n-level syntactical models together with their possible world realisation and a path nominal function δ for each such model.*

1. *Any conjunctive normal form of the form*

$$\mu_0 = \varepsilon = \bigwedge_{q \in \mathbb{Q}} \pm q$$

is a syntactical level 0 model. For each such ε let $\mathbf{n}_\varepsilon = (t_\varepsilon, h)$ be a corresponding level 0 model, with $t_\varepsilon \in S$ and $h(t_\varepsilon, q) = 1$ iff $\varepsilon \vdash q$. Assume $\varepsilon \neq \varepsilon' \Rightarrow t_\varepsilon \neq t_{\varepsilon'}$. Let $\delta(t_\varepsilon) = \varepsilon$.

2. *The syntactical models of level 1 have the form*

$$\mu = \mu_1 = \varepsilon_1 \wedge \bigwedge_{\varepsilon \in I} \Diamond\varepsilon \wedge \bigwedge_{\eta \in J} \mathbb{P}\eta \wedge \bigwedge_{\varepsilon \notin I} \neg\Diamond\varepsilon \wedge \bigwedge_{\eta \notin J} \neg\mathbb{P}\eta$$

where ε_1 is a level 0 syntactical model and ε, η range over level 0 syntactical models and J, I are sets of level 0 syntactical models.

The corresponding level 1 reduced model realisation of μ_1 is defined as follows

$$\mathbf{n}_\mu = (S_\mu, \rho_1, \rho_2, a_\mu, h)$$

where

$$S_\mu = \{a_\mu\} \cup \{t_\varepsilon \mid \varepsilon \in I\} \cup \{s_\varepsilon \mid \varepsilon \in J\}.$$

We assume $a_\mu, t_\varepsilon, s_\varepsilon$ are all different elements of S.

We let $a_\mu \rho_1 t_\varepsilon, \varepsilon \in I$ and $a_\mu \rho_2 s_\varepsilon, \varepsilon \in J$.

Let

$$h(a, q) = 1 \textit{ iff } \varepsilon_1 \vDash q$$
$$h(t_\varepsilon, q) = 1 \textit{ iff } \varepsilon \vDash q$$
$$h(s_\varepsilon, q) = 1 \textit{ iff } \varepsilon \vDash q$$

Let

$$\delta(t_\varepsilon) = \varepsilon$$
$$\delta(s_\varepsilon) = \varepsilon$$
$$\delta(a_\mu) = \mu$$

3. *Assume the syntactical models are defined for any level $\leq m$. Also assume that for any such syntactical model μ a realisation model $\mathbf{n}_\mu = (S_\mu, \rho_1, \rho_2, a_\mu, h)$ is also defined with*

$$\mu_1 \neq \mu_2 \Rightarrow S_{\mu_1} \cap S_{\mu_2} = \varnothing$$

Let I, J be any sets of syntactical models of level $\leq m$. Assume at least one model in $I \cup J$ is of level m and not of any level $< m$.
Define the level $m + 1$ model μ as follows

$$\mu = \varepsilon_\mu \wedge \bigwedge_{\alpha \in I} \Diamond\alpha \wedge \bigwedge_{\beta \in J} \mathbb{P}\beta \wedge \bigwedge_{\alpha \notin I} \neg\Diamond\alpha \wedge \bigwedge_{\beta \notin J} \neg\mathbb{P}\beta.$$

Define the level $m + 1$ realisation model for μ as follows:

$$\mathbf{n}_\mu = (S_\mu, \rho_1, \rho_2, a_\mu, h).$$

For each $\alpha \in I, \beta \in J$ let $\mathbf{n}_\alpha = (S^I_\alpha, \rho^\alpha_1, \rho^\alpha_2, a_\alpha, h_\alpha)$ and $\mathbf{n}_\beta = (S^J_\beta, \rho^\beta_1, \rho^\beta_2, a_\beta, h_\beta)$ be their respective realisation models. Assume all S^I_α, S^J_β are pairwise disjoint. Also assume $\delta_\alpha, \delta_\beta$ are given on $\mathbf{n}_\alpha, \mathbf{n}_\beta$. Let $S_\mu = \{a_\mu\} \cup \bigcup_{\alpha \in I} S^I_\alpha \cup \bigcup_{\beta \in J} S^J_\beta$ where a_μ is a new label from S.
Define

$$a_\mu \rho_1 a_\alpha, \alpha \in I$$
$$a_\mu \rho_2 a_\beta, \beta \in J.$$

$\rho_i, i = 1, 2$ is the same as ρ^α_i on S^I_α, for $\alpha \in I$, and the same as ρ^β_i on S^J_β for $\beta \in J$.
Let $h(a_\mu, q) = 1$ iff $\varepsilon_\mu \vDash q$ and let h be h_α on S^I_α and h_β on S^J_β.
Let δ_μ be δ_α on S^I_α, δ_β on S^J_β and let $\delta(a_\mu) = \mu$.

Lemma 5.33. *Let μ be a syntactical model and let $\mathbf{n}_\mu$ be its realisation as defined in Definition 5.32. Let δ_μ be its path nominal system as defined in Definition 5.32.*

Then δ_μ is indeed a path nominal function according to Definition 5.16 and Lemma 5.25 holds for δ_μ.

Proof. Simple, by induction.

Lemma 5.34. *Let $\mathbf{m} = (S, \rho_1, \rho_2, a, h)$ be a reduced model of level n for $\mathbb{Q}$. Let μ be a syntactical model for $\mathbb{Q}$. Assume that $a \vDash \mu$. Let $\mathbf{n}_\mu$ be the realisation model for μ. Then $\mathbf{m}$ is isomorphic to $\mathbf{n}_\mu$.*

Proof. By induction on the level of $\mathbf{m}$.

Definition 5.35 (Axiom system for KR). *Consider the following axioms for* **KR**.

1. *Modal K axioms and rules for* $\mathbb{P}$ *and* $\Diamond$.
2. $\mathbb{P}A \to \Diamond A$ *for* A *not containing* $\mathbb{P}$.
3. *For each syntactical model* μ, *let* $\mathbb{RA}(\mu)$ *be the axiom as defined on page 151 in Part 3 of Remark 5.30. Then our axiom is* $\mu \to \mathbb{RA}(\mu)$.

Remark 5.36. Note that by axiom (2) $\mathbb{P}q \to \Diamond q$ is a theorem for atomic q. Thus $\mu = q \wedge \mathbb{P}q \wedge \neg\Diamond q$ is an inconsistent level 1 syntactical model for $\{q\}$. This does not affect axiom 3 because μ is in the antecedent!

Remark 5.37. Let us comment on axiom 2 of our logic:

1. If A is a wff without $\mathbb{P}$, then $\mathbb{P}A \to \Diamond A$. This is not true for arbitrary A. Consider Figure 5.27.

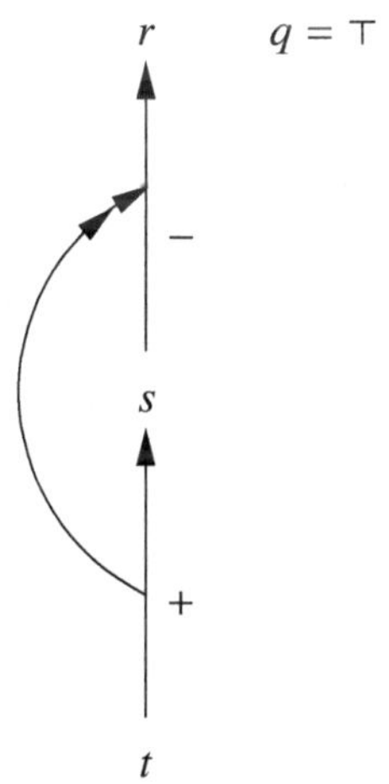

Fig. 5.27

In this figure, $s \to r$ is off, so $\Diamond\mathbb{P}q$ is false at t, however, $\mathbb{P}\mathbb{P}q$ is true at t, because the double arrow $(t \to s) \twoheadrightarrow (s \to r)$ is triggered by $\mathbb{P}$ and so it activates the connection $s \to r$ and hence $s \vDash \mathbb{P}q$.

Remark 5.38. The perceptive reader will have noticed that our axiomatisation of **KR** is a 'brute-force' axiom system. Our 'axioms' simply describe syntactically the correct semantics. Perhaps the reader would have expected a nicer axiom system. This is not possible. We shall offer an explanation by giving an example.

Consider ordinary modal logic **K** with a single modality $\Diamond$. Its models have the form $(S, R, a), a \in S, R \subseteq S \times S$ and R is not necessarily reflexive. The truth condition for $\Diamond$ is (1).

1. $t \vDash \Diamond A$ iff for some s, tRs and $s \vDash A$.

Let us now change the semantics a bit. Consider models of the form (S, E, R, a, h), with $E \subseteq S$. We change the truth condition to be

2.
$$t \vDash \Diamond A \text{ iff } \begin{cases} t \notin E \text{ and for some } s, tRs \text{ and } s \vDash A \\ \text{and} \\ t \in E \text{ and } t \vDash A \text{ or for some } s, tRs \text{ and } s \vDash A \end{cases}$$

Thus the set E tells us that we want reflexivity.

Now define a family of finite models (S, E, R, a, h) as follows.

Take any finite model $\mathbf{m} = (S, R, a, h)$. Code this model by a Turing machine. Apply an effective procedure to extract a set $E_\mathbf{m} \subseteq S$, and define the new model $\mathbf{n} = (S, E_\mathbf{m}, R, a, h)$.

Let the logic $\mathbf{K}_R$ be defined by the set of all such models $\mathbf{n}$.

Now I ask the reader: how can you axiomatise this logic without syntactially describing the models $\mathbf{n}$?

If the definition of $\mathbf{n}$ is simple we can probably do it nicely but otherwise brute force is needed.

For example, we can let

$$t \in E \text{ iff } \exists x(xRt \wedge \neg xRx).$$

Then maybe we can axiomatise this logic nicely.

In the case of **KR** we allow double arrows to go anywhere and we can switch connections on and off. This is too free to control nicely, so brute force is called for. Our difficulty is compounded by the fact that we do not have any modality or other special connective controlling the double arrows, so our only means to describe their behaviour is through the coding of the changes they cause in the model, hence the need for a brute force axiomatisation.

Here are some simple conditions on reactivity which can make for nicer logics.

Next step reactivity
If $(x \to y) \twoheadrightarrow (w \to z) \in R_2$ then $y = w$. I.e. we are allowed double arrows only in the form of Figure 5.28

Forward look reactivity
The target of a subsequent arrow must be above the target of a previous arrow. Namely if

$$(x \to y) \twoheadrightarrow (u \to v) \in R_2$$

and

$$(x' \to y') \twoheadrightarrow (u' \to v') \in R_2$$

and for some $m \geq 1$

$$x\rho^m x'$$

Then for some $n \geq 0$

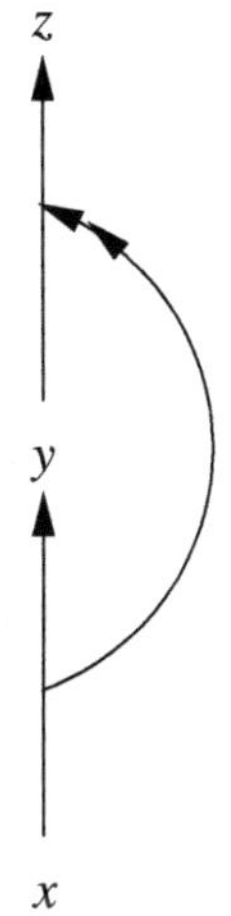

Fig. 5.28

$$u \rho^n u'$$

In words: a later arrow may not 'hit' an earlier target.

Figure 5.29 illustrates the idea.

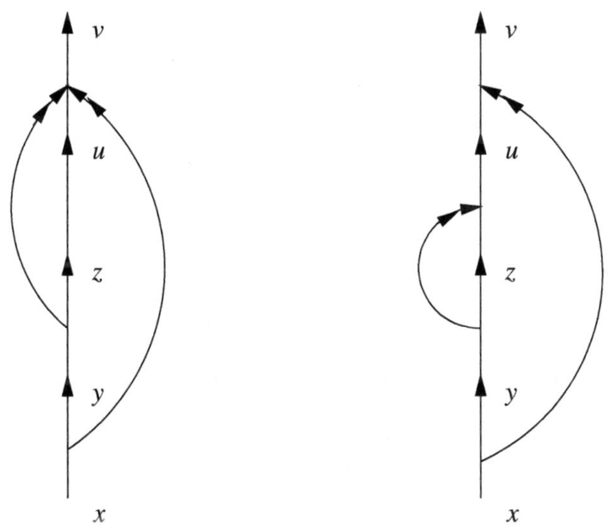

Fig. 5.29

5.3.2 The canonical model

We now construct a canonical finite reduced model for two modalities $\mathbb{P}$ and $\Diamond$ and then convert it to a **KR** model. We shall construct a finite model for any consistent formula A. This will also give us a finite model property and decidability.

Our starting point is to construct a very specific bi-modal finite model with two accessibility relations, one for $\Box$ and the other for $\oslash$. Additional axioms will allow us to convert this special model into a reactive one.

We need some definitions.

1. A wff A is said to be of level n if n is the maximal number of nested $\{\mathbb{P}, \Diamond\}$ in A.
2. Let $\mathbb{Q} = \{q_1, \ldots, q_k\}$ be a set of atoms, then $\mathbb{Q}^n$ is the set of all wffs based on $\mathbb{Q}$ of level $\leq n$.

Let A be a consistent wff. We want to construct a finite model for A. Let $q_1, \ldots, q_k$ be all atomic formulas appearing in A and let n be the maximal number of nested modal operators $\{\mathbb{P}, \Diamond\}$ in A.

Step 1
We first extend A to a theory Θ containing A satisfying the following conditions:

1. For any wff B built up from $\{q_1, \ldots, q_k\}$ and containing no more than n nested modalities we have $B \in \Theta$ or $\neg B \in \Theta$.
2. Θ is **KR** consistent. We shall construct a finite model for Θ.

Step 2a
Let $\mathbb{P}D \in \Theta$.

We say E_D of level $\leq n - 1$ is maximal for $\mathbb{P}D \in \Theta$, iff $\mathbb{P}(D \wedge E_D) \in \Theta$ and whenever $\mathbb{P}X \in \Theta$ and X is of level $\leq n-1$ and $x \vdash D \wedge E_D$, we also have $D \wedge E_D \vdash X$.

We now show that such an E_D exists.

Lemma 5.39. *Let $\mathbb{P}Y \in \Theta$ and let X be any wff of level $\leq n-2$. Then either $\mathbb{P}(Y \wedge X) \in \Theta$ or $\mathbb{P}(Y \wedge \neg X) \in \Theta$.*

Proof. If neither is in Θ then $\neg\mathbb{P}(Y \wedge X) \wedge \neg\mathbb{P}(Y \wedge \neg X)$ is in Θ. Therefore

$$\mathbb{P}(Y \to \neg X) \wedge \mathbb{P}(Y \to X) \in \Theta$$

and hence $\mathbb{P}\neg Y \in \Theta$ contradicting the consistency of Θ.

From the lemma it follows that we can inductively add $\pm X_{m+1}$ to $\mathbb{P}(Y \wedge \pm X_1 \wedge \ldots \wedge \pm X_m)$ and hence such a maximal E_D exists for any $\mathbb{P}D \in \Theta$.

This is so because assume we added Y_i such that $\mathbb{P}(Y \wedge \bigwedge Y_i) \in \Theta$ and for every Z of level $\leq n - 1$, there exists an i such that $Y_i = Z$ or $Y_i = \neg Z$.

Now assume $\mathbb{P}U \in \Theta$ and $U \vdash Y \wedge \bigwedge Y_i$.

Since U is of level $\leq n - 1$ then for some Y_i, either $Y_i = U$ or $Y_i = \neg U$. If $Y_i = U$ then $Y \wedge \bigwedge_i Y_i \vdash U$ as required.

If $Y_i = \neg U$ then we get that $U \vdash \neg U$, i.e. $U \vdash \bot$ and so $\mathbb{P}\bot \in \Theta$. This is not possible.

Let $\mathbb{P}(D^i \wedge E^i_D) \in \Theta$ be the entire list of all such maximal wffs, $i = 1, \ldots, r_n$. Note that the index 'n' in 'r_n' is the level n of the theory Θ. In the sequel we shall use theories of level $n - m, m = 1, 2, \ldots$ and we shall respectively use $r_{n-m}, m = 1, 2 \ldots$. Similarly we shall introduce below $s_n, s_{n-1} \ldots$

Step 2b

Similarly suppose we have that $\Diamond B \in \Theta$, then there exists a maximal E_B such that $\Diamond(B \wedge E_B) \in \Theta$.

We need another lemma.

Lemma 5.40. *Let $\Diamond B \in \Theta$ and let x be any wff of level $\leq n-1$. Then either $\Diamond(B \wedge X) \in \Theta$ or $\Diamond(B \wedge \neg X) \in \Theta$.*

Proof. Otherwise

$$\neg\Diamond(B \wedge X) \wedge \neg\Diamond(B \wedge \neg X) \in \Theta$$

hence $\Box(B \to X) \wedge \Box(B \to \neg X) \in \Theta$ and hence $\Box\neg B \in \Theta$ contradicting the consistency of Θ.

We can thus let $\Diamond(B_j \wedge E_{B_j}) \in \Theta$ enumerate all the wffs of this form which are maximal and are in Θ, for $j = r_n + 1, \ldots, r_n + s_n$.

Lemma 5.41. *If $\Diamond X \in \Theta$ and $\Diamond Y \in \Theta$ and X and Y are maximal then either $X \equiv Y$ or ($X \nvdash Y$ and $Y \nvdash X$).*

Proof. If $Y \vdash X$ then X and Y are consistent. We get by maximality of X that $X \vdash Y$, hence $X \equiv Y$. Similarly for $\mathbb{P}X, \mathbb{P}Y$.

The above two step considerations prove the following.

Lemma 5.42. *Let $\mathbb{Q}$ be finite and let Θ be any complete theory of the language $\mathbb{Q}^n$.*

Then there exist $D_1^{n-1}, \ldots, D_{r_n}^{n-1}, B_{r_n+1}^{n-1}, \ldots, B_{r_n+s_n}^{n-1}$ of level $\leq n - 1$ such that the following hold

1. $\mathbb{P}D_i^{n-1} \in \Theta, i = 1 \ldots r_n$
2. $\Diamond B_j^{n-1} \in \Theta, j = r_n + 1, \ldots, r_n + s_n$
   a) $\mathbb{P}C_k^{a-1} \wedge \Diamond C_k^{n-1} \in \Theta,\ s_n + 1 \leq k \leq w_n$
3. For any D such that $\mathbb{P}D \in \Theta$, there exists an i such that $D_i^{n-1} \vdash D$.
4. For any B such that $\Diamond B \in \Theta$, there exists a j such that $B_j^{n-1} \vdash B$.
5. For any two non-equivalent $Y, X \in \{D_i^{n-1}, B_j^{n-1}\}$ we have $Y \nvdash X$ and $X \nvdash Y$.

Step 3

Consider the situation of Lemma 5.42.

1. For each $\mathbb{P}D_i^n \in \Theta$ let $\Delta_i^{n-1} = \{D_i^{n-1}\} \cup \{u \mid U$ of level $\leq n - 1$ and $\varnothing U \in \Theta\}$. This set is consistent and can be extended to a complete theory Θ_i^{n-1} in the language $\mathbb{Q}^{n-1}$.
2. Similarly for any $\Diamond B_j^{n-1} \in \Theta$ we let $\Delta_j^{n-1} = \{B_j^{n-1}\} \cup \{U \mid U$ of level $\leq n - 1$ and $\Box U \in \Theta\}$.

   This set is consistent and can be extended to a complete and consistent theory Θ_j^{n-1} of $\mathbb{Q}^{n-1}$.

Step 4
We start with a complete and consistent theory Θ of $\mathbb{Q}^n$ and construct a finite tree model for it as follows. $(S, \rho_\diamond, \rho_\mathbb{P}, 0)$ is the tree. We first define S and $R = \varphi_\diamond \cup \varphi_\mathbb{P}$ in steps.

Construction step 4.0
Let $0 \in S$ and let $\Delta_0 = \Theta$.

Construction step 4.1
Consider the situation described in Lemma 5.42 for Θ. Let $(0, i), i = 1, \ldots, r_n + s_n$ be new elements which we put into S and let

$$\Delta_{(0,i)} = \Theta_i^{n-1}, i = 1, \ldots, r_n$$
$$\Delta_{(0,j)} = \Theta_j^{n-1}, j = r_n + 1, \ldots, r_n + s_n$$

Construction step 4.m for $1 < m \leq n$
Assume S contains a sequence $\mathbf{t} = (0, x_1, \ldots, x_m)$ and that a theory $\Theta^{n-m} = \Delta^{n-m}_{(0,x_1,\ldots,x_m)}$ has been defined.

We repeat the construction of step 4.1 for the theory Θ^{n-m} using the indices r_{n-m} and s_{n-m} and define respective sequences

$$\mathbf{t} * (i), i = 1, \ldots, r_{n-m}$$

and

$$\mathbf{t} * (j), = r_{n-m} + 1, \ldots, r_{n-m} + s_{n-m}$$

and put them in S. We also define theories $\Delta^{n-m-1}_{\mathbf{t}*(i)}$ and $\Delta^{n-m-1}_{\mathbf{t}*(j)}$ using Θ_i^{n-m} and Θ_j^{n-m} respectively as done in Step 4.1.

The construction stops at $m = n - 1$. We get a set S of sequences $\mathbf{t} \in S$. The set forms a tree under the relation $\mathbf{t}R\mathbf{s}$ iff $\exists x(\mathbf{s} = \mathbf{t} * (x))$. With each $\mathbf{t}$ of length m we have a theory $\Delta_\mathbf{t}^{n-m}$.

For $\mathbf{t} = (0, x_1, \ldots, x_m)$ let $\mathbf{t}\rho_\mathbb{P}\mathbf{s}$ hold when $\mathbf{s} = \mathbf{t} * (i)$ for $1 \leq i \leq r_{n-m}$ (i.e. $\mathbf{s}$ was created because of $\mathbb{P} * D_i^{n-m}$ in $\Delta_\mathbf{t}^{n-m}$). $\mathbf{t}\rho_\diamond\mathbf{s}$ iff $\mathbf{s} = \mathbf{t} * (j)$ for $r_{n-m} + 1 \leq j \leq s_{n-m}$ (i.e. s was created because of $\diamond B_j^{n-m} \in \Theta$).

Define an assignment h as follows: $\mathbf{t} \vDash q$ iff $q \in \Delta_\mathbf{t}$.

Lemma 5.43. *In the model* $(S, \rho_\diamond, \rho_\mathbb{P}, a, h)$ *we have for any wff of level* $\leq n - m$ *and any* $\mathbf{t}$ *of length* m

$$\mathbf{t} \vDash A \textit{ iff } A \in \Delta_\mathbf{t}.$$

Proof. By induction on A.

For atoms this holds.

Consider the case of $\mathbb{P}A$.

1. Assume $\mathbb{P}A \in \Delta_\mathbf{t}^{n-m}$. Then for some D_i^{n-m-1} we have $D_i^{n-m-1} \vdash A$ and $\mathbb{P}D_i^{n-m-1} \in \Delta_\mathbf{t}^{n-m}$. Hence $D_i^{n-m-1} \in \Delta^{n-m-1}_{\mathbf{t}*(i)}$. By the induction hypothesis $\mathbf{t} * (i) \vDash D_i^{n-m-1}$ and therefore $\mathbf{t} * (i) \vDash A$.

2. Assume $\oslash A \in \Delta_{\mathbf{t}}^{n-m}$. Then by construction for any i, $1 \leq i \leq r_{n-m}$ we have $A \in \Delta_{\mathbf{t}*(i)}^{n-m-1}$ and so $\mathbf{t} * (i) \vDash A$, and hence $\mathbf{t} \vDash \oslash A$.

The proof for $\lozenge$ is similar.

We have proved weak completeness for the logic with $\lozenge$ and $\mathbb{P}$ using the axioms of **K** for each.

We have $R = \rho_\lozenge \cup \rho_\mathbb{P}$ and we need to convert this model to a model of **KR** with R_1 and R_2.

We now need to use our additional axioms $\mathbb{RA}$ to determine how $\rho_\mathbb{P}$ correctly derives a set of double arrows R_2.

Remark 5.44. We make some observations about the model $(S, \rho_\lozenge, \rho_\mathbb{P}, 0, h)$ which we constructed.

Let $\mathbf{t}_m = (0, x_1, \ldots, x_m)$ be a point in the tree. Let $\mathbf{t}_i = (0, x_1, \ldots, x_i), i \leq m$. We know that $\mathbf{t}_{i+1} = \mathbf{t}_i * (x_{i+1})$.

1. If $x_{i+1} \leq r_i$ then $\mathbb{P}D_{x_i}^{n-i-1} \in \Delta_{\mathbf{t}_i}^{n-i}$ and $D_{x_i}^{n-i-1} \in \Delta_{\mathbf{t}_{i+1}}^{n-i-1}$. We write $x_i \xrightarrow{\mathbb{P}} x_{i+1}$.
2. If $r_i < x_{i+1} \leq s_i$ then $\lozenge B_{x_i}^{n-i-1} \in \Delta_{\mathbf{t}_i}^{n-i}$ and $B_{x_i}^{n-i-1} \in \Delta_{\mathbf{t}_{i+1}}^{n-i-1}$. We write $x_i \xrightarrow{\lozenge} x_{i+1}$.

Let E_{x_i} be the key formula $D_{x_i}^{n-i-1}$ or $B_{x_i}^{n-i-1}$ as the case may be. Then clearly at Δ_0^n the following holds

$$\beta_{\mathbf{t}_m} = M_1(E_{x_1} \wedge M_2(E_{x_2} \wedge \ldots M_{m-1}(E_{x_{m-1}} \wedge M_m E_{x_m})) \ldots)$$

where $M_i \in \{\mathbb{P}, \lozenge\}$ and $M_i = \mathbb{P}$ if $n \leq x_i \leq r_i$ and $M_i = \lozenge$ if $r_i + 1 \leq x_i \leq s_i$.

We write $x_i \xrightarrow{M_i} x_{i+1}$.

Clearly by the completeness theorem Lemma 5.43, we also have $0 \vDash \beta_{\mathbf{t}_m}$.

Following Lemma 5.41, $\beta_{\mathbf{t}_m}$ uniquely characterises the path leading to $\mathbf{t}_m$ in the model. Therefore $\delta_x^* = E_x$, as defined in Lemma 5.42, can act as path nominals for the points $\mathbf{t}$ via their validity at 0.

Theorem 5.45 (Completeness theorem for KR). **KR** *of Definition 5.35 is complete for the proposed semantics of Definition 5.12.*

Proof. Let A be a consistent formula. We constructed a reduced finite bi-modal model **m** for A as in Section 5.3.2. By Lemma 5.43, A holds in this model. By Remark 5.30 and Lemma 5.31, **m** can be converted to a **KR** model in which A holds.

Theorem 5.46 (Decidability and finite model property). **KR** *has the finite model property and is decidable.*

Proof. From Theorem 5.45.

5.4 Concluding remarks

We saw that reactive models are very useful and the idea of reactivity is applicable in many areas. The basic system **KR** of modal reactivity turned out to be not so easily axiomatisable and this calls for possibly some simplifying assumptions on the reactive double arrows, possibly emerging from applications.

So this is one direction of future research.

Another direction is to look at some of the new additional modal connectives and use them to introduce reactivity. We chose $\Diamond$ and $\mathbb{P}$, where $\Diamond$ ignores reactivity and $\mathbb{P}$ uses it. This is the most direct and natural connective. However, we can approach reactivity 'sideways' as can be seen from Figure 5.30.

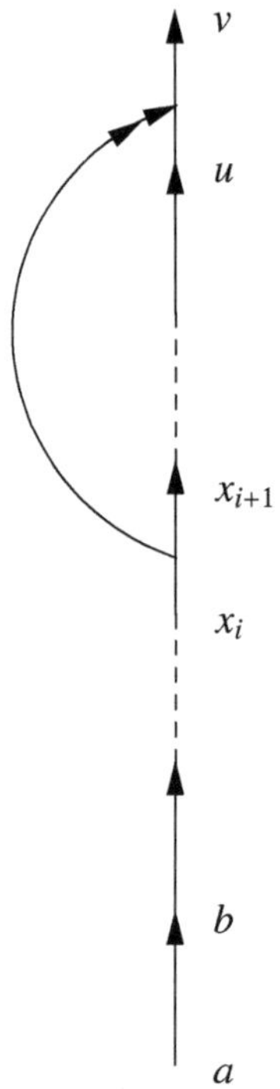

Fig. 5.30

Since our models are trees, the node x_{i+1} completely identifies the arc $x_i \to x_{i+1}$. Similarly, the node v completely identifies $u \to v$. Thus if we add a new binary relation R_3 on S with vR_3x reading

- There is a double arrow of the form ((predecessor of x) $\to x$) $\twoheadrightarrow$ ((predecessor of v) $\to v$)

We can map all double arrows this way.

So in a reactive model, R_2 can be represented by R_3. So we can have a new modality, say $\Diamond'$, for double arrows. We can add a special proposition **on** and have

$$v \vDash \Diamond'\mathbf{on} \text{ iff } \exists x(vR_3x \wedge x \vDash \mathbf{on})$$

This device may give us some scope for simplifying the axiomatisation but we still need to represent the switch nature of the double arrows. We still have to count how many 'hits' v receives, and it is this part which makes the axiomatisation complicated.

We invite the reader to use the classical translation of modal logic into classical logic. In classical logic we have the full power of the classical quantifiers and still it is not simple to express that $\{y \mid xR_3y\}$ is odd.

Furthermore, $\Diamond$ can ignore the double arrows and so we need to count the 'hits' from arcs not ignored by $\Diamond$. 'Ignored by $\Diamond$' implies 'motion' along the arcs and this is foreign to traditional modal logic or to classical logic.

My guess is that axiomatising with the aid of $\Diamond'$ or any additional tricks would still be very complicated.

Of course good restrictions on the reactivity can simplify the system considerably, and with good luck the simplified system may still be interesting and applicable. One such restriction is *next step reactivity*, mentioned in Remark 5.38.

Proposition 5.47. *Any next step reactivity logic system is equivalent to a bi-modal system with one modality implying the other.*

Proof. Figure 5.31 explains the idea of the proof.

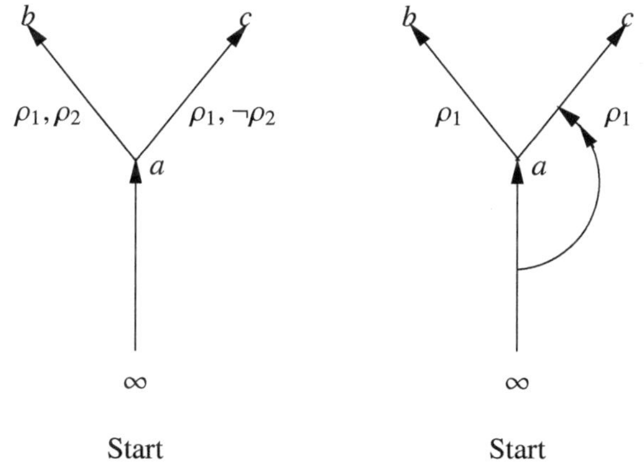

We have $\rho_2 \subseteq \rho_1$.

Fig. 5.31

So that is simple enough! It is also still more powerful. Theorem 5.2 still holds.

So we need to find good conditions on R_1 and R_2 of the semantics! Indeed, there is still lots to be done.

Remark 5.48. The idea of modality $\Diamond'$ and its corresponding relation R_3 has other unexpected consequences. Consider the situation in Figure 5.32. This is an interesting situation where we have a higher-level double arrow of the form $(b \to c) \twoheadrightarrow ((a \to b) \twoheadrightarrow (c \to d))$.

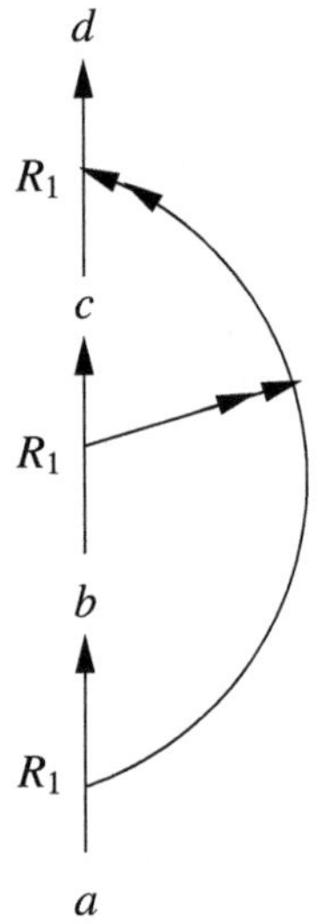

Fig. 5.32

If we use the relation R_3 and the modality $\Diamond'$, we get the situation in Figure 5.33. Now we have a situation with ordinary double arrows, so we can use a new relation R_4 and a modality $\Diamond''$ to eliminate all double arrows, as in Figure 5.34.

We must remember that yR_4x indicates double arrows from the arc ((the R_1 predecessor of x) $\to x$) to the arc ((the R_3 predecessor of y) $\to y$). So when we have many relations R, a better notation would be as follows:

Use R_{R_1,R_1} instead of R_3.
Use $R_{(R_1,(R_1,R_1))}$ instead of R_4.

The conclusion from this example is that higher-level double arrows can be reduced to multiple modalities. Looking at it the other way round (in the spirit of Proposition 5.47), multimodal logic can be reduced (given the right conditions) to a higher level reactive modal logic with one modality.

These investigations we postpone unitl later.

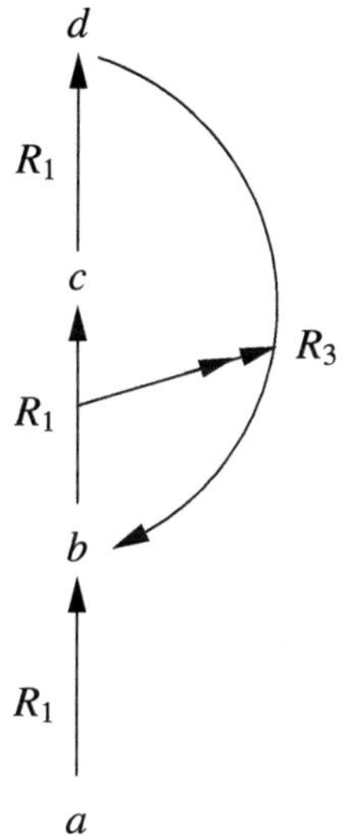

Fig. 5.33

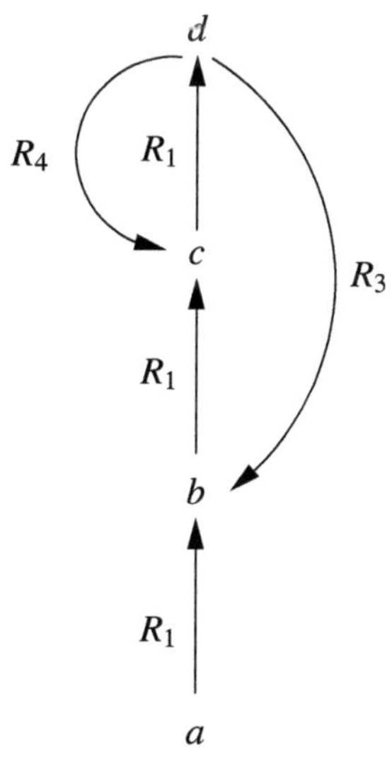

Fig. 5.34

6

Modal Logics of Reactive Frames

6.1 Introduction

Reactiveness.

The reactive idea is very simple. Given a system with states and the possibility of transitions between states, we can naturally imagine a path beginning at an initial state and moving along the path following allowed transitions. If our starting point is s_0, and the path is $s_0 \dots s_n$, then the system is an ordinary non-reactive system if the options available at s_n (i.e. which states t we can go to from s_n) do not depend on the path $s_0 \dots s_n$ (i.e. do not depend on how we got to s_n). Otherwise if there is such dependence then the system is *reactive*.

A simple example would be to consider as worlds the configurations on a chess board and the allowed moves. It is clear that this system is reactive in the above sense. To be able to castle one must not have moved either the the king or the rook, it is not enough that they are in their original positions. Moving the king or the rook corresponds to a higher order state transition, changing its nature. - It seems that the simple idea of taking existing systems and making them reactive in certain ways has many applications. Indeed, there are applications of the reactive idea in such diverse areas as modal logic, preferential non-monotonic logic, inheritance systems, context free grammars, automata theory, deontic logic and contrary-to-duty and argumentation and other networks, see other chapters in this book and papers [14, 66, 17, 37, 64, 75, 64, 72, 86].

One can take a reactive system and turn it into an ordinary system by taking the new states as the paths. This is true, but from the point of view of applications there is a serious loss of information, as the applicability of the reactive system may come from the manner in which the change occurs along the path. In any specific application, the states have meaning, the transitions have meaning and the paths have meaning. Therefore the changes in the system as we go along a path can have very important meaning in the context, which enhances the usability of the model.

D.M. Gabbay, *Reactive Kripke Semantics*, Cognitive Technologies,
DOI 10.1007/978-3-642-41389-6_6, © Springer-Verlag Berlin Heidelberg 2013

Reactive-switch graphs.

The ideas and concepts involved are best explained through examples.

Figure 6.1 presents a simple transition system.

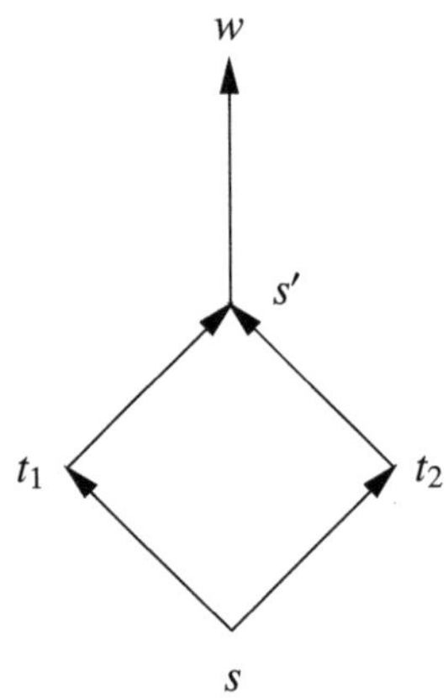

Fig. 6.1

We show only the transition options, without details of what causes the transitions. We have the transitions:

$$\begin{aligned} s &\to t_1 \\ s &\to t_2 \\ t_1 &\to s' \\ t_2 &\to s' \\ s' &\to w. \end{aligned}$$

The system is non-deterministic. Such systems are widely applicable in many areas and their exact meaning depends on the application area in which they are used.

It could be part of an automaton table (where the input letter causing the transition is not shown) or a set of context-free rules, or even part of a Kripke model where $x \to y$ means y is accessible from x.

A **reactive graph** (or reactive transition system) is a set of sequences of elements of a set, the **paths**, that contains all the sequences of length one and is closed under prefixes. It is clear it generalizes the static notion of a graph in which the set of possible extensions of a path depends only on its endpoint (see Proposition 6.23).

A system becomes reactive when the transition table changes as we move through the graph. We can make the above system reactive by, for example, saying that if we start in s, when we reach s' we see w if and only if we pass through t_1. A natural way of depicting this situation can be found in Digure 6.2.

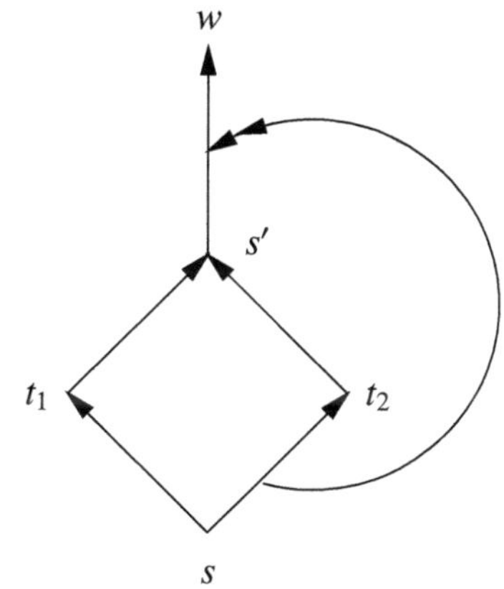

Fig. 6.2

The double arrow from the edge $s \to t_2$ to the edge $s' \to w$ indicates that if we go from s to t_2 then the connection $s' \to w$ is no longer available. The double arrow expresses the dependence on the path by codifying the transition's effects on the system configuration. We can make double arrows act other double arrows, getting higher and higher levels of dependence.

A **switch graph** (introduced in [2]) is such a graph enriched with higher order arrows, called switches. This **local** way of presenting (generating) the relational dynamics, as can be proved, is enough to generate all reactive graphs. There are many possible variations on how to define it. In the mentioned result a switch is an arrow from an edge to another edge or a switch, and it may be a connecting or a disconnecting switch.

Reactive modal logic.

As we observed above, a reactive system can be seen from a static point of view by considering their paths as the actual states.

Figure 6.3 shows the conversion of our example:

The points in Figure 6.3 are the paths of Figure 6.2. So from the point st_1s', there is a transition to $st_1s'w$ and from the point st_2s' there is no transition to $st_2s'w$. That the accessible points vary when we move is already a property of a static graph; what reactivity adds is the possibility of the points accessible from a point to be different depending on how we got there. In this sense Figure 6.3 is not a proper representation of Figure 6.2. The two paths st_1s' and st_2s' share an endpoint (i.e. the same state s') and this can be important. So we need to indicate this by an equivalence relation. Figure 6.4 does this, through the circle around the equivalent points, this is the unfolded version of Figure 6.2.

We will introduce an interpretation of modal logic over reactive graphs (frames), in a way that generalizes the identification between Kripke frames and static graphs, taking into consideration these last remarks.

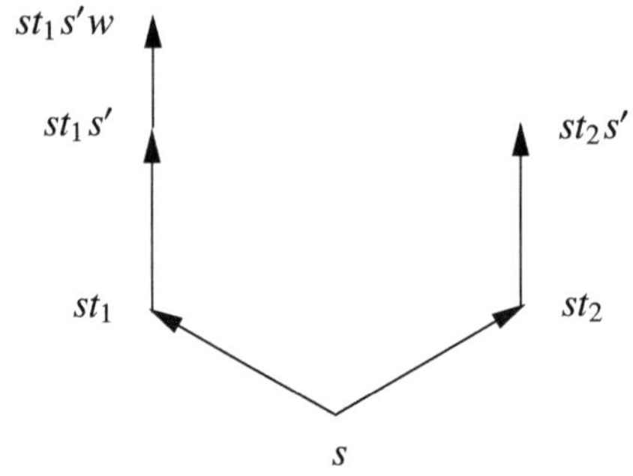

Fig. 6.3

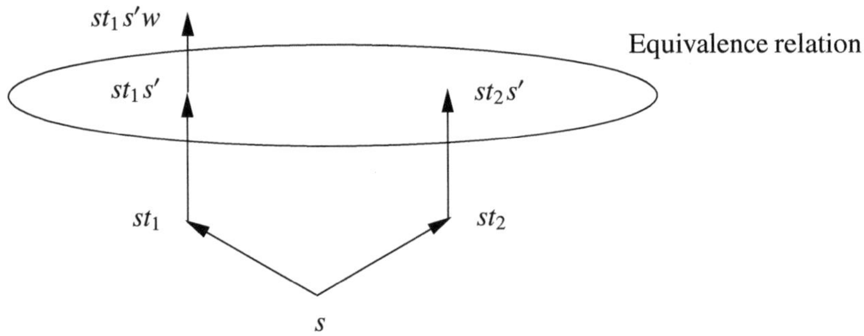

Fig. 6.4

We were naturally led to the choice of a bimodal language, where $\Diamond_R$ deals, as usual, with the dynamics and $\Diamond_P$ with the relation that identifies all the paths with the same endpoint. As we stressed before, reactivity is about the change to the set of accessible points of a given point. $\Diamond_P$ ranges over the various relational states of each point. As we shall see, many reactive properties can be axiomatised by the interaction between $\Diamond_R$ and $\Diamond_P$.

Like the accessibility relation, we let the value of (part of) the variables evolve while we go through the graph. The subset X of variables which we consider fixed, possibly all or none of them, will be a parameter of our logics. So the value of the variables in X depends only on the position in the graph, i.e. the path endpoint. The inclusion of this features in the models gives the results more generality and wider applicability and allows us to understand better the influence of each component of the generalization in the results.

We can now ask the obvious questions

- what axioms on $\Box_P, \Box_R$ correspond to this semantics?
- if we add reflexivity or transitivity or other obvious axioms to $\Box_R$, what are the corresponding conditions on the models?
- what other properties can we express with this language?

As an answer we present a procedure to prove that, given a logic and some reactive properties, the first axiomatises the second. While soundness is proved directly, to prove completeness we will use the static view of reactivity by unfolding the models. This allows us to use the canonical model theorem and other classic techniques. When we unfold the notion of a reactive model we obtain a classic bimodal Kripke model, (W, R, P, V) (where R is the dynamics and P the equivalence relation relating the paths with the same end point), satisfying three additional properties:

- there is a family $I \subseteq W$ picking one element of each P-equivalence class that R-generates the whole graph;
- xRy & xRz & yPz imply that $y = z$;
- the worlds related by P satisfy the same variables in X.

A **shattered** frame is a frame satisfying the first property and if it satisfies also the second we say it is a **coherently shattered (cs)** frame. The completeness proofs are done in two steps:

- we use the canonical model theorem to obtain completeness of a certain subclass of shattered frames corresponding to the reactive one (usually the first-order correspondent of the added axiom)
- we find a truth-preserving model transformation that on the given shattered frame gives a cs-frame with that property.

The procedure proves successful in many cases (see at the end of the chapter table 6.1) but we also present its limitations in the form of an open problem. We hope the techniques prove useful in characterizing logics, possibly in extended languages, expressing reactivity properties coming from research areas where reactivity is being applied, or properties suggested by switch graph research.

6.2 Reactive models

Definition 6.1. The modal language $\mathcal{L}_r$ is defined by

$$\varphi \;=\; p \mid \neg\varphi \mid \varphi_1 \wedge \varphi_2 \mid \Diamond_R\varphi \mid \Diamond_P\varphi\,.$$

where $p \in \Pi$ is a propositional variable. The other connectives: $\top$, $\bot$, $\vee$, $\rightarrow$, $\leftrightarrow$, $\Box_R$ and $\Box_P$ are introduced as the usual abbreviations.

Definition 6.2. A **path** over a set W is any finite sequence of points from W. A **prefix** of a path $w_0w_1 \dots w_n$ is any path of the form $w_0w_1 \dots w_k$ for $k \leq n$. Given a path $\lambda = w_0w_1 \dots w_n$, we let $t(\lambda) = w_n$ and n be its length.

A **reactive frame** is a pair (W, Δ), where W is a non-empty set and Δ is a set of paths over W that

- contains all one-element paths, i.e. W,
- is closed under taking prefixes.

Given $X \subseteq \Pi$, an X**-reactive model** is a triple $\mathcal{M} = (W, \Delta, \nu)$, where (W, Δ) is a reactive frame and ν is a function $\nu : \Pi \to 2^{\Delta}$ such that for $p \in X$ and $\lambda w, \lambda' w \in \Delta$ we have $\lambda w \in \nu(p)$ iff $\lambda' w \in \nu(p)$. X corresponds to the subset of variables that are fixed while we move, i.e. the value of which is determined by the current world.

Given a X-reactive model $\mathcal{M}$, for every $\lambda \in \Delta$ and every $\mathcal{L}_r$-formula φ, we define the notion 'φ **is true at** λ **in** $\mathcal{M}$ ($\mathcal{M}, \lambda \models_X \varphi$)' inductively as follows:

- $\mathcal{M}, \lambda \models_X p$ iff $\lambda \in \nu(p)$ for variables p,
- $\mathcal{M}, \lambda \models_X \neg\varphi$ iff $\mathcal{M}, \lambda \not\models_X \varphi$,
- $\mathcal{M}, \lambda \models_X \varphi_1 \wedge \varphi_2$ iff $\mathcal{M}, \lambda \models_X \varphi_1$ and $\mathcal{M}, \lambda \models_X \varphi_2$,
- $\mathcal{M}, \lambda \models_X \Diamond_R \varphi$ iff there is $w \in W$ such that $\lambda w \in \Delta$ and $\mathcal{M}, \lambda w \models_X \varphi$,
- $\mathcal{M}, \lambda \models_X \Diamond_P \varphi$ iff there is $\gamma \in \Delta$ such that $t(\gamma) = t(\lambda)$ and $\mathcal{M}, \gamma \models_X \varphi$.

Notice that if $X = \Pi$, then ν can be seen as a function from Π to 2^W and the first line of the definition of $\models_X$ can be equivalently replaced by $\mathcal{M}, \lambda \models_\Pi p$ iff $t(\lambda) \in \nu(p)$.

We say that φ **is true in** $\mathcal{M}$ iff $\mathcal{M}, \lambda \models_X \varphi$ for every $\lambda \in \Delta$. We say that φ **is** X**-valid in** a reactive frame if it is true in every X-reactive model over it.

When $X = \Pi$ we may omit X.

Definition 6.3. Given a reactive frame $F = (W, \Delta)$ we define $\sim_F \subseteq \Delta^2$ as:

$$\lambda \sim_F \gamma \text{ iff } t(\lambda) = t(\gamma) \ \&\ \forall \beta \in W^* \lambda\beta \in \Delta \leftrightarrow \gamma\beta \in \Delta.$$

We may omit F when it is clear from the context. We write $[\lambda]$ to refer to the equivalence class of $\lambda \in \Delta$.

Remark 6.4. We will only be interested in studying properties of X-reactive frames, that is, logics valid in the whole class of X-reactive models over reactive frames. Otherwise the notion of $\sim$ could be adapted to the context of X-reactive models and many of the following results would be valid in its model version.

Proposition 6.5. *Let*(W, Δ) *be a reactive frame.*

1. *If* $\lambda \sim \gamma$ *and* $(W, \Delta, \nu), \lambda \models_\Pi \varphi$ *then* $(W, \Delta, \nu), \gamma \models_\Pi \varphi$.
2. *If* $\lambda, \gamma \in \Delta$ *and* $\lambda \nsim \gamma$ *then there exists* φ *and* ν *s.t.* $(W, \Delta, \nu), \lambda \models_\Pi \varphi$ *and* $(W, \Delta, \nu), \gamma \models_\Pi \neg\varphi$.

Proof. 1. Let us prove it by induction on the structure of φ:

- if φ is a variable this is trivial since $t(\lambda) = t(\gamma)$.
- if $\varphi = \neg\psi$ then $\mathcal{M}, \lambda \not\models_\Pi \psi$ and so by IH and symmetry of $\sim$, $\mathcal{M}, \gamma \not\models_\Pi \psi$ thus $\mathcal{M}, \gamma \models_\Pi \neg\psi = \varphi$. The $\varphi = \varphi_1 \wedge \varphi_2$ case is trivial.
- if $\varphi = \Diamond_R \psi$ then it exists $w \in W$ s.t. $\lambda w \in \Delta$ and $\mathcal{M}, \lambda w \models_\Pi \psi$. It is clear that we also have $\lambda w \sim \gamma w \in \Delta$ and so by IH $\mathcal{M}, \gamma w \models_\Pi \psi$ hence $\mathcal{M}, \gamma \models_\Pi \Diamond_R \psi = \varphi$.
- the case $\varphi = \Diamond_P \psi$ also comes from $t(\lambda) = t(\gamma)$.

2. Suppose $t(\lambda) = t(\gamma)$ and there exists $\beta = w_1 \dots w_n$ s.t. we do not have $\lambda\beta \in \Delta$ iff $\gamma\beta \in \Delta$. Without loss of generality let us assume $\lambda\beta \in \Delta$ and $\gamma\beta \notin \Delta$ and pick a valuation ν that distinguishes all w_i, i.e., let $p_i \in \Pi$ with $1 \leq i \leq n$ and $\alpha \in \nu(p_i)$ iff $t(\alpha) = w_i$. Let $\varphi = \Diamond_R \varphi_\beta$ where φ_β is defined recursively by:

$$\varphi_{w_n} = p_n$$
$$\varphi_{w_i\lambda} = p_i \wedge \Diamond_R \varphi_\lambda.$$

It is clear that $(W, \Delta, \nu), \lambda \models_\Pi \varphi$ but $(W, \Delta, \nu), \gamma \not\models_\Pi \varphi$.
If $t(\lambda) \neq t(\gamma)$ we pick a valuation that distinguishes them.

Definition 6.6. $L_{r,X}$ is the logic of all reactive frames:

$$L_{r,X} = \{\varphi : \varphi \text{ is X-valid in every reactive frame}\}.$$

6.2.1 Reactiveness unfolded

Definition 6.7. 1. A **shattered frame** is a bimodal frame (W, R, P) such that P is an equivalence relation over W. Given a shattered frame, we say $I \subseteq W$ is an **initial family** if it picks one element from each P-class that R-generates the whole frame.
2. A **cs-frame** (coherently shattered) is a shattered frame (W, R, P) that admits an initial family and that is **coherent**, i.e., such that for all $w, w', w'' \in W$, if wRw', wRw'' and $w'Pw''$ then $w' = w''$.
3. An X-**shattered model** is a Kripke model $\mathcal{M} = (W, R, P, V)$ over a shattered frame (W, R, P) for which V is X-**admissible**, i.e. for all $w, w' \in W$, if wPw' then

$$w \in V(p) \text{ iff } w' \in V(p) \text{ for all } p \in X.$$

We say φ is X-**valid** in a shattered frame if it is true in every X-shattered model over it.
4. An X-**cs-model** is an X-shattered model over a cs-frame.

Notice that in order for the restrictions on the valuations in a Π-shattered model (W, R, P, V) to correspond to the restrictions on a general frame (W, R, P, A), where A is the boolean algebra generated by the equivalence classes of P, A has to be closed under m_R (and m_P). As we shall see in Proposition 6.23 the cs-frames that satisfy this requirement are those coming from unfolding static reactive frames. But in this case there is no restriction to deal with!

Definition 6.8. $L_{cs,X}$ is the logic of cs-frames:

$$L_{cs,X} = \{\varphi : \varphi \text{ is } X\text{-valid in every cs-frame}\}.$$

Remark 6.9. It is straightforward to see that every reactive frame $F = (W, \Delta)$ can be regarded as a cs-frame $F_{cs} = (\Delta, R^\Delta, P^\Delta)$ where W is an initial family and

- $\lambda R^\Delta \gamma$ iff there is some $w \in W$ such that $\gamma = \lambda w$,

- $\lambda P^\Delta \gamma$ iff $t(\lambda) = t(\gamma)$.

We call F_{cs} the **unfolding** of F. Furthermore it is easy to see that there is a bijective correspondence between X-reactive models $\mathcal{M} = (W, \Delta, v)$ and X-cs-models $\mathcal{M}_{cs} = (\Delta, R^\Delta, P^\Delta, v)$ that preserves valuations, i.e. for every $\lambda \in \Delta$ and every $\mathcal{L}_r$-formula φ,

$$\mathcal{M}, \lambda \models_X \varphi \quad \text{iff} \quad \mathcal{M}_{cs}, \lambda \models_X \varphi,$$

and so we have $L_{cs,X} \subseteq L_{r,X}$.

The converse is also true.

Proposition 6.10. *Let (W, R, P) be a cs-frame with I as initial family. There is a reactive frame s.t. there is a bijective correspondence between X-cs-models over the first and X-reactive models over the second that preserves valuations.*

Proof. We call an R-path $\lambda = w_0 \dots w_n$ an **I-initial path** if $w_0 \in I$. For every such I-initial path λ, let

$$l_I(\lambda) = i_0 \dots i_n$$

for the unique sequence of $i_j \in I$ with $w_j P i_j$ for $1 \le j \le n$ (see Def. 6.7).

Moreover l_I is an injective function from the set of all $\mathcal{M}$-initial paths into paths over I, that is, if $l_I(\lambda) = l_I(\gamma)$ then $\lambda = \gamma$. Indeed, as l clearly preserves the length of a path, we can do induction on the length n of $\lambda = w_0 \dots w_n$, $\gamma = v_0 \dots v_n$. If $n = 0$ then $w_0 = l_I(w_0) = l_I(v_0) = v_0$ follows. Now suppose that $l_I(w_0 \dots w_n w_{n+1}) = l_I(v_0 \dots v_n v_{n+1}) = i_1 \dots i_n i_{n+1}$ for some $i_j \in I$, $1 \le j \le n+1$. Then $l_I(w_0 \dots w_n) = l_I(v_0 \dots v_n) = i_1 \dots i_n$ and $w_{n+1} P i_{n+1} P v_{n+1}$. By the IH, we have $w_j = v_j$ for $1 \le j \le n$. Therefore not only $w_n R w_{n+1}$, but also $w_n R v_{n+1}$. So $w_{n+1} = v_{n+1}$ follows by coherence.

Let $\Delta^I = \{l_I(\lambda) : \lambda \text{ is an } I\text{-initial path}\}$. (I, Δ^I) is clearly a reactive frame, we call it a **folding** of (W, R, P).

Given an X-cs-model $\mathcal{M} = (W, R, P, V)$ (with I as an initial family) let $\mathcal{M}' = (I, \Delta^I, v)$ with

$$v(p) = \{l_I(\lambda) \in \Delta^I : t(\lambda) \in V(p)\}, \text{ for each variable } p.$$

It is straightforward to see that $\mathcal{M}'$ is an X-reactive model. We will show that $\mathcal{M} \models_X \varphi$ iff $\mathcal{M}' \models_X \varphi$.

According to Remark 6.9 we can regard $\mathcal{M}'$ as a X-cs-model $\mathcal{M}'_{cs}$ preserving valuations. Hence, to conclude, it is enough to prove that $\mathcal{M}$ is a bounded morphic image of $\mathcal{M}'_{cs}$:

We define a function $f : \Delta^I \to W$ by taking

$$f(l_I(\lambda)) = t(\lambda).$$

which is well defined since l_I is injective. We claim that f is a surjective bounded morphism from $\mathcal{M}'_{cs}$ onto $\mathcal{M}$:

- f is surjective since by Def. 6.7(1) we have that I R-generates the whole frame

- p-morphism in R. First, if $l_I(\lambda), l_I(\lambda)i \in \Delta^I$ then there is some $w \in W$ such that $t(\lambda)Rw$ and $l_I(\lambda)i = l_I(\lambda w)$. So $t(\lambda)Rw = f(l_I(\lambda)i)$. Second, if $f(l_I(\lambda)) = t(\lambda)Rw$ then $l_I(\lambda w) \in \Delta^I$ and $f(l_I(\lambda w)) = w$.
- p-morphism in P. First, if $t(l_I(\lambda)) = t(l_I(\gamma))$ then $t(\lambda)Pt(\gamma)$. Second, if $f(l_I(\lambda)) = t(\lambda)Pw$ then, by Def. 6.7(1), there is some $\mathcal{M}$-initial path γ such that $t(\gamma) = w$. Then $l_I(\gamma) \in \Delta^I$ and $t(l_I(\lambda)) = t(l_I(\gamma))$ follows.
- p-morphism in V: $l_I(\lambda) \in v$ iff $t(\lambda) = f(l_I(\lambda)) \in V(p)$.

Corollary 6.11. $L_{cs,X} = L_{r,X}$.

Proof. From remark 6.9 we get one direction. For the other, suppose that $\varphi \notin L_{cs,X}$, that is, there is some X-cs-model $\mathcal{M} = (W, R, P, V)$ having I as initial family and such that $\mathcal{M} \not\models_X \varphi$. $\mathcal{M}'$ is an X-reactive model and $\mathcal{M}' \not\models_X \varphi$ follows from the previous proposition.

Given a reactive frame we can obtain a cs-frame by unfolding (and vice versa by folding) where each X-model over the first corresponds to an X-model over the second and modal satisfaction is preserved.

With shattered frames we are back in the classic setting even if, in the case of $X \neq \emptyset$, we are not at the frame level. There are restrictions on the models over the correspondent Kripke frames; the situation is closer to the case of general frames.

Folding and unfolding will be the bridge from reactiveness to this classical setting and will allow us to use some known techniques in the study of the axiomatisation of logics over reactive frames. The most immediate consequence of this connection is that the logics formed by formulas valid over reactive frames and cs-frames coincide.

6.3 Axiomatisations

In this section we will deal with axiomatising logics over some classes of reactive frames.

Definition 6.12. Let $L \cup L' \cup \{\varphi\}$ be a set of formulas.

- $L \oplus L'$ is the closure by the rules of modus ponens and necessity of the set
$$\{\psi : \psi \text{ is a propositional instance of } \varphi \in L \cup L'\}.$$
If L, L' have only one element we may omit the curly brackets.
- $L +_X \varphi$ is the closure by the rules of modus ponens and necessity of the set
$$L \cup \{\psi : \psi \text{ is the result of substituting the variables in } \varphi \text{ by variables in } X\}.$$

Let
$$L_X = K_R \oplus S5_P +_X p \leftrightarrow \Box_P p,$$
meaning that L_X is the closure by the rules of modus ponens and necessity (for $\Box_R$ and $\Box_P$) of the set containing

$$p \leftrightarrow \Box_P p$$

for every propositional variable $p \in X$, the substitution instances of all propositional tautologies and the following axioms:

1. $\Box_R(p \to q) \to (\Box_R p \to \Box_R q)$
2. $\Box_P(p \to q) \to (\Box_P p \to \Box_P q)$
3. $\Box_P p \to p$
4. $\Box_P p \to \Box_P \Box_P p$
5. $p \to \Box_P \Diamond_P p$

It is clear that $L_X \subseteq L_{cs,X} = L_{r,X}$ since all the L_X axioms are sound with respect to shattered frames. Notice that the $\Diamond_R$ fragment of L_X is just K_R, i.e. K. It is obvious that $K \subseteq L_X$ and from the fact that every Kripke model is a trivial X-reactive model we get the other inclusion.

We want to show that L_X is complete with respect to all reactive frames. For this we will prove that L_X's canonical model is an X-shattered model and then prove that any such model is the bounded image of a generated subframe of an X-cs-frame, hence concluding that $L_X = L_{cs,X} = L_{r,X}$.

If $X \neq \emptyset$ then L_X is not closed under structural substitution. One can define the canonical model of a normal logic L in the usual way, $\mathcal{M}_L = (W^L, R^L, P^L, V^L)$ where:

$$\begin{aligned}
&W^L = \{s : s \text{ is } L - \text{theory}\}\\
&sR^L t \text{ iff } \{\varphi : \Box_R\varphi \in s\} \subseteq t \text{ iff } \{\Diamond_R\varphi : \varphi \in r\} \subseteq s\\
&sP^L t \text{ iff } \{\varphi : \Box_P\varphi \in s\} \subseteq t \text{ iff } \{\Diamond_P\varphi : \varphi \in r\} \subseteq s\\
&V^L(p) = \{s \in W^{L_X} : p \in s\}.
\end{aligned}$$

and prove the well-known truth lemma about it:

$$\{\varphi : \mathcal{M}_{L_X} \models \varphi\} = L_X$$

even if the logic in question is not closed under the rule of substitution, see e.g. [99].

Proposition 6.13. *If a logic L contains $p \leftrightarrow \Box_P p$ for all $p \in X \subseteq \Pi$ then given $vP^{\mathcal{M}_L}w$ we have that:*

$$v \in V^{\mathcal{M}_L}(p) \;\Rightarrow\; w \in V^{\mathcal{M}_L}(p) \textit{ for all } p \in X.$$

Proof. If $sP^{\mathcal{M}_L}t$ and $p \in s$ then $\Box_P p \in s$ and so $p \in t$.

Clearly, if $P^{\mathcal{M}_L}$ is a equivalence class then all the worlds related by $P^{\mathcal{M}_L}$ satisfy the same variables in X.

Corollary 6.14. *$\mathcal{M}_{L_X}$ is an X-shattered model and so*

$$\{\, \varphi : \varphi \textit{ is } X\textit{-valid in every shattered frame} \,\} = L_X.$$

In other words we have that L_X is sound and complete with respect to the class of X-models over shattered frames. Next, by showing that L_X is also complete with respect to its subclass of cs-frames, we conclude that the axioms generating L_X axiomatise the minimal logic over reactive frames.

Theorem 6.15. $L_X = L_{r,X}$

Proof. Since $L_{r,X} = L_{cs,X}$ it is equivalent to prove that $L_X = L_{cs,X}$.

Every cs-frame is also a shattered frame, hence we have that $L_{cs,X} \subseteq L_X$.

Let $\varphi \notin L_X$ then there exists some X-shattered model $\mathcal{M} = (W, R, P, V)$ that does not satisfy φ.

We define $\mathcal{N} = (W', R', P', V')$, where:

$$W' = \{(z, y) : yPz\} \cup W \times \{*\}$$
$$(x, i)R'(y, j) \text{ iff } i = * \,\&\, x = y \text{ or } i \neq * \,\&\, xRy \,\&\, y = j$$
$$(x, i)P'(y, j) \text{ iff } i = j \,\&\, (i = * \to x = y)$$
$$(x, i) \in V'(p) \text{ iff } x \in V(p).$$

$\mathcal{N}$ is an X-cs-model and $\mathcal{M}$ is a bounded morphic image of the submodel generated by $W'' = W' - W \times \{*\}$ (and therefore $\varphi \notin L_{cs,X}$):

- let I be a family of elements of W' picking one element from each P-class. It is clear that $W \times \{*\} \subseteq I$ and so I generates $\mathcal{N}$.
  V' is an X-admissible valuation: $(x, i)P'(y, j)$ imply $i = j$ hence xPy and therefore if $p \in X$,

  $$(x, i) \in V'(p) \text{ iff } x \in V(p) \text{ iff } y \in V(p) \text{ iff } (y, j) \in V'(p).$$

  Let us check that it is coherent. Given $(x, i), (y, j), (z, k) \in W'$. If $(y, j)P'(z, k)$ then $j = k$. Let us assume also $(x, i)R'(y, j)$ and $(x, i)R'(z, k)$. If $i = *$ then $x = y = z$ and if $i \neq *$ then $y = j = k = z$. So $(y, j) = (z, k)$.

- let $f : W'' \to W$ be defined by: $f((x, i)) = x$.

  It is straightforward to see that f is surjective, let us check that it is actually a p-morphism:
  - f is a p-morphism in R
    - $(x, i)R'(y, j)$ then $f((x, i)) = xRy = f((y, j))$.
    - $f((x, i))Ry$ then $(x, i)R'(y, y)$.
  - f is a p-morphism in P
    - $(x, i)P'(y, j)$ then $i = j$ and so, by construction, we have $f((x, i)) = xPy = f((y, j))$.
    - $f((x, i))Py$ then $(x, i)P'(y, i)$
  - f is a p-morphism in V
    $(x, y) \in V'(p)$ iff $f((x, i)) = x \in V'(p)$.

We will call this way of generating an X-cs-model from a given X-shattered model the *blow up trick* and it is inspired by the standard extensions used in [100].

If we consider the basic properties in graphs like reflexivity, symmetry or transitivity, we see that there are many ways of generalising them to the reactive graph level. These properties refer to the accessibility of points without referring to the changes in their relational state since it is always the same. In the reactive case this is not true and they may mean different things. Reflexivity can mean for example that we can always access a loop without any change to the set of accessible points or we may require anything else. All the mentioned properties have similar variants, which are the *generalised* notions of the originals.

Just as the (static) graph properties are studied in classic modal logic, we dedicate the rest of the chapter to the study of logics of the subclasses of reactive frames satisfying some of these properties.

Our strategy will be the same as for the general case. Given a class of reactive frames and a set of candidate axioms, we first check whether they *correspond* to the reactive frame property in question (in some cases we only have soundness). Then we prove completeness in steps:

- we establish that the logic they originate is complete with respect to a class of shattered frames with a certain property by analysing its canonical model;
- we check that its subclass of cs-frames validates exactly the same formulas by showing the existence of a transformation from shattered frames to cs-frames (using the blow up trick) that preserves the class we are considering and showing the first one is a bounded morphic image of a generated submodel of the second;
- we show that folding these cs-frames originates reactive frames with the required property, thus obtaining the result by applying Proposition 6.10, as in Corollary 6.11.

6.3.1 Reflexivity and transitivity

Now we will study some subclasses of reactive frames obtained by imposing properties that generalize the notions of reflexivity and transitivity in the static case and axiomatise them.

Let us introduce some variations of the blow up trick that will work for all the subsequent cases.

Proposition 6.16. *Given* φ *and some* X*-shattered model* $\mathcal{M} = (W, R, P, V)$ *we define* $t_i(\mathcal{M}) = (W_i, R_i, P_i, V_i)$ *for* $i = 1, 2, 3, 4$*:*

1.
- $W_1 = \{(x, (y, 0)), (x, (y, 1)) : xPy\} \cup W \times \{*\}$
- $(x, i)R_1(y, j)$ *iff* $i = * \;\&\; x = y$ *or*
  $(i = (x', a) \;\&\; j = (y', b) \;\&$
  $(b = |a - 1| \;\&\; y = t'(x, y')$ *or* $i = j \;\&\; y = t'(x, x)))$
- $(x, i)P_1(y, j)$ *iff* $i = j \;\&\; (i = 1 \rightarrow x = y)$
- $(z, i) \in V_1(p)$ *iff* $z \in V(p)$.

2. • $W_2 = \{(t^n(c), n, c, d, i) : cPd; i = 0, 1, *; n < \omega\}$
   • $(a, b, c, d, i)R_2(a', b', c', d', i')$ *iff* $(c, d) = (c', d')$ &
   $(i = * \;\&\; (a, b) = (a', b')$ *or*
   $i \neq * \;\&\; aRa' \;\&$
   $(a = t(a') \;\&\; b = b' + 1 \;\&\; i = i'$ *or*
   $a' = c' = d' \;\&\; b' = 0 \;\&\; i = |i' - 1|))$
   • $w = (a, b, c, d, i)P_2(a', b', c', d', i') = w'$ *iff* $(d, i) = (d', i')$ &
   $(i = * \rightarrow w = w')$
   • $(a, b, c, d, i) \in V_2(p)$ *iff* $a \in V(p)$

   $(t^0(w) = w)$
3. • $W_3 = \{((x, y), n), (x, n) : xPy, n < \omega\}$
   • $(x, i)R_3(y, j)$ *iff* $y = (y_1, y_2)$ &
   $(x \in W \;\&\; (x = y_1 \;\&\; i = j$ *or* $j > i \;\&\; y_1 = y_2)$ *or*
   $x = (x_1, x_2) \;\&\; j > i \;\&\; x_1 R y_1 = y_2)$
   • $(x, i)P_3(y, j)$ *iff* $i = j \;\&\; (x \in W \;\&\; x = y$ *or* $x = (x_1, w) \;\&\; y = (y_1, w))$
   • $(x, i) \in V_3(p)$ *iff* $x \in W \;\&\; x \in V(P)$ *or* $x = (x_1, x_2) \;\&\; x_1 \in V(p)$.
4. • $W_4 = W_3, P_4 = P_3 \;\&\; V_4 = V_3$
   • $(x, i)R_4(y, j)$ *iff* $y = (y_1, y_2)$ &
   $(x \in W \;\&\; (x = y_1 \;\&\; i = j$ *or* $j > i \;\&\; y_1 = y_2)$ *or*
   $x = (x_1, x_2) \;\&\; j > i \;\&\; t'(x_1, y_2) = y_1)$

   where t, t' are defined by
   • $t(w) = v$ *s.t.* $vRwPv$ *if there is such a* v*, otherwise it is undefined,*
   • $t'(w, v) = v'$*, s.t.* $wRv'Pv$ *and* $wRv \rightarrow v' = v$ *if there is such a* v'*, and is undefined otherwise.*

$t_i(\mathcal{M})$ is an X-cs-model and $\mathcal{M}$ is a bounded morphic image of a generated submodel of $t_i(\mathcal{M})$ for $i = 1, 2, 3, 4$.

Remark 6.17. Any of these constructions could have been used in Theorem 6.15 instead of $\mathcal{N}$.

Proof. 1. $t_1(\mathcal{M})$ is an X-cs-model and $\mathcal{M}$ is a bounded morphic image of the submodel of $t_1(\mathcal{M})$ generated by $W_1' = \{(x, (y, 0)), (x, (y, 1)) : xPy\}$:

- the existence of an initial family and that V_1 is an X-admissible valuation is shown just like in Theorem 6.15.
  Let us check that it is coherent. Given $(x, i), (y, j), (z, k) \in W_1$. If $(y, j)P_1(z, k)$ then $j = k$. Let us assume that we also have $(x, i)R_1(y, j)$ and $(x, i)R_1(z, k)$. If $i = *$ we get $x = y = z$. If $i = (z, a)$ then $j = (w, b) = k$. If $a = b$ then $y = t(x, x) = z$ and if $a \neq b$ then $y = t(x, w) = z$. In any case $(y, j) = (z, k)$.
- let $f : W_1' \rightarrow W$ be defined by: $f((x, i)) = x$. It is straightforward to see that f is a surjective function, let us see that it is actually a p-morphism:
  - f is a p-morphism in R
    · $(x, (x', i))R_1(y, (y', j))$ then $y = t'(x, y')$ or $y = t'(x, x)$. In both cases $f((x, i)) = xRy = f((y, j))$,
    · $f((x, (t, a)))Ry$ then we have that $t'(x, y) = y$,
    hence $(x, (t, a))R_1(y, (y, |a - 1|))$;

- f is a p-morphism in P
  - $(x, i)P_1(y, j)$ then $i = j$ and so $f((x, i)) = xPy = f((y, j))$,
  - $f((x, i))Py$ then $(x, i)P_1(y, i)$;
- f is a p-morphism in V, is trivial since by definition $(x, i) \in V_1(p)$ iff $f((x, i)) = x \in V$.

2. $t_2(\mathcal{M})$ is an X-cs-model and $\mathcal{M}$ is a bounded morphic image of the submodel of $t_2(\mathcal{M})$ generated by $W_2' = \{(t^n(c), n, c, d, i) : cPd; i = 0, 1; n < \omega\}$:
   - it is easy to verify that V_2 is admissible and that any choice of I, picking one element from each P-class and containing $\{(t^n(c), n, c, d, *) : cPd; n < \omega\}$, works as initial family.
     Let us check that it is coherent. Given $w = (a, b, c, d, i)$, $w' = (a', b', c', d', i')$ and $w'' = (a'', b'', c'', d'', i'')$ in W_2. If $w'P'w''$ then $(d', i') = (d'', i'')$. Let us assume that we also have wR_2w' and wR_2w'' so $(c', d') = (c, d) = (c'', d'')$. If $i = *$ we get $(a, b) = (a', b') = (a'', b'')$. Otherwise, if either $i = i'$ or $i = i''$ then $i = i' = i''$, so $b' = b - 1 = b''$ and $a = t^{b'}(c) = t^{b''}(c) = a''$. If $i \neq i'$ then $b = b' = 0$, $a' = c' = d' = d'' = c'' = a''$. Hence, in any case, $w' = w''$.
   - let $f : W_2' \to W$ be defined by $f((a, b, c, d, i)) = a$.
     It is straightforward to see that f is a surjective function, let us see that it is actually a p-morphism:
     The condition in the valuation is trivial as before;
     - f is a p-morphism in R
       - $w = (a, b, c, d, i)R_2(a', b', c', d', i') = w'$ then $f((x, i)) = aRb = f((y, j))$,
       - $f((a, b, c, d, i))Ry$ then $(a, b, c, d, i)R_2(y, 0, y, y, |i - 1|)$;
     - f is a p-morphism in P
       - $w = (a, b, c, d, i)P_2(a', b', c', d', i') = w'$ then $aPcPd = d'Pc'Pa'$. Thus, by transitivity of P, we obtain $f(w) = aPa' = f(w')$,
       - $f((a, b, c, d, i))Py$ then $(a, b, c, d, i)P_2(y, 0, y, d, i)$.
3. $t_3(\mathcal{M})$ is an X-cs-model and $\mathcal{M}$ is a bounded morphic image of the submodel of $t_3(\mathcal{M})$ generated by $W_3' = \{((x, y), n) : xPy, n < \omega\}$:
   - the existence of an initial family and the admissibility of V_3 are shown as before (in this case the generator is $\{(x, n) : n < \omega\}$).
     Let us check that it satisfies coherence. Given $(x, i), (y, j), (z, k) \in W_3$. Let us assume that we have $(x, i)R_3(y, j)$, $(x, i)R_3(z, k)$ and $(y, j)P_3(z, k)$ then $y = (y_1, w)$, $z = (z_1, w)$ and $j = k$. So, either $i = j = k$, in which case, $x \in W$ and $y_1 = x = z_1$; or $i < j = k$ and $y_1 = w = z_1$.
   - let $f : W_3' \to W$ be defined by: $f((x, y), i)) = x$.
     It is straightforward to see that f is a function, let us see that it is also a p-morphism (the condition in V is handled as before):
     - f is a p-morphism in R
       - $((x_1, x_2), i)R_3((y_1, y_2), j)$ then
         $$f(((x_1, x_2), i)) = x_1Ry_1 = f(((y_1, y_2), j)),$$
       - $f(((x_1, x_2), i))Ry$ then $((x_1, x_2), i)R_3((y, y), j)$ for any $j > i$;

- f is a p-morphism in P
  - $((x_1, x_2), i)P_3((y_1, y_2), j)$ then

$$f(((x_1, x_2), i)) = x_1 P x_2 = y_2 P y_1 = f(((y_1, y_2), j)),$$

  - $f((x_1, x_2), i)Py$ then $((x_1, x_2), i)P_3((y, x_2), i)$.

4. Let us consider the morphism as in 3., we just need to check the conditions involving R_4:
   - initiality is shown as in 3.
     Coherence: Let us assume that we have $(x, i)R_4(y, j)$, $(x, i)R_4(z, k)$ and $(y, j)P_4(z, k)$. Then $y = (y_1, w)$, $z = (z_1, w)$ and $j = k$. If $x \in W$ then, if $i = j = k$ then $x = y_1 = z_1$ and if $i < j = k$ then $y_1 = w = z_1$. Otherwise $x = (x_1, x_2)$ and $y_1 = t(x_1, w) = z_1$.
   - $((x_1, x_2), i)R_4((y_1, y_2), j)$ then $y_1 = t'(x_1, y_2)$ hence

$$f(((x_1, x_2), i)) = x_1 R y_1 = f(((y_1, y_2), j)),$$

   - $f(((x_1, x_2), i))Ry$ then $t'(x_1, y) = y$ thus $((x_1, x_2), i)R_4((y, y), j)$ for any $j > i$.

We start by analysing the subclasses of reactive frames (or models) that correspond to the usual axioms for reflexivity and transitivity (with one operator).

Let us prove a lemma that will be useful throughout the chapter:

Lemma 6.18. *Given a cs-frame (W, R, P) admitting as initial family I. Let γs and $\gamma' s$ be two I-initial paths, then*

$$l_I(\gamma s) \sim_{(I, \Delta_I)} l_I(\gamma' s).$$

Proof. $l_I(\gamma s)\alpha \in \Delta_I$ iff there is a β s.t. $\gamma s\beta$ is an I-initial path (iff $\gamma' s\beta$ is an I-initial path) and $l_I(\beta) = \alpha$ iff $l_I(\gamma' s)\alpha \in \Delta_I$. Since clearly $t(l_I(\gamma s)) = l_I(s) = t(l_I(\gamma' s))$ we get $l_I(\gamma s) \sim_{(I, \Delta_I)} l_I(\gamma' s)$.

Proposition 6.19. *1. A reactive frame (W, Δ) (Π)-validates $p \rightarrow \Diamond_R p$ iff it is* **strongly reflexive**, *i.e. satisfies:*

$$\lambda w \in \Delta \rightarrow \lambda w \sim_{(I, \Delta_I)} \lambda ww \in \Delta.$$

Let $L_X^T = L_X \oplus p \rightarrow \Diamond_R p$.

2. *L_Π^T is sound and complete with respect to the class of (all Π-reactive models over) strongly reflexive frames.*
3. *If $X \neq \Pi$, L_X^T is not sound and complete with respect to (all X-models over) any class of reactive frames.*

Proof. 1.

- Given a strongly reflexive Π-reactive model $\mathcal{M} = (W, \Delta, v)$ and $\lambda w \in \Delta$ s.t. $\mathcal{M}, \lambda w \models_\Pi \varphi$. Thus $\lambda w \sim \lambda ww$ and, by Proposition 6.5, $\mathcal{M}, \lambda ww \models_\Pi \varphi$ and so $\mathcal{M}, \lambda w \models_\Pi \Diamond_R \varphi$.

- Given a non-strongly reactive frame (W, Δ). There exists some $\lambda w \in \Delta$ s.t. $\lambda w \nsim \lambda ww$, so we have two cases.
  Either $\lambda ww \notin \Delta$ so we can take $p \in \Pi$ and pick v s.t. $\lambda w \in v(p)$ and $\lambda ww' \notin v(p)$ for all $\lambda ww' \in \Delta$ in which case
  $$(W, \Delta, v), \lambda w \models p \wedge \neg \Diamond_R p.$$
  Or there exists $\beta = w_1 \dots w_n$ s.t. we don't have $\lambda w\beta \in \Delta \leftrightarrow \lambda ww\beta \in \Delta$. We consider $w_0 = t(\lambda) = t(\gamma)$, $p_0, \dots, p_n$ and pick v s.t. $\alpha \in v(p_i)$ iff $t(\alpha) = w_i$. Let $\varphi = \Diamond_R \varphi_\beta$ be as defined in Proposition (6.5.2). If $\lambda w\beta \in \Delta$ but $\lambda ww\beta \notin \Delta$ then λw satisfies $\psi = (p_0 \wedge \varphi) \vee \neg p_0$ and λww does not. If $\lambda w\beta \notin \Delta$ but $\lambda ww\beta \in \Delta$ then λw satisfies $\neg\psi = \neg(p_0 \wedge \varphi) \wedge p_0$ and λww does not. Since no $\lambda ww' \in \Delta$ with $w' \neq w$ satisfies p_0 and both ψ and $\neg\psi$ imply p_0 (which is only satisfied at paths with w as end point), we know that either $(W, \Delta, v), \lambda w \not\models \psi \to \Diamond_R \psi$ or $(W, \Delta, v), \lambda w \not\models \neg\psi \to \Diamond_R \neg\psi$.

Imposing the usual axiom for reflexivity forces a very strong notion of reflexivity in reactive frames. Strong reflexivity imposes that, no matter which path we have covered, we can always loop without any change to the accessible worlds.

2. Soundness of L_Π^T comes easily from the soundness of the new axiom which has just been established.

Let us prove that $L_\Pi^T = \{\varphi : \varphi$ is valid in every strongly reflexive reactive frame$\}$:

- using Proposition 6.13 it is easy to check that the canonical model for L_Π^T is a Π-shattered model (W, R, P, V) where R is reflexive. Since every cs-frame is also a shattered frame we have that:
  $$\begin{aligned} L_\Pi^T &= \{\varphi : \varphi \text{ is true in every } \Pi\text{-shattered model } (W, R, P, v) \text{ where } R \text{ is reflexive}\} \\ &\subseteq \{\varphi : \varphi \text{ is true in every } \Pi\text{-cs model } (W, R, P, v) \text{ where } R \text{ is reflexive}\}. \end{aligned}$$
- t_1 preserves R-reflexivity:
  Given (x, i) the case $i = *$ is trivial and if $i = (x', a)$ we have
  $$(x, (x', a))R_2(t'(x, x), (x', a)) = (x, (x', a))$$
  since $t'(x, x) = x$. Hence
  $$L_\Pi^T = \{\varphi : \varphi \text{ is true in every } \Pi\text{-cs model } (W, R, P, v) \text{ where } R \text{ is reflexive}\}.$$
- given an R-reflexive cs-frame $F = (W, R, P)$ with initial family I, (I, Δ^I) is a strongly reflexive reactive frame:
  Let $\lambda w \in \Delta^I$, so there is $\gamma \in W^*$ and $s \in W$ s.t. γs is an I-initial path and $l_I(\gamma s) = \lambda w$. Since F is R-reflexive we have that γss is also an I-initial path and clearly $l_I(\gamma ss) = \lambda ww \in \Delta$. Applying lemma 6.18 we conclude that $\lambda w \sim \lambda ww$.

3. Let $X \subsetneq \Pi$ and $\mathcal{F}$ be a class of reactive frames.

If there is a reactive frame $F = (W, \Delta) \in \mathcal{F}$ then let $\lambda \in \Delta$ and $p \in \Pi - X$. We pick v s.t. $\lambda \in v(p)$ and $\lambda w \notin v(p)$ for all $\lambda w \in \Delta$. So

$$(W, \Delta, v), \lambda w \models p \wedge \neg \Diamond_R p.$$

If $\mathcal{F}$ is the empty class it validates $\bot$. Since there are Π-reactive models over strongly reflexive frames and $L_X^T \subseteq L_\Pi^T$ we conclude that

$$L_X^T \subsetneq \{\varphi : \varphi \text{ is valid in all } F \in \mathcal{F}\} \ni \bot.$$

Notice that to have reactive frame completeness we have to impose that $X = \Pi$. This will also happen in the next case.

Proposition 6.20. *1. A reactive frame* (W, Δ) *(*Π*-)validates* $\Diamond_R \Diamond_R p \to \Diamond_R p$ *iff it is* **strongly transitive***, i.e. satisfies:*

$$\lambda ww'w'' \in \Delta \to \lambda ww'w'' \sim \lambda ww'' \in \Delta.$$

Let $L_X^4 = L_X \oplus \Diamond_R \Diamond_R p \to \Diamond_R p$.

2. L_Π^4 *is sound and complete with respect to the class of* Π*-reactive models over strongly transitive frames.*
3. *If* $X \neq \Pi$, L_X^4 *is not sound and complete with respect to (all* X*-models over) any class of reactive frames.*

Proof. 1.

- Given a strongly transitive reactive model $\mathcal{M} = (W, \Delta, v)$ and $\lambda w \in \Delta$ s.t. $\mathcal{M}, \lambda w \models \Diamond_R \Diamond_R \varphi$ then there are $w', w'' \in W$ s.t. $\lambda ww'w'' \in \Delta$ and $\mathcal{M}, \lambda ww'w'' \models \varphi$. Thus $\lambda ww'w'' \sim \lambda ww'' \in \Delta$ and, by Proposition 6.5, $\mathcal{M}, \lambda ww'' \models \varphi$ hence $\mathcal{M}, \lambda w \models \Diamond_R \varphi$.
- Given a non-strongly transitive reactive frame (W, Δ). There exists some $\lambda ww'w'' \in \Delta$ s.t. $\lambda ww'w'' \nsim \lambda ww''$ then we have two cases:
  Either $\lambda ww'' \notin \Delta$ which implies $w'' \neq w'$. Hence, we can take $p \in \Pi$ and pick v s.t. $\lambda ww'w'' \in v(p)$ and $\lambda wv \notin v(p)$ for all $\lambda wv \in \Delta$, in which case

  $$\lambda w \models \Diamond_R \Diamond_R p \wedge \neg \Diamond_R p.$$

  Or there exists β s.t. we don't have $\lambda ww'w''\beta \in \Delta \leftrightarrow \lambda ww''\beta \in \Delta$. In this case we pick $w_0 = w''$ and define ψ as in Proposition 6.19.1. As before, we conclude that either $(W, \Delta, v), \lambda w \not\models \Diamond_R \Diamond_R \psi \to \Diamond_R \psi$ or $(W, \Delta, v), \lambda w \not\models \Diamond_R \Diamond_R \neg\psi \to \Diamond_R \neg\psi$.

Similarly to the case of strong reflexivity, strong transitivity imposes that, regardless of the path we have covered, every world accessible in two steps is accessible in one and the set of accessible worlds is the same in both cases.

2. Soundness of L_Π^4 comes easily from the soundness of the new axiom which has just been established.

Let us prove that $L_\Pi^4 = \{\varphi : \varphi \text{ is valid in every strongly reflexive reactive frame}\}$:

- using Proposition 6.13 it is easy to check that the canonical model for L_X^4 is a shattered model (W, R, P, V) where R is transitive.

- t_3 preserves R-transitivity:
  If $(x_1, i_1)R_3(x_2, i_2)R_3(x_3, i_3)$ then $x_k = (x_{k,1}, x_{k,2})$ for $k = 2, 3$, $i_1 \leq i_2 < i_3$ and $x_{3,1} = x_{3,2}$. If $x_1 \in W$ then immediately we conclude $(x_1, i_1)R_3(x_3, i_3)$. If $x_1 = (x_{1,1}, x_{1,2})$ then $x_{1,1}Rx_{2,1}Rx_{3,1}$. Hence $x_{1,1}Rx_{3,1} = x_{3,2}$ and $(x_1, i_1)R_3(x_3, i_3)$.
- given an R-transitive cs-frame with initial family I, (I, Δ^I) is a strongly transitive reactive frame:
  Let $\lambda ww'w'' \in \Delta^I$, so there is $\gamma \in W^*$ and $s, s', s'' \in W$ s.t. $\gamma ss's''$ is an I-initial path and $l_I(\gamma ss's'') = \lambda ww'w''$. Since F is R-transitive we have that $\gamma ss''$ is also an I-initial path and clearly $l_I(\gamma ss'') = \lambda ww'' \in \Delta$. Applying Lemma 6.18 we conclude that $\lambda ww'w'' \sim \lambda ww''$.

3. Let $X \subsetneq \Pi$ and $\mathcal{F}$ be a class of reactive frames.
If there is a reactive frame $F = (W, \Delta) \in \mathcal{F}$ with a path of length three, $w_0w_1w_2$, let v be s.t. $w_0w_1w_2 \in v(p)$ and $w_0v \notin v(p)$ for all $w_0v \in \Delta$, then $(W, \Delta, v), w_0 \models \Diamond_R\Diamond_R p \wedge \neg\Diamond_R p$.
Given a reactive frame (W, Δ) s.t. there are no $w_0, w_1, w_2 \in W$ s.t. $w_0w_1w_2 \in \Delta$ then it validates $\Diamond_R\Box_R\bot$ but $\Diamond_R\Box_R\bot \notin L^T_X$ since

$$L^4_X \subseteq L^4_\Pi \subseteq \{\varphi : (\{a\}, \{a\}^*, v) \models \varphi \ \& \ v(p) = \{a\}^* \text{ for all } p \in \Pi\} \not\ni \Diamond_R\Box_R\bot.$$

So if there is no frame with a path of length three in $\mathcal{F}$ (in particular if $\mathcal{F}$ is empty) we have

$$L^4_X \subsetneq \{\varphi : \varphi \text{ is valid in all } F \in \mathcal{F}\}.$$

Let us consider some variants of these axioms and see that they axiomatise other generalized notions of reflexivity and transitivity.

Proposition 6.21. *1. A reactive frame (W, Δ) X-validates $p \to \Diamond_R\Diamond_P p$ iff it is* **outwardly reflexive**, *i.e. satisfies:*

$$\lambda w \in \Delta \to \lambda ww \in \Delta.$$

$L^{T_o}_X = L_X \oplus p \to \Diamond_R\Diamond_P p$ is sound and complete with respect to the class of X-reactive models over outwardly reflexive frames.

2. Let $L^{T_i}_X = L_X \oplus p \to \Diamond_P\Diamond_R p$. If $X = \Pi$, $L^{T_i}_\Pi$ is sound and complete with respect to the class of Π-reactive models over **inwardly reflexive** *frames, i.e. satisfying:*

$$\lambda w \in \Delta \to \exists\lambda' \ \lambda w \sim \lambda' ww \in \Delta.$$

If $X \subsetneq \Pi$, $L^{T_i}_X$ is not sound with respect to (all X-models over) any class of reactive frames.

3. A reactive frame (W, Δ) X-validates $\Box_P p \to \Diamond_P\Diamond_R p$ iff is **lightly reflexive**, *i.e. satisfies:*

$$w \in W \to \exists\lambda \ \lambda ww$$

$L^{T_l}_X = L_X \oplus \Box_P p \to \Diamond_P\Diamond_R p$ is sound and complete with respect to the class of X-reactive models over inwardly reflexive frames.

It is clear that strong reflexivity implies inward and outward reflexivity and both imply light reflexivity.

Proof.

1. Given an X-reactive model $\mathcal{M} = (W, \Delta, v)$ over an outwardly reflexive reactive frame and $\lambda w \in \Delta$ s.t. $\mathcal{M}, \lambda w \models \varphi$ then since $\lambda ww \in \Delta$ and $t(\lambda w) = t(\lambda ww)$ we have $\mathcal{M}, \lambda ww \models \Diamond_P \varphi$ and so $\mathcal{M}, \lambda w \models \Diamond_R \Diamond_P \varphi$.

If a reactive frame (W, Δ) is not outwardly reflexive then there exists some $\lambda w \in \Delta$ s.t. $\lambda ww \notin \Delta$. Picking a $p \in \Pi$ and choosing a v s.t. $\lambda' \in v(p)$ iff $t(\lambda') = w$, now we have that $(W, \Delta, v), \lambda w \models p \wedge \neg \Diamond_R \Diamond_P p$.

Soundness of $L_X^{T_o}$ comes easily from the soundness of the new axiom which has just been established.

Using the same strategy as before we will obtain the equality:

$$L_X^{T_o} = \{\varphi : \varphi \text{ is } X\text{-valid in every outwardly reflexive reactive frame}\}.$$

- $q \to \Diamond_R \Diamond_P q$ is a Sahlqvist formula and so the canonical frame (W, R, P) for $L_X^{T_o}$ satisfies:
$$\forall w \exists w' w R w' P w \tag{a}$$
since it is the first-order correspondent to the new axiom.
- t_1 preserves property (a):
Given (x, i), if $i = *$ then $(x, i) R_1 (x, i) P_1 (x, i)$ and if $i = (x', a)$ then by property (a) there exists y s.t. $xRyPx$ thus $t'(x, x)$ is defined and $(x, i) R_1 (t'(x, x), i) P_1 (x, i)$.
- Let (W, R, P) be a cs-frame satisfying property (a) and I an initial family. (I, Δ^I) is an outward reflexive reactive frame:
Let $\lambda w \in \Delta^I$ then there are $\gamma \in W^*$ and $s \in W$ s.t. γs is an I-initial path and $l_I(\gamma s) = \lambda w$. By property (a) there exists $t \in W$ s.t. $sRtPs$ (which implies tPw) thus γst is also an I-initial path and $l_I(\gamma st) = \lambda ww \in \Delta^I$.

2. To establish soundness we have just to check whether the added axiom is sound: given a Π-reactive model $\mathcal{M} = (W, \Delta, v)$ over an inwardly reflexive reactive frame and $\lambda w \in \Delta$ s.t. $\mathcal{M}, \lambda w \models \varphi$ then since there exists λ' s.t. $\lambda' ww \in \Delta$ and $\lambda w \sim \lambda' ww$ by Proposition 6.5 we have $\mathcal{M}, \lambda' ww \models \varphi$ and since $t(\lambda w) = t(\lambda' w)$ we have that $\mathcal{M}, \lambda w \models \Diamond_P \Diamond_R \varphi$.

For completeness we proceed as before:

- $p \to \Diamond_P \Diamond_R q$ is a Sahlqvist formula and so the canonical frame $F = (W, R, P)$ for $L_\Pi^{T_i}$ satisfies:
$$\forall w \exists w' w P w' R w \tag{b}$$
since it is its first-order correspondent.
- t_2 preserves property (b): given $w = (a, b, c, d, i) \in W_2$, if $i = *$ then $wP_2 wR_2 w$ and if $i \neq *$ then $(a, b, c, d, i) P_2 (t(a), b + 1, c, d, i) R_2 (a, b, c, d, i)$.
- let (W, R, P) be a cs-frame satisfying property (b) and I an initial family. (I, Δ^I) is an inwardly reflexive reactive frame:
Let $\lambda w \in \Delta^I$, so there is $\gamma \in W^*$ and $s \in W$ s.t. γs is an I-initial path and $l_I(\gamma s) = \lambda w$. By property (b) there exists $t \in W$ s.t. $tRsPt$. From the initiality of I we know that there is some γ' s.t. $\gamma' t$ is an I-initial path, thus $\gamma' ts$ is also an I-initial path and $l_I(\gamma st) = l_I(\gamma) ww = \lambda' ww \in \Delta^I$. Applying Lemma 6.18 we conclude that $\lambda w \sim \lambda' ww$.

So

$$L_{\Pi}^{T_i} = \{\varphi : \varphi \text{ is valid in every inwardly reflexive reactive frame}\}.$$

Given $X \subsetneq \Pi$ and a reactive frame (W, Δ) there is an X-reactive model $\mathcal{M} = (W, \Delta, v)$, $\lambda \in \Delta$ and φ s.t. $\mathcal{M}, \lambda \not\models \varphi \to \Diamond_R \varphi$: given $\lambda w \in \Delta$ and some $p \in \Pi - X$, let v s.t. $\lambda w \in v(p)$ and $\lambda' ww \notin v(p)$ for all $\lambda' ww \in \Delta$. So

$$(W, \Delta, v), \lambda w \models p \wedge \neg \Diamond_R \Diamond p.$$

3. Let $\mathcal{M} = (W, \Delta, v)$ be an X-reactive model over a lightly reflexive reactive frame and $\lambda w \in \Delta$ s.t. $\mathcal{M}, \lambda w \models \Box_P \varphi$. Since there exists λ' s.t. $\lambda' ww \in \Delta$ and $t(\lambda w) = t(\lambda' ww)$ we have $\mathcal{M}, \lambda' ww \models \varphi$ and so $\mathcal{M}, \lambda' w \models \Diamond_R \varphi$ hence $\mathcal{M}, \lambda w \models \Diamond_P \Diamond_R \varphi$.

If a reactive frame (W, Δ) is not lightly reflexive then there exists some $w \in W$ for which there is no $\lambda \in \Delta$ s.t. $\lambda ww \in \Delta$. So we pick a $p \in \Pi$ and choose a v s.t. $\lambda' \in v(p)$ iff $t(\lambda') = w$. Now we have that $(W, \Delta, v), w \models \Box_P p \wedge \neg \Diamond_P \Diamond_R p$.

Soundness of $L_X^{T_l}$ comes easily from the soundness of the new axiom which has just been established.

For completeness, as before:

- $\Box p \to \Diamond_P \Diamond_R q$ is a Sahlqvist formula and (just as before) the canonical model (W, R, P) for $L_X^{T_o}$ satisfies:

$$\forall w \exists w' w'' \; wPw'Rw''Pw. \qquad \text{(c)}$$

  since it is its first-order correspondent.
- t_1 preserves property (c): given $(x, i) \in W_1$, if $i = 1$ then

$$(x, i)P_1(x, i)R_1(x, i)P_1(x, i)$$

  and if $i = (x', a)$ then by property (c) there exist y, z s.t. $xPyRzPx$ thus $t'(y, y)$ is defined and

$$(x, i)P_1(y, i)R_1(t(y, y), i)P_1(x, i).$$

- let (W, R, P) be a cs-frame satisfying property (c) and I an initial family. (I, Δ^I) is a lightly reflexive reactive frame:
  Let $\lambda w \in \Delta^I$ then there are s, t s.t. $l_I(s) = w = l_I(t)$ and sRt. By initiality of I there is an I-initial path ending in s, γs, making γst also an I-initial path, hence $l_I(\gamma)ww = \lambda' ww \in \Delta$.

So

$$L_X^{T_l} = \{\varphi : \varphi \text{ is } X\text{-valid in every lightly reflexive reactive frame}\}.$$

Proposition 6.22. *1. A reactive frame (W, Δ) X-validates $\Diamond_R \Diamond_R p \to \Diamond_R \Diamond_P p$ iff it is* **left transitive**, *i.e. satisfies:*

$$\lambda ww'w'' \in \Delta \to \lambda ww'' \in \Delta$$

$L_X^{4_l} = L_X \oplus \Diamond_R \Diamond_R p \to \Diamond_R \Diamond_P p$ is sound and complete with respect to the class of X-reactive models over left transitive reactive frames.

2. *A reactive frame* (W, Δ) *X-validates* $\Diamond_R\Diamond_R p \to \Diamond_P\Diamond_R\Diamond_P p$ *iff it is* **middle transitive**, *i.e. satisfies:*

$$\lambda ww'w'' \in \Delta \to \exists \lambda' \; \lambda' ww'' \in \Delta$$

$L_X^{4_m} = L_X \oplus \Diamond_R\Diamond_R p \to \Diamond_P\Diamond_R\Diamond_P p$ *is sound and complete with respect to the class of X-reactive models over middle transitive reactive frames.*

3. *Let* $L_X^{4_r} = L_X \oplus \Diamond_R\Diamond_R p \to \Diamond_P\Diamond_R p$. $L_\Pi^{4_r}$ *is sound and complete with respect to the class of* Π*-reactive models over* **right transitive** *reactive frames, i.e. satisfying:*

$$\lambda ww'w'' \in \Delta \to \exists \lambda' \; \lambda ww'w'' \sim \lambda' ww'' \in \Delta.$$

If $X \subsetneq \Pi$, $L_X^{4_r}$ *is not sound and complete with respect to (all X-models over) any class of reactive frames.*

4. *A reactive frame* (W, Δ) *X-validates* $\Diamond_R\Diamond_P\Diamond_R p \to \Diamond_R\Diamond_P p$ *iff it is* **globally left transitive**, *i.e. satisfies:*

$$\lambda ww', \lambda' w'w'' \in \Delta \to \lambda ww'' \in \Delta.$$

$L_X^{4_{gl}} = L_X \oplus \Diamond_R\Diamond_P\Diamond_R p \to \Diamond_R\Diamond_P p$ *is sound and complete with respect to the class of X-reactive models over globally left transitive reactive frames.*

5. *A reactive frame* (W, Δ) *X-validates* $\Diamond_R\Diamond_P\Diamond_R p \to \Diamond_P\Diamond_R\Diamond_P p$ *iff it is* **globally middle transitive**, *i.e. satisfies:*

$$\lambda ww', \lambda' w'w'' \in \Delta \to \exists \lambda'' \; \lambda'' ww'' \in \Delta.$$

$L_X^{4_{gm}} = L_X \oplus \Diamond_R\Diamond_P\Diamond_R p \to \Diamond_P\Diamond_R\Diamond_P p$ *is sound and complete with respect to the class of X-reactive models over globally transitive reactive frames.*

6. *Let* $L_X^{4_{gr}} = L_X \oplus \Diamond_R\Diamond_P\Diamond_R p \to \Diamond_P\Diamond_R p$. $L_\Pi^{4_{gr}}$ *is sound and complete with respect to the class of* Π*-reactive models over* **globally right transitive** *reactive frames, i.e. satisfies:*

$$\lambda ww', \lambda' w'w'' \in \Delta \to \exists \lambda'' \; \lambda' w'w'' \sim \lambda'' ww'' \in \Delta.$$

If $X \subsetneq \Pi$, $L_X^{4_{gr}}$ *is not sound and complete with respect to (all X-models over) any class of reactive frames.*

It is clear that strong transitivity implies left, middle and right transitivity. Both left and right transitivity imply middle transitivity and each of them is implied by its global version.

Proof. 1. Given an X-reactive model $\mathcal{M} = (W, \Delta, v)$ over a left transitive reactive frame and $\lambda w \in \Delta$ s.t.

$$\mathcal{M}, \lambda w \models \Diamond_R\Diamond_R\varphi,$$

i.e. there are w', w'' s.t. $\lambda ww'w'' \in \Delta$ and $\mathcal{M}, \lambda ww'w'' \models \varphi$. By left transitivity we get that $\lambda ww'' \in \Delta$ and since $t(\lambda ww'w'') = t(\lambda ww'')$ we have

$$\mathcal{M}, \lambda w \models \Diamond_R \Diamond_P \varphi.$$

If a reactive frame (W, Δ) is not left transitive then there exists $\lambda ww'w'' \in \Delta$ s.t. $\lambda ww'' \notin \Delta$. So we pick a $p \in \Pi$ and choose a v s.t. $\gamma \in v(p)$ iff $t(\gamma) = w''$. Now we have that

$$(W, \Delta, v), \lambda w \models \Diamond_R \Diamond_R p \wedge \neg \Diamond_R \Diamond_P p.$$

Soundness of $L_X^{4_l}$ follows from the soundness of the new axiom which has just been established.

Let us prove that $L_X^{4_l} = \{\varphi : \varphi \text{ is } X\text{-valid in every left transitive reactive frame}\}$:

- $\Diamond_R \Diamond_R p \rightarrow \Diamond_R \Diamond_P p$ is a Sahlqvist formula and so the canonical model (W, R, P) for $L_X^{4_l}$ satisfies:
$$\forall tvw \exists w' \; tRvRw \rightarrow tRw'Pw \qquad \text{(LT)}$$
since it is its first-order correspondent.
- t_4 preserves property (LT):
Given $(x_k, i_k) \in W_4$ for $k = 1, 2, 3$ s.t. $(x_1, i_1)R_4(x_2, i_2)R_4(x_3, i_3)$. So $i_1 \leq i_2 < i_3$, $x_k = (x_{k,1}, x_{k,2})$ for $k = 2, 3$. If $x_1 \in W$ we have that
$$(x_1, i_1)R_4((x_{3,2}, x_{3,2}), i_3)P_4(x_3, i_3).$$
If $x_1 = (x_{1,1}, x_{1,2})$ then $x_{1,1}Rx_{2,1}Rx_{3,1}$ hence, by property (LT), there exist x s.t. $x_{1,1}RxPx_{3,1}(Px_{3,2})$. Thus $t'(x_{1,1}, x_{3,2})$ is defined and
$$(x_1, i_1)R_4((t'(x_{1,1}, x_{3,2})), i_3)P_4(x_3, i_3).$$
- let (W, R, P) be a cs-frame satisfying property (LT) and I an initial family. (I, Δ^I) is a left transitive reactive frame:
Let $\lambda ww'w'' \in I$ then there are some $\gamma \in W^*$ and $s, s', s'' \in W$ s.t. $\gamma ss's''$ is an I-initial path and $l_I(\gamma ss's'') = \lambda ww'w''$. By (LT) there exists some $t \in W$ s.t. $sRtPs''Pw$ and so γst is an I-initial path. Hence $l_I(\gamma st) = \lambda ww'' \in \Delta^I$.

2. Given an X-reactive model $\mathcal{M} = (W, \Delta, v)$ over a middle transitive reactive frame and $\lambda w \in \Delta$ s.t.

$$\mathcal{M}, \lambda w \models \Diamond_R \Diamond_R \varphi,$$

i.e. there are w', w'' s.t. $\lambda ww'w'' \in \Delta$ and $\mathcal{M}, \lambda ww'w'' \models \varphi$. By middle transitivity, there exists λ' s.t. $\lambda'ww'' \in \Delta$. Since $t(\lambda ww'w'') = t(\lambda'ww'')$ and $t(\lambda w) = t(\lambda' w)$ we have

$$\mathcal{M}, \lambda w \models \Diamond_P \Diamond_R \Diamond_P \varphi.$$

If a reactive frame (W, Δ) is not middle transitive then there exists some $\lambda ww'w'' \in \Delta$ and no λ' s.t. $\lambda'ww'' \in \Delta$. So we pick a $p \in \Pi$ and choose a v s.t. $\gamma \in v(p)$ iff $t(\gamma) = w''$. We obtain

$$(W, \Delta, v), \lambda w \models \Diamond_R \Diamond_R p \wedge \neg \Diamond_P \Diamond_R \Diamond_P p.$$

Soundness of $L_X^{4_l}$ follows from the soundness of the new axiom which has just been established.

We prove completeness as before:

- $\Diamond_R\Diamond_R p \to \Diamond_P\Diamond_R\Diamond_P p$ is a Sahlqvist formula and so the canonical model (W, R, P) for $L_X^{4_m}$ satisfies:
$$\forall tvw \exists t'w'\ tRvRw \to tPt'Rw'Pw \qquad \text{(MT)}$$
since it is its first-order correspondent.
- t_4 preserves property property (MT):
Given $(x_k, i_k) \in W_4$ for $k = 1, 2, 3$ s.t. $(x_1, i_1)R_4(x_2, i_2)R_4(x_3, i_3)$. So $i_1 \le i_2 < i_3$, $x_k = (x_{k,1}, x_{k,2})$ for $k = 2, 3$. If $x_1 \in W$ we have that
$$(x_1, i_1)P_4(x_1, i_1)R_4((x_{3,2}, x_{3,2}), i_3)P_4(x_3, i_3).$$
If $x_1 = (x_{1,1}, x_{1,2})$ then $x_{1,1}Rx_{2,1}Rx_{3,1}$ hence, by property (MT), there are x, x' s.t. $x_{1,1}PxRx'Px_{3,1}(Px_{3,2})$ and so $t'(x, x_{3,2})$ is defined and
$$(x_1, i_1)P_4((x, x_{1,2}), i_1)R_4((t'(x, x_{3,2})), i_3)P_4(x_3, i_3).$$
- let (W, R, P) be a cs-frame satisfying property (MT) and I an initial family. (I, Δ^I) is a middle transitive reactive frame:
Let $\lambda ww'w'' \in I$ then there are some $\gamma \in W^*$ and $s, s', s'' \in W$ s.t. $\gamma ss's''$ is an initial path and $l_I(\gamma ss's'') = \lambda ww'w''$. From property (MT) it follows that there are some $t, t'' \in W$ s.t. $sPtRt''Ps''$. By initiality of I there is an I-initial path $\gamma't$, making $\gamma'tt''$ also an I-initial path thus $l_I(\gamma'tt'') = l_I(\gamma')ww'' = \lambda'ww'' \in \Delta^I$.

So $L_X^{4_m} = \{\varphi : \varphi$ is X-valid in every middle transitive reactive frame$\}$.

3. To establish soundness of $L_\Pi^{4_r}$ it is enough to verify soundness of the added axiom: given a Π-reactive model $\mathcal{M} = (W, \Delta, \nu)$ over a right transitive reactive frame and $\lambda w \in \Delta$ s.t.
$$\mathcal{M}, \lambda w \models \Diamond_R\Diamond_R\varphi,$$
so there are w', w'' s.t. $\lambda ww'w'' \in \Delta$ and $\mathcal{M}, \lambda ww'w'' \models \varphi$. By right transitivity there is $\lambda'ww'' \in \Delta$ s.t. $\lambda ww'w'' \sim \lambda'ww''$ and so, by Proposition 6.5, $\mathcal{M}, \lambda'ww'' \models \varphi$. Thus
$$\mathcal{M}, \lambda w \models \Diamond_P\Diamond_R\varphi.$$

For completeness:

- $\Diamond_R\Diamond_R p \to \Diamond_P\Diamond_R p$ is a Sahlqvist formula and so the canonical model (W, R, P) for $L_\Pi^{4_r}$ satisfies:
$$\forall tvw \exists t'\ tRvRw \to tPt'Rw \qquad \text{(RT)}$$
since it is its first-order correspondent.
- t_3 preserves property (RT):
Given $(x_k, i_k) \in W_3$ for $k = 1, 2, 3$ s.t. $(x_1, i_1)R_3(x_2, i_2)R_3(x_3, i_3)$. So $i_1 \le i_2 < i_3$, $x_k = (x_{k,1}, x_{k,2})$ for $k = 2, 3$ and $x_{3,1} = x_{3,2}$. If $x_1 \in W$ then immediately we conclude that $(x_1, i_1)P_3(x_1, i_1)R_3(x_3, i_3)$. If $x_1 = (x_{1,1}, x_{1,2})$ then $x_{1,1}Rx_{2,1}Rx_{3,1} = x_{3,2}$ so by property (RT) there exists an x s.t. $x_{1,1}PxRx_{3,1}$ and so $(x_1, i_1)P_3((x, x_{1,2}), i_1)R_3(x_3, i_3)$.

- let (W, R, P) be a cs-frame satisfying property (RT) and I an initial family. (I, Δ^I) is a right transitive reactive frame:
  Let $\lambda ww'w'' \in \Delta^I$ then there is an I-initial path $\gamma s_1 s_2 s_3$ ($s_i \in W$) s.t. $l_I(\gamma s_1 s_2 s_3) = \lambda ww'w''$. By property (RT) there exists $s \in W$ s.t. $s_1 P s R s_3$. From the initiality of I we know that there is some γ' s.t. $\gamma' s$ is an I-initial path and so $\gamma' s s_3$ is also an I-initial path. Applying Lemma 6.18 we conclude that $\Delta^I \ni l_I(\gamma' st) = l_I(\gamma')ww'' = \lambda' ww'' \sim \lambda ww'w''$.

Hence,

$$L_{\Pi}^{4_r} = \{\varphi : \varphi \text{ is } \Pi\text{-valid in every right transitive reactive frame}\}.$$

Given $X \subsetneq \Pi$ and a (non-empty) class of reactive frames $\mathcal{F}$:

- if there is a reactive frame $(W, \Delta) \in \mathcal{F}$ with a path $\lambda w_0 w_1 w_2$, where $w_0 \neq w_1$, we pick $p \in \Pi - X$ and v s.t. $\gamma \in v(p)$ iff $\gamma = \lambda w_0 w_1 w_2$. Thus

$$(W, \Delta, v), \lambda w_0 \models \Diamond_R \Diamond_R p \wedge \neg \Diamond_P \Diamond_R p.$$

- if $\mathcal{F}$ contains only reactive frames with paths of length greater than 2 of the form $\lambda www'$ then it validates $\Diamond_R(\varphi \wedge \Diamond_R \top) \to \Diamond_P \varphi$. Consider that $L_X^{4_r} \subseteq L_{\Pi}^{4_r}$ and the right transitive Π-reactive model: $\mathcal{M} = (\{0, 1\}, \{0, 1, 01, 011\}, v)$ s.t. $\lambda \in v(p)$ iff $t(\lambda) = 1$. It is clear that $\mathcal{M}, 0 \models \Diamond_R(p \wedge \Diamond_R \top) \wedge \neg \Diamond_P p$ and so:

$$L_X^{4_r} \subsetneq \{\varphi : \varphi \text{ is valid in all } F \in \mathcal{F}\}.$$

- if $\mathcal{F}$ contains only reactive frames with paths of length smaller than 3, following the same reasoning as in Proposition 6.20.3, together with the fact that the strong transitive reactive frame used there - $(\{a\}, \{a\}^*)$ - is also a right transitive reactive frame, we conclude that $L_X^{4_r}$ is not complete with respect to $\mathcal{F}$.

4. Given an X-reactive model $\mathcal{M} = (W, \Delta, v)$ over a globally left transitive reactive frame and $\lambda w \in \Delta$ s.t.

$$\mathcal{M}, \lambda w \models \Diamond_R \Diamond_P \Diamond_R \varphi,$$

i.e. there are $w', w'' \in W$ and λ' s.t. $\lambda ww', \lambda' w'w'' \in \Delta$ and $\mathcal{M}, \lambda' w'w'' \models \varphi$. By left transitivity, there is $\lambda ww'' \in \Delta$ and since $t(\lambda ww'') = t(\lambda' w'w'')$ we have

$$\mathcal{M}, \lambda w \models \Diamond_R \Diamond_P \varphi.$$

If a reactive frame (W, Δ) is not globally left transitive then there exists some $\lambda ww'$, $\lambda' w'w'' \in \Delta$ s.t. $\lambda ww'' \notin \Delta$. So we pick a $p \in \Pi$ and choose a v s.t. $\gamma \in v(p)$ iff $t(\gamma) = w''$. Hence

$$(W, \Delta, v), \lambda w \models \Diamond_R \Diamond_P \Diamond_R p \wedge \neg \Diamond_R \Diamond_P p.$$

Soundness of $L_X^{4_l}$ comes easily from the soundness of the new axiom which has just been established.

We establish completeness as before:

- $\Diamond_R\Diamond_P\Diamond_R p \to \Diamond_R\Diamond_P p$ is a Sahlqvist formula and so the canonical model (W, R, P) for $L_X^{4_{gl}}$ satisfies:
$$\forall tvv'w\exists w' \; tRvPv'Rw \to tRw'Pw \tag{LLT}$$
since it is its first-order correspondent.
- t_4 preserves property (LLT):
Given $(x_k, i_k) \in W_4$ for $k = 1, 2, 3, 4$ s.t. $(x_1, i_1)R_4(x_2, i_2)P_4(x_3, i_3)R_4(x_4, i_4)$. So $i_1 \leq i_2 = i_3 < i_4$, $x_k = (x_{k,1}, x_{k,2})$ for $k = 2, 3, 4$. If $x_1 \in W$ we have that
$$(x_1, i_1)R_4((x_{4,2}, x_{4,2}), i_4)P_4(x_4, i_4).$$
If $x_1 = (x_{1,1}, x_{1,2})$ then $x_{1,1}Rx_{2,1}Px_{3,1}Rx_{4,1}$ hence, by property (LLT), there exists x s.t. $x_{1,1}RxPx_{4,1}(Px_{4,1})$ and so $t'(x_{1,1}, x_{4,2})$ is defined and
$$(x_1, i_1)R_4((t'(x_{1,1}, x_{4,2})), i_3)P_4(x_4, i_4).$$
- given a cs-frame (W, R, P) satisfying property (LLT) with initial family I, (I, Δ^I) is a left transitive reactive frame: Let $\lambda ww', \lambda' w'w'' \in \Delta^I$ then there are I-initial paths $\gamma s_1 s_2$ and $\gamma' s_3 s_4$ ($s_i \in W$) s.t. $l_I(\gamma s_1 s_2) = \lambda ww'$ and $l_I(\gamma' s_3 s_4) = \lambda' w'w''$ (so $s_1 R s_2 P s_3 R s_4$). By property (LLT) there exists $s \in W$ s.t. $s_1 R s P s_4$. So $\gamma s_1 s$ is also an I-initial path and $l_I(\gamma s_1 s) = \lambda ww'' \in \Delta^I$.

Hence
$$L_X^{4_{gl}} = \{\varphi : \varphi \text{ is } X\text{-valid in every left transitive reactive frame}\}.$$

5. Given an X-reactive model $\mathcal{M} = (W, \Delta, v)$ over a transitive reactive frame and $\lambda w \in \Delta$ s.t.
$$\mathcal{M}, \lambda w \models \Diamond_R\Diamond_P\Diamond_R\varphi,$$
i.e. there are $w', w'' \in W$ and λ' s.t. $\lambda ww', \lambda' w'w'' \in \Delta$ and $\mathcal{M}, \lambda' w'w'' \models \varphi$. By globally middle transitivity, there is λ'' s.t. $\lambda'' ww'' \in \Delta$, thus
$$\mathcal{M}, \lambda w \models \Diamond_P\Diamond_R\Diamond_P\varphi.$$

If a reactive frame (W, Δ) is not globally middle transitive then there exists some $\lambda ww', \lambda' w'w'' \in \Delta$ s.t. for all λ'' we have $\lambda'' ww' \notin \Delta$. So we pick $p \in \Pi$ and v s.t. $\gamma \in v(p)$ iff $t(\gamma) = w''$. Hence
$$(W, \Delta, v), \lambda w \models \Diamond_R\Diamond_P\Diamond_R p \wedge \neg\Diamond_P\Diamond_R\Diamond_P p.$$
Soundness of $L_X^{4_l}$ comes easily from the soundness of the new axiom which has just been established.

Completeness:

- $\Diamond_R\Diamond_P\Diamond_R p \to \Diamond_P\Diamond_R\Diamond_P p$ is a Sahlqvist formula and so the canonical model (W, R, P) for $L_X^{4_l}$ satisfies:
$$\forall t, v, v', w\exists t', w' \; tRvPv'Rw \to tPt'Rw'Pw \tag{GM}$$
since it is the first-order correspondent to the new axiom.

- t_4 preserves property (GM):
  Given $(x_k, i_k) \in W_4$ for $k = 1, 2, 3, 4$ s.t. $(x_1, i_1)R_4(x_2, i_2)P_4(x_3, i_3)R_4(x_4, i_4)$. So $i_1 \leq i_2 = i_3 < i_4$, $x_k = (x_{k,1}, x_{k,2})$ for $k = 2, 3, 4$. If $x_1 \in W$ we have that

$$(x_1, i_1)P_4(x_1, i_1)R_4((x_{4,2}, x_{4,2}), i_4)P_4(x_4, i_4).$$

  If $x_1 = (x_{1,1}, x_{1,2})$ then $x_{1,1}Rx_{2,1}Px_{3,1}Rx_{4,1}$ hence, by property (GM), there is x, x' s.t. $x_{1,1}PxRx'Px_{4,1}$ and so $t'(x, x_{4,2})$ is defined and

$$(x_1, i_1)P_4((x, x_{1,2}), i_1)R_4((t'(x, x_{4,2})), i_3)P_4(x_4, i_4).$$

- given a cs-frame (W, R, P) satisfying property (GM) with initial family I, (I, Δ^I) is a globally middle transitive reactive frame:
  Let $\lambda ww', \lambda' w' w'' \in \Delta^I$ then there are I-initial paths $\gamma s_1 s_2$ and $\gamma' s_3 s_4$ ($s_i \in W$) s.t. $l_I(\gamma s_1 s_2) = \lambda ww'$ and $l_I(\gamma' s_3 s_4) = \lambda' w' w''$ (in particular $s_1 R s_2 P s_3 R s_4$). By property (GM) there exists $t, t' \in W$ s.t. $s_1 PtRt' R s_4$. By initiality of I, there is some γ'' s.t. $\gamma'' t$ is an I-initial path and so $\gamma'' tt'$ is also an I-initial path and $l_I(\gamma'' tt') = \lambda'' ww'' \in \Delta^I$.

Hence $L_X^{4_{gm}} = \{\varphi : \varphi$ is X-valid in every globally middle transitive reactive frame$\}$.

6. To establish soundness of $L_\Pi^{4_{gr}}$ it is enough to verify soundness of the added axiom: given a Π-reactive model $\mathcal{M} = (W, \Delta, v)$ over a globally right transitive reactive frame and $\lambda w \in \Delta$ s.t.

$$\mathcal{M}, \lambda w \models \Diamond_R \Diamond_P \Diamond_R \varphi,$$

i.e. there are $w', w'' \in W$ and λ' s.t. $\lambda ww', \lambda' w' w'' \in \Delta$ and $\mathcal{M}, \lambda' w' w'' \models \varphi$. By globally right transitivity we get that there is λ'' s.t. $\lambda'' ww'' \sim \lambda' w' w''$, so by Proposition 6.5 we get $\mathcal{M}, \lambda'' ww'' \models \varphi$ and so

$$\mathcal{M}, \lambda w \models \Diamond_P \Diamond_R \varphi.$$

For completeness:

- $\Diamond_R \Diamond_P \Diamond_R p \to \Diamond_P \Diamond_R p$ is a Sahlqvist formula and so the canonical model (W, R, P) for $L_X^{4_{gr}}$ satisfies:

$$\forall tvv'w \exists t' \; tRvPv'Rw \to tPt'Rw \tag{GR}$$

  since it is its first-order correspondent.
- t_3 preserves property (GR):
  Given $(x_k, i_k) \in W_3$ for $k = 1, 2, 3, 4$ s.t. $(x_1, i_1)R_3(x_2, i_2)P_3(x_3, i_3)R_3(x_4, i_4)$. So $i_1 \leq i_2 = i_3 < i_4$, $x_k = (x_{k,1}, x_{k,2})$ for $k = 2, 3, 4$ and $x_{4,1} = x_{4,2}$. If $x_1 \in W$ then immediately we conclude $(x_1, i_1)P_3(x_1, i_1)R_3(x_4, i_4)$. If $x_1 = (x_{1,1}, x_{1,2})$ then $x_{1,1}Rx_{2,1}Px_{3,1}Rx_{4,1} = x_{4,2}$ so by property (GR) there exists an x s.t. $x_{1,1}PxRx_{4,1}$ and so $(x_1, i_1)P_3((x, x_{1,2}), i_1)R_3(x_4, i_4)$.
- given a cs-frame (W, R, P) satisfying property (GR) with initial family I, (I, Δ^I) is a globally right transitive reactive frame:
  Let $\lambda ww', \lambda' w' w'' \in \Delta^I$ then there are I-initial paths $\gamma s_1 s_2$ and $\gamma' s_3 s_4$ ($s_i \in W$)

s.t. $l_I(\gamma s_1 s_2) = \lambda ww'$ and $l_I(\gamma' s_3 s_4) = \lambda' w' w''$ (so $s_1 R s_2 P s_3 R s_4$). By property (GR) there exists $s \in W$ s.t. $s_1 P s R s_4$. By initiality of I, there is some γ'' s.t. $\gamma'' s$ is an I-initial path and so $\gamma'' s s_4$is an I-initial path, thus $l_I(\gamma'' s s_4) = \lambda'' ww'' \in \Delta$. Applying Lemma 6.18 we conclude that $\lambda' w' w'' \sim \lambda'' ww''$.

So

$$L_\Pi^{4_{gr}} = \{\varphi : \varphi \text{ is valid in every globally right transitive reactive frame}\}.$$

Given $X \subsetneq \Pi$, $L_X^{4_{gr}}$ is not sound and complete for any class of reactive frames: It follows from the proof in 3. for $L_X^{4_r}$. In the first case we can also conclude that the frame does not validate $\Diamond_R \Diamond_P \Diamond_R p \to \Diamond_P \Diamond_R p$ thus $L_X^{4_{rl}} \subset L_X^{4_r}$ is not sound in relation to any class of frames containing a reactive frame with a path of the form $\lambda w_0 w_1 w_2$ with $w_0 \neq w_1$. Nor is it complete with respect to the other two cases considered since both reactive frames used in the proof are also globally right transitive.

6.3.2 Static and quasi-static frames

Proposition 6.23. *We have that* $(\Diamond_P \Diamond_R p \to \Diamond_R \Diamond_P p) \leftrightarrow (\Diamond_R \Box_P p \to \Box_P \Diamond_R p) \in L_X$ *and* $L_X^S = L_X \oplus \Diamond_P \Diamond_R p \to \Diamond_R \Diamond_P p = L_X \oplus \Diamond_R \Box_P p \to \Box_P \Diamond_R p$ *is sound and complete with respect to the class of X-reactive models over* **static** *reactive frames, i.e. satisfying:*

$$\lambda w, \lambda w' \in \Delta \to \lambda w \sim \lambda' w$$

Notice that if $X = \Pi$, the new axioms are equivalent to $\Diamond_P p \to p$ and that if we impose this axiom instead we get a result limited to this case. This tell us that in the other cases where we have this kind of restriction there may be better axiomatisations valid for all X.

Proof. Let $com_{PR}^r = \Diamond_P \Diamond_R p \to \Diamond_R \Diamond_P p$ and $chr_{PR} = \Diamond_R \Box_P p \to \Box_P \Diamond_R p$.

- Using the equality $L_{r,X} = L_X$:
  - chr_{PR} implies com_{PR}^r
    $\mathcal{M}, \lambda w \models \Diamond_P \Diamond_R \varphi$, since $\varphi \to \Box_P \Diamond_P \varphi$, we have $\mathcal{M}, \lambda w \models \Diamond_P \Diamond_R \Box_P \Diamond_P \varphi$. Applying chr_{PR} we get $\mathcal{M}, \lambda w \models \Diamond_P \Box_P \Diamond_R \Diamond_P \varphi$ and, again from $\psi \to \Box_P \Diamond_P \psi$ (equivalent to $\Diamond_P \Box_P \psi \to \psi$), we obtain $\mathcal{M}, \lambda w \models \Diamond_R \Diamond_P \varphi$.
  - com_{PR}^r implies chr_{PR}
    $\mathcal{M}, \lambda w \models \Diamond_P \Box_R \varphi$, since $\varphi \to \Box_P \Diamond_P \varphi$, we have $\mathcal{M}, \lambda w \models \Diamond_P \Box_R \Box_P \Diamond_P \varphi$. Applying com_{PR}^r we get $\mathcal{M}, \lambda w \models \Diamond_P \Box_P \Box_R \Diamond_P \varphi$ and, again from $\Diamond_P \Box_P \psi \to \psi$, we obtain $\mathcal{M}, \lambda w \models \Box_R \Diamond_P \varphi$.
- Soundness of L_X^S comes easily from the soundness of the new axiom:
  Given an X-reactive model $\mathcal{M} = (W, \Delta, v)$ over a static frame and $\lambda w \in \Delta$ s.t. $\mathcal{M}, \lambda w \models \Diamond_P \Diamond_R \varphi$ then there exists $\lambda' ww' \in \Delta$ s.t. $\mathcal{M}, \lambda' ww' \models \varphi$ thus $\lambda ww' \in \Delta$ (since $\lambda w \sim \lambda' w$) and $\mathcal{M}, \lambda w \models \Diamond_R \Diamond_P \varphi$ (since $t(\lambda ww') = t(\lambda' ww')$).
- For completeness we proceed as before:

- $\Diamond_P \Diamond_R p \to \Diamond_R \Diamond_P p$ is a Sahlqvist formula and so the canonical model $F = (W, R, P)$ for L^S_X satisfies PR-commutativity:

$$\forall xyz \exists y' \; xPyRz \to xRy'Pz$$

  since it is the first-order correspondent to the new axiom.
- PR-commutativity is preserved by t_1:
  Let $(x, i), (y, j), (z, k) \in W'$ s.t. $(x, i)P'(y, j)R'(z, k)$. If $i = *$ then $(x, i) = (y, j)$ and $(x, i)R'(z, k)P'(z, k)$. If $i \neq *$ then $xPyRz$, $i = j = (x', a)$ and $k = (z', b)$. So there exists y' s.t. $xRy'Pz$, and so $t'(x, v)$ is defined for any v s.t. vPz. Either $i = j = k$, so $t'(x, x)$ is defined and $(x, i)R'(t'(x, x), k)P'(z, k)$, or $i = j \neq k$, so $t'(x, z')$ is defined and $(x, i)R'(t'(x, z'), k)P'(z, k)$.

- Given a PR-commutative cs-frame with initial family I then (I, Δ^I) is a static reactive frame:
  Let $\lambda w, \lambda' w \in \Delta^I$ so there are βb_1 and $\beta' b'_1$ I-initial paths s.t. $l_I(\beta b_1) = \lambda w$ and $l_I(\beta' b'_1) = \lambda' w$. Let us see that for every $\gamma = w_1 \ldots w_n$ we have $\beta b_1 \gamma$ iff there exist $\gamma' = v_1 \ldots v_n$ s.t. $\beta' b_1 \gamma'$ and $v_i P w_i$ for $i = 1, \ldots, n$. By induction on n (we do it only in one direction, the other is the same):
  - $n = 1$
    $\beta b_1 w_1$ is an I-initial path so by PR-commutativity there exists v_1 s.t. $b_2 R v_1 P w_1$ and so $\beta' b_2 v_1$ is an I-initial path.
  - $n + 1$
    $\beta b_1 w_1 \ldots w_n w_{n+1}$ is an I-initial path so by PR-commutativity there exists v_{n+1} s.t. $v_n R v_{n+1} P w_{n+1}$ and so $\beta' b_2 v_1$ is an I-initial path.

  Hence we conclude that $\lambda w \sim \lambda' w$.

A static frame is a reactive frame that does not react, that is, the accessible worlds depend only on the current world and not on how you got there. It is clear that from such a reactive frame (W, Δ) we can obtain a classic Kripke frame (W, R) where

$$R = \{(w, v) : wv \in \Delta\} = \{(w, v) : \lambda wv \in \Delta\}.$$

It is straightforward to see that in such a reactive frame all the variants of transitivity and reflexivity on reactive frames coincide with the usual notions on the correspondigt Kripke frame.

It is easy to see that a Kripke model over a general frame (W, R, P, A), where P is an equivalence relation, is a Π-shattered model iff $A \subseteq 2^W$ – the boolean algebra generated by the P equivalence classes – is closed for the operators: $m_R(X) = \{x \in W : \exists y \in W \text{ s.t. } xRy\}$ and $m_P(X) = \{x \in W : \exists y \in W \text{ s.t. } xPy\}$. As P is an equivalence relation, A is trivially closed under m_P. It is easy to see that A being closed under m_R means that if a world is in $m_R(X)$ so must be all of its P-class, which corresponds to PR-commutativity. From the point of view of reactivity, the use of (shattered) general frames to deal with the restrictions over the valuations (even if only in the case of $X = \Pi$) does not help, it demands a very strict interaction between R and P, which corresponds, at the reactive level to asking it not to react! See [120] for an equivalent presentation of Π-shattered models (with only $\Diamond_R$) and its relation with general frames.

Proposition 6.24. *Let* $L_X^{qS} = L_X \oplus \Diamond_R\Diamond_P p \rightarrow \Diamond_P\Diamond_R p$.

L_Π^{qS} *is sound and complete with respect to the class of (*Π*-shattered models over)* **quasi-static** *reactive frames, i.e. that satisfy:*

$$\lambda ww', \lambda' w' \in \Delta \rightarrow \exists \lambda'' \; \lambda' w' \sim \lambda'' ww' \in \Delta.$$

If $X \subsetneq \Pi$, L_X^{qS} *is not sound and complete with respect to (all* X*-models over) any class of reactive frames.*

Proof. • Soundness:
Given a Π-reactive model $\mathcal{M} = (W, \Delta, v)$ over a quasi-static reactive frame and $\lambda w \in \Delta$ s.t. $\mathcal{M}, \lambda w \models \Diamond_R\Diamond_P\varphi$ then there exists $\lambda ww', \lambda' w' \in \Delta$ s.t. $\mathcal{M}, \lambda' w' \models \varphi$ thus $\lambda' w' \sim \lambda'' ww' \in \Delta$ and $\mathcal{M}, \lambda'' ww' \models \varphi$ (by Lemma 6.5). Hence $\mathcal{M}, \lambda w \models \Diamond_P\Diamond_R\varphi$ (since $t(\lambda ww') = t(\lambda' ww')$).

• Completeness:
$\Diamond_R\Diamond_P p \rightarrow \Diamond_P\Diamond_R p$ is a Sahlqvist formula and so the canonical frame $F = (W, R, P)$ for L_X^{qS} satisfies RP commutativity:

$$\forall xyz \exists y' \; xRyPz \rightarrow xPy'Rz$$

since it is the first-order correspondent to the new axiom.

Given an RP-commutative shattered frame $M = (W, R, P, V)$, let $M' = (W', R', P', V')$ be defined by:
- $W' = (W^+)^3 \cup (W^+)^2 \times \{*\}$
- $(a, b, c)R'(a', b', c')$ iff $c = *$ & $a = a'$ or
  $c \neq * \neq c'$ & $o(a)Ro(a')$ &
  ($a \in W$ & $a' = b' = c'$ or $a \notin W$ &
  ($b' \neq a' \neq c'$ & $r(a) = a'$ or $r(a) = b'$ & $a' = c'$ or $r(a) = c'$ & $a' = b'$))
- $(a, b, c)P'(a', b', c')$ iff $(b, c) = (b', c')$ & $o(a)Po(a')$
- $(a, b, c) \in V'(p)$ iff $o(a) \in V(p)$,

where $o(w_1 \ldots w_n) = w_1$, $t(w_1 \ldots w_n) = w_n$ and $r(w_1 \ldots w_n) = w_2 \ldots w_n$.
We have that (W', R', P') is an RP-commutative cs-frame and M' is a bounded morphic image of M:
- (W', R', P') is a RP-commutative cs-frame. Let $(a, b, c)R'(a', b', c')P'(a'', b'', c'')$. If $c = *$ then $a = a'$ and $o(a) = o(a')Po(a'')$ thus

$$(a, b, c)P'(a'', b, c)R'(a'', b'', c'').$$

  If $c \neq * \neq c'$ then $o(a)Ro(a')Po(a'')$ so there exists w s.t. $o(a)PwRo(a'')$. We have three possibilities:
  either $b' = b'' \neq a'' \neq c'' = c'$ and so $(a, b, c)P'(wa'', b, c)R'(a'', b'', c'')$;
  or $a'' = b''$ thus

$$(a, b, c)P'(wc'', b, c)R'(a'', b'', c'');$$

  or $a'' = c''$ and

$$(a, b, c)P'(wb'', b, c)R'(a'', b'', c'').$$

- Clearly V' is admissible and any choice of I picking one element from each P-class and containing $\{(a, a, *) : a \in W^+\}$ works as an initial family.
- Let us check that it is coherent. Let $(a, b, c)R'(a', b', c')$, $(a, b, c)R'(a'', b'', c'')$ and $(a', b', c')P'(a'', b'', c'')$. From $(a', b', c')P'(a'', b'', c'')$ we conclude that $(b', c') = (b'', c'')$. If $c = *$ then $a' = a = a''$. If $c \neq * \neq c'$, then either $a \in W$ and $a' = b' = c' = c'' = b'' = a''$; or $a \notin W$, in which case we have three subcases: either $b' \neq a' \neq c'$ and so $r(a) = a' = a''$; or $a' = b'$, thus $r(a) = c' = c''$ and $a'' = b'' = b' = a'$; or $a' = c'$ and $r(a) = b' = b''$ so $a'' = c'' = c' = a'$. In any case, $(a', b', c') = (a'', b'', c'')$.
- Let $f : (W^+)^3 \to W$ be defined by $f((a, b, c)) = o(a)$.
  It is straightforward to see that f is surjective, let us see that it is actually a p-morphism:
  The condition in the valuation is trivial as before;
  · f is a p-morphism in R
    · $(a, b, c)R'(a', b', c')$ then $f((a, b, c)) = o(a)Ro(a') = f((a', b', c'))$,
    · $f((a, b, c))Rw$,
      if $a \in W$ then $(a, b, c)R'(w, w, w)$ otherwise $(a, b, c)R'(w, r(a), w)$;
  · f is a p-morphism in P
    · $(a, b, c)P'(a', b', c')$ then $f((a, b, c)) = o(a)Po(a') = f((a', b', c'))$,
    · $f((a, b, c))Pw$ then $(a, b, c)P'(w, b, c)$.

• Given a PR-commutative cs-frame with initial family I then (I, Δ^I) is a such a reactive frame:
  Let $\lambda ww', \lambda' w' \in \Delta^I$, so there is $\gamma, \gamma' \in W^*$ and $s, s', s'' \in W$ s.t. $\gamma ss'$ and $\gamma' s''$ are I-initial paths, $l_I(\gamma ss') = \lambda ww'$ and $l_I(\gamma' s'') = \lambda' w'$. So $sRs'Ps''$, thus there exists a t s.t. $sPtRs''$. Let γ'' be s.t. $\gamma'' t$ is an I-initial path, $\gamma'' ts''$ is I-initial too. Furthermore, using Lemma 6.18, we conclude that $\lambda w' \sim \lambda'' ww' = l_I(\gamma'' ts'')$.

It is clear that if a reactive frame is static then it is also quasi-static. Furthermore it is interesting to notice that if a frame is static all the variants of transitivity and reflexivity coincide.

6.3.3 Open problem — symmetry

Our method seems not to be as fruitful with the notions of generalized symmetry. We are unable to prove reactive frame completeness. To prove that it is sound with respect to a certain class of reactive frames and that it is complete with respect to the corresponding class of shattered frames is straightforward as before. However completeness cannot be proved in the same way. We just cannot simply pass from the shattered to cs-frames. Let us look just at the case of strong symmetry.

Conjecture 6.25. Let $L^B_X = L_X \oplus p \to \Box_R \Diamond_R p$. L^B_Π is sound and complete with respect to the class of reactive frames that are **strongly symmetric**, i.e. satisfying:

$$\lambda ww' \in \Delta \to \lambda w \sim \lambda ww'w \in \Delta.$$

Still we have not been able to prove it is not complete either. We present the positive result as a conjecture and add a brief explanation on how our attempts failed.

We present an R-reflexive shattered frame that cannot be transformed into a cs-frame using the classic backward truth-preserving transformations.

If a shattered frame (W, R, P) with $R \neq \emptyset$ satisfies for all w there is w' s.t. wPw', there is no w'' satisfying $w'Rw''$ or $w''Rw'$, then it does not admit an initial family.

Let us assume the contrary, let I be its initial family and

$$A = \{w : \text{ there is no } w' \text{ s.t. } wRw' \text{ or } w'Rw\}.$$

Since I R-generates W, the isolated points must be in I, so $A \subseteq I$. From the fact that I picks only one element from each P-class and for every w there is $w' \in A$ s.t. wPw' we have that $I = A$. As there is an element in $W - A$ I does not R-generate all W.

It is also clear that a frame that has such a frame as a generated subframe cannot have an initial family. Or else the elements of the initial family of the bigger frame, present in the smaller one, would be an initial family to the latter.

Furthermore, the operation of taking pre-images, ultrafilter extensions and ultrapowers preserves this (bad) property.

The following shattered frame $(\{0, 1\}, \{(1, 1)\}, \{0, 1\}^2)$ is R-symmetric and satisfies the bad property. Hence classical ways of generating new models from old preserving modal satisfaction do not allow us to find a general recipe to convert the relevant shattered frames into the corresponding cs-frames. In particular no variation of blow up will work.

This may not be a limitation on the method, instead it may be that the conjecture is not true and that the missing axiom would restrict us to a class of shattered frames that do not have this property. We have not been able to prove either way.

In the presence of (strong) reflexivity this problem disappears and the blow up method works:

Proposition 6.26. *Let* $L_X^{TB} = L_X^T \oplus p \rightarrow \Box_R \Diamond_R p = L_X^B \oplus p \rightarrow \Diamond_R p$. L_Π^{TB} *is sound and complete with respect to the class of strongly symmetric reactive frames that are also strongly reflexive.*

Proof. Soundness is obtained just as before by checking that $p \rightarrow \Box_R \Diamond_R p$ is sound:

Given a strongly symmetric reactive model $\mathcal{M} = (W, \Delta, v)$ and $\lambda w \in \Delta$ s.t. $\mathcal{M}, \lambda w \models \varphi$ then given v s.t. $\lambda wv \in \Delta$ then $\lambda wv \sim \lambda wvw \in \Delta$ so, by Proposition 6.5, $\mathcal{M}, \lambda wvw \models \varphi$ thus $\mathcal{M}, \lambda wv \models \Diamond_R \varphi$. Hence $\mathcal{M}, \lambda w \models \Box_R \Diamond_R \varphi$

In this case we are able to prove completeness by applying the blow up method:

Given an R-symmetric and R-reflexive shattered frame $M = (W, R, P, V)$, let $M' = (W', R', P', V)$ be defined by:

- $W' = \bigcup_{n<\omega} W^{2n+1} \times \bigcup_{n<\omega} W^{2n+1}$
- $(x, y)R'(x', y')$ iff $x = x'$ or
  $(y = y' \ \& \ wRw' \ \& \ (x = z \ \& \ x' = z' \text{ or } x' = z \ \& \ x = z') \ \&$
  $z = (w, \bar{v}) \ \& \ z' = (w', w', w, \bar{v}))$

- $(x, y)P'(x', y')$ iff $y = y'$ & $x = (w, \bar{v})$ & $x' = (w', \bar{v})$ & wPw'
- $((x_1, \ldots, x_k), w) \in V'(p)$ iff $x_1 \in V(p)$.
  where $\bar{v} = v_1, \ldots, v_k$.

We have that (W', R', P') is a cs-frame (where R' is clearly symmetric and reflexive) and M' is a bounded morphic image of M:

- clearly V' is admissible and any choice of I picking one element from each P-class and containing $\{(x, y) \in W' : x = y\}$ works as an initial family.
- let us check that it is coherent. Given $(x, y), (x', y'), (x'', y'') \in W'$. If $(x', y')P'(x'', y'')$ then $x' = (v, \bar{v})$, $x'' = (v', \bar{v})$ (so $|x'| = |x''|$) with wPw' and $y' = y''$. Let us assume that we also have $(x, y)R'(x', y')$ and $(x, y)R'(x'', y'')$.
  If $x = x'$ and $x \neq x''$ then $|x'| = |x| \neq |x''|$ which contradicts $(x', y')P'(x'', y'')$, so $x = x''$. The same applies if $x = x''$ and $x \neq x'$.
  If $x \neq x'$ and $x' \neq x''$ then either $x = (w, \bar{v})$ and $x' = (w', w', w, \bar{v}) = x''$, or $x = (w', w', w, \bar{v})$ and $x' = (w, \bar{v}) = x''$.
  In any case: $(x', y') = (x'', y'')$.
- let $f : W' \to W$ be defined by $f(((w, \bar{x}), y)) = w$.
  It is straightforward to see that f is surjective, let us see that it is actually a p-morphism:
  The condition in the valuation is trivial as before;
  - f is a p-morphism in R
    - $((w, \bar{x}), y)R'((w', \bar{x}), y')$ then either $w = w'$ or wRw', in any case $f(((w', \bar{x}), y)) = wRw' = f(((w', \bar{x}), y'))$.
    - $f(((w, \bar{x}), v))Rw'$ then $(((w, \bar{x}), v))R'(((w', w', w, \bar{x}), v))$;
  - f is a p-morphism in P
    - $((w, \bar{x}), y)P'((w', \bar{x}), y')$ then $f(((w', \bar{x}), y)) = wPw' = f(((w', \bar{x}), y'))$.
    - $f(((w, \bar{x}), v))Pw'$ then $(((w, \bar{x}), v))P'(((w', w, \bar{x}), v))$.

    In Proposition 6.19 we have checked that given an R-reflexive cs-frame with initial family I then (I, Δ^I) is a strongly reflexive reactive frame. Let us see now that if the cs-frame is also R-symmetric then (I, Δ^I) is also strongly symmetric:

    Let $\lambda ww' \in \Delta^I$, so there exists $\gamma \in W^*$ and $s, s' \in W$ s.t. $\gamma ss'$ is an I-initial path and $l_I(\gamma ss') = \lambda ww'$. Since F is R-symmetric we have that $\gamma ss's$ is also an I-initial path and clearly $l_I(\gamma ss's) = \lambda ww'w \in \Delta$. Applying Lemma 6.18 we conclude that $\lambda w \sim \lambda ww'w$.

This result tells us that if the above conjecture is not true, the formula valid in all strong symmetric reactive frames that is not in L_X^B must be an L_X-consequence of $p \to \Diamond_R p$.

6.4 Results and final comments

Table 6.1 summarizes the main results presented in this chapter. One can read along its lines the correspondence between logic, reactive frame property (including its dependence on X) and the shattered frame unfolded property.

For example, in the first line we see that the logic L_X corresponds to all shattered frames and to the whole class of (X-models over) reactive frames, hence being the smallest (minimal) "reactive" logic. In general there is a dependence on X, for instance in the sixth line, we see that the completeness of $L_X^T = L \oplus p \to \Diamond p$ with respect to the class of (all X-models over) strong reflexive reactive frames requires that $X = \Pi$, i.e. that all variables have to be fixed. We do not know whether this dependence can be avoided by a more inspired choice of axiom as in the case of static reactive frames, see Proposition 6.23.

In Section 6.3.3 we stated that the question of axiomatizing the generalized notions of symmetry is still open. The same is true of many other completeness results. It is not clear that the combination of the axioms corresponding to certain reactive properties will correspond to the combination of these properties. This is the case in the classical setting when we add reflexivity to transitivity or symmetry. It may be that the reactive properties interact and these classes satisfy formulas not captured by the axioms. It certainly does not follow from the results we presented above. For that to happen the (blow up) transformation would have to preserve the combination of all properties in question. The fact is that we have been unable to find a transformation that would work for all cases, in particular when considering generalizations of different properties. In this sense our choice of transformations is very weak. By contrast, in the strong symmetry plus strong reflexivity case we see how gathering properties may also be helpful. The strong symmetry case alone is worse behaved that the combined one.

In this chapter we have used classical tools to study completeness in a new interpretation of modal logic. We have been successful in studying some of its subsystems although we are aware that it may fail (see Section 6.3.3) and we should not expect it to work in every case. Furthermore it would be interesting to find some new methods that would allow a more direct way of studying these logics, leading to a reactivisation of modal techniques.

logic	X	reactive property		unfolded property
$L = K_R \oplus S5_P +_X p \leftrightarrow \Box_P p$	any		minimal	
$L^S_\Pi = L_X \oplus \Diamond_P p \to p$	Π			
$L^S_X = L_X \oplus \Diamond_P \Diamond_R p \to \Diamond_R \Diamond_P p$	any	$\lambda w, \lambda w' \in \Delta \to \lambda w \sim \lambda' w$	static	$\forall xyz \exists y'\ xPyRz \to xRy'Pz$
$L^S_X = L_X \oplus \Diamond_R \Box_P p \to \Box_P \Diamond_R p$	any			
$L^{qS}_X = L_X \oplus \Diamond_R \Diamond_P p \to \Diamond_P \Diamond_R p$	Π	$\lambda ww', \lambda' w' \in \Delta \to \exists \lambda''\ \lambda' w' \sim \lambda'' ww' \in \Delta$	q-static	$\forall xyz \exists x'\ xRyPz \to xPx'Ry$
$L^T_X = L \oplus p \to \Diamond p$	Π	$\lambda w \in \Delta \to \lambda w \sim \lambda ww \in \Delta$	gener-	R-reflexive
$L^{T_o}_X = L \oplus p \to \Diamond_R \Diamond_P p$	any	$\lambda w \in \Delta \to \lambda ww \in \Delta$	alising	$\forall w \exists w' wRw'Pw$
$L^{T_i}_X = L \oplus p \to \Diamond_P \Diamond_R p$	Π	$\lambda w \in \Delta \to \exists \lambda'\ \lambda w \sim \lambda' ww \in \Delta$	reflex-	$\forall w \exists w' wPw'Rw$
$L^{T_l}_X = L \oplus \Box_P p \to \Diamond_P \Diamond_R p$	any	$w \in W \to \exists \lambda\ \lambda ww$	ivity	$\forall w \exists w' w''\ wPw'Rw''Pw$
$L^4_X = L \oplus \Diamond_R \Diamond_R p \to \Diamond_R p$	Π	$\lambda ww'w'' \in \Delta \to \lambda ww'w'' \sim \lambda ww'' \in \Delta$		R-transitive
$L^{4_l}_X = L \oplus \Diamond_R \Diamond_R p \to \Diamond_R \Diamond_P p$	any	$\lambda ww'w'' \in \Delta \to \lambda ww'' \in \Delta$	general-	$\forall tvw \exists w'\ tRvRw \to tRw'Pw$
$L^{4_m}_X = L \oplus \Diamond_R \Diamond_R p \to \Diamond_P \Diamond_R \Diamond_P p$	any	$\lambda ww'w'' \in \Delta \to \exists \lambda'\ \lambda' ww'' \in \Delta$	ising	$\forall tvw \exists t'w'\ tRvRw \to tPt'Rw'Pw$
$L^{4_r}_X = L \oplus \Diamond_R \Diamond_R p \to \Diamond_P \Diamond_R p$	Π	$\lambda ww'w'' \in \Delta \to \exists \lambda'\ \lambda ww'w'' \sim \lambda' ww'' \in \Delta$		$\forall tvw \exists t'\ tRvRw \to tPt'Rw$
$L^{4_{gl}}_X = L \oplus \Diamond_R \Diamond_P \Diamond_R p \to \Diamond_R \Diamond_P p$	any	$\lambda ww', \lambda' w'w'' \in \Delta \to \lambda ww'' \in \Delta$	transi-	$\forall tvv'w \exists w'\ tRvPv'Rw \to tRw'Pw$
$L^{4_{gm}}_X = L \oplus \Diamond_R \Diamond_P \Diamond_R p \to \Diamond_P \Diamond_R \Diamond_P p$	any	$\lambda ww', \lambda' w'w'' \in \Delta \to \exists \lambda''\ \lambda'' ww'' \in \Delta$	tivity	$\forall tvv'w \exists t'w'\ tRvPv'Rw \to tPt'Rw'Pw$
$L^{4_{gr}}_X = L \oplus \Diamond_R \Diamond_P \Diamond_R p \to \Diamond_P \Diamond_R p$	Π	$\lambda ww', \lambda' w'w'' \in \Delta \to \exists \lambda''\ \lambda' w'w'' \sim \lambda'' ww'' \in \Delta$		$\forall tvv'w \exists t'\ tRvPv'Rw \to tPt'Rw$
$L^{TB}_X = L^T_X \oplus p \to \Box_R \Diamond_R p = L^B_X \oplus p \to \Diamond_R p$	Π	$\lambda ww' \in \Delta \to \lambda w \sim \lambda ww \in \Delta\ \&\ \lambda w \sim \lambda ww'w \in \Delta$	ref.+trans.	R-reflexive and R-symmetric

Table 6.1: Table of results

7

Global View on Reactivity: Switch Graphs and their Logics

7.1 Introduction

In computer science the word reactivity has been used to denote systems that react to their environment and are not meant to terminate, as coined by Pnueli and Harel in [104]. In this chapter the word has a different meaning, reactive systems are history-dependent relational structures, where the accessibility relation is determined not only by the point where one is, but also by the previous transitions. This concept was introduced by us in 2004, see [67].

In Chapter 2 we introduced the idea of enriching graph-based structures with arrows of a new type, called double arrows. Double arrows, instead of connecting points, connect arrows with arrows or other double arrows, see Figure 7.1. The idea is that this new kind of arrows can represent the dependence of the state of the targeted arrow (or double arrow) upon the crossing of the arrow in its origin. In this first presentation the double arrows would simply change the targeted arrow's state. Let us see how it works by playing with the example in Figure 7.1. We represent the fact that an arrow is off by drawing its body as a dotted line. Let us see the effect that crossing some of its edges has. As shown in Figure 7.2, when we cross the edges (a,b) (left) or (c,d) (right), the arrows that are in the scope of the double arrows coming out of them, become off or on if they are on or off respectively (that is, their state changes). This process is cumulative; after crossing (a,b) we can also cross (b,c) and the effects are determined by the new state of the double arrows, see Figure 7.3. These ideas were presented using suggestive motivational cases, for example in Figure 7.4 we see how these new arrows can represent a classical inheritance networks case.

The general idea that a relational structure may vary when one moves through it and the enriched kind of frames that came along with it fuelled publications in many areas. Indeed, there are applications of the reactive idea in such diverse areas as modal logic, preferential non-monotonic logic, inheritance systems, context-free grammars, automata theory, deontic logic and contrary-to-duty, argumentation and other networks, see papers [16, 37, 65, 75, 72, 73, 68, 130]. Some of this material is included in the rest of this book. For example when one adds these kind of double

D.M. Gabbay, *Reactive Kripke Semantics*, Cognitive Technologies,
DOI 10.1007/978-3-642-41389-6_7, © Springer-Verlag Berlin Heidelberg 2013

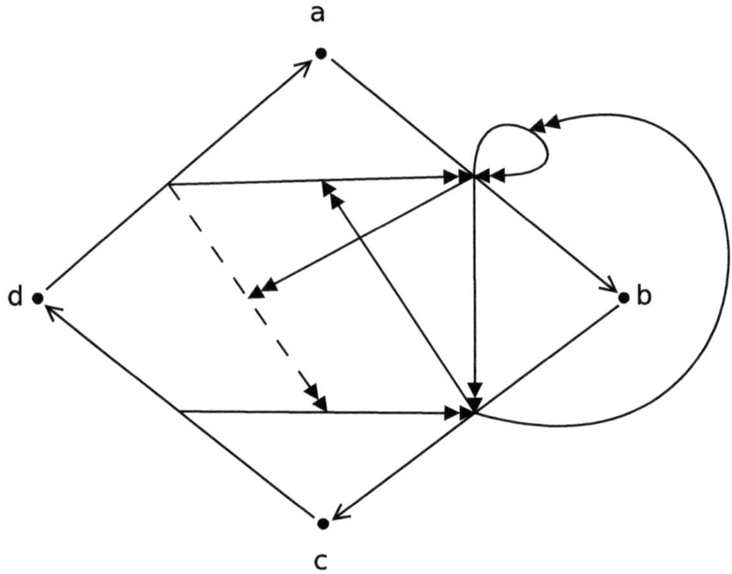

Fig. 7.1: An enriched graph.

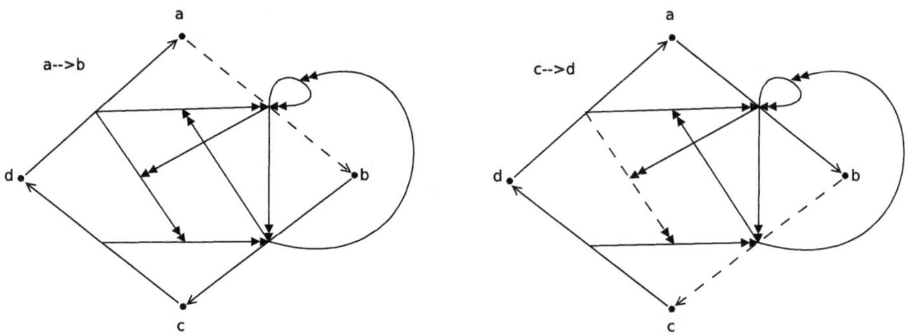

Fig. 7.2: The effect of crossing edges.

arrows to the structure of an automaton, one allows it to modify its transition relation while reading a sentence, that is, one makes it reactive. This alternative paradigm competes with non-determinism in the task of obtaining automata with the minimal number of states accepting a language. Indeed, the following theorem (Proposition 6.1, [37] and Proposition 8.8 of Chapter 8 of this book) is proven:

> If A is a deterministic automaton with k^n states, it has an equivalent reactive automaton $R(A)$ with kn states.

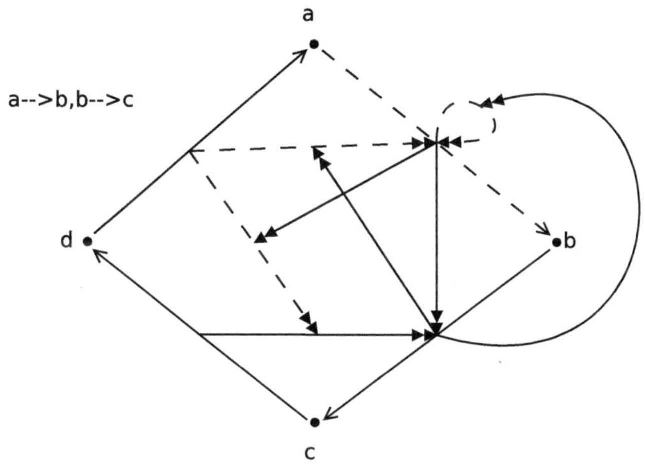

Fig. 7.3: The effect of crossing edges.

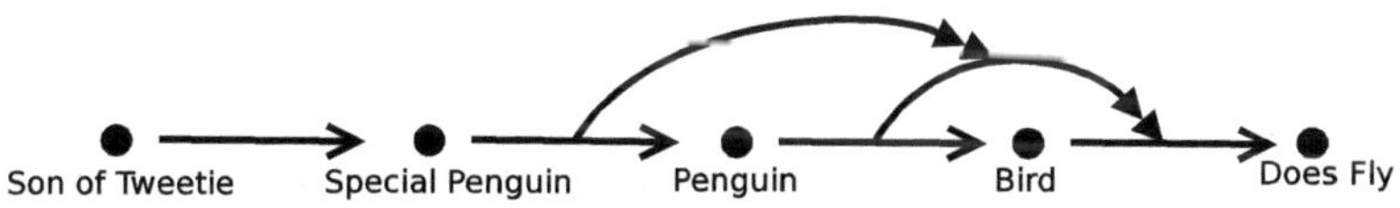

Fig. 7.4: A classical inheritance networks example. We can represent simple exceptions: birds do fly, but although penguins are birds they do not fly. And we can also represent higher order exceptions (exceptions to exceptions): even though son of Tweety is a special penguin (so also a penguin) he does fly.

Another interesting application of this kind of enriched graphs (though not using the dynamical counterpart) can be found in [130], where Dung's abstract argumentation theory is extended, incorporating meta-level argumentation-based reasoning about possibly conflicting preferences between arguments.

Changing Kripke structures

One can say that the relational semantics of modal logic already encompasses change. In fact, one can consider (or access) different worlds, and the propositional truths, given by the propositional variables' valuation, may change. Also the accessible worlds may change with these transitions. Yet, the truths at a given world are 'still'. In a Kripke model both propositional truths and the accessible worlds are fixed for each world. One can take it a step further and let these vary. In many situations it makes sense to consider semantics such that when certain operators are evaluated, the model, where the formula is being evaluated, changes. Therefore, the interpretation of a formula in the scope of a modal operator is given by a general condition of the type:

$$\mathfrak{M}, x \models \Diamond\varphi \text{ if } \mathfrak{M}', x' \models \varphi$$

where x' is a point in a new model $\mathfrak{M}'$. Indeed there are various examples of such approaches, e.g.:

- in dynamic epistemic logics with agents' public announcements [155];
- in sabotage logics edges can be deleted [154];
- in memory logics one may keep the information that a certain world was visited, adding it to the memory of the model [11];
- in hyper-modalities the meaning of the modal operators depends on where in the formula they occur [69];
- in product logics one may think that while moving along one direction the valuation of the remaining hyper-plan is changing, for example modelling valuation change in time if that direction is a time flow [70].

Local view: reactive logics, making Kripke reactive

In [67] Kripke structures are made reactive. Gabbay introduces a semantics based on Kripke frames enriched with double arrows, where the interpretation of a formula is changed by the action of the double arrows. The dependence is on where one has been before, that is, neither relation nor propositional variables' values change with the clock ticking but they react when and because we move. Moreover, the changes are sensitive to the way we got to the current world. Subsequently it is proven that this semantics strictly generalises the Kripke semantics, in fact, we may have classes of these frames originating logics that are not closed under substitution. It is not claimed that the classical Kripke frames cannot cope with these types of change, the matter is more about how to incorporate these meta-level notions into the models and which language to consider in order to reason about them. In [62, 126] a more abstract notion of reactive Kripke frames is introduced. Whereas in [67] the changes in the accessibility relation are the ones produced by the action of the double arrows, in the abstract notion of reactive Kripke frames these changes are given. In reality, in the usual semantics of modal logic the only information important to the value of a modal formula is the set of successors at each moment, i.e the **local** accessibility relation. Therefore the notion of reactive Kripke frame boils down to a set of admissible sequences of points, that is, the set of admissible paths. One can picture the initial accessibility relation by considering paths of size two, and its evolution is encoded in the one-step prolongment relation on bigger paths. The semantics presented in [67] is generalised over these abstract structures, allowing the valuation to vary along the paths, and the language is enriched with an extra operator relating paths that have the same endpoint. This is similar to what it is done in the branching-time logic with quantification over branches in [159]. Finally results of soundness and completeness are presented characterising some properties of these frames (generalising some familiar properties of 'static' Kripke frames: reflexivity, symmetry and transitivity). Furthermore some results regarding the decidability of the resulting logics are obtained. Let us look at a concrete case and see some examples of what can be expressed in the considered language.

Example 7.1. Let us consider the situation of a traveller with a budget. The set of his possible moves depends on whether he has enough money to make them (to pay tolls, fuel, train or flight tickets), furthermore his actual moves also determine his future possibilities. So the paths of the corresponding reactive frame are the sequences of cities he can visit with a certain budget. The formulas are interpreted over these paths. Let $\Diamond_R$ stand for the dynamics operator, that is, corresponding to the accessibility relation, and $\Diamond_P$ to the relation identifying the paths with the same endpoint. So, $\Diamond_R \varphi$ means that after the current path we can access a city such that the resulting path satisfies φ. $\Diamond_P \varphi$ means that there is a path to the current world satisfying φ. Let us consider that the propositional symbols p_b and p_w correspond to the predicate of being able to buy bread and wine respectively, and let m be true if there is still some money left. See Table 7.1 for examples of what can be said.

Modal language	Natural language
$\neg m \to \Box_R \bot$	If the traveller has no money left then he cannot move
$\Diamond_P(\Box_R p_b \wedge \Diamond_R \top)$	There is a path to the current city, after which the traveller is not blocked and he has enough money to buy bread in every city he can access
$(p_w \wedge \Box_R p_w) \to \Box_P p_b$	If the traveller can buy wine now, and at any immediate next stop, then he is always able to buy bread in the current city regardless of how he got there

Table 7.1: Possible statements in the considered modal semantics.

The truth values of the sentences in Table 7.1 depend on the particular valuation we pick, but we noticed above, with this language we can capture interesting structural aspects of these frames (see Chapter 6).

7.1.1 Global view: Switch graphs and their frames

While the notion of reactive frame contains exactly the necessary information to generalise the usual Kripke semantics to the reactive case, it ignores the state of the global accessibility relation. In order to model this global dependence we consider the concept of reactive graphs. Intuitively a reactive graph consists of a graph that may change its configuration when a certain edge is crossed.

As in the case of reactive frames, we define a reactive graph to be a set of admissible sequences of edges. Clearly, from such a set one can extract the evolution of the whole relational structure while transversing the graph edges. For any admissible sequence λ, the relational state of the reactive graph after λ is given by

$$R_\lambda = \{(w, w') : \lambda(w, w') \text{ is an admissible sequence of edges}\}.$$

At this point a natural question arises: can all these relational behaviours be encoded by double arrows? In order to answer this question we introduce the concept of a switch graph.

Switch graphs

A switch graph is a graph enriched with two kinds of double arrows, the connecting and the disconnecting switches. As their names suggest, when the origin of a connecting/disconnecting switch is crossed its target is connected/disconnected.

Given a set W, a switch is either an edge $s \in W^2$ (switch of level 0, neither connecting or disconnecting) or a triple $s = (a, s', *)$ (of level $n > 0$), where

- $a \in W^2$ in the edge that triggers its action,
- s' is the targeted switch (of level $n - 1$)
- $* \in \{\bullet, \circ\}$ says whether it is a connecting (black circle) or disconnecting (white circle) switch.

The type of the switches of level 0 is ϵ (the empty sequence) and of $s = (a, s', *)$ is $\sigma *$ where σ is the type of s'. We use the following notation to refer to switches more easily:

- $(ab, \epsilon) = (a, b)$,
- $(v_1 v_2, \ldots, v_{2n+1} v_{2n+2}, a, *_1 \ldots *_{n+1}) =$
  $((v_1, v_2), (v_3 v_4, \ldots, v_{2n+1} v_{2n+2}, *_1 \ldots *_n), *_{n+1})$.

In graphical representations we use white headed arrows to represent the disconnecting switches and black headed arrows to represent the connecting ones. Let us see an example of a situation where the dynamical restrictions are easily represented by these structures.

Example 7.2. Switches can easily grasp the fact that certain resources are finite, that is, one can use them a finite number of times. Depending on the meaning of the accessibility relation (e.g. crossing a bridge, driving a road, taking a pill from a tablet, printing pages, asking a person for a cigarette, etc) the switch configuration presented in Figure 7.5 represents the fact that a particular action can be taken exactly $k > 1$ number of times. The set of switches is given by $\{(a, b), (ab, \ldots, ab, \circ\bullet^{k-1})\}$. For $k = 0$ we would have (a, b) and $(ab, ab, \circ)$ on, and would not need more switches.

The main result of section 7.2 answers the above question, we prove that any reactive graph can be generated by a switch graph. We also provide an example where switches are used to represent the dynamical restrictions required for the solution of the mutual exclusion problem.

Switch reactive hybrid logic

In order to reason about switches and their actions we introduce an interpretation of a hybrid modal language over Kripke frames generated by switch graphs dynamics. Usually nominals are valid in exactly one point in each model. In the reactive setting they are valid in exactly one point for each of the components representing different reactive moments, with different relational states. The general idea is as follows:

- Instead of having an operator relating the different relational states of a point, we use nominals to identify them.

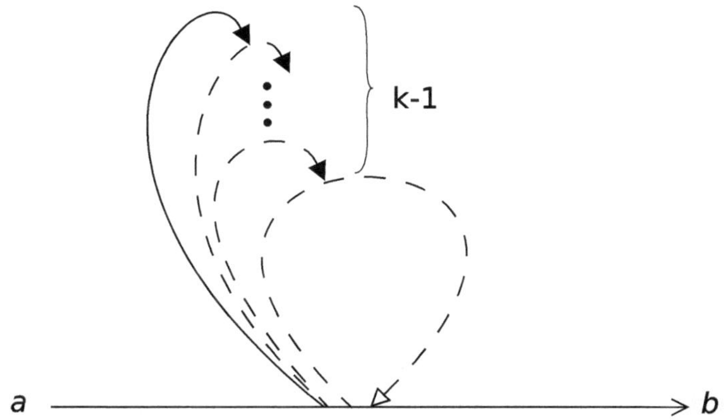

Fig. 7.5: The edge (a, b) can be crossed exactly $k > 1$ times.

- Also, using the fact that double arrows are relations (of different arities) over the carrier, we consider the corresponding modal operators. To each type of switches we consider the associated $2n + 2$ relation

$$R^{\sigma} = \{(w_1, w_2, \ldots, w_{2n+1} w_{2n+2}) : (w_1 w_2, \ldots, w_{2n+1} w_{2n+1}, \sigma) \textbf{ is on.}\}$$

  for $|\sigma| = n$, and its corresponding $2n + 1$-ary modal operator $\Diamond_{\sigma}$.
- Moreover the use of the hybrid operator @ allows us to have a global view over the switch configuration at each moment.
- To the fragment of the language introduced above, that allows us to talk about the switches' state in each moment, we add the modal operator $\not\circ$ relating the different states of the switch graph, being the real dynamics operator (corresponding to $\Diamond_R$ in the reactive Kripke frames).

In Section 7.3 we give an axiomatisation of the switches' evolution by capturing the interaction between these components but for now let us look at some examples of what we may express in this language.

Example 7.3. Let us consider a non-local version of Example 7.1. Instead of considering a single traveller, who can be at only one place at a time, let us consider the same problem but with a truck company (or group of travellers) with a common budget. Clearly each move affects all the subsequent possible moves. We know from the result proven in Chapter 7.2 that any such reactive dynamics can be expressed by switches. If we consider a switch graph generating this dynamics and the language described above we may express the local interdependencies explicitly, e.g.

$$@_a \wedge \triangleleft\!|_{\circ}(b, c, d)$$

means that the fact that a truck in a goes to b implies that no truck in c can go to d. That is

$$@_a \not\Diamond(b \rightarrow @_c \neg \not\Diamond d).$$

In the case of Example 7.2 dealing with bounded resources, where we allow only certain kinds of switches:

$$@_a(\Diamond\!|_{\epsilon} b \wedge \Diamond\!|_{\circ}(b, a, b))$$

means that (a, b) can be crossed exactly once, $@_a \not\Diamond(b \rightarrow @_a \neg \not\Diamond b)$,

$$@_a(\Diamond\!|_{\epsilon} b \wedge \neg \diamond|_{\circ}(b, a, b) \wedge \neg \diamond|_{\circ\bullet}(b, a, b, a, b) \wedge \Diamond\!|_{\circ\bullet\bullet}(b, a, b, a, b, a, b))$$

means that (a, b) can be crossed up to three times, $@_a \not\Diamond(b \rightarrow @_a \not\Diamond(b \rightarrow @_a \neg \not\Diamond b))$, and so on.

We do not claim that this is the most appropriate language to reason about all cases of reactivity. For instance, in Examples 7.1 and 7.3 we can envisage a language that could explicitly reason about the cost of each move and the remaining budget after it. And in the case of Example 7.2, where only some shapes of switches are allowed, the language could be simplified. Still, the fact is that all reactive systems can be generated by switches and this language seems adequate to express the local dynamic dependencies of each move, imposed by the switches. We hope that it represents a kind of skeleton for the various possibilities. In the conclusion of section 7.3 we discuss possible extensions of this language with operators of the kind we find in CTL and CTL^*. The main result of this section is the proof of the usual hybrid completeness result in this dynamic context. An important fact to retain is that in each moment the whole future relational dynamics is coded in the switches, so it may be that some properties (that depend only on the worlds - corresponding to pure formulas, having no propositional symbol that is not a nominal) can be derived by reasoning locally, using only the information contained in the switches configuration.

7.2 Switch graphs

A graph is the abstract representation of a binary relation between objects. There are many graph-based structures, and often, to read (some of) the information represented in them, one needs to travel across their vertices following their edges, e.g. to check whether two points are connected or to interpret modal formulas in a Kripke model. In the usual notion of graph the vertices that are accessible from a vertex are fixed.

When we modify the notion of graph allowing the accessible vertices to depend on the sequence of edges we have crossed, we obtain a reactive graph. So, to each sequence of crossed edges corresponds a (relational) state of the graph, where the edges that are available are the ones that can prolong the current sequence. The information in such a graph boils down to the sequences of edges (or actions) over a certain set.

A direct way of representing such a graph would be to draw the tree given by the admissible sequences of edges, and, perhaps, to draw at each of its nodes the state of the graph at that point. Although this idea of having different points representing the relational state of the same point is useful (in [62] we used it to obtain a relation with classical modal logic), it does not offer an easy reading of the changes that crossing an edge implies. As stated in the introduction, the concept of reactivity was initially introduced using an enrichment of graphs with new kinds of arrows representing the local effects of traversing an edge in the global relation. The representation offered by these structures seems to be much more interesting as it truly grasps the flavour of reactivity. Furthermore, we prove that these new multi-level-arrowed structures are enough to code all the relational dynamics (Theorem 7.11).

The simplest effect crossing an edge can have on the accessibility relation is to turn on or off a connection. These elementary changes can be represented by drawing an arrow from the crossed edge to the edge representing the connection being altered. Crossing an edge does not have necessarily always the same effects. To represent these changes we use arrows from the edges to the arrows representing the elementary effects, and so on, obtaining infinite levels of arrows.

In this context we refer to the arrows as switches. The edges connecting points are 0-level switches and the switches connecting 0-level and n-level switches are $n + 1$-level switches. The switches of level greater than 0 can be of two kinds, the connecting and disconnecting ones. A set equipped with a set of switches is called a switch graph.

The state of the relation after a certain sequence of actions in a switch graph takes into account its layered structure. The switches of level greater than 0 do not represent the accessibility relation between points but how this relation changes after each action, corresponding to the 0-level arrows. When we cross an edge we turn on/off the switches that are the targets of the connecting/disconnecting switches coming out of it.

In Figure 7.6 we can see the switch graph representing the reactive graph with set of points $\{a, b\}$, and admitting as set of actions the set given by the following regular expression (with alphabet $\{a, b\}^2$):

$$((a, a) + (a, b))^*(b, b)((a, b) + (b, b))^*(a, a)((a, a) + (b, b))^*.$$

We see that once (b, b) is crossed, the disconnecting switch coming from (a, a) to (a, b) becomes connected. Thus, after crossing (b, b) and (a, a), (a, b) can no longer be crossed.

Many other kinds of switches, having different effects, could be considered. These seem to be the most primitive ones, since they incorporate the basic actions in graph reactivity, connecting and disconnecting edges. We shall see that this is enough to represent all reactive graphs. That is: any reactive behaviour can be decomposed into this kind of local actions.

7.2.1 Reactive by Switch

Definition 7.4. A *reactive graph* is a pair (W, Δ), where

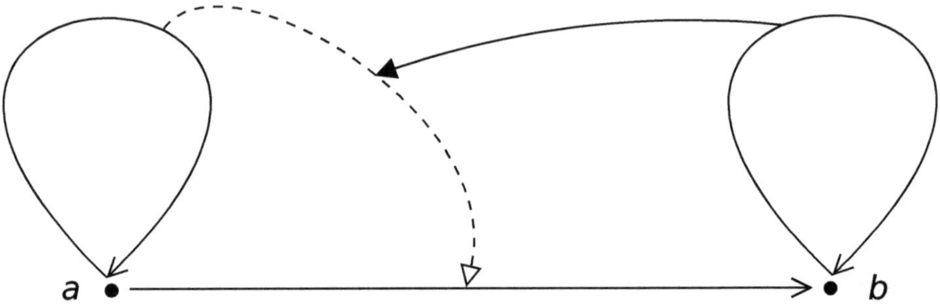

Fig. 7.6: The white pointed arrows represent the connecting switches, the black arrows the disconnecting ones and the dashed line the fact that the switch is initially disconnected.

- W is a non-empty set, the set of *worlds*, and
- Δ, the *behaviour*, is the set of admissible sequences of edges, i.e., a subset of $(W \times W)^*$ closed under prefixes containing ϵ (the empty sequence is always admissible).

The state of the accessibility relation after an admissible sequence of edges λ being covered is given by

$$R_\lambda = \{(w, w') : \lambda(w, w') \in \Delta\}.$$

We now introduce the enriched notion of graph, the switch graphs, and formalise in which sense they can be used to represent (or generate) reactive graphs highlighting the effects that moving around the base graph has on the accessibility relation.

Definition 7.5. For a non-empty set W and $n < \omega$, the set $\mathcal{A}_n(W)$ of *switches* over W of level n is defined as:

- $\mathcal{A}_0(W) = W \times W$ (of level 0),
- $\mathcal{A}_{n+1}(W) = (W \times W) \times \mathcal{A}_n(W) \times \{\bullet, \circ\}$ (of level n).

A switch with $\bullet$ ($\circ$) as its third component is called a *connecting (disconnecting) switch*.

The set of all switches on W is defined by taking $\mathcal{A}(W) = \bigcup_{n<\omega} \mathcal{A}_n(W)$.

Definition 7.6. A *switch graph* is a pair (W, R), where

- W is a non-empty set, the set of *worlds*, and
- $R \subseteq \mathcal{A}(W)$ is the set of *switches*.

We say that (W, R) has *reactivity of level n* if $R \subseteq \bigcup_{i \leq n} \mathcal{A}_i(W)$.

Remark 7.7. In the graphical representation we use normal arrows for the 0-level switches, a black (white) pointed arrow for the connecting (disconnecting) switches and switches that are off are drawn with a dashed line.

Here we introduce some notation to refer to switches in an easier fashion:

Definition 7.8. $(v_1v_2, \ldots, v_{2n-1}v_{2n}, a, \sigma)$ for $\sigma = s_1 \ldots s_n \in \{\bullet, \circ\}^n$, $n < \omega$ and $a \in \mathcal{A}(W)$ is defined as:

- $(a, \epsilon) = a$,
- $(v_1v_2, \ldots, v_{2n+1}v_{2n+2}, a, s_1 \ldots s_{n+1}) = ((v_1, v_2), (v_3v_4, \ldots, v_{2n+1}v_{2n+2}, a, s_1 \ldots s_n), s_{n+1})$.

We say that $(v_1v_2, \ldots, v_{2n-1}v_{2n}, \sigma)$ is a switch of type σ.
For example,

$$(w_1w_2, w_3w_4, w_5w_6, \circ\bullet\bullet)_p = \Big((w_1, w_2), ((w_2, w_3), ((w_3, w_4), (w_4, w_5), (w_5, w_6)\circ), \bullet), \bullet\Big)$$

is of type $\circ \bullet \bullet$.

Definition 7.9. Given a switch graph $S = (W, R)$, the *behaviour of* S is the smallest set, Δ_S, such that:

- $\epsilon \in \Delta_S$,
- If $\alpha \in \Delta_S^n$ then $\alpha(w, w') \in \Delta_S$ for all $(w, w') \in R_\alpha$.
  Here $R_\alpha \subseteq \mathcal{A}(W)$ is the *switch state* of our switch graph after crossing the sequence of edges α:
  - $R_\epsilon = R$, the initial state,
  - $R_{\alpha(w,w')} = (R_\alpha - \{a : ((w, w'), a, \circ) \in R_\alpha\}) \cup \{a : ((w, w'), a, \bullet) \in R_\alpha\}$.

  (W, Δ_S) is the reactive graph generated by S and

$$\Delta_S = \{R_\alpha : \alpha \in \Delta_S\}$$

of *switch states* of S.

Notice that the reactive level of R_α is the same for all α. The highest order arrows can never be turned off, nor can higher order ones be introduced, since a switch that acts over a switch of order n has order $n + 1$.

Remark 7.10. When we informally introduced the switches' dynamics we did not specify what should happen when a connecting and a disconnecting switch act simultaneously over the same switch. In the previous definition the convention is that the connecting action prevails. We could have opted instead for:

- the disconnecting switch would prevail
  $R_{\alpha(w,w')} = (R_\alpha \cup \{a : ((w, w'), a, \bullet) \in R_\alpha\}) - \{a : ((w, w'), a, \circ) \in R_\alpha\}$,
- or that it would depend on the state of the target, by, for example, always changing its state
  $R_{\alpha(w,w')} = (R_\alpha - \{a : ((w, w'), a, \circ) \in R_\alpha\}) \cup \{a : ((w, w'), a, \bullet) \in R_\alpha \;\&\; a \notin R_\alpha\}$.

We are interested in studying how expressive these structures are regarding the generation of reactive graphs. It is easy to see that given a reactive graph $\mathfrak{F} = (W, \Delta)$ the switch graph $\mathfrak{G} = (W, R^\circ \cup R^\bullet)$ where

$$R^{\bullet} = \{(w_1 w_2, \ldots, w_{2|\sigma|+1} w_{2|\sigma|+2}, \sigma) \;:\; \sigma \in \{\bullet\}^*,\ w_1 \ldots w_{2|\sigma|+2} \in \Delta\},$$
$$R^{\circ} = \{(vv', w_1 w_2, \ldots, w_{2|\sigma|+1} w_{2|\sigma|+2}, \sigma\circ) \;:\; \sigma \in \{\bullet\}^*,\ v, v', w_i \in W\},$$

generates $\mathfrak{F}$ *if the connecting switches prevail.* For the other options there does not seem to be such a direct way of coding the reactive behaviour. We can then ask whether these options are relevant for this goal. The theorem shows that if one allows unbounded levels, these options do not limit the switch graphs' expressivity. Furthermore, it presents a general construction such that given any reactive graph we obtain a switch graph that, regardless of the chosen option, generates the given reactive graph.

Theorem 7.11. *Any reactive graph can be generated by a switch graph, furthermore this switch graph can be chosen such that no connecting and disconnecting switches ever act simultaneously over the same switch.*

Proof. Given a reactive graph (W, Δ), we define a relation $C \subseteq \{\bullet, \circ\}^* \times (W \times W)^*$ by taking

$$C(\sigma, \alpha) \quad \text{iff either the number of } \circ\text{s in } \sigma \text{ is even and } \alpha \in \Delta, \text{ or the number of } \circ\text{s in } \sigma \text{ is odd and } \alpha \notin \Delta.$$

This definition clearly implies the following:

Lemma 7.12. $C(\sigma\bullet, \alpha) \leftrightarrow \neg C(\sigma\circ, \alpha)$.

Now let

$$R = \{(w_1 w_2, \ldots, w_{2|\sigma|+1} w_{2|\sigma|+2}, \sigma) \;:\; C(\sigma, (w_1, w_2) \ldots (w_{2|\sigma|+1}, w_{2|\sigma|+2})),\ \sigma \in \{\bullet, \circ\}^*,\ w_i \in W \text{ and } w_1 \ldots w_{2|\sigma|} \in \Delta\}.$$

We claim that, for every $\alpha \in \Delta_S$, the following holds:

Lemma 7.13. *For every* $(\beta, \sigma) \in \mathcal{A}(W)$, $(\beta, \sigma) \in R_\alpha \iff C(\sigma, \alpha\beta)$.

We prove the lemma by induction on α. For $\alpha = \epsilon$ this is just the definition of R. Now suppose that $\alpha(w, w') \in \Delta_S$ and Lemma 7.13 holds for α. Then, by Lemma 7.12, we have that

$$((w, w')\beta, \sigma\bullet) \in R_\alpha \iff ((w, w')\beta, \sigma\circ) \notin R_\alpha. \tag{7.1}$$

Now by (7.1) we have:

$$\text{for every } (\beta, \sigma) \in \mathcal{A}(W), \quad (\beta, \sigma) \in R_{\alpha(w,w')} \iff ((w, w')\beta, \sigma\bullet) \in R_\alpha. \tag{7.2}$$

Now we can show Lemma 7.13 for $\alpha(w, w')$:

$$(\beta, \sigma) \in R_{\alpha(w,w')}$$

iff (by (7.2))

$$((w, w')\beta, \sigma\bullet) \in R_\alpha$$

iff (by the IH)

$$C(\sigma\bullet, \alpha(w,w')\beta)$$

iff

$$C(\sigma, \alpha(w,w')\beta).$$

Now we can complete the proof of the theorem as follows. First, $\epsilon \in \Delta \cap \Delta_S$. Otherwise, $\alpha(w,w') \in \Delta_S$ iff $\alpha \in \Delta_S$ and $(w,w') \in R_\alpha$ iff (by Lemma 7.13) $\alpha \in \Delta_S$ and $C(\epsilon, \alpha(w,w'))$ iff $\alpha \in \Delta_S$ and $\alpha(w,w') \in \Delta$ iff $\alpha(w,w') \in \Delta$.

Remark 7.14. Notice that the condition that $w_1 \dots w_{2|\sigma|} \in \Delta$ in the definition of R is not necessary but avoids the inclusion of switches that will have no part in the dynamics. Indeed it is easy to see that with regard to the behaviour of a switch graph only such switches matter.

7.2.2 Modelling multiple agents or processes

Reactive/Switch graphs can model situations where the accessibility relations change when an edge is crossed, without any limitation on the number of individuals going through the edges. In fact a reactive graph may be used to design a particular interaction between many agents going around in a graph.

In a k-agents (processes, individuals, etc.) setting, the switch configuration ceases to be the only relevant information, we need to keep track of each agent's position. The k-behaviour of a switch graph with set of initial configurations $C \subseteq W^k$ is the set of allowed sequences of moves when the k agents start in positions in C. Let us formalise this notion.

Definition 7.15. *Given a reactive graph $\mathcal{R} = (W, \Delta)$ and a number k of agents wandering about in the graph we define:*

- *$(W^k)^{\%}$ is the subset of $(W^k)^+$ formed by the sequences where each element differs from its successor in only one component, that is:*
  - *$W^k \in (W^k)^{\%}$,*
  - *if $\alpha(w_1, \dots, w_i, \dots, w_k) \in (W^k)^{\%}$ then $\alpha(w_1, \dots, w_i, \dots, w_k)[i \to w] \in (W^k)^{\%}$ for any $1 \leq i \leq n$, where*

$$\alpha(w_1, \dots, w_i, \dots, w_k)[i \to w] = \alpha(w_1, \dots, w_i, \dots, w_k)(w_1, \dots, w, \dots, w_k).$$

- *$\mathcal{E} : (W^k)^{\%} \to (W \times W)^+$ is defined as:*
  - *if $\alpha \in W^k$ then $\mathcal{E}(\alpha) = \epsilon$,*
  - *if $\alpha = \alpha'(w_1, \dots, w_i, \dots, w_k)[i \to w] \in (W^k)^{\%}$ then*

$$\mathcal{E}(\alpha) = \mathcal{E}(\alpha(w_1, \dots, w_i, \dots, w_k))(w_i, w).$$

Let $C \subseteq W^k$ be a set of initial allowed configurations for the k agents, C generated k-behaviour is:

$$\Delta_C^k = \{\alpha \in (W^k)^{\%} : \mathcal{E}(\alpha) \in \Delta^k \text{ and there exists } \gamma \in C \text{ that is a prefix of } \alpha\}.$$

Switch graphs can be useful in modelling programs or protocols, capturing the intended interaction between the entities involved. The design of the constraints can be directly imposed by strategically locating the appropriate switches. Ideally, this double perspective on the dynamics would allow that the verification of such properties over a switch graph S, departing from a given configuration γ, could either be extracted from the layout of the switches of S or exhaustively checked over Δ_C^k.

Example 7.16. Let us consider the mutual exclusion problem, taken from [108]. There we can find a model-based approach to the verification of the required properties in a given system, presented as the solution to this problem. The idea is to code the intended properties in CTL and verify whether the transition system associated with the solution satisfies them. Here we lay the basis for a different approach.

As we mentioned in the introduction, in [37] by adding higher order arrows to the structure of an automaton, and thus allowing its transition table to change while it is reading a sentence, the authors achieved an exponential reduction in the minimal number of states needed to accept a language. Therefore, this way of representing systems may have an impact on model checking, where the state explosion problem is a serious drawback. The switches' configuration at each point determines all the future dynamics, coding the interdependence of actions. Of course if one obtains fewer states when considering the switch graph corresponding to a certain protocol instead of considering its associated transition system, this is not because some information was thrown away. The fact is that this information is coded in a different form, hopefully in a more intuitive and accessible way. It is the extra expressivity, given by the higher order arrows, that allows us to identify different states with the same point by associating them with different switch configurations. Moreover, if one finds an appropriate way to extract the corresponding switch graph of a program or protocol, one may expect that the verification of some properties can be reduced (by means of some intermediary reasoning) to a simpler verification over the switches. Although we do not develop the verification part, in the next section we introduce a language to reason about switches and their effects. We believe that this language can be extended to a point where we can reason about both the switch view and the usual transition system view, and derive enough knowledge about their interaction in such a way that in order to verify some properties at one level it is enough to guarantee some related properties in the other and vice versa. Nevertheless, here, we concentrate on showing how switches can be integrated in transition systems to improve their modelling expressivity. We understand that in this case the switches are associated into conditional commands and changes with the auxiliary variables used in these conditions. However, no systematic knowledge about that connection was obtained yet. We will limit ourselves to present examples of switch graphs 'associated' (in an informal and intuitive way) with protocols that constitute possible solutions to the problem and discuss them.

The mutual exclusion problem as presented in Chapter 3 of [108] is:

> *When concurrent processes share a resource (such as a file on a disk or a database entry), it may be necessary to ensure that they do not have access to it at the same time. Several processes simultaneously editing the same*

file would not be desirable. We therefore identify certain critical sections of each process's code and arrange that only one process can be in its critical section at a time. The critical section should include all the access to the shared resource (though it should be as small as possible so that no unnecessary exclusion takes place). The problem we are faced with is to find a protocol for determining which process is allowed to enter its critical section at which time. Once we have found one which we think works, we verify our solution by checking that it has some expected properties, such as the following ones:

***Safety**: The protocol allows only one process to be in its critical section at any time.*

This safety property is not enough, since a protocol which permanently excludes every process from its critical section would be safe, but not very useful. Therefore, we should also require:

***Liveness**: Whenever any process wants to enter its critical section, it will eventually be permitted to do so.*

***Non-blocking:** A process can always request to enter its critical section.*

Some rather crude protocols might work on the basis that they cycle through the processes, making each one in turn enter its critical section. Since it might naturally be the case that some of them request accesses to the shared resource more than others, we should make sure our protocol has the property:

***No strict sequencing:** Processes need not enter their critical section in strict sequence.*

We model N processes, each of which is in its non-critical state (n), or trying to enter its critical state (t), or in its critical state (c). Each individual process undergoes transitions in the cycle $n \to t \to c \to n \to ...$, see in Figure 7.7[1] the case $N = 2$, but the two processes interleave with each other.

One immediate solution (for $N = 2$) one can think of, is to use switches to directly impose the condition that when one process gets to its critical area the other cannot – using $(t_1c_1, t_2c_2, \circ)$ and $(t_2c_2, t_1c_1, \circ)$ – and to remove this restriction when it returns to the non-critical state – using $(c_1n_1, t_2c_2, \bullet)$ and $(c_2n_2, t_1c_1, \bullet)$, see Figure 7.8. The two processes start off in their non-critical states as indicated by the incoming edges with no source. Either of them may now move to its trying state, but only one of them can ever make a transition at a time (asynchronous interleaving). The problem here is that nothing prevents one of the processes from getting stuck at the trying state while the other one continuously accesses the critical area, so liveness fails.

A way to guarantee liveness is, when a process requires permission to move to the critical state (moves to state t), to allow the other process to require access to the critical area only once before the first process accesses it. In Figure 7.9 we

[1] In the graphical representation one finds numbers juxtaposed with the letters, this is due to a limitation of the application used, thus they should be seen as being subscripts. Outside the figures we will use the proper form to avoid the text with their intense use becoming unreadable. E.g. $t1$ in the figure corresponds to t_1 in the text.

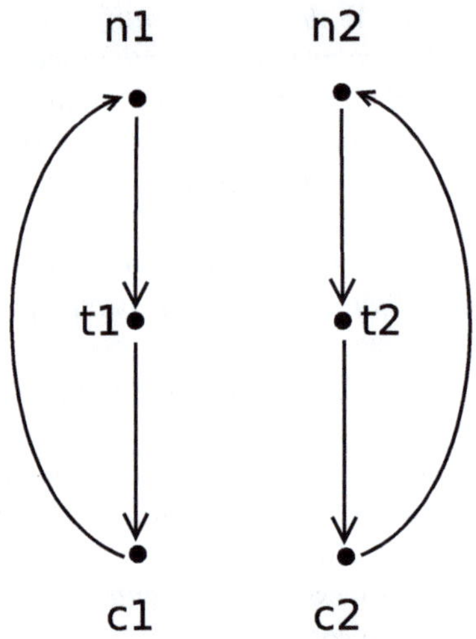

Fig. 7.7: The mutual exclusion problem: combining processes sharing resources, two processes scenario.

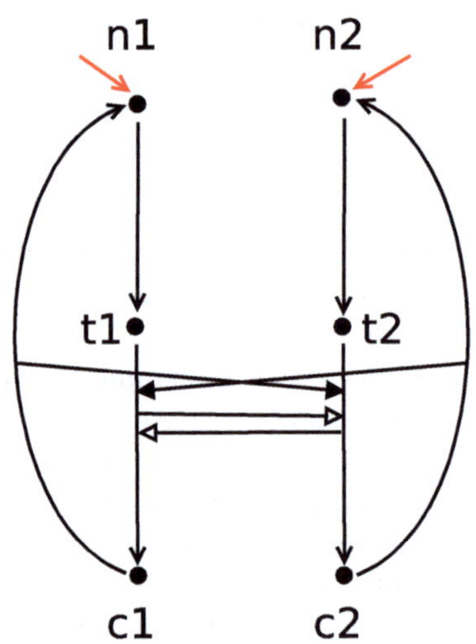

Fig. 7.8: $N = 2$: liveness fails.

can see that the extra switches do exactly this. The situation is symmetric so lets us check liveness for the first process, that is, if it moves to t_1, it will eventually be able to move to c_1. If process 1 moves to t_1, it turns on $(c_2n_2, n_2t_2, \circ)$ (by the action of $(n_1t_1, c_2n_2, n_2t_2, \circ\bullet)$), which guarantees that if process 2 does the whole cycle once, then (since $(c_2n_2, n_2t_2, \circ)$ is on) it can only try again if process 2 passes by c_2, $(t_1c_1, n_2t_2, \bullet)$ and $(t_1c_1, c_2n_2, n_2t_2, \circ\circ)$ remove that restriction. But this is done by restricting process 2's ability to *require* access to its critical section, thus failing the non-blocking constraint.

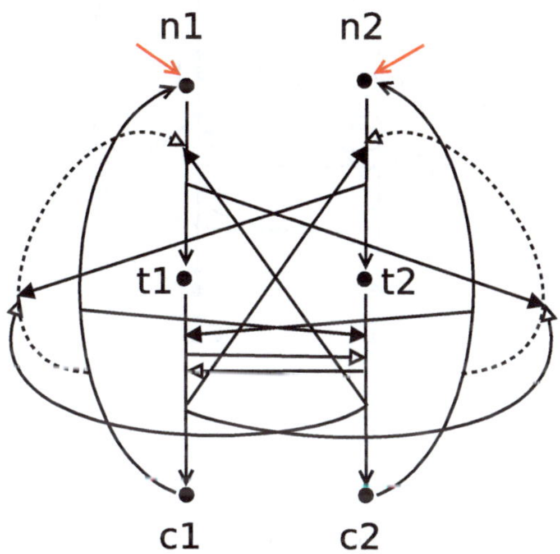

Fig. 7.9: $N = 2$: non-blocking fails.

This is easily solved in Figure 7.10, where the limitation is forced only at the last stage. One can see that the two switch graphs are really similar, the difference is that the disconnected edge is (t_2, c_2) instead of (n_2, t_2). That is, instead of using the following set of switches

$$\{(n_it_i, c_jn_j, n_jt_j, \circ\bullet), (t_ic_i, n_jt_j, \bullet), (t_ic_i, c_jn_j, n_jt_j, \circ\circ) : i, j \in \{1, 2\}, i \neq j\},$$

we use

$$\{(n_it_i, c_jn_j, t_jc_j, \circ\bullet), (t_ic_i, t_jc_j, \bullet), (t_ic_i, c_jn_j, t_jc_j, \circ\circ) : i, j \in \{1, 2\}, i \neq j\}.$$

So when one process requires access to the critical state, the other may access it only once, though being still able to require access to it, until the first's access is granted (other levels could be as easily set). Thus this solution complies with all the problem requirements.

Hence, the advantage of the switch graph representation is that we can force/check the properties at the meta-level. We directly observe the effect that each transition has on the global accessibility relation. This strongly suggests how these solutions could be implemented by programming. Though the other direction (from programs to switch graphs) is not considered here, it seems a fundamental step in evaluating these structures' potential.

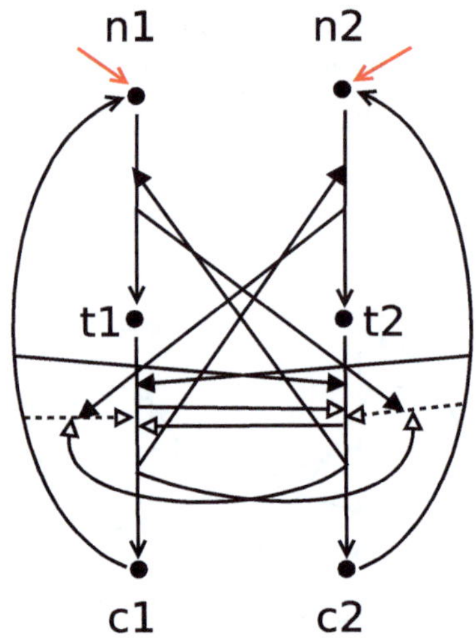

Fig. 7.10: $N = 2$: a complete solution.

Till now we tried solutions where each process runs around on different connected components of the presented switch graph, but there is no reason for it to be so. If we allow more processes to run around the solutions in 7.9 and 7.10, then both liveness and non-blocking properties would clearly be lost. Let us look at some solutions with only one connected component where all processes run, thus saving on the number of points required in the graph.

In Figure 7.11 we have a general straightforward solution for the N processes case, but again liveness is not guaranteed. Again nothing forbids a process (or a group of $N' < N$ processes) to keep accessing the critical area making it impossible for other processes to do it. There does not seem to be a way of avoiding this with this base graph (without losing the non-blocking property), at least if we do not use multiple connections between each pair of points, which are not allowed in our definition of switch graphs. Of course it would be easy to consider labelled switch graphs where this was allowed but it would be closer to considering different connected components.

Instead, in order to guarantee the other properties we will consider a slightly more complex base graph. We use a number of 'trying' states equal to the number of processes. We learned the kind of restrictions needed from Figure 7.10 and we can see in Figure 7.12 a switch graph using the same idea for avoiding the loss of liveness. Since here both processes share the same graph point for non-critical and critical sections the way it is imposed changes slightly. First we need to guarantee we do not have more than one process in the same point representing 'trying to get to the critical section', by having $(nt_i, nt_i, \circ)$ and $(t_i c, nt_i, \bullet)$, for $i = 1, 2$, controlling that. Initially we have both $(cn, t_1 c, \bullet)$ and $(cn, t_2 c, \bullet)$ on. One way to avoid the loss of liveness is to have that when a process moves to t_i then $(cn, t_j c, \bullet)$ (for $j = 3 - i$)

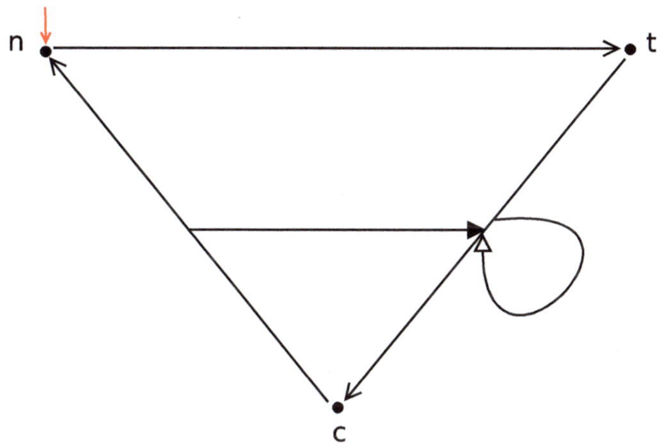

Fig. 7.11: N arbitrary: safety and no strict sequence only.

becomes off, that is we have the switch $(nt_i, cn, t_jc, \bullet\circ)$. Clearly when that process moves out of t_i the restriction is not needed anymore, and so we have $(t_ic, cn, t_jc, \bullet\bullet)$. This constitutes a complete solution for $N = 2$ but for $N > 2$ it loses liveness and non-blocking. If we want to accommodate more processes we can simply add more trying states, as many as the number of processes being considered, and have the same switch structure between every pair t_i, t_j for $i \neq j$. This means that a solution for arbitrary N is given by $S_N = (W_N, R_N)$ with $W_N = \{n, c, t_1, \ldots, t_N\}$ and

$$\begin{aligned} R_N = \{&(n, t_i), (t_i, c), (c, n), \\ &(nt_i, nt_i, \circ), (t_ic, nt_i, \bullet), (cn, t_ic, \bullet), (cn, t_jc, \bullet), (t_ic, t_jc, \circ), \\ &(nt_i, cn, t_jc, \bullet\circ), (t_ic, cn, t_jc, \bullet\bullet) : 1 \leq i, j \leq N, i \neq j\} \end{aligned}$$

Clearly for $N > 2$, S_N is not as easily visualised, but this is a cost to pay also when drawing the transition system associated with the protocol for mutual exclusion with various processes. We believe that it is still impressive that the switch structure is easily definable for all N, and that it is clear from the $N = 2$ case analysis why it works in the general case.

7.2.3 Comments and remarks

Level of Reactivity

Given that all reactive graphs can be represented by switch graphs, the level of reactivity of the representation is of obvious practical importance.

Given a reactive graph, it is easy to see that the minimal reactive level for a switch graph representing it, depends on the choice of dynamics we discussed above. Consider $R = (\{a, b\}, \varDelta)$, where $\varDelta = \{\alpha : \lambda = (a, a)^n$ or $\alpha = (a, a)^{2n}(a, b), n <$

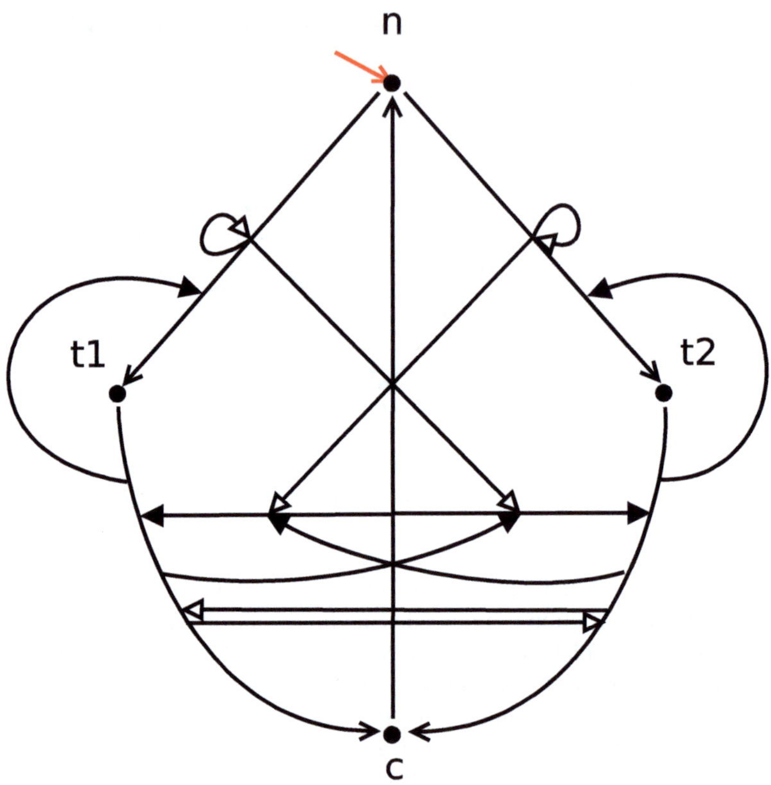

Fig. 7.12: N arbitrary: safety, liveness, no strict sequencing and non-blocking for $N = 2$.

$\omega\}$. Using the definition where the connecting switches win we cannot find a finite reactivity level switch graph that represents it. However, using the alternating one it can be represented by a switch graph of reactivity level 1:

$$(\{a, b\}, \{(a, a), (a, b), ((a, a), (a, b), \circ), ((a, a), (a, b), \bullet)\}).$$

It is also easy to devise a reactive graph that cannot be represented by a finite level graph for any of the options we mentioned above, nor with any switches with 'recursive behaviour'. Let f be a non-recursive binary sequence and

$R = (\{a, b\}, \Delta)$, where $\Delta = \{\alpha : \alpha = (a, a)^n$ or $\alpha = (a, a)^k(a, b), n < \omega, f(k) = 1\}$.

Question 7.17. For each of the considered options what is the set of reactive graphs generated by switch graphs of reactivity level $n \in \omega$?

Further extensions

We have seen how arrows can represent more than the basic transitions, they can represent the transitions between states of the accessibility relation itself. These new

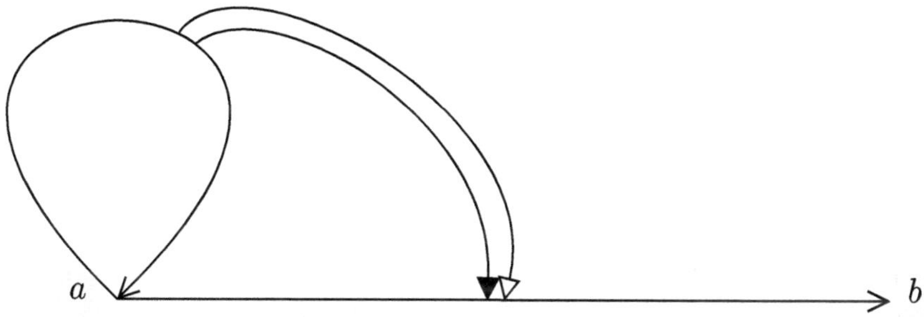

Fig. 7.13: Alternating dynamics example with the following behaviour: $\Delta = \{\alpha : \lambda = (a,a)^n$ or $\alpha = (a,a)^{2n}(a,b), n < \omega\}$.

arrows are a natural extension of the concept of graph corresponding to a kind of meta-transition in the above sense. They provide an explicit way of expressing the meta-level graph's notion of reactivity, in a way that allows an immediate (and complete) reading of the effects of crossing an edge.

In Section 7.2.2 we considered multiple entities moving through the graph, the effects of each action were independent of who was doing it and no synchronous movements were allowed. Other types of dependences can be considered:

- Dependence on the identity of the individual entities could be modelled by having different relations for each entity, or groups of entities.
- Synchronous movements could also be represented by special relations. Of course switches can connect arrows regardless of the relation they represent.

This leads to the question:

Question 7.18. Is the switches formalism enough to represent these more general behaviours?

Model Checking

Above we gave an example of how a mutual exclusion protocol may be coded by a switch graph. Specifically, we model at the 'Kripke level', how the action of one agent (i.e. the transition between worlds/states) precludes the execution of an action (i.e. transition) of another agent. In this case switches are associated with changes to the auxiliary variables keeping the critical area safe. It is clear that the fact that we directly express that when one process accesses the critical area, all the others will not be able to access it (by having the appropriate switch in place, and no other switch acting on it), may give enough evidence to guaranty safety, without the need to check the whole state space.
It may be that depending on the property in question, the most effective way of verifying that a certain system satisfiesit involves considering different types of models associated with that system, and checking a stronger property that entails it. Hopefully switch graphs are among the structures which are useful for this. In the future

we hope to find more meaningful examples that allow us to understand the general case.

7.3 Reactive hybrid switch logics

In the previous section we have shown that switch graphs are suitable structures to embody and represent the reactive paradigm. In this section we introduce a logic that allows us to reason about switches and their effects and prove completeness. A basic feature of such a logic would be to be able to express that if a certain switch is on, then after such a move the state of a certain switch will be such.

In the introduction we gave some motivation for the language used. On one hand, given a switch graph (W, R), for each switch type $\sigma \in \{\circ, \bullet\}^*$, the set of σ-switches forms a $|\sigma| + 2$-ary relation over W and there is the $|\sigma| + 1$-ary modal correspondent, $\diamond|_{\sigma}$. On the other hand, similarly to $\Diamond_R$ in [62], there is the operator corresponding to its behaviour, $\not\diamond$. The goal is to relate the two, expressing how the switches determine the behaviour.

The logics we considered in [62] were suitable to talk about local effects of reactivity, but here we are dealing with its global effects and at the same time we feel the need to to explicitly refer to specific states. Ordinary modal logic's lack of mechanisms for dealing with states explicitly is a recognised weakness. In fact the idea of adding variables that are used to name worlds dates back to the pioneering work of Prior [138, 137] and especially of Bull [34](see [97]). Recently the study of this idea gained popularity and its development became an autonomous subfield of modal logic, called *hybrid logic*[2].

Hybrid languages are a very simple extension of modal ones. A special set of propositional variables, called nominals, is added and used to name worlds. Nominals are true at exactly one world in any model. Many operators related to nominals have been studied. A particularly simple one is @ that allows one to jump to particular worlds, in the sense that, for each nominal i, $@_i\varphi$ is true if φ is true in the world named i, which suits our need to think globally.

We consider models constructed from switch graphs in such a way that for each admissible sequence of edges we get a usual hybrid model based on the Kripke frame given by the switches' state at that point. Of course each nominal must be true in the same world in all these models but all the other variables may change.

This language allows us to express some strong reactive assertions. Consider the formula

$$@_i \not\diamond(j \to @_i \neg \not\diamond j).$$

This says that if we cross the edge from the world named i to the world named j, that connection will be turned off. Another way to say this is to say that there is a disconnecting switch from that edge to itself:

$$@_i \diamond|_{\circ}(j, i, j).$$

[2] For more details see [9, 8, 23].

We obtain completeness for the introduced logic adapting the usual Henkin style proof of completeness for hybrid logics, see [24, 26]. This further justifies the choice of language by showing that it is appropriate to capture the switches' dynamics.

7.3.1 Switch Models

Definition 7.19. We consider the *reactive switch similarity type*

$$s = (\{\not\Diamond\} \cup \{\lhd\!|_{\sigma} : \sigma \in \{\circ, \bullet\}^*\}, \rho),$$

where $\rho(\not\Diamond) = 1$ and $\rho(\lhd\!|_{\sigma}) = 2|\sigma| + 1$. Thus we define the hybrid modal language $\mathcal{H}_s(@)$, defined by

$$\varphi ::= i \mid p \mid \neg\varphi \mid \varphi_1 \wedge \varphi_2 \mid \not\Diamond\varphi \mid \lhd\!|_{\sigma}(\varphi_1, \ldots, \varphi_{2|\sigma|+1}) \mid @_i\varphi,$$

where $p \in \Pi$, $i \in NOM$ and $\sigma \in \{\circ, \bullet\}^*$. The other connectives: $\top$, $\bot$, $\vee$, $\rightarrow$, $\leftrightarrow$, $\not\Box$ and $\Box\!|_{\sigma}$ are introduced as the usual abbreviations.

Given $\lambda \in (NOM \times NOM)^*$ we define the abbreviation $\not\Diamond^{\lambda}\varphi$ by recursion:

- $\not\Diamond^{\epsilon}\varphi = \varphi$,
- $\not\Diamond^{(i,j)\lambda}\varphi = @_i \not\Diamond(j \wedge \not\Diamond^{\lambda}\varphi)$ (clearly $\not\Diamond^{(i,j)\lambda}\varphi = \not\Diamond^{(i,j)} \not\Diamond^{\lambda}\varphi$),

and $\not\Box^{\lambda}\varphi = \neg \not\Diamond^{\lambda}\neg\varphi$.

Definition 7.20. Given a switch graph $S = (W, R)$ and using the dynamics defined in 7.9 we generate, for each $\lambda \in \Delta_S$, a reactive graph: $S_{\lambda} = (W, R_{\lambda})$.

A switch frame is the Kripke frame given by the disjoint union of all these switch graphs, where σ switches in R_{λ} give rise to local $(2|\sigma| + 2)$-ary relations and with a global accessibility relation connecting w in S_{λ} and w' in $S_{\lambda(w,w')}$, see Figure 7.14.

Formally, the **switch frame** over S is $\mathfrak{F}_S = (W \times \Delta_S, \check{R}_S)$ where

- $\check{R}_S = \{\check{R}^{\sigma}_{\lambda}\}_{\sigma \in \{\circ,\bullet\}^*, \lambda \in \Delta_S}$ and
- $((w_1, \lambda), \ldots, (w_{2n+2}, \lambda)) \in \check{R}^{\sigma}_{\lambda}$ iff $(w_1 w_2, \ldots, w_{2n+1} w_{2n+2}, \sigma) \in R_{\lambda}$.

A *switch model* over $\mathfrak{F}_S$ is a pair $\mathfrak{M} = (\mathfrak{F}, v)$, where v is a function

$$v : \Pi \cup NOM \rightarrow 2^{\Delta_S \times W}$$

such that for $s \in NOM$ we have $V(s) = \{w\} \times \Delta_S$ for some $w \in W$.

Given a switch model $\mathfrak{M} = (W \times \Delta_S, R_S, v)$, for every $(w, \lambda) \in W \times \Delta_S$ and every L-formula φ, we define the notion φ *is true at* (w, λ) *in* $\mathfrak{M}$ ($\mathfrak{M}, (w, \lambda) \models \varphi$) inductively as follows:

- $\mathfrak{M}, (w, \lambda) \models p$ iff $(w, \lambda) \in v(p)$ for variables p,
- $\mathfrak{M}, (w, \lambda) \models s$ iff $(w, \lambda) \in v(s)$ for nominals s,
- $\mathfrak{M}, (w, \lambda) \models \neg\varphi$ iff $\mathfrak{M}, (w, \lambda) \not\models \varphi$,
- $\mathfrak{M}, (w, \lambda) \models \varphi_1 \wedge \varphi_2$ iff $\mathfrak{M}, (w, \lambda) \models \varphi_1$ and $\mathfrak{M}, (w, \lambda) \models \varphi_2$,

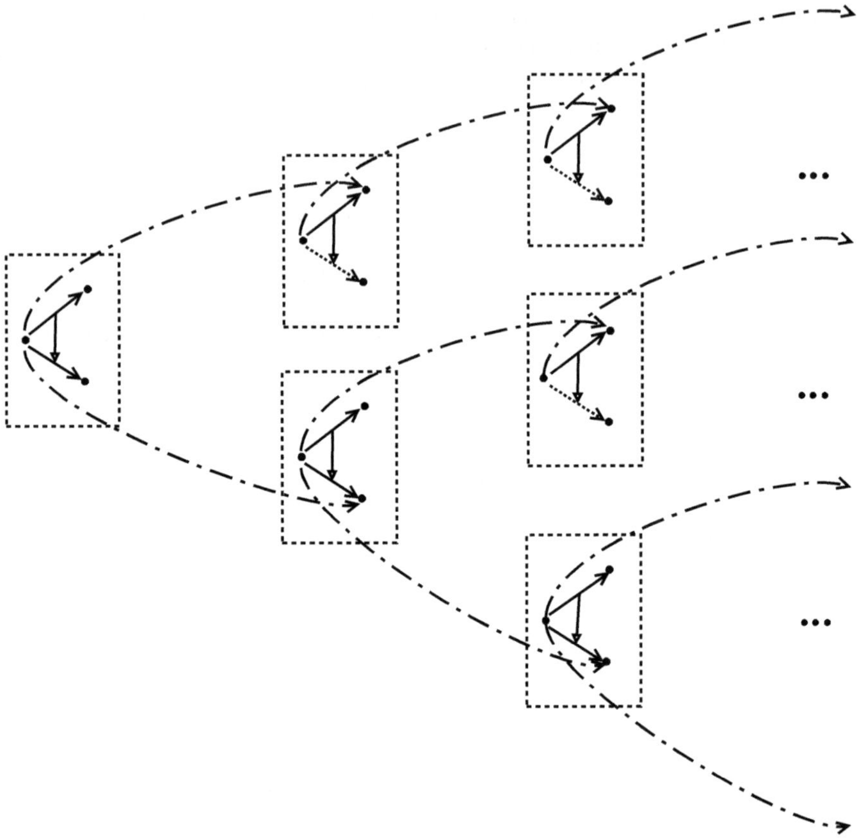

Fig. 7.14: A representation of a switch frame. The various components are delimited by a dashed line, the switches of that component are represented by the arrows as before and the (reactive) transitions by the line formed by lines and dots.

- $\mathfrak{M}, (w, \lambda) \models \not\triangleright\varphi$ iff there is $w' \in W$ such that $\lambda(w, w') \in \Delta$ and $\mathfrak{M}, (w', \lambda(w, w')) \models \varphi$,
  Notice that $\lambda(w, w')$ iff $(w, w') = (w, w', \epsilon) \in R_\lambda$ iff $(w, w') \in \check{R}^{\epsilon}_{\lambda}$,
- $\mathfrak{M}, (w, \lambda) \models @_i\varphi$ iff $\mathfrak{M}, (w', \lambda) \models \varphi$ for $v(i) = \Delta_S \times \{w'\}$,
- $\mathfrak{M}, (w, \lambda) \models \triangleleft_\sigma(\varphi_1, \ldots, \varphi_{2n+1})$ iff there are $v_1 \ldots v_{2n+1} \in W$ such that $(w, v_1, \ldots, v_{2n}, v_{2n+1}) \in \check{R}^{\sigma}_{\lambda}$ and $\mathfrak{M}, (\lambda, v_i) \models \varphi_i$.

We say that φ *is true in* $\mathfrak{M}$ iff $\mathfrak{M}, (w, \lambda) \models \varphi$ for every $(w, \lambda) \in \Delta_S \times W$. We say that φ *is valid in* a switch frame if it is true in every switch model over it.

7.3.2 Axiomatising

In this section we prove that the axiomatisation presented in Figure 7.15 generates all the $\mathcal{H}_s(@)$-formulas that are valid in all switch frames. Furthermore, we obtain the usual hybrid automatic completeness for pure axioms. The following axiomatisation is the natural adaptation of the ones for the standard hybrid logics given in [25, 26].

L_s	
Axioms:	
CT	All classical tautologies
$K^l_{\Box\!\|_\sigma}$	$\Box\!\|_\sigma(\varphi_1,\ldots,\varphi \to \psi,\ldots,\varphi_{2n+1}) \to$ $(\Box\!\|_\sigma(\varphi_1,\ldots,\varphi,\ldots,\varphi_{2n+1}) \to \Box\!\|_\sigma(\varphi_1,\ldots,\psi,\ldots,\varphi_{2n+1}))$ where l is the component including $\varphi \to \psi$ is.
$K_@$	$@_i(\varphi \to \psi) \to (@_i\varphi \to @_i\psi)$
$Selfdual_@$	$@_i\varphi \leftrightarrow \neg @_i\neg\varphi$
$Ref_@$	$@_i i$
$Agree$	$@_i @_j\varphi \leftrightarrow @_j\varphi$
$Intro$	$i \to (\varphi \leftrightarrow @_i\varphi)$
$Sync$	$\not\Diamond i \leftrightarrow \lhd\!\|_\epsilon i$
$Det_{\not\Diamond}$	$\not\Diamond(i \wedge \varphi) \to \not\Box(i \to \varphi)$
$K_{\not\Box}$	$\not\Box(\varphi \to \psi) \to (\not\Box\varphi \to \not\Box\psi)$
Dyn	$\not\Diamond^{(i,j)} @_{i_1} \Diamond\|_\sigma(i_2,\ldots,i_{2n+2}) \leftrightarrow$ $(@_{i_1} \Diamond\|_\sigma(i_2,\ldots,i_{2n+2}) \wedge \neg @_i \Diamond\|_{\sigma\circ}(j,i_1 \ldots,i_{2n+2}) \vee @_i \Diamond\|_{\sigma\bullet}(j,i_1 \ldots,i_{2n+2}))$
Rules:	
MP	If $\vdash \varphi$ and $\vdash \varphi \to \psi$ then $\vdash \psi$
$Subst$	If $\vdash \varphi$ then $\vdash \varphi^\theta$
$Gen_{\Box\!\|_\sigma}$	If $\vdash \varphi$ then $\vdash \Box\!\|_\sigma(\bot,\ldots,\bot,\varphi,\bot,\ldots,\bot)$
$Gen_@$	If $\vdash \varphi$ then $\vdash @_i\varphi$
$Gen_{\not\Box}$	If $\vdash \varphi$ then $\vdash \not\Box\varphi$
$Name$	If $\vdash @_i\varphi$ and i does not occur in φ then $\vdash \varphi$
$Paste_{\not\Diamond^\lambda \Diamond\|_\sigma}$	If $\vdash \not\Diamond^\lambda @_i \Diamond\|_\sigma(j_1,\ldots,j_{2n+1}) \wedge \not\Diamond^\lambda(\bigwedge_{1\geq l\geq 2n+1} @_{j_l}\varphi_l) \to \psi$ and $j_m \neq i$ does not occur in φ_l or ψ then $\vdash \not\Diamond^\lambda @_i \Diamond\|_\sigma(\varphi_1,\ldots,\varphi_{2n+1}) \to \psi$
$Paste'_{\not\Diamond^\lambda \not\Diamond}$	If $\vdash \not\Diamond^\lambda @_i \not\Diamond j \wedge \not\Diamond^\lambda \not\Diamond @_j\varphi \to \psi$ and $j \neq i$ does not occur in φ or ψ then $\vdash \not\Diamond^\lambda @_i \not\Diamond\varphi \to \psi$

Fig. 7.15: The L_s axiomatisation.

Remark 7.21. Notice that if we considered a different dynamics for the switches, *Dyn* would have to be different. For example the alternatives in 7.10 correspond to:

- the disconnecting switch prevails
  $\not\Diamond^{(i,j)} @_{i_1} \Diamond|_\sigma(i_2,\ldots,i_{2n+2}) \leftrightarrow$
  $\neg @_i \Diamond|_{\sigma\circ}(j,i_1 \ldots,i_{2n+2}) \wedge (@_{i_1} \Diamond|_\sigma(i_2,\ldots,i_{2n+2}) \vee @_i \Diamond|_{\sigma\bullet}(j,i_1 \ldots,i_{2n+2}))$,
- always change state
  $\not\Diamond^{(i,j)} @_{i_1} \Diamond|_\sigma(i_2,\ldots,i_{2n+2}) \leftrightarrow$

$$(@_{i_1} \Diamond|_\sigma(i_2,\ldots,i_{2n+2}) \wedge \neg @_i \Diamond|_{\sigma\circ}(j,i_1 \ldots,i_{2n+2}) \vee$$
$$\neg @_{i_1} \Diamond|_\sigma(i_2,\ldots,i_{2n+2}) \wedge @_i \Diamond|_{\sigma\bullet}(j,i_1 \ldots,i_{2n+2})).$$

Here is the theorem we shall prove:

Theorem 7.22. *Let Λ be a set of pure $\mathcal{H}_s(@)$ formulas. A set of $\mathcal{H}_s(@)$ formulas Σ is $L_s + \Lambda$-consistent iff Σ is satisfiable in a model satisfying the frame properties defined by Γ. Here $L_s + \Lambda$ is the above axiomatisation extended with the axioms of Λ.*

Lemma 7.23. *The following are derivable*

- $K^{-1}_{@}$: $\vdash (@_i\varphi \to @_i\psi) \to @_i(\varphi \to \psi)$,
- *Nom:* $\vdash @_i j \to (@_i\varphi \to @_j\varphi)$,
- *Sym:* $\vdash @_i j \to @_j i$,
- *Name′: If* $\vdash i \to \varphi$ *then* $\vdash \varphi$ *where* i *does not occur in* φ,
- $Back_{\lhd|_\sigma}$: $\lhd|_\sigma(\varphi_1, \ldots, @_i\varphi_k, \ldots, \varphi_{2n+1}) \to @_i\varphi_k$,
- $Bridge_{\lhd|_\sigma}$: $@_i \diamond|_\sigma(j_1, \ldots, j_{2n+1}) \wedge (\bigwedge_{1 \geq l \geq 2n+1} @_{j_l}\varphi_l) \to @_i \diamond|_\sigma(\varphi_1, \ldots, \varphi_{2n+1})$.

Proof. The first 4 are proved in [26].

- $Back_{\not\lhd}$

$$\cfrac{\cfrac{\cfrac{\overline{@_i @_j\varphi \leftrightarrow @_u @_j\varphi}\ \textit{Agree, Agree}}{@_u \diamond|_\sigma(\varphi_1, \ldots, j, \ldots, \varphi_{2n+1}) \wedge @_i @_j\varphi \to @_u @_j\varphi}}{@_u \diamond|_\sigma(\varphi_1, \ldots, @_j\varphi, \ldots, \varphi_{2n+1}) \to @_u @_j\varphi}\ Paste_{\lhd|_\sigma}}{\lhd|_\sigma(\varphi_1, \ldots, @_j\varphi, \ldots, \varphi_{2n+1}) \to @_j\varphi}\ K^{-1}_{@}, Name$$

- $Bridge_{\not\lhd^\sigma}$ by induction on σ, $\sigma = \epsilon$ is trivial and

$$\cfrac{\cfrac{\cfrac{\cfrac{\overline{\Box|_\epsilon(\neg\varphi \to \neg j) \to (\Box|_\epsilon\neg\varphi \to \Box|_\epsilon\neg j)}\ K_{\Box|_\epsilon}}{\Box|_\epsilon\neg\varphi \wedge \lhd|_\epsilon j \to \lhd|_\epsilon(j \wedge \neg\varphi)}\ CT \qquad \overline{j \wedge \neg\varphi \to @_j\neg\varphi}\ Int_@, CT}{\Box|_\epsilon\neg\varphi \wedge \lhd|_\epsilon j \to \lhd|_\epsilon @_j\neg\varphi}\ CT \qquad \overline{\lhd|_\epsilon @_j\neg\varphi \to @_j\neg\varphi}\ Back}{@_i(\Box|_\epsilon\neg\varphi \wedge \lhd|_\epsilon j) \to @_i @_j\neg\varphi}\ Gen_@, K_@}{\cfrac{@_i\Box|_\epsilon\neg\varphi \wedge @_i \diamond|_\epsilon j \to @_j\neg\varphi}{@_i \diamond|_\epsilon j \wedge @_j\varphi \to @_i \diamond|_\epsilon\varphi}\ CT}\ K_@, CT, Agree, Selfdual_@$$

The proof for arbitrary σ's consists of the repetition of this derivation applied to each of the $2|\sigma| - 1$ coordinates, where each time an argument eats a conjunct.

Lemma 7.24. *The following are derivable:*

- $\not\Box^{(i,j)\lambda}\varphi \leftrightarrow @_i \not\Box(j \to \not\Box^\lambda\varphi)$,
- $Gen_{\not\Box^\lambda}$: *If* $\vdash \varphi$ *then* $\vdash \not\Box^\lambda\varphi$,
- $K_{\not\Box^\lambda}$: $\not\Box^\lambda(\varphi \to \psi) \to (\not\Box^\lambda\varphi \to \not\Box^\lambda\psi)$,
- $Det_{\not\lhd^\lambda}$: $\not\lhd^\lambda\varphi \leftrightarrow \not\Box^\lambda\varphi \wedge \not\lhd^\lambda\top$,
- $Gen'_{\not\lhd^\lambda}$: $\vdash \varphi \to \psi$ *then* $\vdash \not\lhd^\lambda\varphi \to \not\lhd^\lambda\varphi$.

Furthermore from $Gen_{\not\Box^\lambda}$ *and* $K_{\not\Box^\lambda}$ *we easily get* $/\!\diamond^\lambda(\varphi \wedge \psi) \leftrightarrow /\!\diamond^\lambda\varphi \wedge /\!\diamond^\lambda\psi$ *and* $\not\lhd^\lambda(\varphi \vee \psi) \leftrightarrow \not\lhd^\lambda\varphi \vee \not\lhd^\lambda\psi$.

Proof. • $\not\Box^{(i,j)\lambda}\varphi = \neg @_i \not\Diamond(j\wedge \not\Diamond^{\lambda}\neg\varphi)$ which is equivalent to $@_i \not\Box(j \to \neg \not\Diamond^{\lambda}\neg\varphi) = @_i \not\Box(j \to \not\Box^{\lambda}\varphi)$.

- $Gen_{\not\Box^{\lambda}}$ by induction on λ, ϵ is trivial and

$$\dfrac{\dfrac{\dfrac{\dfrac{\varphi}{\not\Box^{\lambda}\varphi}\,IH}{j \to \not\Box^{\lambda}\varphi}\,CT}{\not\Box(j \to \not\Box^{\lambda}\varphi)}\,Gen_{\not\Box}}{@_i \not\Box(j \to \not\Box^{\lambda}\varphi)}\,Gen_{@}$$

- $K_{\not\Box^{\lambda}}$ by induction on λ, $\lambda = \epsilon$ is trivial and

$$\dfrac{\dfrac{\dfrac{\dfrac{\varphi}{\not\Box^{(i,j)\lambda}(\varphi \to \psi)}\,IH}{@_i \not\Box(j \to (\not\Box^{\lambda}\varphi \to \not\Box^{\lambda}\psi))}\,CT}{@_i \not\Box(j \to \not\Box^{\lambda}\varphi) \to @_i \not\Box(j \to \not\Box^{\lambda}\psi)}\,K_{\not\Box}, K_{@}}{\not\Box^{(i,j)\lambda}\varphi \to \not\Box^{(i,j)\lambda}\psi}\,def$$

- $Det_{\not\Diamond^{(i,j)\lambda}}$ by induction on λ, $\lambda = \epsilon$ is trivial and

$$\dfrac{\dfrac{}{\not\Box^{(i,j)\lambda}\varphi\wedge \not\Diamond^{(i,j)\lambda}\top \to \not\Diamond^{(i,j)\lambda}\varphi}\,K_{\not\Box}^{(i,j)\lambda}, CT \qquad \dfrac{\dfrac{\dfrac{}{\not\Diamond^{(i,j)\lambda}\varphi \to @_i \not\Diamond(j \wedge [\not\Box^{\lambda}\varphi\wedge \not\Diamond^{\lambda}\top])}\,Def, IH}{\not\Diamond^{(i,j)\lambda}\varphi \to @_i \not\Diamond(j\wedge \not\Box^{\lambda}\varphi)}\,IH, CT}{\not\Diamond^{(i,j)\lambda}\varphi \to \not\Box^{(i,j)\lambda}\varphi\wedge \not\Diamond^{(i,j)\lambda}\top}\,CT}{\not\Diamond^{(i,j)\lambda}\varphi \leftrightarrow \not\Box^{(i,j)\lambda}\varphi\wedge \not\Diamond^{(i,j)\lambda}\top}\,CT$$

- $Gen'_{\not\Diamond^{\lambda}}$ by induction on λ, $\lambda = \epsilon$ is trivial and

$$\dfrac{\dfrac{\dfrac{\dfrac{\varphi \to \psi}{\not\Box^{\lambda}(\varphi \to \psi)}\,Gen_{\not\Box^{\lambda}}}{\not\Diamond^{\lambda}\top\wedge \not\Box^{\lambda}(\varphi \to \psi)}\,CT}{\not\Box^{\lambda}\varphi\wedge \not\Diamond^{\lambda}\top \to \not\Box^{\lambda}\psi\wedge \not\Diamond^{\lambda}\top}\,K_{\not\Box^{\lambda}}, CT}{\not\Diamond^{\lambda}\varphi \to \not\Diamond^{\lambda}\psi}\,Det_{\not\Diamond^{\lambda}}$$

Definition 7.25. *Let Σ be a set of $\mathcal{H}_s(@)$-formulas:*

- *Σ is named if one of its elements is a nominal.*
- *Σ is $\triangleleft|_{\sigma}$-saturated if for all σ' and $\not\Diamond^{\lambda}@_i\triangleleft|_{\sigma'}(\varphi_1,\ldots,\varphi_{2n+1}) \in \Sigma$ there are nominals $j_1,\ldots,j_{2n+1}$ such that $\not\Diamond^{\lambda}@_i \Diamond|_{\sigma'}(j_1,\ldots,j_{2n+1}) \in \Sigma$ and $\not\Diamond^{\lambda}@_{j_k}\varphi_k \in \Sigma$, $k = 1,\ldots,2n+1$.*
- *Σ is $\not\Diamond$-saturated if for all $\not\Diamond^{\lambda}@_i \not\Diamond\varphi \in \Sigma$ there is a nominal j such that $\not\Diamond^{\lambda}@_i \not\Diamond(j \wedge \varphi) = \not\Diamond^{\lambda(i,j)}\varphi \in \Sigma$.*

Lemma 7.26. *(Lindenbaum Lemma). Every $L_s + \Gamma$-consistent set of formulas can be extended to a named, $\not\Diamond$-saturated and $\triangleleft|_{\sigma}$-saturated MCS, by adding countably many new nominals to the language.*

Proof. Let $(i_n)_{n<\omega}$ be an enumeration of the new nominals and $(\varphi_n)_{n<\omega}$ an enumeration of the formulas in the extended language.

We define $\Sigma^0 = \Sigma \cup i_0$, *Name′* guarantees that it is consistent.

If $\Sigma^n \cup \{\varphi_n\}$ is inconsistent then $\Sigma^{n+1} = \Sigma^n$. Otherwise:

1. $\Sigma^{n+1} = \Sigma^n \cup \{\varphi_n\}$ if φ_n is not of the form $/\!\!\diamond^{\lambda} @_i \diamond\!|_\sigma(\psi_1, \ldots, \psi_{2k+1})$ or $/\!\!\diamond^{\lambda} @_i\ /\!\!\diamond\psi$,
2. $\Sigma^{n+1} = \Sigma^n \cup \{\varphi_n\} \cup \{/\!\!\diamond^{\lambda} @_i \diamond\!|_\sigma(i_m, \ldots, i_{m+2n+1})\} \cup \{/\!\!\diamond^{\lambda} @_{i_{m+l}} \psi_l : 1 \leq l \leq 2k+1\}$ if it is of the form $/\!\!\diamond^{\lambda} @_i \diamond\!|_\sigma(\psi_1, \ldots, \psi_{2k+1})$,
3. $\Sigma^{n+1} = \Sigma^n \cup \{\varphi_n\} \cup \{/\!\!\diamond^{\lambda} @_i\ /\!\!\diamond(i_m \wedge \psi)\}$ if it is of the form $/\!\!\diamond^{\lambda} @_i\ /\!\!\diamond\psi$.

Here i_m is the first new nominal that does not occur in Σ^n or φ_n.
Let $\Sigma^\omega = \bigcup_{n<\omega} \Sigma^n$. Then $\Sigma \subseteq \Sigma^\omega$ and Σ^ω is named, $\diamond\!|_\sigma$-saturated, $/\!\!\diamond$-saturated, maximal and consistent. The only non-trivial step is in 2, and here consistency is guaranteed by $Paste_{\not\diamond^{\lambda}\diamond|_\sigma}$ and $Paste'_{\not\diamond^{\lambda}\not\diamond}$.

Definition 7.27. *(Henkin model from Γ). Let Γ be a maximal consistent set of $\mathcal{H}_s(@)$ formulas. For all nominals i, let $|i| = \{j : @_i j \in \Gamma\}$. Then $\mathfrak{M}_\Gamma = (W, \Delta, T, v)$ is given by:*

$$
\begin{aligned}
&W = \{|i| : i \text{ is a nominal}\},\\
&\Delta = \{\ll \lambda : /\!\!\diamond^{\lambda} \top \in \Gamma\},\\
&\check{R} = \{T^{\sigma}_{\ll\lambda}\}_{\sigma,\ll\lambda},\\
&\check{R}^{\sigma}_{\ll\lambda}(|i_1|, \ldots, |i_{2n+2}|) \text{ iff } /\!\!\diamond^{\lambda} @_{i_1} \diamond\!|_\sigma(i_2, \ldots, i_{2n+2}) \in \Gamma,\\
&v(p) = \{(|i|, \ll \lambda) : /\!\!\diamond^{\lambda} @_i p \in \Gamma\},\\
&v(i) = \{(|i|, \ll \lambda) : \ll \lambda \in \Delta\},
\end{aligned}
$$

where $\ll \epsilon = \epsilon$ and $\ll \lambda(i, j) = \ll \lambda(|i|, |j|)$. That $\mathfrak{M}_\Gamma$ is well defined follows from Ref, Sym, Nom and $Gen'_{\not\diamond|_\lambda}$.

Lemma 7.28. $/\!\!\diamond^{\lambda(i,j)}\varphi = /\!\!\diamond^{\lambda} @_i\ /\!\!\diamond(j \wedge \varphi)$.

Proof. By induction on λ:

For $\lambda = \epsilon$ the two formulas coincide trivially and for $\lambda = (s, t)\gamma$

$$/\!\!\diamond^{(s,t)\gamma(i,j)}\varphi = /\!\!\diamond^{(s,t)}\ /\!\!\diamond^{\gamma(i,j)}\varphi =^{IH} /\!\!\diamond^{(s,t)}\ /\!\!\diamond^{\gamma} @_i\ /\!\!\diamond(j \wedge \varphi) = /\!\!\diamond^{(s,t)\gamma} @_i\ /\!\!\diamond(j \wedge \varphi).$$

Lemma 7.29. *$\mathfrak{M}_\Gamma$ is a switch model where $(W \times \Delta, \check{R})$ is the switch frame generated by $S = (W, R)$ such that*

$$R = \{(|i_1||i_2|, \ldots, |i_{2n+1}||i_{2n+2}|, \sigma) : (|i_1|, |i_2|, \ldots, |i_{2n+1}|, |i_{2n+2}|) \in \check{R}^{\sigma}_{\epsilon}\}.$$

Proof. Let Δ_S and $R_S = \{R^{\sigma}_{\lambda}\}$ as defined in 7.9. The proof that $\Delta = \Delta_S$ and $\check{R}^{\sigma}_{\lambda} = R^{\sigma}_{\lambda}$ is done by induction on the length of $\ll \lambda$.

The case of $\ll \lambda = \epsilon$ is trivial since it is in both Δ ($\top \in \Gamma$ by CT) and Δ_S, furthermore $\check{R}^{\sigma}_{\epsilon} = R^{\sigma}_{\epsilon}$ by definition.

The induction step ($\ll \lambda \rightarrow \ll \lambda(i, j)$):
$\ll \lambda(|i|, |j|) \in \Delta_S$ iff
$\ll \lambda \in \Delta_S$ and $(|i|, |j|) \in R_\lambda$ iff $\ll \lambda \in \Delta_S$ and $R^{\epsilon}_{\ll\lambda}(|i|, |j|)$ iff (IH)
$/\!\!\diamond^{\lambda}\top \in \Gamma$ and $\check{R}^{\epsilon}_{\ll\lambda}(|i|, |j|)$ iff $/\!\!\diamond^{\lambda}@_i\ /\!\!\diamond(j \wedge \top) = /\!\!\diamond^{\lambda(i,j)}\top \in \Gamma$ iff
$\ll \lambda(|i|, |j|) \in \Delta$.

Also,
$(|i_1||i_2|, \ldots, |i_{2n+1}||i_{2n+2}|, \sigma) \in R_{\ll\lambda(i,j)}$ iff
$(|i_1||i_2|, \ldots, |i_{2n+1}||i_{2n+2}|, \sigma) \in R_{\ll\lambda}$ and $(|i||j|, |i_1||i_2|, \ldots, |i_{2n+1}||i_{2n+2}|, \sigma\circ) \notin R_{\ll\lambda}$
or
$(|i||j|, |i_1||i_2|, \ldots, |i_{2n+1}||i_{2n+2}|, \sigma\bullet) \in R_{\ll\lambda}$
iff (IH)
$\check{R}^{\sigma}_{\lambda}(|i_1|, |i_2|, \ldots, |i_{2n+1}|, |i_{2n+2}|, \sigma)$ and $\check{R}^{\sigma\circ}_{\lambda}(|i|, |j|, |i_1|, |i_2|, \ldots, |i_{2n+1}|, |i_{2n+2}|)$
or
$\check{R}^{\sigma\bullet}_{\lambda}(|i|, |j|, |i_1|, |i_2|, \ldots, |i_{2n+1}|, |i_{2n+2}|)$
iff
$/\!\!\diamond^{\lambda}@_{i_1} \diamond|_{\sigma}(i_2, \ldots, i_{2n+2}) \in \Gamma$ and $/\!\!\diamond^{\lambda}@_i \diamond|_{\sigma\circ}(j, i_1 \ldots, i_{2n+2}) \notin \Gamma$
or
$/\!\!\diamond^{\lambda}@_i \diamond|_{\sigma\bullet}(j, i_1 \ldots, i_{2n+2}) \subset \Gamma$
iff
$/\!\!\diamond^{\lambda}@_{i_1} \diamond|_{\sigma}(i_2, \ldots, i_{2n+2}) \wedge\ /\!\!\diamond^{\lambda}\neg @_i \diamond|_{\sigma\circ}(j, i_1 \ldots, i_{2n+2}) \vee\ /\!\!\diamond^{\lambda}@_i \diamond|_{\sigma\bullet}(j, i_1 \ldots, i_{2n+2}) \in$
Γ
iff (Dyn, $Gen'_{\not\diamond^{\lambda}}$ and Lemma 7.24)
$/\!\!\diamond^{\lambda}\ /\!\!\diamond^{(i,j)}@_{i_1} \diamond|_{\sigma}(i_2, \ldots, i_{2n+2}) \in \Gamma$
iff (Lemma 7.28)
$/\!\!\diamond^{\lambda(i,j)}@_{i_1} \diamond|_{\sigma}(i_2, \ldots, i_{2n+2}) \in \Gamma$.

Lemma 7.30 (Truth lemma). *For all $/\!\!\diamond$-saturated and $\triangleleft|_\sigma$-saturated $L_s + \Lambda$-MCSs Γ, nominals i and formulas φ,*

$$\mathfrak{M}, (|i|, \ll \lambda) \models \varphi \textit{ iff } /\!\!\diamond^{\lambda}@_i\varphi \in \Gamma.$$

Proof. By induction on the length of φ.

- propositional symbols and nominals by definition;
- $\varphi = \neg\psi$
  $\mathfrak{M}_\Gamma, (|i|, \ll \lambda) \models \neg\psi$ iff $\mathfrak{M}_\Gamma, (|i|, \ll \lambda) \not\models \psi$ and $\mathfrak{M}_\Gamma, (|i|, \ll \lambda) \models \top$
  iff (IH)
  $/\!\!\diamond^{\lambda}@_i\psi \notin \Gamma$ and $/\!\!\diamond^{\lambda}@_i\top \in \Gamma$ iff $\neg\ /\!\!\diamond^{\lambda}@_i\psi \wedge\ /\!\!\diamond^{\lambda}\top \in \Gamma$
  iff ($Selfdual_@$, $Gen'_{\not\diamond^{\lambda}}$ and $Det_{\not\diamond^{\lambda}}$)
  $/\!\!\diamond^{\lambda}@_i\neg\psi$;
- $\varphi = \psi_1 \rightarrow \psi_2$ as in the previous case we apply $K_@$, $K_@^{-1}$, $Gen'_{\not\diamond^{\lambda}}$ and $Det_{\not\diamond^{\lambda}}$;
- $\varphi = @_j\psi$ apply $Agree$ and $Gen'_{\not\diamond^{\lambda}}$;

- $\varphi = \triangleleft_\sigma \psi$;
  If $\mathfrak{M}_\Gamma, (|i|, \ll \lambda) \models \triangleleft_\sigma(\psi_1, \ldots, \psi_{2n+1})$ then there are $|j_1|, \ldots, |j_{2n+1}|$ such that

  $$\check{R}^\sigma_{\ll\lambda}(|i|, |j_1|, \ldots, |j_{2n+1}|)$$

  and $\mathfrak{M}_\Gamma, (|j_l|, \ll \lambda) \models \psi_l$. By definition $/\!\!\diamond^\lambda @_i \diamond|_\sigma(j_1, \ldots, i_{2n+1}) \in \Gamma$ and (IH) $@_{j_l}\psi_l \in \Gamma$ for $1 \leq l \leq 2n+1$. Thus, using $Bridge_{\triangleleft_\sigma}$, $Gen'_{\not\square^\lambda}$ and Lemma 7.24, we have that

  $$/\!\!\diamond^\lambda @_i \diamond|_\sigma(\varphi_1, \ldots, \varphi_{2n+1}) \in \Gamma.$$

  Conversely, suppose $/\!\!\diamond^\lambda @_i \diamond|_\sigma(\varphi_1, \ldots, \varphi_{2n+1}) \in \Gamma$, then by $/\!\!\diamond^\lambda$-saturation there are $j_1, \ldots, j_{2n+1}$ such that $/\!\!\diamond^\lambda @_i \diamond|_\sigma(j_1, \ldots, j_{2n+1}) \in \Gamma$ and $/\!\!\diamond^\lambda @_{j_l}\psi_l \in \Gamma$ for $1 \leq l \leq 2n+1$. By IH $\mathfrak{M}_\Gamma, (|j_l|, \ll \lambda) \models \psi_l$ and by definition $\check{R}^\sigma_{\ll\lambda}(|i|, |j_1|, \ldots, |j_{2n+1}|)$. Hence,

  $$\mathfrak{M}_\Gamma, (|i|, \ll \lambda) \models \triangleleft_\sigma(\psi_1, \ldots, \psi_{2n+1}).$$

- $\varphi = /\!\!\diamond\psi$
  $\mathfrak{M}_\Gamma, (|i|, \ll \lambda) \models /\!\!\diamond\varphi$ iff
  there exists $|j|$ such that $\ll \lambda(|i|, |j|) \in \Delta$ such that $\mathfrak{M}_\Gamma, (|j|, \ll \lambda(|i|, |j|)) \models /\!\!\diamond\varphi$ iff (I.H.)
  there exists $|j|$ such that $/\!\!\diamond^{\lambda(i,j)} @_j\varphi = /\!\!\diamond^\lambda @_i\ /\!\!\diamond(j \wedge @_j\varphi) \in \Gamma$ iff (using $/\!\!\diamond$-saturation and CT) $/\!\!\diamond^\lambda @_i\ /\!\!\diamond\varphi \in \Gamma$.

Lemma 7.31 (Frame lemma). *For all $/\!\!\diamond$-saturated and $\triangleleft_\sigma$-saturated $L_s + \Lambda$-MCSs, $\mathfrak{M}_\Gamma$ satisfies the switch frame properties defined by Λ.*

Proof. It follows from the fact that $\Lambda \subseteq \Gamma$ contains $\not\square^\lambda @_i\varphi$ where φ is an instance of an element of Λ from $Subst$, $Gen_@$ and $Gen_{\not\square^\lambda}$.

We can now prove Theorem 7.22.

Proof. (Of Theorem 7.22) Suppose Σ is $L_s + \Lambda$-consistent. By Lemma 7.26, Σ can be extended to a named, $/\!\!\diamond$-saturated and $\triangleleft_\sigma$-saturated $L_s + \Lambda$-MCSs Γ. Let $i \in \Sigma$. By Lemma 7.30 we have $\mathfrak{M}_\Gamma, (|i|, \ll \epsilon) \models \Sigma$. By Lemma 7.31, $\mathfrak{M}_\Gamma$ satisfies all required frame properties.

7.3.3 Comments and remarks

Small switch frames

The fact that the propositional valuation is allowed to change at each move implies that switch frames based on switch graphs with infinite behaviour have infinite components. Nevertheless, for a complete view on the switch dynamics it would be enough to have one component for each reachable switch configuration. Let $A_S = \{R_\lambda : \lambda \in \Delta_S\}$ be the set of switch configurations that are obtained in S's dynamics. One can then consider the Kripke frame $\mathfrak{F}'_S = (W \times A, \check{R}'_S)$ (the small switch frame over S) where

- $\check{R}'_S = \{\hat{R}^\sigma\}_{\sigma \in \{\circ,\bullet\}^*}$,
- $((w_1, R), \ldots, (w_{2n+2}, R)) \in \check{R}^\sigma_\lambda$ iff $(w_1 w_2, \ldots, w_{2n+1} w_{2n+2}, \sigma) \in R$ and
- (w, R_λ) connects (reactively) with $(w', R_{\lambda w w'})$.

So everything is the same apart from the fact that we do not always get a new component from a move, the tree structure disappears. It is easy to see that $\mathfrak{F}'_S$ is finite if and only if S is finite. See Figure 7.16 for an example.

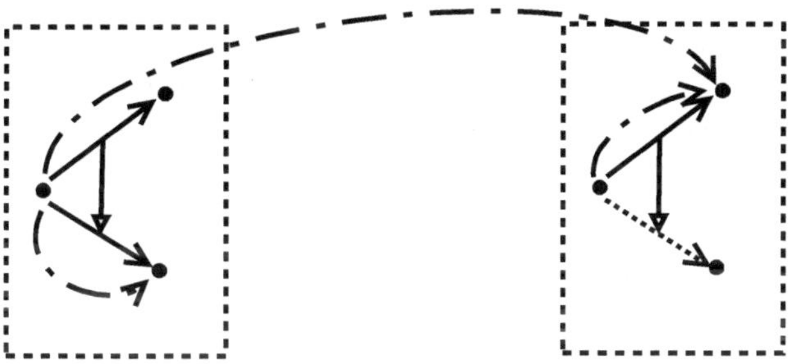

Fig. 7.16: Small switch frame of $\{\{a, b, c\}, \{(a, b), (a, c), (ab, cd, \circ)\}\}$.

Let us consider a semantics of $\mathcal{H}_s(@)$ over the small switch frames trivially adapted from Definition 7.20 (only the interpretation of $/\!\!\diamond$ changes), and call them small switch models. If the set of propositional symbols is empty ($\Pi = \emptyset$) these two classes of frames trivially generate the same logic. Also, when, e.g. for model checking purposes, everything in the model is finite, including the number of propositional variables involved, one would consider similarly a finite representation by having one component for each configuration of the switches and distribution of the relevant propositional variables.

Moreover, it is interesting to notice that even if the classes of small and the original switch frames are not immediately reducible to one another they yield the same logic.

It is trivial to define a model over $\mathfrak{F}_S$ from a model over $\mathfrak{F}'_S$ preserving modal truths, since we just have to copy the valuation of each configuration and paste it where it appears, so

$$\{\varphi \in \mathcal{H}_s(@) : \varphi \text{ is valid over switch frames}\} \subseteq$$
$$\{\varphi \in \mathcal{H}_s(@) : \varphi \text{ is valid over small switch frames}\}.$$

The converse is not so immediate since we cannot fit two different configurations in the same component of a small switch model. However, it is not hard to see that these two classes define the same logic. The idea of the proof is the following:

Given φ and a switch model $(W \times \Delta_S, R_S, v), (w, \epsilon) \models \varphi$ (we can assume $\lambda = \epsilon$ since the past clearly does not interfere with the evaluation of φ). We then consider a switch graph S' which is a copy of S plus some switches of a high enough level so that they do not interfere with the evaluation of φ and during the evaluation of φ the new switches' configuration changes so that we can accommodate the v. Let k be the higher level of the switches referred to in φ; it is clear we can add new points and switches such that at each crossing of an edge a different switch, of level higher than k, changes its state. For example one could add for each admissible sequence of edges covered in the interpretation of $\not\Diamond$, $\lambda = (w_1 w_2) \dots (w_{2n-1} w_{2n})$, a point w_λ and the switch

$$(w_1 w_2, \dots, w_{2n-1} w_{2n}, [w_\lambda w_\lambda]^k, \bullet^{n-1+k})$$

where $[w_\lambda w_\lambda]^k$ stands for $k+1$-occurrences of $w_\lambda w_\lambda$ with a comma dividing each pair of occurrences, thus guaranteeing a different component for each sequence of edges and therefore that we may define the appropriate valuation to each of them.

Decidability and f.m.p.

Given a satisfiable formula φ, that is, such that for some $(w, \lambda) \in W \times \Delta_S$ and valuation μ, we have:

$$(W \times \Delta_S, R_S, \mu), (w, \lambda) \models \varphi.$$

Clearly this is determined by just a part of the whole model. As we have noticed above we can start by throwing out the past, that is, assuming that $\lambda = \epsilon$. It is also clear that only the fragment corresponding to the paths bounded by the $\not\Diamond$-modal nesting-depth of φ, $md_{\not\Diamond}(\varphi)$, matters. Furthermore, the level of the relevant switches is bounded by $n + md_{\not\Diamond}(\varphi)$, where n is the level of the highest switch in the formula. Thus, the satisfiability of φ is reduced to checking the satisfiability over certain Kripke models corresponding to that kind of fragment. Of course, if W is infinite these fragments are infinite. The most direct way to finitise such a fragment would be to adapt the filtration method by identifying only the points that satisfy the same relevant formulas in all of the fragment components.

We start by noticing that this method fails over general switch models. Consider the following switch graph

$$S = (\omega, \{(0, n), (0n, nn, \circ) : n < \omega\}),$$

see Figure 7.17, and $\not\Diamond \not\Diamond \top \in \epsilon \subseteq \mathcal{H}_s(@)$.

Given a model over the switch frame generated by S, it is clear that no two points will be identified since $(n, 0n)' \models' \neg \not\Diamond \not\Diamond \top$ and $(n, 0m)' \models' \not\Diamond \not\Diamond \top$ for $n \neq m$.

We were also unable to adapt the game-based argument to establish the PSPACE upper bound for the satisfiability problem of $\mathcal{H}(@)$, see [10, 24]. The problem is, when adding a new world to a particular component, how to update the other components without falling into a infinite loop of verifications.

The study of the logics corresponding to carefully chosen subclasses may result in more treatable problems and may lead to a better understanding of the general case. One interesting example is the class of switch graphs representing that each edge can only be crossed a certain number of times, modelling the finite durability of certain resources, e.g. roads, bridges, product stocks, etc. The class of switches required to model this is very simple (see Example 7.2), which may be reflected

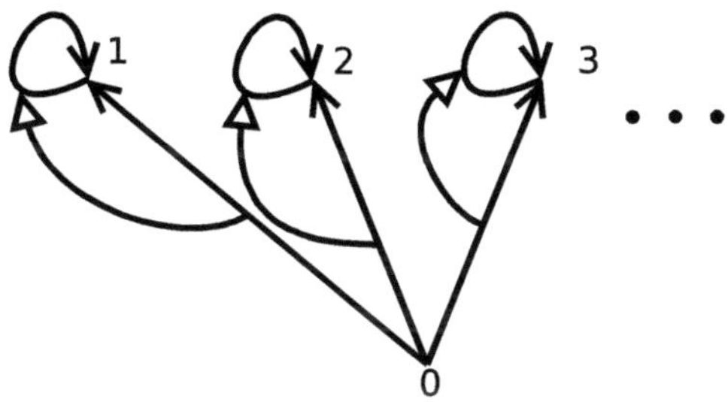

Fig. 7.17: A switch graph modelling a filtration problem.

positively in an easier understanding of certain properties of the originating logics, e.g. the finite model property.

Relation to Reactive Logics

There are no obvious connections (e.g. translations) between these switch logics and the ones studied in [62]. Even if we can extract a reactive frame from a switch graph by considering its local behaviour (its paths without jumps), the $\Diamond_P$ operator allows a global access over all the reactive states of a point that cannot be mimicked by the operators considered here. The two approaches are two transversal extensions of classical modal logic with different expressivity. Applications may assert which should be developed or even determine the necessity of joining the two views.

Further Extensions

Many other hybrid operators have been studied and it would be interesting to investigate how they would play in this context. A particularly natural extension of these logics would be to enrich $\mathcal{H}_s(@)$ with computational tree logic's (CTL) operators ([27]) or even μ-calculus ([145]). This would greatly reinforce our ability to reason about the behaviour of a switch graph and increase its usability. The way this would be done is not completely clear and it would depend on one's particular interests.

If we want to reason about k agents acting in the graph (in Section 7.2.3 there are such examples) and they are not allowed to jump, we are not interested in the whole set of sequences of edges, but only in the ones that the agents may cross, which are determined by their position at each moment. To express these restrictions, it may be necessary to include in the language the possibility of referring explicitly to the distribution of the agents in the graph. In this direction, a possibility would be to model each agent's location by variables, maybe even to introduce another sort of variables to deal with it like we did with nominals. In the language used we would be able to express for example that each agent can be only in one point at a time:

$$@_i(a \wedge \neg j) \rightarrow @_j \neg a,$$

considering a to be an agent variable. Also, to express their movement, we could associate a specific modal operator $\not\Box_a$ with each agent, and say:

$$@_i(a \wedge \neg j \wedge \not\Diamond j) \rightarrow \not\Box_a(j \rightarrow (@_j a \wedge @_i \neg a)).$$

8

Reactive Automata

8.1 Introduction and background

This chapter offers a reactive point of view for automata and shows that using reactive arrows can reduce the number of states of an automaton. Within the framework of reactive automata we provide some examples where such a reduction is striking.

Membership for regular languages, i.e. the problem of testing whether a word belongs to the language, is traditionally done with the help of a finite automaton representing the language, which is equivalent due to Kleene's Theorem (see for example [106]). When the automaton is deterministic the acceptance of a word of length m can be tested in linear time using $O(kn)$ space, where k is the size of the alphabet and n the number of states of the automaton. If the automaton is non-deterministic the alternatives are to simulate an equivalent deterministic automaton or to transform it into a deterministic automaton. The first option leads to $O(mn)$ membership time with space proportional to that of the non-deterministic automaton. The second option yields linear membership but at the cost of the determinisation which can be time and space exponential in the number of states. For related algorithms, see [4] or [124] and references therein. Both solutions are indeed implemented in the many variant of `grep` software aimed at locating regular motifs in texts.

The possible exponential blow up of automata when they are determinised is a major concern for applications. In this chapter we outline a solution based on extra arcs called reactive links in so-called reactive automata. Their size seems to be restrained even with a small number of reactive links.

The notion of reactive automata was introduced in [68] and [66], as part of a general reactive methodology (see also [17, 78, 89, 64, 76, 72]). The basic idea is that a reactive system is a system that changes when it is used, that is, it dynamically changes during its execution as a reaction to the manner in which it is being utilised. A practical example of this is any machine whose components change under stress or overuse. To give some such examples, consider:

1. an old bridge that might collapse if used by heavy lorries,
2. a fuse,

D.M. Gabbay, *Reactive Kripke Semantics*, Cognitive Technologies,
DOI 10.1007/978-3-642-41389-6_8, © Springer-Verlag Berlin Heidelberg 2013

3. a computer that crashes if overheated.

A reactive system has to be distinguished from a time-dependent system as it is not dependent on an objective clock.

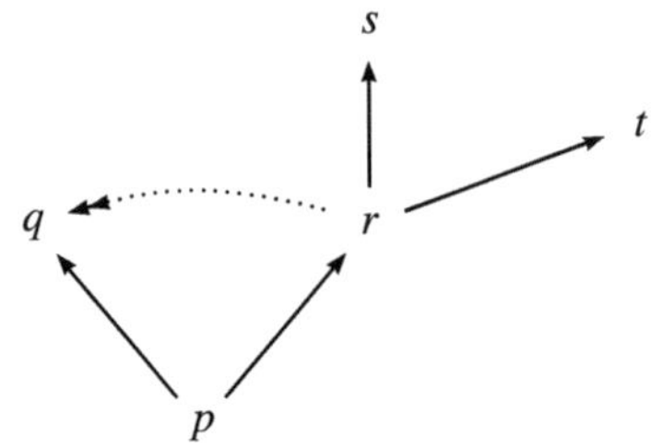

Fig. 8.1: Police road blocks: the signal along the dotted double-headed arrow from point r tells point q that it need not block the road at q.

Police road blocks

A bank robbery took place at 5pm just before bank closure time. The police institute road blocks to catch the getaway car. If a car is spotted crashing through road block r, a signal is sent to other road blocks along the route to remain alert, while another signal is sent to road blocks not along the route to be lifted (see Figure 8.1).

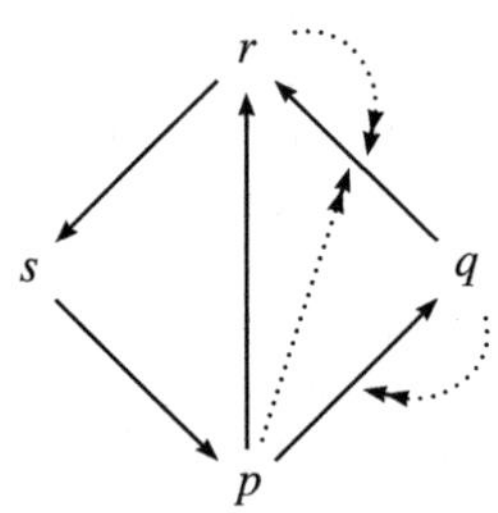

Fig. 8.2: A reactive graph.

Reactive graph

A simple theoretical example of a reactive system is the reactive graph of Figure 8.2. The single-headed arrows are directional links from vertices. The double-headed arrows are links from vertices to single arrows. The expression “using the graph” can mean in this context “traversing the graph.” Let us start from node p and move along the single-headed arrows to other nodes. As we exit p to go either to q or to r along the single-headed arrows the double-headed arrow originating at p fires and disconnects the single-headed arrow (q,r) at which it points. Thus, by exiting node p to go to node q, the arrow from q to r is cancelled. Upon continuing from r to s a signal is again sent to the arrow from q to r, which is then reactivated. Here, "use" means "walking through the graph" and the "change" or "reaction" is executed by the double-head arrows, switching single-head arrows *on* and *off*.

Algebraically the graph can be described as a set of nodes $\{p,q,r,s\}$, with the single-headed arrows written as pairs $\{(p,q),(p,r),(q,r),(r,s),(s,p)\}$ and the double-headed arrows as $\{(p,(q,r)),(q,(p,q)),(r,(r,q))\}$.

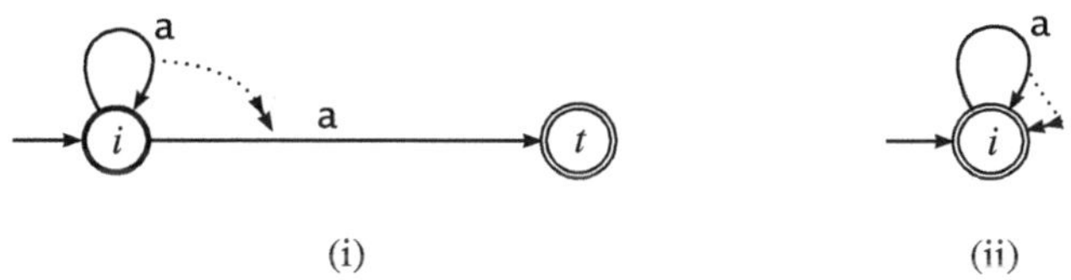

Fig. 8.3: Two reactive automata accepting the language a^{2n+1}. All arcs are initially active. (i) The automaton uses a reactive edge-to-edge link, which cancels or activates the arc from the initial state i to the terminal state t. (ii) The automaton uses a reactive edge-to-state link, which flips the status of state i alternately to terminal and non-terminal.

Reactive automaton

Figure 8.3 displays an example of a reactive automaton. Consider the automaton (ii), which has one state i. The transition is indicated as above by the single-headed arrow and reactivity by the double-headed arrow. For simplicity we assume we have only one letter a. Assume i is the initial state as well as the terminal state. Upon receiving a letter a the machine stays at i, and can accept or continue. The status of i as a terminal state is then cancelled by the reactive arrow. Upon receiving a second letter a the machine can no longer accept the word aa as i is no longer a terminal state. Instead if the machine receives a third letter a state i is reactivated as a terminal state and the machine can accept the word a^3. Obviously this reactive automaton accepts words of the form a^{2n+1}, $n \geq 0$, only. To implement such an acceptor without reactivity we would need more states (at least two for this example). We shall see in Section 8.3 that any reactive automaton can be simulated by a non-deterministic automaton. However the saving we get through reactivity is in the complexity of states. Further sections of the article consolidate this idea.

The above examples show that there are several ways of defining reactive links: from state to state (Figure 8.1), state to edge (Figure 8.2), edge to state (Figure 8.3(ii)), or edge to edge (Figure 8.3(i)); in the rest, we choose the most general reactivity: edge-to-edge reaction.

The value of our approach requires some clarification and discussion. Suppose we have an algorithm A involving some parameters π. If we offer a variant algorithm A' which reduces the number of parameters in π but introduces some new parameters π', then we might question the value of the new A', especially if the overall complexity remains the same. A reasonable objection can be made that no gain was demonstrated. This surely might be a valid objection in an abstract setting: if A and A' have no intuitive meaning and are mere combinatorial variations of each other, then we have indeed gained nothing. Instead, in our case the reactive arrows have a meaning, are part of a general methodology with meaningful applications in many areas, and have a human intuitive appeal. So just on the grounds of meaning and intuition our contribution stands. But there is more in favour of the approach.

When an automaton is used in some applied area, its states mean something in the application. The transitions, in response to input, mirror some natural behaviour in the application. Our reactive arrows are fully compatible with the automaton transitions. All they say is similar to: if you make transition 1, then do not execute transition 2 and here is transition 3 instead! So this compatibility implies that reactive arrows can carry something that has a natural meaning in the application and therefore one may be able to devise a new simpler reactive automaton that does the same job with less complexity. It is true that when we prove a general reduction theorem we cannot perhaps show a real reduction in size, but in many cases the application area and the nature of the job to be done allow us to simplify. This is because the reactive arrows have a natural meaning beyond their mere combinatorial use.

Take for example the reduction from k^n to kn states (see Section 8.5). Our formal proof embeds the k^n states into the paths of a $k \times n$ matrix. Formally this matrix is arbitrary and meaningless. But in an application it may present itself naturally, and because of that serious simplifications may be possible.

This type of competition between algorithms A and A' mentioned above is not new. Look for example at the classical propositional calculus. This is an NP-complete system and many problems in diverse areas are translated into it. It has several methodologies for theorem proving. Consider two of them: resolution, which is disjunction-negation based, and say the tableaux type, which is implication-conjunction based. Which one is to be used depends on the application. For colouring problems or Sudoku, resolution is better, but for rule-based problems tableaux are better. We have elsewhere developed reactive tableaux [89] and their use can simplify the matter. If we realise during the computation that certain rules are no longer needed we can block them. So, an algorithm for reducing the number of disjunctions in favour of implications can be useful in an appropriate application.

In the next section we define the notion of a reactive automaton and prove in Section 8.3 that the expressive power of automata is unchanged by adding reactive links. In Section 8.4 we show how to reduce the number of states of an ordinary non-deterministic automaton by adding reactive links. Section 8.5 is the analogue for deterministic automata and shows an exponential decrease in the number of states provided enough reactive links are added to the structure. Examples of reactive automata given in Section 8.6 do not have this drawback because they have only a linear number of reactive links while they are logarithmically smaller than the minimal automata associated with their accepted language. Some remarks and open questions are stated in the conclusion.

8.2 Reactive automata

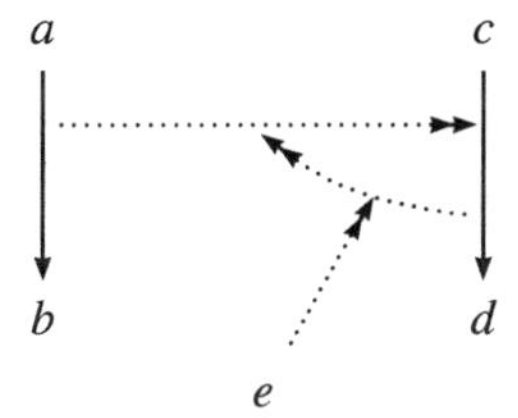

Fig. 8.4: Different types of reactive arrows.

We first explain our reactive concepts, then go to the formal definition.

Concepts

Let us fix the letter σ. A transition (by σ) from state a to state b can be "on" or can be "off". We write it as (on, a, b) or as (off, a, b), or just as (a, b), leaving the on/off status to context. We also have reactive arrows, pictured as dotted arrows with a double head. The typical situations are depicted in Figure 8.4.

When the automaton of Figure 8.4 moves from a to b it sends a signal to the arrow (c, d). This signal can be of three types:

1. $(+, (a, b), (c, d))$ switches (c, d) to *on* if it is *off* and leaves it *on* if it is *on*.
2. $(-, (a, b), (c, d))$ switches (c, d) to *off* if it is *on* and leaves it *off* if it is *off*.
3. $(\pm, (a, b), (c, d))$ is a proper switch: if (c, d) is *off* it changes it to *on*, and if it is *on* it changes it to *off*.

The reactive arrow itself (from (a, b) to (c, d)) can be on or off, which is denoted by $(\text{on}, (?, (a, b), (c, d)))$ or $(\text{off}, (?, (a, b), (c, c)))$, where ? can be + or −, or ±.

Similarly we have a reactive arrow from (c, d) to the reactive arrow from (a, b) to (c, d). It can also be +, −, or ±, which for example is denoted by $(+, (c, d), (+, (a, b)$ $(c, d)))$.

A reactive arrow can emanate from a state. For example in the diagram $(+, e, (+, (c, d), (+, (a, b), (c, d))))$ is emanating from e.

An ordinary non-deterministic automaton $\mathcal{A} = (S, i, \Sigma, F, R)$ includes a set of states, S, an initial state $i \in S$, and a non-empty subset $F \subseteq S$ of terminal states. The automaton receives words on the alphabet Σ. For every $\sigma \in \Sigma$ and state $p \in S$ the relation $R(p, \sigma, q)$ indicates that if the automaton is in state p and receives the letter σ then it can non-deterministically shift to state q.

Definition 8.1 (Switch Reactive Transformation). *Let $R \subseteq S \times \Sigma \times S$ be the transition relation of an (ordinary) automaton $\mathcal{A}$. Let $\mathbf{T}^+$, $\mathbf{T}^-$ be two subsets of $S \times \Sigma \times S \times S \times \Sigma \times S$; they are composed of pairs of the form $((p, \sigma, q), (r, \tau, s))$ where $\sigma, \tau \in \Sigma$, $p, q, r, s \in S$, and $(p, \sigma, q) \in R$.*

We define a transformation $(p, \sigma, q) \longrightarrow R^{(p,\sigma,q)}$ for $(p, \sigma, q) \in R$ using the sets $\mathbf{T}^+$ and $\mathbf{T}^-$ as follows:

$$R^{(p,\sigma,q)} = (R \setminus \{(r, \tau, s) \mid (r, \tau, s) \in R \text{ and } ((p, \sigma, q), (r, \tau, s)) \in \mathbf{T}^-\}) \\ \cup \{(r, \tau, s) \mid (r, \tau, s) \in R \text{ and } ((p, \sigma, q), (r, \tau, s)) \in \mathbf{T}^+\}$$

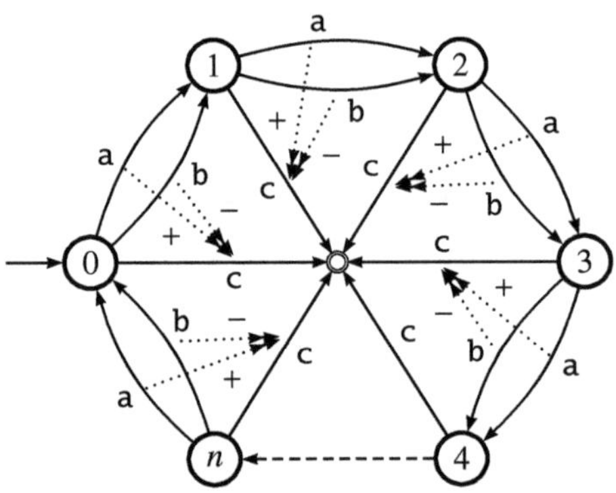

Fig. 8.5: Reactive automaton on the alphabet $A = \{\mathsf{a}, \mathsf{b}, \mathsf{c}\}$ accepting the language $A^*\mathsf{a}A^n\mathsf{c}$. The automaton is deterministic. Any a-transition activates a c-arc from its original state to the centre, which is the only terminal state. Any b-transition switches off such an arc. If a run over a word stops in the terminal state (the last step is a c-transition), the $(n + 1)$th letter before the end of the word must be a, so the word belongs to the language, and conversely.

Definition 8.2 (Switch Reactive Automaton).

1. *A reactive automaton is an ordinary non-deterministic automaton with a switch reactive transformation, i.e. a triple $\mathcal{R} = (\mathcal{A}, \mathbf{T}^+, \mathbf{T}^-)$ which defines the switch reactive transformation above.*
2. *Let $\sigma_1\sigma_2 \ldots \sigma_n$ be a word on the alphabet Σ. We define the notion of a (non-deterministic) run of $\mathcal{R}$ over $\sigma_1\sigma_2 \ldots \sigma_n$. The run is a sequence of pairs (p_k, R_k), $k = 0, \ldots, n$, defined as follows:*

   Step 0. *We start with the pair $(p_0, R_0) = (i, R)$ from the automaton $\mathcal{A} = (S, i, \Sigma, F, R)$.*

   Step $k > 0$. Assume pairs $(p_0, R_0), (p_1, R_1), \ldots, (p_{k-1}, R_{k-1})$ have been defined as the result of a run over $\sigma_1\sigma_2 \ldots \sigma_{k-1}$. Then, state p_k is such that $(p_{k-1}, \sigma_k, p_k) \in R_{k-1}$ and $R_k = R_{k-1}^{(p_{k-1},\sigma_k,p_k)}$.

3. *We say that the reactive automaton $\mathcal{R}$ accepts the word $\sigma_1\sigma_2\ldots\sigma_n$ if there is a run of the automaton over this word that ends with $p_n \in F$.*

Figure 8.5 shows a reactive automaton that changes its arcs going to the unique terminal state. The automaton is deterministic and has linear size, $O(n)$. The automaton accepts the language $A^*\mathrm{a}A^n\mathrm{c}$ over the alphabet $A = \{\mathrm{a}, \mathrm{b}, \mathrm{c}\}$. It is known that the minimal deterministic (ordinary) automaton accepting the same language has $2^{n+1}+1$ states (see [106]).

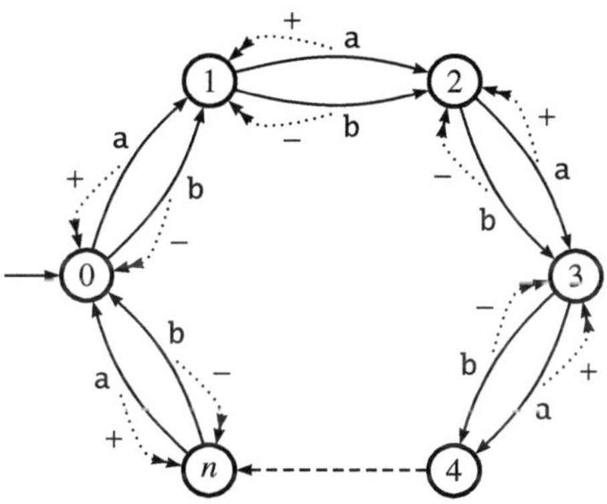

Fig. 8.6: Reactive deterministic automaton on the alphabet $A = \{\mathrm{a}, \mathrm{b}\}$ accepting the language $A^*\mathrm{a}A^n$ similar to the language of the automaton of Figure 8.5. Initially, the automaton has no terminal state and it uses edge-to-state reaction. Any a-transition makes its starting state a terminal state. Any b-transition transforms its starting state into a non-terminal state. If the run over a word stops in a terminal state, since the state has been set as a terminal state by an a-transition on the nth letter before, the word belongs to the language, and conversely.

Figure 8.6 shows another reactive automaton which changes only its terminal states and accepts a language similar to that of Figure 8.5. The definition of a reactive automaton changing its terminal states is a simple adaptation of the above definition.

Higher level reactive links

For the results of Section 8.5 we need to use higher order links. These arrows emanating from arrows to arrows, etc., are inductively defined to any level. We define the concepts now.

Definition 8.3. *Let S be the states and Σ be the alphabet of the automaton.*

1. *A level $0 \longrightarrow 0$ arrow has the form (s_1, σ, s_2) where $s_1, s_2 \in S$ and $\sigma \in \Sigma$.*
2. *Let α and β be arrows of levels ℓ_α and ℓ_β respectively. Then $\alpha \longrightarrow \beta$ is an arrow of level $\ell_\alpha \longrightarrow \ell_\beta$.*

Figure 8.7 shows a reactive automaton having reactive arrows of different levels. Arrow 1 has level $\ell_1 = [(0 \longrightarrow 0) \longrightarrow (0 \longrightarrow 0)]$. *Arrow 3 is of level* $\ell_2 = [((0 \longrightarrow 0) \longrightarrow (0 \longrightarrow 0)) \longrightarrow ((0 \longrightarrow 0) \longrightarrow (0 \longrightarrow 0))]$. *Arrow 4 has level* $\ell_1 \longrightarrow \ell_2$.

When we move from a to b we pass through the connection $a \longrightarrow b$. *So arrow 1 is "asked to act" and sends a signal to the arrow* $b \longrightarrow c$. *Since arrow 1 was asked to act, in turn it asks arrow 4 to act and send a signal to arrow 3. Thus, the effect of using the arrow* $a \longrightarrow b$ *is transmitted to arrows 1, 4, and 3.*

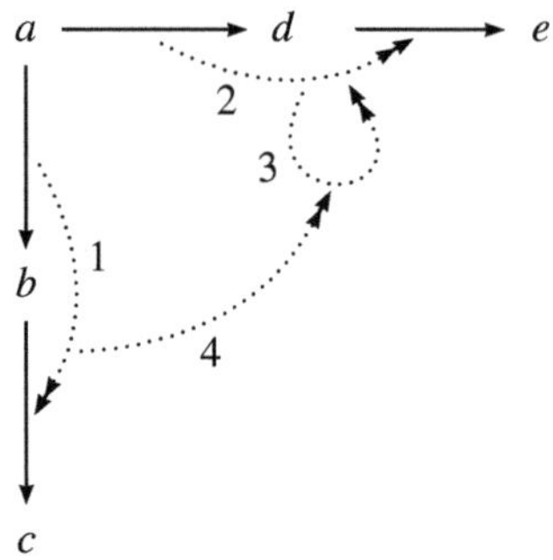

Fig. 8.7: Reactive arrows of different levels.

To describe reactions as in the above example, we need to give a formal inductive definition.

Definition 8.4. *Let* S *be a set of states,* Σ *the alphabet, and* T *a set of higher level transitions as defined in Definition 8.3. Assume that if* $\alpha \longrightarrow \beta \in T$ *then* $\alpha \in T$ *and* $\beta \in T$. *Let* τ *be a function from* T *to* $\{0, 1\}$: $\tau(\alpha) = 1$ *says that* α *is on (or live), and* $\tau(\alpha) = 0$ *says that* α *is off (or not on). (Note that being "on" is not equivalent to being "asked to act".) Assume* $(s, \sigma, t) \in T$ *is on. We now calculate by induction the cumulative effect of the transition from* s *to* t *that changes the* τ *value of some arrows.*

The inductive algorithm *Assume* (s, σ, t) *is on.*

Step 0 Make a move along the arc from s *to* t. *This starts a chain reaction as follows. Let* $\gamma = [(s, \sigma, t) \longrightarrow \beta]$. *We say that* γ *is "asked to act" at step 0. Two cases:* γ *may be on or off. If it is off nothing happens. If it is on it will act. So at this step 0 let* $\tau^0 = \tau$ *and let the arrows which are asked to act at step 0 all be arrows* γ *of the form* $[(s, \sigma, t) \longrightarrow \beta]$.

Step 1 Let $\gamma = [(s, \sigma, t) \longrightarrow \beta]$ *be any arrow that was asked to act at step 0 such that it is on. So* γ *is now ready to act at step 1. If* β *is on according to* τ^0, *then make* β *off according to* τ^1, *that is, set* $\tau(\beta) = 0$. *If* β *is off according to* τ^0, *then make* β *on according to* τ^1. *This defines* τ^1.

Also if $\epsilon = (\gamma \longrightarrow \delta)$ *then since* γ *is acting at step 1,* ϵ *is asked to act at step 1. It will act at stage 2 if* ϵ *is on at step 2. The above gives us* τ^1 *and gives us a list of arrows that are asked to act at step 1. We can now go to step 2.*

Step $n+1$ We assume we have defined τ^n at step n and have identified at step n those arrows $\gamma = \beta \longrightarrow \delta$ which were "asked to act" at step n. We now start the process of stage $n+1$. For each $\gamma = \beta \longrightarrow \delta$ which was asked to act at step n and which is on according to τ^n, we look at β. If β is on according to τ^n it is turned off for τ^{n+1} and if it is off it is turned on for τ^{n+1}. We also look at arrows η of the form $\gamma \longrightarrow \epsilon$ and ask them to act at step $n+1$. We now have τ^{n+1} and a list of arrows asked to act at step $n+1$, so we can carry on to step $n+2$. Note that the level of the arrows η which we ask to act goes up at every stage, and so this will prevent looping — at some step no arrow will be asked to act. At some step m there will be no arrow η being asked to act unless the graph is infinite. Note that we may have loops in the reactivity arrows but whenever an arrow is asked to act it must emanate from another arrow acting already and so its level is higher. So after m steps the process will stop and then let $\tau_{s,\sigma,t} = \tau^m$, for the first value m for which there is no change, that is, $\tau^m = \tau^{m+1}$. Thus we defined the following notion by induction: given τ and (s, σ, τ) which is on according to τ we defined $\tau_{(s,\sigma,\tau)}$.

8.3 Reactivity and non-reactivity

We focus on switch reactive automata and show that their expressive power is identical to that of ordinary automata. The proof can be adapted to automata with the various types of reactive links described in previous sections.

Theorem 8.5. *Any switch reactive deterministic or non-deterministic automaton is equivalent to an (ordinary) deterministic or non-deterministic automaton, respectively.*

Proof. Let $\mathcal{R} = (\mathcal{A}, \mathbf{T}^+, \mathbf{T}^-)$. We define the automaton $\mathcal{B} = (\hat{S}, (i, R), \Sigma, F, \hat{R})$ whose set of states $\hat{S}$ is composed of pairs of the form (x, R') where $x \in S$ and R' is related to R via the switch reactive transformation using $\mathbf{T}^+$ and $\mathbf{T}^-$. The transition relation $\hat{R}$ of $\mathcal{B}$ is defined using $\mathbf{T}^+$ and $\mathbf{T}^-$ as follows: $((x_1, R_1), \sigma, (x_2, R_2)) \in \hat{R}$ iff $(x_1, \sigma, x_2) \in R_1$ and $R_2 = R_1^{(x_1,\sigma,x_2)}$. Then $\mathcal{B} = (\hat{S}, i, \Sigma, F, \hat{R})$.

It is straightforward to see that a run of $\mathcal{R}$ on $\sigma_1\sigma_2 \ldots \sigma_n$ corresponds to a run of $\mathcal{B}$ on the same word and vice versa.

Remark

We saw that a switch reactive automaton actually starts as an ordinary automaton $\mathcal{A} = (S, i, \Sigma, F, R)$ and then changes into different automata while receiving the input. The changes can affect either the transition relation or the terminal states or both, as shown in the previous examples.

8.4 Saving states of non-deterministic automata using reactive links

Reactive links can be used to reduce the number of states of (ordinary) automata. In the section we show how it works in general for non-deterministic automata (deterministic automata are considered in the next section). It shows the power of reactive links provided we add enough of them to the new structure. Bounding linearly the number of reactive arcs to get the same result is an open question.

Let $\mathcal{A}$ be a non-deterministic automaton with kn states on the alphabet Σ. Denote its states by pairs (s_i, e_j), $i = 1, \ldots, k$, $j = 1, \ldots, n$.

Let $S = \{s_1, \ldots, s_k, e_1, \ldots, e_n\}$. We define a reactive automaton $\mathbb{R}(\mathcal{A})$ on the set of states S that has the same behaviour as $\mathcal{A}$.

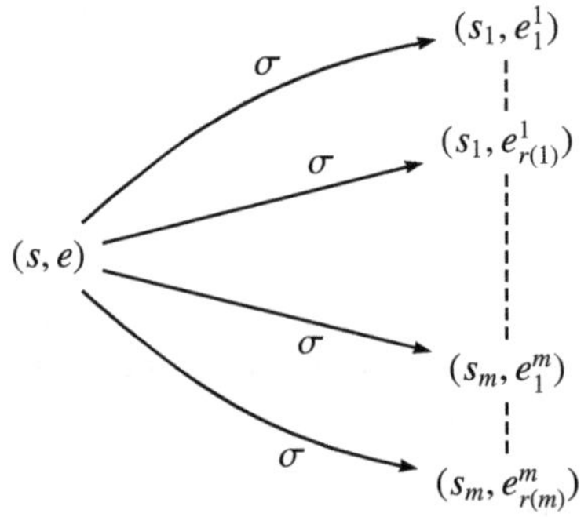

Fig. 8.8: Edges emanating from state (s, e) in the non-deterministic automaton $\mathcal{A}$, and their target states.

Basic idea

We first explain the basic idea. Let σ be a fixed letter and let (s, e) be a fixed state. Record the σ-transitions from (s, e) as in Figure 8.8.

We organise the non-deterministic targets according to the lexicographic order of pairs (s, e). We "break" the non-deterministic choice of edges by first choosing the s component and then the corresponding e component of pairs.

Figure 8.9 illustrates the states of $\mathbb{R}(\mathcal{A})$. The arrow from e to s represents the pair (s, e). The automaton $\mathcal{A}$ being in state (s, e) is represented by the automaton $\mathbb{R}(\mathcal{A})$ being in state e with the arrow from e to s being the only active arrow.

Suppose we see the letter σ and want to move from (s, e) to one of its targets, say (s_i, e^i_j), $1 \le j \le r(i)$. We want to end up in state e^i_j of $\mathbb{R}(\mathcal{A})$ with the only active arrow being the arrow from e^i_j to s_i. To do that, we first move along the arrow from e to s. This move activates all (non-deterministic) connections from s to s_i, $i = 1, \ldots, m$

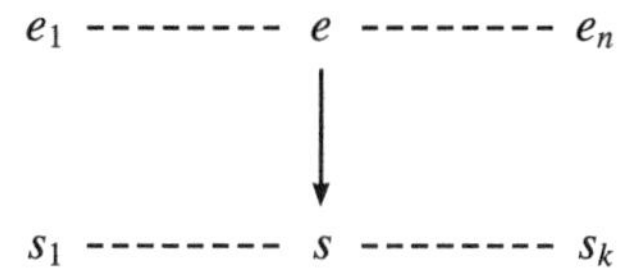

Fig. 8.9: An arrow from e to s in the reactive automaton $\mathbb{R}(\mathcal{A})$ represents the state (s, e) of the automaton $\mathcal{A}$.

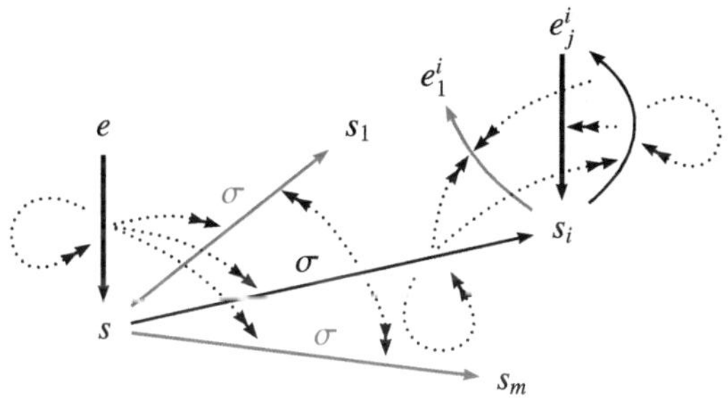

Fig. 8.10: Simulation of a transition in the automaton $\mathcal{A}$ from state (s, e) to state (s_i, e^i_j) by arcs and reactive edges of the reactive automaton $\mathbb{R}(\mathcal{A})$: from state e to state e^i_j.

(see Figure 8.10). We also deactivate the arrow from e to s. We then move non-deterministically along one of the newly activated links labelled σ, say from s to s_i. This is a choice of the $(s_i, .)$ group. Immediately as we move along the arrow from s to s_i we activate (connect) the links to $e^i_1, \ldots, e^i_{r(i)}$ emanating from s_i and disconnect all arrows emanating from s including the arrow from s to s_i itself. Now we are at s_i with arrows leading to e^i_j, $1 \leq j \leq r(i)$. We move non-deterministically along one of them to reach, say e^i_j. Immediately as we move along the arrow from s_i to e^i_j we disconnect all arrows emanating from s_i and activate the arrow from e^i_j to s_i. This leads to state e^i_j with the only activated link being from e^i_j to s_i. Figure 8.10 shows the connections we need.

Note that the same procedure works for the deterministic case as well, but we show later that a better result can be obtained in the deterministic case. Also note that the move in $\mathcal{A}$ from (s, e) to (s_i, e^i_j) is simulated by three moves in $\mathbb{R}(\mathcal{A})$ and that we use only reactive arrows of the second level, i.e., $(0 \to 0) \to (0 \to 0)$. For the result of Section 5 we need higher level arrows.

With the above procedure the simulation requires only a switch automaton.

The reactive automaton $\mathbb{R}(\mathcal{A})$ is defined by the states and connections as defined above for each σ and each (s, e). For each edge $((s, e), \sigma, (s_i, e^i_j))$ of $\mathcal{A}$, $1 \leq j \leq r(i)$, the automaton $\mathbb{R}(\mathcal{A})$ contains the following links:

(e, s)	active at start
(s, e)	non-active
$((e, s), (e, s))$	active
(s, s_i)	not active, $1 \leq i \leq m$
$((e, s), (s, s_i))$	active
$((s, s_i), (s, s_\ell))$	active, $1 \leq i, \ell \leq m$
$((s, s_i), (s_i, e^i_j))$	active, $1 \leq i \leq m$, $1 \leq j \leq r(i)$
(s_i, e^i_j)	not active, $1 \leq i \leq m$, $1 \leq j \leq r(i)$
(e^i_j, s_i)	not active, $1 \leq i \leq m$, $1 \leq j \leq r(i)$
$((s_i, e^i_j), (s_i, e^i_\ell))$	active, $1 \leq j, \ell \leq r(i)$
$((s_i, e^i_j), (e^i_j, s_i))$	active, $1 \leq j, \ell \leq r(i)$.

The above transformation of the non-deterministic automaton $\mathcal{A}$ to the switch automaton $\mathbb{R}(\mathcal{A})$ proves the next statement.

Lemma 8.6. *Associate the double condition $\mathbb{R}(\mathcal{A})$ is in state e and (e, s) is the only active arrow in $\mathbb{R}(\mathcal{A})$ with the state (s, e) of $\mathcal{A}$. Set the arrow (e, s) as terminal in $\mathbb{R}(\mathcal{A})$ whenever state (s, e) is terminal in $\mathcal{A}$. Then the automaton $\mathcal{A}$ accepts w iff $\mathbb{R}(\mathcal{A})$ accepts it.*

Corollary 8.7. *If w is recognised by $\mathcal{A}$ with kn states then it is recognised by $\mathbb{R}(\mathcal{A})$ with $k + n$ states. If $\mathcal{A}$ is deterministic, so is $\mathbb{R}(\mathcal{A})$.*

Can the reduction process be repeated?

Let $\mathcal{A}$ be a deterministic automaton having knm states. Can we collapse it to a reactive automaton with $k + n + m$ states?

Consider the states $\{s_1, \ldots, s_k, e_1, \ldots, e_n, a_1, \ldots, a_m\}$. We can represent the states of $\mathcal{A}$ as triples (s, e, a). The question is, if we use a reactive automaton representing the "(s, e)" part as before, then in the same spirit we need to represent $((s, e), a)$ as in Figure 8.11.

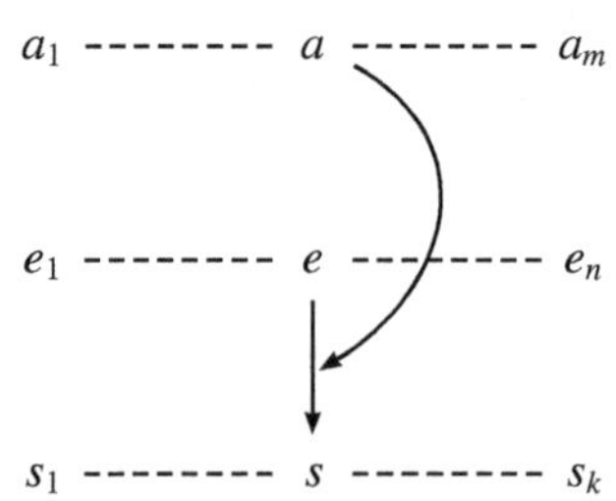

Fig. 8.11: Possible representation of state (s, e, a), among the knm states of the non-deterministic automaton $\mathcal{A}$, by arrows in the reactive automaton $\mathbb{R}(\mathcal{A})$.

We need an arrow from a to the arrow (e, s). This is *not* an arrow to e but really to the arrow (e, s) because the arrow is associated with the pair (s, e). However, this cannot be done as these arrows must move from state to state. Indeed, there is another solution to reduce the number of states when the automaton $\mathcal{A}$ is deterministic, as shown in the next section.

8.5 Reducing the number of states of deterministic automata

On deterministic automata reactive links are even more powerful in reducing their number of states. This section shows that a deterministic automaton with k^n states can be reduced to an equivalent deterministic reactive automaton with $k \cdot n$ states by adding enough reactive links. Bounding linearly the number of them to get the same result, as in the previous section, is an open question.

Let $\mathcal{A}$ be a deterministic automaton. Let σ be a fixed letter. Let f_σ be the transition table associated with σ ($f_\sigma(x) = y$ means that state y is the next state after x under the action of the input σ).

Assume $m = k^n$. We represent the states x of $\mathcal{A}$ as vectors $x = (x_1, \ldots, x_n)$, where $x_i \in \{a^i_1, \ldots, a^i_k\}$ for each i. The $k \times n$-matrix $\mathbb{A} = \{a^i_j, i = 1, \ldots, k, j = 1, \ldots, n\}$ comprises pairwise distinct elements.

We now define a deterministic reactive automaton $\mathbb{R}(\mathcal{A})$ from $\mathcal{A}$. The states of $\mathbb{R}(\mathcal{A})$ are the $k \times n$ elements

$$a^i_j, i = 1, \ldots, n, j = 1, \ldots, k.$$

Let R be the set of connections in the automaton $\mathbb{R}(\mathcal{A})$. We write $(x, y) \in R$ to mean that there is a connection from x to y. It may be *on* or it may be *off*. This is not indicated since it changes as we move from state to state. Double arrows are written as $(?, x, y)$ as mentioned above. Again these may be *on* or *off*, depending on the context.

We need some more notation. Let $x = (x_1, \ldots, x_n)$ and $y = (y_1, \ldots, y_n)$ be two vectors ($x_i, y_i \in \{a^i_1, \ldots, a^i_k\}$) for which $f_\sigma(x) = y$. We need to represent the move from x to y in the reactive automaton with states a^i_j, $i = 1, \ldots, n$, $j = 1, \ldots, k$. To do this, we have to encode states $x = (x_1, \ldots, x_n)$ in the reactive automaton in some way. We use paths in the $k \times n$-matrix $\mathbb{A}$. A path through the matrix (from top to bottom) corresponds to a vector $x = (x_1, \ldots, x_n)$ for which $x_i \in \{a^i_1, \ldots, a^i_k\}$. We can represent paths by connecting arrows.

Figure 8.12 shows two paths for a 3×3-matrix. Let $x = (a^1_1, a^2_2, a^3_3)$ and $y = (a^1_3, a^2_2, a^3_1)$. The active arrows are (a^1_1, a^2_2), (a^2_2, a^3_3), (a^1_3, a^2_2), and (a^2_2, a^3_1). If we rely on active arrows only to represent triplets, then we are also activating two more paths, namely (a^1_1, a^2_2, a^3_1) and (a^1_3, a^2_2, a^3_3). So we have to do it differently.

In Figure 8.12 the triplet (a^1_1, a^2_2, a^3_3) is characterised both by the edge (a^1_2, a^2_2) and the reactive arrow $((a^1_2, a^2_2), (a^2_2, a^3_3))$; (a^1_3, a^2_2, a^3_1) is characterised both by (a^1_3, a^2_2) and $((a^1_3, a^2_2), (a^2_2, a^3_1))$. In fact, the two dashed arrows alone contain all the necessary information. Each dashed arrow represents a unique triplet.

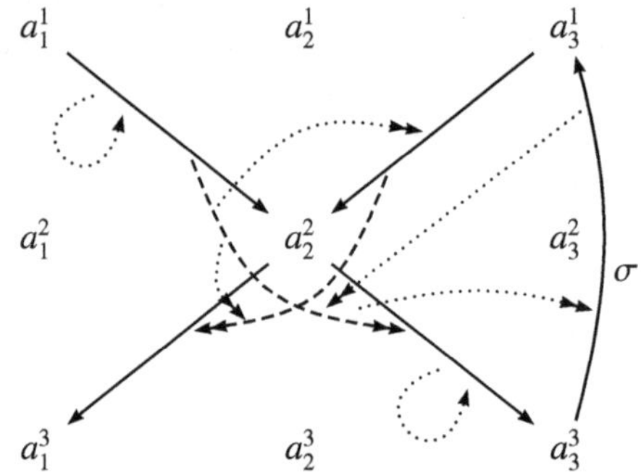

Fig. 8.12: Two paths in a (3×3)-matrix representing two states of a deterministic automaton with 3^3 states with the help of the dashed reactive arrows.

Now suppose $x = (a_1^1, a_2^2, a_3^3)$ represents a state in a deterministic automaton with 3^3 states. Suppose the transition from x by σ goes to $y = (a_3^1, a_2^2, a_1^3)$. How can we represent that? We start with the active edge (a_1^1, a_2^2) and the active dashed arrow $((a_1^1, a_2^2), (a_2^2, a_3^3))$. This means that the automaton $\mathbb{R}(\mathcal{A})$ is in state x and we are at a_1^1 in the matrix. Deterministically, we can move only to a_2^2. This activates the edge (a_2^2, a_3^3) via the dashed arrow that uniquely represents the triplet $x = (a_1^1, a_2^2, a_3^3)$. Now we can activate $y = (a_3^1, a_2^2, a_1^3)$ as follows. We send two signals to activate the edge (a_3^1, a_2^2) and the arrow $((a_3^1, a_2^2), (a_2^2, a_1^3))$. We also send a signal to activate the σ-edge. Once we get to a_3^3 we follow the edge labelled by σ to a_3^1 (first element of y) and deactivate x, that is, deactivate (a_1^1, a_2^2), (a_2^2, a_3^3), and $((a_1^1, a_2^2), (a_2^2, a_3^3))$.

The basic idea is that we encoded a triple (a_1, a_2, a_3) by edges and arrows as shown in Figure 8.13 (left). To encode a 4-tuple (a_1, a_2, a_3, a_4) we do it as in Figure 8.13 (right). To encode an n-tuple, we cascade the construction of the previous figure, which is displayed in Figure 8.14, as well as a σ-transition.

The above construction reduces the number of states of the initial automaton from k^n to kn. It replaces them by reactive arrows but sending signals through arrows may be cheaper than having states. It is obvious that one equivalent presentation of a problem may be cheaper than another. [1]

The conclusion of the section is in the next statement.

Proposition 8.8. *If $\mathcal{A}$ is deterministic automaton with k^n states, it has an equivalent reactive automaton $\mathbb{R}(\mathcal{A})$ with $k \cdot n$ states.*

---

[1] A typical example is the polynomial evaluation:

$$f(x) = 1 + 2x + 3x^2 + x^3 = 1 + x(2 + x(3 + x)),$$

for which the second expression is cheaper to evaluate.

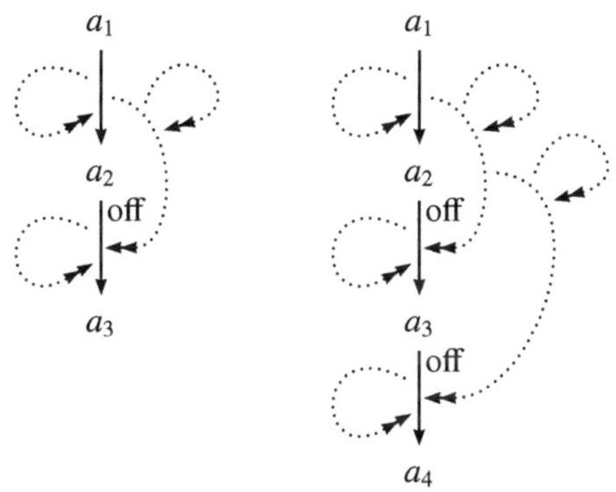

Fig. 8.13: Representing a triple (left) and a 4-tuple (right). The bottom dotted arrow encodes the triple (a_1, a_2, a_3) or the four element vector (a_1, a_2, a_3, a_4), respectively. Only deactivated edges are indicated as off.

8.6 Examples of size reduction

The use of reactive links in automata can dramatically reduce the size of a deterministic automaton accepting a given regular language. The two previous sections show that this is possible for deterministic as well as non-deterministic automata but with a large number of reactive arcs. Instead, in the next examples, reactivity keeps the total size of automata as small as the size of their non-deterministic equivalent but without losing determinism. In these examples determinism without reactivity leads to an exponential blow up of the number of states and then of the total size of the automata.

The first example corresponds to the finite language of words in which each letter of the alphabet appears at most once (see Figure 8.15). The principle is that loops on the initial state are cancelled by a reactive link immediately after being used. States of a deterministic automaton for the language have to store the set of letters already treated and therefore the minimal automaton has at least an exponential number of states. Instead the total size of the reactive automaton accepting the language is $O(k)$ on a k-letter alphabet.

The automaton of Figure 8.16 accepts all the $k!$ permutations of letters. To do this we add a path of length k from the initial state to the unique terminal state. Compared with the automaton of the previous example, the aim is to count the number of letters read in the initial state. Each loop on the initial state has an additional reactive link that activates its associated ε-arc on the path. So, the terminal state can be reached only if all ε-arcs are activated. We view the automaton as deterministic because its light non-determinism due to ε-arcs can be remove by considering a special symbol marking the end of words. The size of the reactive automaton is $O(k)$, which contrasts with the $O(2^k)$ size of the minimal automaton accepting the language.

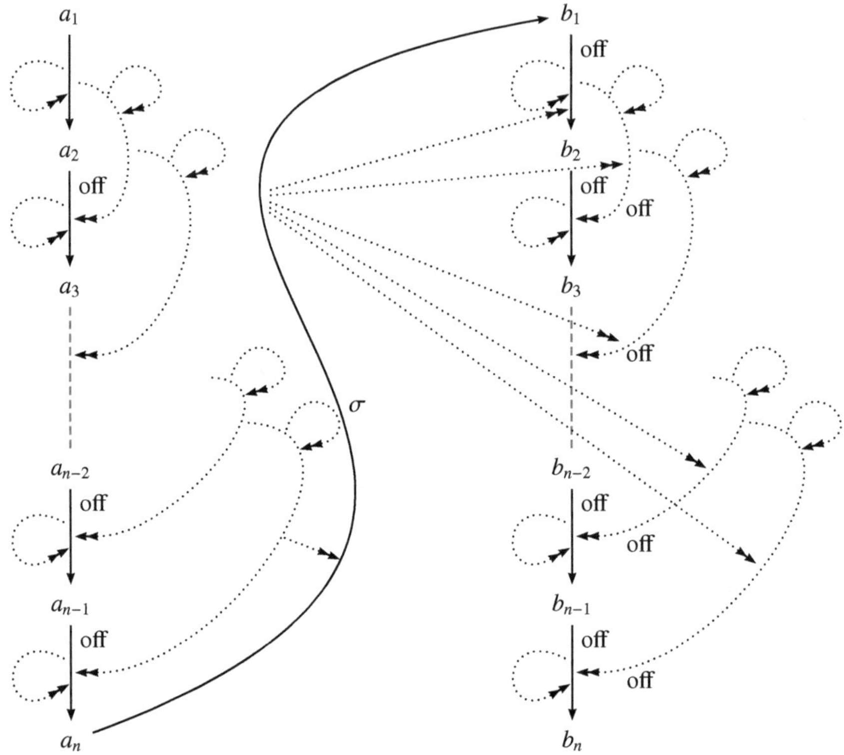

Fig. 8.14: Transition from the n-tuple $(a_1, a_2, \ldots, a_n)$ to the n-tuple $(b_1, b_2, \ldots, b_n)$. Only deactivated edges and arrows are indicated as off. Loops are permanently on.

The language $A^*\mathsf{a}A^n$ is accepted by the non-deterministic automaton of Figure 8.17, which has $n+2$ states and $O(n)$ total size. The non-determinism appears in the initial state only. It is known that the minimal deterministic automaton accepting the same language has 2^{n+1} states (see [106]) and also $O(2^n)$ total size, while the equivalent reactive automaton of Figure 8.6 has only also $n+1$ states. It is noticeable that it has $O(n)$ total size despite the addition of reactive links.

The last remarkable example concerns the language $A^{\leq n}(\mathsf{a}A^n)^*$ on the alphabet $A = \{\mathsf{a}, \mathsf{b}\}$. Figure 8.18 displays a non-deterministic automaton accepting it. It is non-deterministic because all its $n+1$ states are initial states. Figure 8.19 shows an equivalent reactive automaton, which, as above, may be considered as deterministic since ε-arcs are useful only at the end of the input word. Reactive links are from b-transitions and cancel their associated ε-arc to the terminal state. The number of states is $n+2$ and the total size is $O(n)$. This is to be compared with the result of Béal

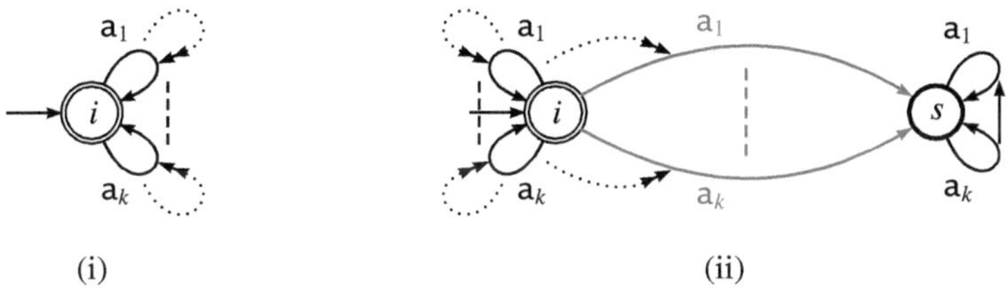

Fig. 8.15: Two deterministic reactive automata accepting the set of strings in which each letter of the alphabet $\{a_1, a_2, \ldots, a_k\}$ appears at most once. All loop arcs are initially active. Loops on state i are made inactive after their first use. (i) Incomplete version. (ii) Complete version: only arcs from i to s are initially inactive and become active after the first use of their corresponding loop on state i.

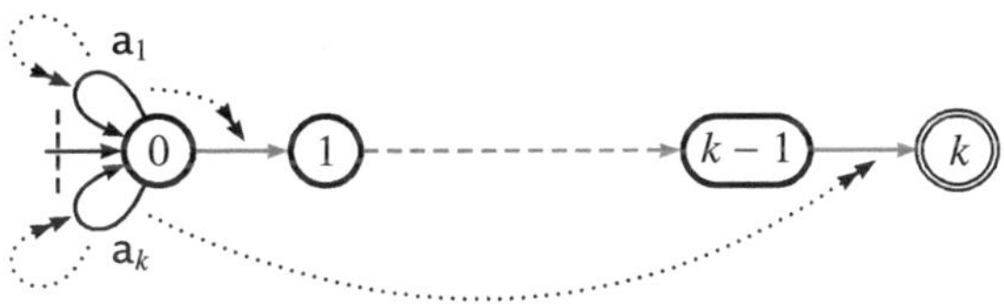

Fig. 8.16: A deterministic reactive automaton accepting the set of strings that are permutations of the letters $a_1, a_2, \ldots, a_k$. All loops on the initial state are initially active and the ε-arcs are inactive. One reactive link for letter a_i cancels its respective loop while the second activates its associated ε-arc.

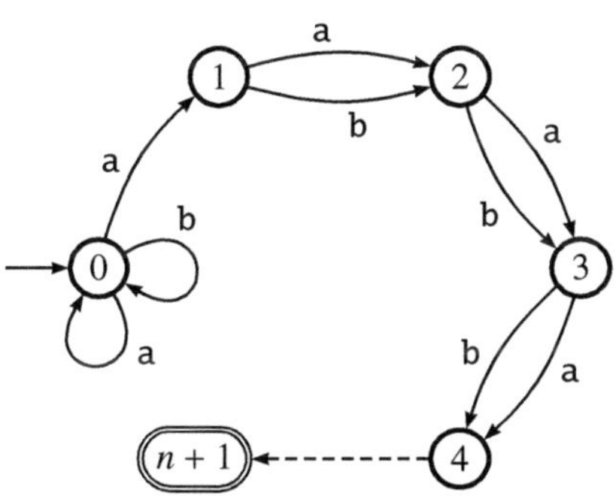

Fig. 8.17: Non-deterministic automaton on the alphabet $A = \{a, b\}$ accepting the language A^*aA^n. Its equivalent reactive automaton of Figure 8.6 has size of the same order.

et al. [21], which shows that the minimal deterministic automaton for this language has an exponential number of states.

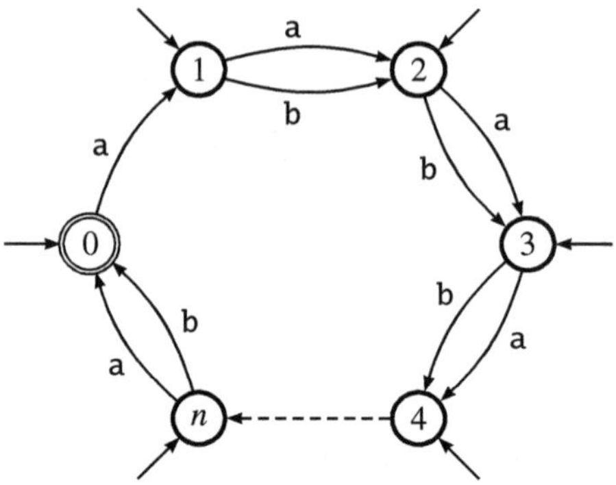

Fig. 8.18: Non-deterministic automaton on the alphabet $A = \{\mathtt{a}, \mathtt{b}\}$ accepting the language $A^{\leq n}(\mathtt{a}A^n)^*$. All states are initial states.

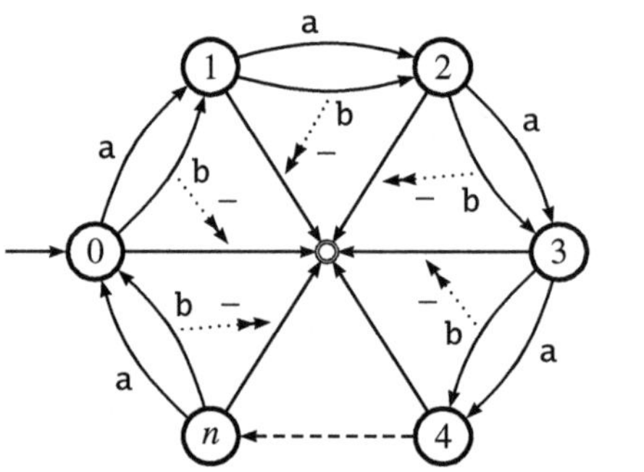

Fig. 8.19: Reactive deterministic automaton on the alphabet $A = \{\mathtt{a}, \mathtt{b}\}$ accepting the same language $A^{\leq n}(\mathtt{a}A^n)^*$ as the automaton of Figure 8.18. It has only one terminal state, in the center. During a run, at least one ε-arc remains if some positions of letter a in the input word form a non-extendible arithmetic progression of period n. The automaton has only one more state and twice as many arcs as the automaton of Figure 8.18.

8.7 Conclusion

The strength of reactive links in automata comes essentially from the reduction of the size of their implementation. Designing a non-deterministic automaton to solve a Pattern Matching question leads to slow algorithms requiring extra work to be implemented. Instead, the use of reactive links can turn a non-deterministic automaton into a deterministic Pattern Matching machine. This has two simultaneous advantages: little effort at implementation and efficient running time since the basic operation required when parsing a stream of data with an automaton is table lookup to change state, which avoids any other more time-consuming low or high level instruction.

The usefulness of reactive links is obvious when they are used to send signals to part of the machine that is to be used in the rest of the parsing. The signals change the behaviour of the machine in a very smooth way but with deep consequences for the future analysis of the stream.

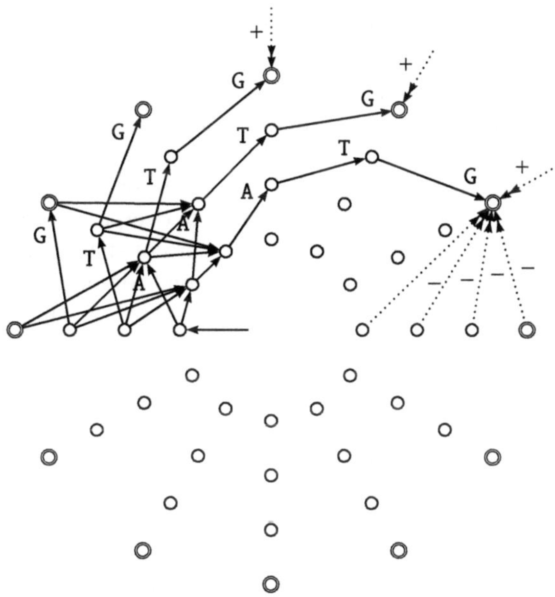

Fig. 8.20: Deterministic reactive automaton accepting the start codon ATG at positions i modulo 12. The automaton is partially drawn. It is composed of four circles of states; the outer circle contains states that are terminal when activated by a reactive link coming from another part of the whole machine. They are deactivated when going to the next spoke. From the inner circle, circles correspond respectively to the prefixes ε, A, AT, and ATG of the codon. Although not shown in the picture all edges have a label that corresponds to the circle of their target. For example, from states of the second outer circle, there is a G-edge that goes to a state of the larger circle, an A-edge that goes to a state of the smaller circle, and all other edges go to a state of the inner circle. The next state is on the next spoke of the wheel. Edges and reactive links from states on two different spokes of the wheel are analogues.

Gene signals

We describe here a possible use of a reactive automaton for fast recognition of coding sequences (potential genes) in genomes or chromosomes. An essential part of the information of these molecules is stored in data bases in the form of DNA sequences on the four-letter alphabet of nucleotides {A, C, G, T}. Some segments of a DNA sequence, called coding genes, encode protein molecules and have specific features. As an example and to simplify the demonstration, we consider only the two following properties: a coding gene is a sequence of codons (words of length 3) that starts at position 0 with the codon ATG; upstream, around position −16 from the position of

A, the DNA sequence contains a promoter segment called the tata-box composed of a sequence similar to TATAAT.

The idea of a reactive automaton for the detection of a start codon based on the simplified description is analogous to that used in previous examples. The automaton is composed of two parts. The first part detects occurrences of the promoter sequence at each position i modulo 12, the distance between the end of the promoter supposedly of length 6 and the end of the start codon. The second part detects occurrences of the start codon at each position j modulo 12. Terminal states in the second part are initially inactive. When an occurrence of the promoter is found by the first part of the automaton, a signal is sent to activate the corresponding terminal state in the second part of the automaton. When the current state overpasses this state it is again inactivated. During a run of the automaton, the current state consists of a pair of states from each part. The state is terminal if its second component is.

Figure 8.20 shows how the second part of the automaton can be designed. The first part is built in the same way. Reactive links are then added as described above.

Open question

In some sense, reactivity competes with non-determinism to get small automata accepting a given language, although they are not antagonistic concepts. The results of Sections 8.4 and 8.5 show that reactivity significantly reduces the number of states of automata, the solution requires in general a large number of extra links, but the significant examples of Section 8.6 raise our hope that this number can indeed be fairly small.

We leave open the question of whether, given a language described by a regular expression of size r, there is a reactive deterministic automaton of size $O(r)$ accepting it.

9

Reactivity and Grammars: An Exploration

9.1 Background and motivation

We consider the relationship between grammars and formal languages, exploring the following idea: Normally, in the process of deriving a word in a language using a grammar, all structures remain fixed except for the intermediate strings, which are changed only by the replacement of substrings. By introducing a more dynamic view of this process, we may allow the grammar to change in various ways as the derivation proceeds, or we may change the notion of applications of a rule to a string, or the intermediate strings may be modified between application of rules. We call these more dynamic approaches to language generation 'reactive grammars' and explore in this chapter a range of such reactivities. Some of these are related to previously introduced notions of generative grammars, others appear to be new. We consider the expressivity of various reactive grammars and also their relationship to each other and to other formalisms including modal logic and deontic logic. Reactivity of computational structures has been explored in other areas, e.g. in Kripke structures and in the general areas of evolvable and adaptive systems.

A standard approach to defining models of computation is to introduce fixed structures over which computation may take place. Such structures include graphs, relations, algebras, grammars, automata, etc. An alternative is to allow the structures themselves to change as computation proceeds. Such computational systems are called 'evolutionary', 'adaptive' or (the term we shall use) 'reactive'. The advantages of this additional flexibility are several: (1) they allow us to model desirable aspects of computation in dynamic settings where we require systems which may reconfigure themselves during computation, (2) they allow us to model the influence of external activity under which computation may have to adapt by changing computational structure, (3) they allow us to re-examine the relationship of a computational model with its semantics and (4) reactive systems allow us to establish relationships with other formalisms for dynamic change such as modal and deontic logics. The semantics may be a function computed, or a language generated or recognised, or a logical expression satisfied (e.g. for modal logics and Kripke models). The extra degree of freedom that reactivity introduces allows us additional mechanisms for

D.M. Gabbay, *Reactive Kripke Semantics*, Cognitive Technologies,

DOI 10.1007/978-3-642-41389-6_9,

generating a required semantics. It is this feature of reactive systems that we consider in this chapter, where we explore ideas of reactivity in grammars and the relationship of these with the generation of formal languages.

For grammars, reactivity is realised through transformations that interact with the derivation steps by which a grammar generates a word in a language. The interaction may for example, at each step of a derivation, control which grammar rules may be applied, or change the notion of application of a rule to a string, or modify rules as they are used, or transform the intermediate strings as they are generated in a derivation. Introducing such reactive elements into the definitions of grammars allows us to re-examine the mathematical structures involved in language recognition and generation, to relate languages according to their reactive structure as well as their production rules, and to model more dynamically changing computational systems.

Before exploring reactivity in grammars, we briefly consider reactivity in the context of Kripke structures and automata. This is related to some of the forms of reactivity in grammars that we introduce later and has already been studied in Chapter 2. Examples drawn from this study illustrate how different forms of reactivity interact with the semantics of these structures.

A Kripke structure has the form (S, R, a), where S is a non-empty set of worlds, $R \subseteq S \times S$ is the accessibility relation and $a \in S$ is the actual world. The semantic evaluation of a formula φ in a world $t \in S$ is inductively defined and involves values of its subformulas in other worlds s related to t. For example, we have:

$$t \models \Box\varphi \text{ iff in all accessible worlds } s \text{ (i.e. such that } tRs) \text{ we have } s \models \varphi.$$

The above is a static view, not a reactive one. The structure does not change as we evaluate. If we take a dynamic view of the evaluation process, we imagine the structure as a graph with a directed relation R (so aRb is viewed as a directed connection $a \longrightarrow b$). When we stand at point t and we want to evaluate $\Box\varphi$, we view this as an invitation to stroll along the arrows (accessibility relation R), go to the accessible worlds and check φ there. If φ holds in all of these worlds we report that $\Box\varphi$ holds at t. Now this point of view allows us to say that the model can change as we stroll through it from world to world. This is the idea of reactivity.

We adopt a similar point of view for automata. An automaton can be viewed either as a mathematical table M which, when given a state t and a letter σ can move to another state s chosen from $s \in M(t, \sigma)$. A more reactive view is to see the automaton as having some internal "will" with which it can change its table M as it reacts to input. So for example if $\{s\} = M(t, \sigma)$ and the automaton gets the input σ three times, then s may become 'exhausted' and the automaton changes to a state $s' \neq s$ when getting the fourth σ. This means that $M(t, \sigma) = \{s'\}$ if σ is a fourth occurrence in the input. See Chapter 8.

We have a similar situation with grammars and also with proofs. We use a rule to make a step in a derivation or a proof and the use of this rule may then change the current state of the system. A simple example of such a change is where the rule deactivates itself on use and is then no longer available unless re-activated later.

Note that the reactive change is in response to the history of use and is not a meta-level (possibly time-dependent) intervention. When the changes involve switching rules/arrows on or off, we represent these graphically by double(-headed) arrows, reserving single(-headed) arrows for the transitions, as in the figure below.

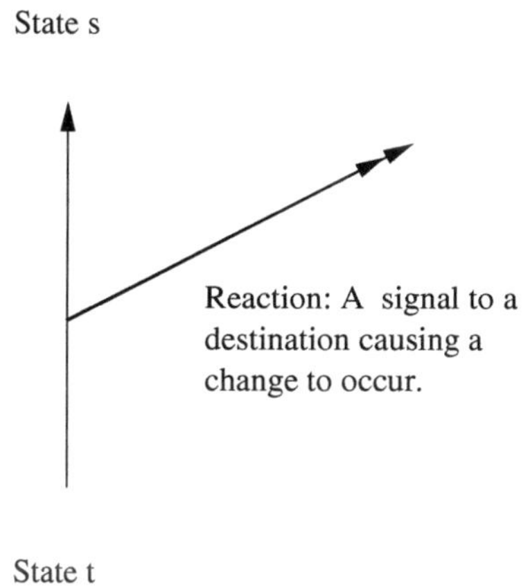

Fig. 9.1: Transition and reaction arrows

Reactivity may be perceived differently from system to system. In some systems, reactivity is perceived as a fault. Consider a washing machine which is continuously being used on the same cycle. If it changes and this overused cycle is no longer available, we see it is as a fault. In some other areas, the reactive point of view is new and offers its own meaning. In normative agent-based systems, the double arrow which changes the system can be viewed as an immediate object-level punishment or response to an agent making a forbidden move. With grammars, reactivity is a formal notion. The grammar rules may, for example, get switched 'on' and 'off' depending on which rules are used. This means that the current rules available at any point depend on the history of which rules have been used up to that point. This is not an unusual point of view for grammars and indeed, in the early years of automata and grammar investigations, many new grammars were introduced adopting a computational point of view yielding variations and generalisations. In this chapter, using the general theme of reactivity and computational variation, we explore a range of reactive grammars, some of which have been considered before; others appear to be new.

Example 9.1 (Kripke semantics). Let us recall Figure 5.2, which we reproduce here for the convenience of the reader as Figure 9.2. It is an example of a Kripke modal logic model.

For the moment, ignore the double arrow. The single arrows indicate accessibility. Described traditionally, we have $S = \{a, b, c, d, e\}$ and $R = \{(a, c), (a, b),$

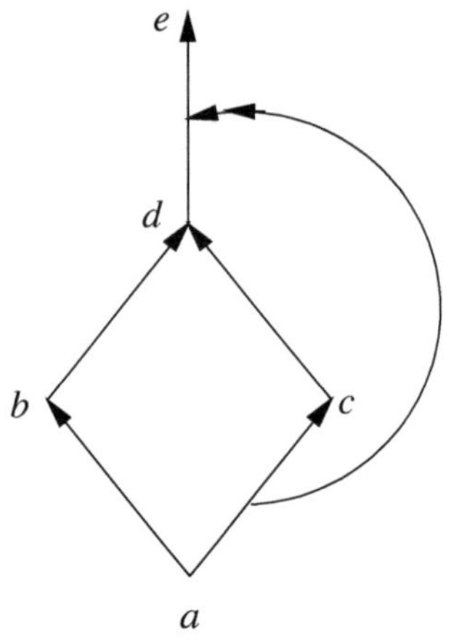

Fig. 9.2: A reactive Kripke structure

$(b,d),(c,d),(d,e)\}$. We also have the double arrow which is really $R' = \{((a,c),(d,e))\}$, i.e. is an arrow whose source and targets are both arrows.

Now consider a formula of modal logic, say

$$A = \Diamond\Diamond\Box\bot.$$

The evaluation process at node a of the model runs as follows:

$$\begin{aligned} a \vDash \Diamond\Diamond\Box\bot \text{ iff } & \exists x \exists y (aRx \wedge xRy \wedge \forall z (yRz \rightarrow z \vDash \bot)) \\ \text{iff } & \exists x \exists y (aRx \wedge xRy \wedge \sim \exists z (yRz)). \end{aligned}$$

This is a static way of looking at the definition. It is a mathematical definition of satisfaction.

Let us adopt a dynamic point of view. We stand at node a and want to check whether $\Diamond\Diamond\Box\bot$ holds at a. We send messages from a to b and from a to c and give them the task to check and report whether $\Diamond\Box\bot$ holds. These messengers in turn send additional messages to d asking whether $\Box\bot$ holds and the people at d send messengers to e to ask whether $\bot$ holds.

The information is passed backwards in the chain and a final decision is made at node a.

Looking at the evaluation process in this way allows for several modifications and practical questions:

1. Are the messengers running in parallel or under any coordination or can one person do the job sequentially?
2. How long does it take to go to the various nodes?
3. How much does it cost to
   a) move from node to node
   b) get a value at a node

4. How can things change or go wrong?

The Kripke model is *reactive* if things can change as we traverse the model. The double arrows indicate how things change.

So a double arrow from $(a \to c)$ to $(d \to e)$ can indicate that when a messenger moves from a to c then a signal is sent to disconnect the arrow from d to e. So the accessible points that a messenger sees standing at point d depend on the path by which the point was reached. If the journey was via b then e is accessible. If the journey was via c, then the connection $d \to e$ is cancelled and the step from d to e is not available. Thus in model $\mathbf{m} = (S, R \cup R', a)$, we do have $a \vDash \Diamond\Diamond\Box\bot$, because there is a way to get to d (via c) and in this case $d \vDash \Box\bot$.

The model $\mathbf{m}$ is equivalent to the model depicted in Figure 9.3. (This is Figure 5.3 of Chapter 5.)

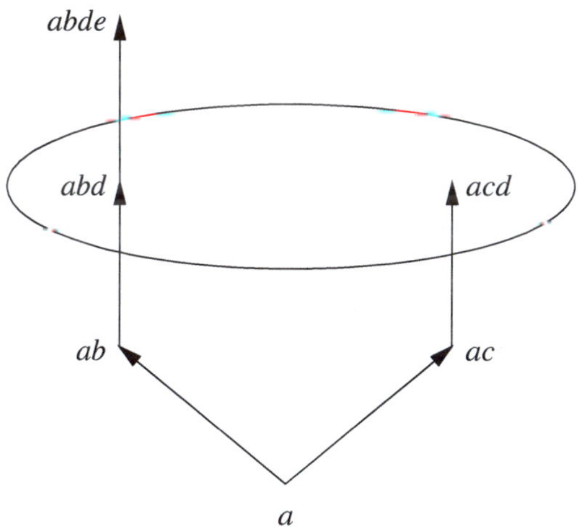

Fig. 9.3

The nodes indicate the paths. The circle around *abd* and *acd* indicates they are really the same point (i.e. for any atomic q, $abd \vDash q$ iff $acd \vDash q$). Such models (with double arrows) are called *reactive models*.

Figure 9.4 is a more complicated example of a reactive model. In Figure 9.4 to the double arrow $((a,c),(d,e))$, we add $R_3 = \{((a,b),((a,c),(d,e)))\}$.

In this model, timing of the movements of the messengers is important. So if the messenger from a to b sets out before the messenger from a to c, then the double arrow $(a,c) \to (d,e)$ will be cancelled and when the messenger from a to c sets out to c, nothing will happen to the connection $d \to e$.

We see here that there is a lot of scope in defining the nature of the models and how reactivity behaves.

What view do we take of the models of Figures 9.2 and 9.4?

There are several options:

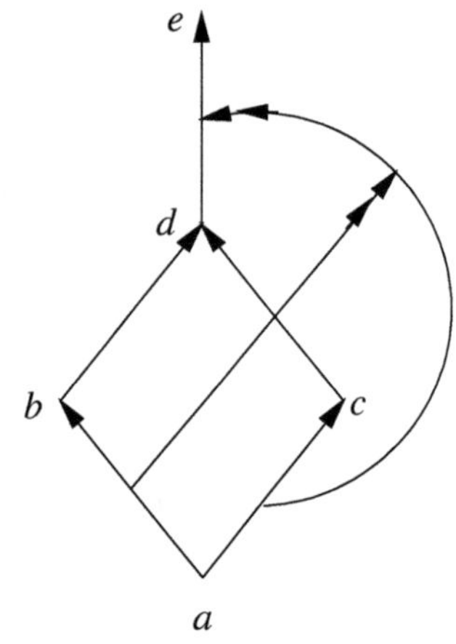

Fig. 9.4

1. Mathematical view, as more complex networks/graphs.
2. Reactive view, as systems which react to algorithms (people) using them (traversing them). They change in a prescribed manner while they are used.

In fact, as we have seen in Chapter 3, reactive models have an increased expressivity, in the following sense:

Theorem 9.2. *There exist modal logics (with a $\Box$) which are complete for a class of reactive models but are not complete for any class of ordinary (Kripke) models.*

Example 9.3 (Finite automata). An automaton is a state machine which responds non-deterministically to inputs and changes state. An input string is accepted by the automaton at its initial state, if after responding to the string as input it ends up in a terminal state.

An automaton can be represented by a multimodal Kripke model. Figure 9.5 is an example of such an automaton with two letter inputs $\{1, 2\}$ and states $\{a, b\}$.

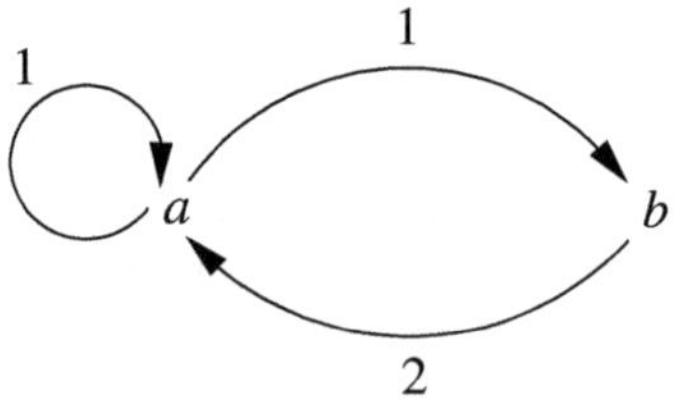

Fig. 9.5: A simple automaton

The initial state is the final state b. Let q be an atom and let q hold only at the final state. The model of Figure 9.5 has two relations $R_1 = \{(a,a),(a,b)\}$ and

$R_2 = \{(b,a)\}$. For any sequence of numbers from $\{1,2\}$ of the form $x_1,\ldots,x_n$ we can check whether the model satisfies $a \vDash \Diamond_{x_1}\Diamond_{x_2}\ldots\Diamond_{x_n}q$. It accepts words of the form $1^{m_1}21^{m_2}2\ldots1^{m_{k-1}}21^{m_k}$ with $m_1,\ldots,m_k \geq 1, k \geq 1$.

If we transfer the reactive idea to automata through the correspondence above, we get that a reactive automaton can change with every move into a new automaton. So given automaton $\mathcal{A}$ at state s, it changes when seeing word σ into a new automaton $\mathcal{A}'$ at state s'.

One conclusion is immediate, any reactive finite automaton is equivalent to another non-reactive ordinary automaton.

We do gain something though, as shown in Chapter 8.

Theorem 9.4. *Every ordinary automaton with n states $n = p_1 \times \ldots \times p_k$ is equivalent to a reactive automaton with m states, $m = p_1 + \ldots + p_k$.*

Thus, the idea of reactivity of this kind does not allow automata to recognise more sets of words but it does allow us to economize on states.

Example 9.5 (An infinite state reactive automaton). The Kripke model point of view of automata does allow for a uniform presentation of automata. Consider an (infinite state) automaton for recognising all words of the form $0^n1^n, n = 0,1,2,\ldots$ ($n = 0$ is the empty word). We need some infinity here, that of a stack getting fuller and fuller and then emptier. We can construct this as a reactive model as in Figure 9.6. The thicker (red) arrows are the transitions of the model; the thinner (blue) arrows are reactivities. Furthermore, all arrows are, by default, self-cancelling, i.e. whenever we traverse an arrow, the arrow is automatically de-activated. Initially, we assume all arrows are off apart from the first upward red arrow from X_0 to X_1. We start walking up from X_0 to X_1. The blue arrows activate the next upward arrow and the return downwards.

Let $q = \Box\bot$. Let X_0 be both the starting point and the terminal point. Then we have that exactly wffs of the form $\Diamond_1^n\Diamond_2^n\Box\bot$ hold at X_0.

Example 9.6 (An infinite state, with infinite branching, automaton). Consider the model of Figure 9.7. This corresponds to an infinite reactive automaton which accepts (or generates) $0^n1^n2^n\ldots(2k+1)^n, n = 0,1,2,\ldots,k = 0,1,2,\ldots$.

There are states X_i for $i = 0,1,\ldots$ and the associated towers of the thicker (red) arrows accept (or generate) the symbols $0,1,2,\ldots$, i.e. traversal of the bottom leftmost red arrow from X_0 to X_1 will accept (or generate) a 0, whereas traversal of the red arrow from X_0 to X_1 in the column above 2 will accept (or generate) a 2. State X_0 is the final state. The thinner (blue) arrows are reactivities that flip the activation status of both red and blue arrows. First note that all arrows are self-cancelling, i.e. as an arrow is traversed its activation status is switched off (and an arrow can only be traversed if its activation status is on). Also, for ease of presentation, we have used a convention that a blue arrow flipping the activation of a red arrow will also flip the activation of all of its associated blue arrows. Thus, for example, the blue arrow labelled (1) will flip the activation of the arrow $X_1 \to X_0$ in the 1 column. Blue arrows

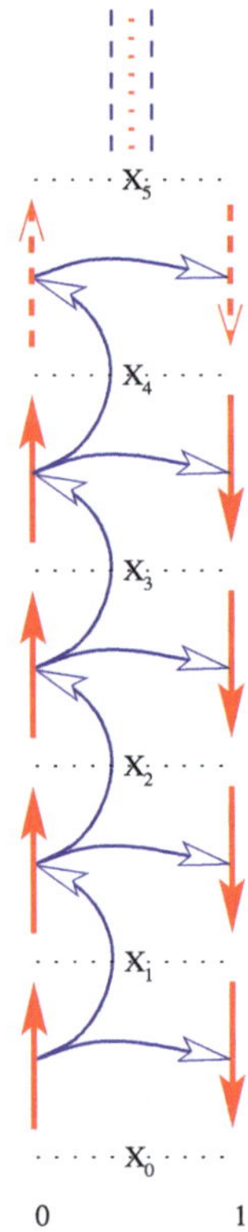

Fig. 9.6: Reactive automaton for $0^n 1^n$

are also used to flip the activation status of other blue arrows as, for example, is the case for the arrow labelled (2). Thus, when the red arrow from X_2 to X_1 is traversed in column 1, the arrow labelled (2) flips the activation status of the blue arrow labelled (3). The activation status of all the arrows is off apart from the $X_0 \to X_1$ arrow in column 0 and its three blue reactivity arrows. The automaton then has the following operational behaviour. It can initially traverse the red $X_0 \to X_1$ arrow accepting 0. As a consequence the red arrows $X_1 \to X_2$ in column 0 and $X_1 \to X_0$ in column 1 are switched on. In addition, the blue arrow from $(X_0, X_1) \to (X_1, X_2)$ is switched on. The automaton is then able to either accept a further 0, moving one place up the leftmost tower, or accept a 1 by moving across and down in the second column (labelled 1). The automaton is thus able to accept, say, a sequence of n 0s. Without loss of generality let $n = 4$ and assume that the next symbol accepted is a 1. The machine thus starts moving down the second column labelled 1. However, in order to accept only words of the form $0^4 1^4 2^4 3^4 \ldots$ it must remember the turnaround point. This is encoded by disabling the reactivity $(X_3, X_4) \to (X_4, X_5)$ in the column labelled 2. Note, of course, that similar switching occurred for all lower transitions, however,

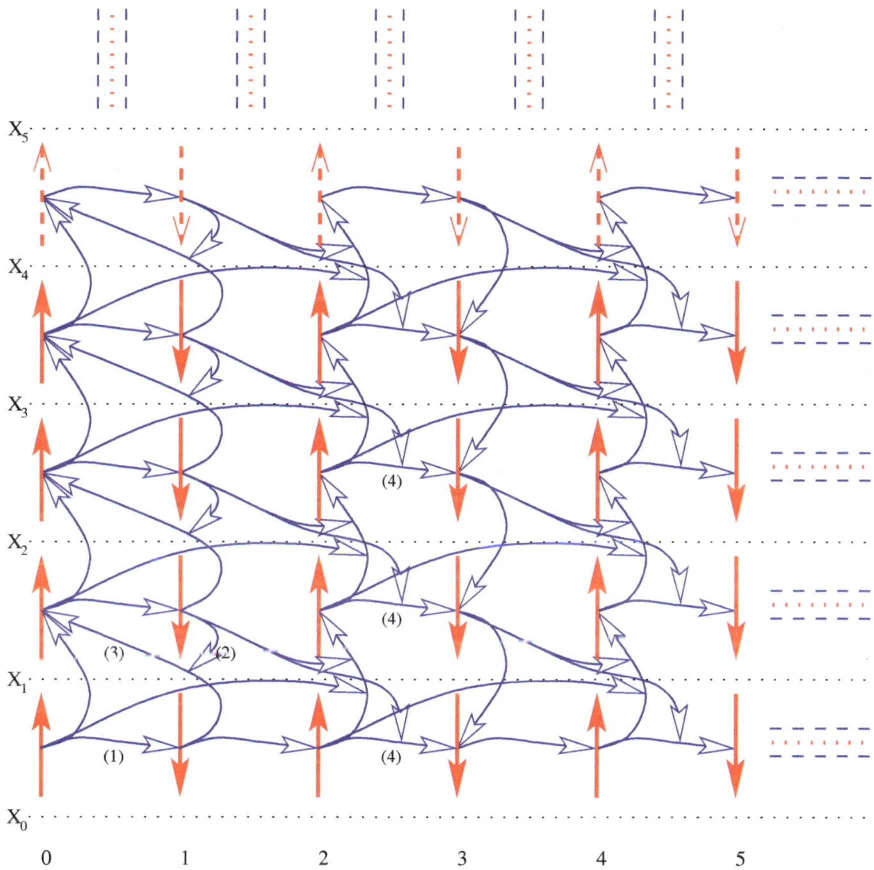

Fig. 9.7: Infinite reactive automaton for $0^n 1^n 2^n 3^n \dots (2k+1)^n$

as we progress down the 1 tower, those reactivities are switched back again with the effect that the automaton will be able to switch, sequentially, the correct number of up-going red arrows in the third column. Note also that all the reactivities labelled (4) will have also been switched so that when the red arrows in the third column are traversed the option of immediately descending in the fourth column is removed up to the turnaround point when the machine is back in state X_4. We thus claim this infinite reactive automaton accepts

$$0^n 1^n 2^n \dots (2k+1)^n, n = 0, 1, 2, \dots, k = 0, 1, 2, \dots$$

9.1.1 Some historical comments

This chapter is a contribution to a continuing exploration that started with the observation that, in the 'Chomsky hierarchy' [39] of formal languages, many languages, particularly those arising in computer science, are not context-free, and so they are present in the hierarchy as instances of context-sensitive languages, or languages requiring full Turing computability. An alternative treatment for presenting and processing many of these languages is to use a context-free grammar together with a mechanism for controlling the choice of derivation step at each point in a derivation. By selecting only certain derivations to be legitimate, various classes of non-context-free languages can be described. Such presentations appear more 'natural' in that they reflect the inherent structure of these languages whose basis is a context-free phrase structure. They also provide alternative (and possibly more efficient) algorithms for processing such languages.

This general approach to grammar is presented in [40] under the name of 'Regulated Rewriting'. The mechanisms for controlling derivations vary from author to author. An early proposal was that of 'matrix grammars' [109] whose rules are grouped in sequences (called 'matrices') and derivation steps iterate through the rules in a sequence. Many other mechanisms have been proposed, amongst which are: indexed grammars [5], programmed grammars [142], regulated grammars [47], grammars with state [118, 129], grammars with control sets [98] and co-operating grammars [128]. Some of these approaches treat the derivation sequences themselves as strings in an additional language (e.g. [98]). Properties of the language of derivation sequences then determine properties of the generated language. Results concerning the equivalence of expressive power of these formalisms can be established in some cases.

In this chapter, we add yet more to the number of such mechanisms! This is not really a case of 'the more, the merrier', rather the opposite. However, our emphasis here is somewhat different, although we will investigate the expressivity of the reactive grammars that we consider. We are interested in the general notion of reactivity in computational systems that may change their behaviour and evolve as computation progresses. There is an increasing focus on the development and analysis of computational systems that exhibit 'evolutionary' behaviour, that is, systems that may either compute normally or may invoke evolutionary changes which can modify the structure of the system itself. For various models of computation, these evolutionary changes take very different forms. For automata consisting of (labelled) transition systems, we have already seen some forms of evolutionary change in examples above. As well as changing the transition relation (i.e. the connectivity of nodes), we may change the labelling, or, if data is stored at nodes, change this data. As a counterpart to reactive automata, here we investigate reactive grammars.

In grammars, normal computation is the process of replacing non-terminals or substrings using the rules of a grammar. The evolutionary steps are the 'reactive elements' which may be introduced into the generation process at various points in the derivation. Reactivity may be present so that (1) at each step of a derivation, we may control which grammar rules may be applied (and this control itself can be of various forms including additional 'reaction rules' which determine the availability of rules), or (2) we may change the notion of application of a rule to a string, or (3) we may modify rules as they are used, or (4) we may transform intermediate strings as they are generated in a derivation, before applying further rules. We consider some of these cases in this chapter. Some of these are equivalent to already introduced forms of grammar, others appear to be new. We also consider examples of what we call 'embedded reactivity' in which substrings of a string can themselves determine changes in either the grammar, the current string or the derivation steps allowed. The extra degree of freedom that reactivity adds allows us to re-examine the mathematical structures involved in language recognition and generation, and to relate languages according to their reactive structure as well as their production rules. We have yet to explore the latter idea.

Reactive grammars have already appeared in applications other than language recognition and generation. For example, in runtime verification of computational systems, reactive rule-based systems arise naturally as in RuleR [20] (see also the related Eagle [19] and MetateM[18] systems). Other non-linguistic areas where non-context-free grammars appear include graph-based problems [134] and modelling biological systems [122].

Our exposition is driven by example – the examples indicate the complex structure of language generation when production rules of grammars are combined with notions of reactivity. We first consider 'switching grammars' and 'switch reactive grammars' in which the use of a grammar rule determines which rules are available for the next derivation step. We then consider 'string transformer grammars' in which each production rule of the grammar has an associated string transformer - a transformation of the string resulting from application of the rule. This is a powerful notion of reactivity and we will show that it can be used to model arbitrary derivation strategies in grammars. Finally, we consider several examples of 'embedded reactivity' in which substrings of a string may determine reactivity. This is all very much an initial exploration of the space of possibilities in this area.

9.2 Switching grammars

We begin with a standard definition of grammar:

Definition 9.7. *A grammar, G, is a four-tuple consisting of a set of non-terminal symbols, N, a set of terminal symbols, Σ, a set of (production) rules, with each rule of the form* $\alpha A\beta \to \gamma$ *for* $A \in N$, $\alpha,\beta,\gamma \in (N \cup \Sigma)^*$, *and an initial non-terminal symbol* $S \in N$.

Recall that a grammar is *context-free* if all the production rules are of the form $A \to \gamma$ where $A \in N$ and $\gamma \in (N\cup\Sigma)^*$, and a grammar is *right regular* if all production rules are of the form $A \to wX$ or $A \to w$ with $w \in \Sigma^*$ and $A, X \in N$. A grammar is *linear* when, for each rule, γ contains at most one non-terminal. A grammar is said to be *ϵ-free* (where ϵ is the empty string) when all production rules have $\gamma \in (N \cup \Sigma)^+$.

To begin the exploration of reactivity in grammars, we introduce a very simple form of reactivity in, what we call, *switching grammars*. Here the application of a production rule can switch production rules of the grammar on or off. These are a special case of programmed grammars [142] where only 'success sets' are associated with production rules. The ϵ-free switching grammars are, effectively, an alternative presentation of (ϵ-free) state grammars [118, 129]. This simple form of reactivity is considered here to illustrate some of the issues which arise in reactive grammars and because it is related to other forms of reactivity that we consider, including 'switch reactive grammars' and 'string transformer grammars'.

Definition 9.8. *A switching grammar is a tuple consisting of a finite set of non-terminal symbols, N, a finite set of terminal symbols, Σ, a set of rule labels, R, a set of R-labelled production rules P, each rule being of the form $r : \alpha A\beta \rightarrow \gamma, \rho$; for $r \in R$, $A \in N$, $\alpha, \beta, \gamma \in (N \cup \Sigma)^*$, $\rho \subseteq R$, together with an initial non-terminal symbol $S \in N$ and an initial rule set $\rho_0 \subseteq R$. A switching grammar is ϵ-free if its production rules have $\gamma \in (N \cup \Sigma)^+$.*

Definition 9.9. *Let $G = \langle N, \Sigma, R, P, S, \rho_0\rangle$ be a switching grammar. G derives the string $s\gamma t$ from $s\alpha A\beta t$ and switches from ρ_i to ρ_j in one step (written as $\langle s\alpha A\beta t, \rho_i\rangle \Rightarrow \langle s\gamma t, \rho_j\rangle$) iff for some $r \in \rho_i$, $r : \alpha A\beta \rightarrow \gamma, \rho_j \subseteq P$. A word $w \in \Sigma^*$ is derived in G iff there is a sequence of single step derivations $\langle w_0, \rho_0\rangle \Rightarrow \langle w_1, \rho_1\rangle \Rightarrow \ldots \Rightarrow \langle w_n, \rho_n\rangle$ such that $w_k \in N \cup \Sigma^*$ and $\rho_k \subseteq R$, for $k \in 0..n$, and $S = w_0$ and $w = w_n$. The language generated by G is the set of its derivable words w.*

Let us look at an example of a switching grammar.

Example 9.10 (A right regular switching grammar). Consider a switching grammar G where $N = \{A\}$, $\Sigma = \{a, b\}$, $R = \{r_0, r_1, r_2, r_3\}$, the production rule set is

$$\begin{aligned}\{\ r_0 &: A \rightarrow a, \{r_2, r_3\},\\ r_1 &: A \rightarrow aA, \{r_2, r_3\},\\ r_2 &: A \rightarrow b, \{r_0, r_1\},\\ r_3 &: A \rightarrow bA, \{r_0, r_1\}\}\end{aligned}$$

with initial non-terminal $S = A$ and initial rule set $\{r_0, r_1\}$. G generates the language $(ab)^*a \cup (ab)^*$.

Note, (a) this is a right regular grammar, and (b) this grammar generates the same language as the (non-switching) right regular grammar $\langle\{A, B\}, \{a, b\}, \{A \rightarrow a, A \rightarrow aB, B \rightarrow b, B \rightarrow bA\}, A\rangle$.

The comment in this example suggests the following result.

Theorem 9.11. *A language is generated by a switching right regular grammar iff it is a regular language.*

Proof. Clearly any right regular grammar can be transformed into a switching right regular grammar. Simply form a production rule set P_R by uniquely labelling each production rule of the regular grammar with a label from R and then make R the initial rule set.

The other direction is established as follows. Let $G = \langle N, \Sigma, R, P, S, \rho_0\rangle$. Create a new set of non-terminal symbols $N_R = \{n_r \mid n \in N \text{ and } r \in R\} \cup \{S_R\}$. Create a new production rule set

$$\begin{aligned} Q_R = {} & \{A_r \to u \mid (r : A \to u, \rho) \in P \text{ for } u \in \Sigma^*\} \cup \\ & \{A_r \to vB_s \mid (r : A \to vB, \rho) \in P \text{ for } v \in \Sigma^* \text{ and } s \in \rho\} \cup \\ & \{S_R \to S_{r_0} \mid r_0 \in \rho_0\}. \end{aligned}$$

By construction, the grammar $G_R = \langle N_R, \Sigma, Q_R, S_R \rangle$ is regular. Furthermore, G_R generates exactly the same set of words as G.

More illustrative of the expressivity of switching grammars is the following example which generates the language $\{a^n b^n c^n \mid n \geq 1\}$. This language is a standard example of a language that cannot be generated by unrestricted rule application in a context-free grammar. There are however simple context-sensitive grammars for this language and, as we shall show below, a range of reactive context-free grammars of various forms which also generate the language. As an example of a context-sensitive grammar for this language, consider the following (non-switching) grammar,

$$\begin{aligned} S &\to aSBC \\ S &\to aBC \\ CB &\to BC \\ aB &\to ab \\ bB &\to bb \\ bC &\to bc \\ cC &\to cc \end{aligned}$$

with initial symbol S. Notice that because of the interchange steps (between Bs and Cs), generating the word $a^n b^n c^n$ requires $O(n^2)$ derivation steps.

Example 9.12 (The language $\{a^n b^n c^n \mid n \geq 1\}$). Consider a switching context-free grammar generating this language:

$$G = \langle \{A, B, C\}, \{a, b, c\}, \{r_0, r_1, r_2, r_3, r_4\}, P, A, \{r_0\} \rangle$$

where

$$P = \left\{ \begin{array}{l} r_0 : A \to BC, \{r_1, r_2\} \\ r_1 : B \to ab, \{r_3\} \\ r_2 : B \to aBb, \{r_4\} \\ r_3 : C \to c, \{\} \\ r_4 : C \to cC, \{r_1, r_2\} \end{array} \right\}$$

The following is an example derivation:

$$\begin{aligned} A, \{r_0\} \Rightarrow BC, \{r_1, r_2\} \Rightarrow aBbC, \{r_4\} \Rightarrow \\ aBbcC, \{r_1, r_2\} \Rightarrow aabbcC, \{r_3\} \Rightarrow aabbcc, \{\} \end{aligned}$$

After the first step replacing A by BC, the grammar reacts by only allowing rules which replace B to be applied. The second step replaces B by aBb and then only allows the rule that expands C by cC. And so on. Thus the language generated is $\{a^n b^n c^n \mid n \geq 1\}$.

Theorem 9.13. *The class of ϵ-free switching context-free grammars is identical to the (ϵ-free) state context-free grammars and hence corresponds to the class of context-sensitive languages.*

See Kasai [118] for the relation between state and context-sensitive grammars. The identity of ϵ-free switching context-free grammars and (ϵ-free) context-free state grammars is trivial.

9.3 Switch reactive grammars

In switch reactive Kripke models, the accessibility of one state from another may be turned 'on' and 'off'. In traversing such a structure only the accessibilities currently 'on' or activated are available. Accessibilities are changed according to additional arcs which determine the reactivity. Such reactive arcs may themselves be turned on and off according to other reactive arcs. This produces a typed hierarchy of arcs, the base arcs being the accessibilities between states.

We now turn to grammars which have an analogous reactivity, extending that of switching grammars. We call these *switch reactive grammars*. It turns out that there is a range of possibilities which we explore here.

The idea is that associated with each production rule in a grammar is a set of 'reaction rules' which determine what rules (production rules and reaction rules) are active (or 'available') at the next step. Reaction rules may themselves have associated reaction rules.

The possibilities here are generated by the following issues:

1. Base rules and reaction rules

We may wish to distinguish the way that rules operate according to whether they are

- base rules, which include the production rules of the grammar, but also may include additional rules for the presentation of grammars,
- reaction rules, whose role is the activation and de-activation of other rules, including reaction rules.

Thus a grammar consists, in general, of a base grammar and a collection of reaction rules.

2. Transient and persistent rules

What happens to an active rule when it is used (or 'fired')? It may

- become de-activated – this is the case of *transient* rules,
- remain activated – this is the case of *persistent* rules.

The transience or persistency of rules may vary either (a) by allowing both forms in a grammar (defining the nature of each rule), or (b) by distinguishing base rules from reaction rules.

3. Operation of reaction rules

There are various possibilities as to how the reaction rules operate: They may

- switch (or 'flip') the activation status of the target rule when the source rule fires – turning on rules that are off, and turning off rules that are on.
- act only 'positively' i.e. turn on target rules that are off and leave target rules on that are on (these may then be turned off by firing, if they are transient).

Let us now look at example of these various cases.

Example 9.14 (Transient base rules, persistent reactivity and switching reaction rules). Consider a grammar for the language $\{a^n b^n \mid n \geq 1\}$. It has production rules:

$$
\begin{aligned}
c_0 &: S \rightarrow AB \\
c_1 &: A \rightarrow a \\
c_2 &: A \rightarrow aA \\
c_3 &: B \rightarrow b \\
c_4 &: B \rightarrow bB
\end{aligned}
$$

The reaction rules are:

$$
\begin{aligned}
r_0 &: c_0 \rightarrow c_1, c_2 \\
r_1 &: c_1 \rightarrow c_2, c_3 \\
r_2 &: c_2 \rightarrow c_1, c_4 \\
r_3 &: c_4 \rightarrow c_1, c_2
\end{aligned}
$$

Let the rules that are initially 'on' be c_0 together with all the reaction rules. Let the base rules be transient and the reaction rules be persistent. Finally, let the reaction rules operate by switching: flipping the on/off status.

Initially, rule c_0 fires (and becomes de-activated). The reaction rule for c_0 is r_0, which thus flips the status of c_1 and c_2, turning both on. At the next step, if c_1 is fired (and is then de-activated), rule c_2 is switched off and rule c_3 switched on, to generate the string ab. Alternatively, if c_2 is fired (and is then de-activated), rule c_1 is switched off and rule c_4 switched on, to generate the string $aAbB$. Then c_1 and c_2 are activated to continue the derivation.

This scheme may be extended to generate the non-context-free language $\{a^n b^n c^n \mid n \geq 1\}$.

The above grammar is directly equivalent to (has the same derivations as) the standard switching grammar:

$$
\begin{aligned}
c_0 &: S \rightarrow AB,\ \{c_0, c_1\} \\
c_1 &: A \rightarrow a,\ \{c_3\} \\
c_2 &: A \rightarrow aA,\ \{c_4\} \\
c_3 &: B \rightarrow b,\ \{\} \\
c_4 &: B \rightarrow bB,\ \{c_1, c_2\}
\end{aligned}
$$

Notice however, that the reaction mechanisms are different. In the case of the switching grammar, at each step, all rules are turned off and only then is the set of rules for the next step turned on.

In this example then, introducing reaction rules does not result in anything more than standard switching grammars.

Let us now consider an example of a different form.

Example 9.15 (All rules persistent, switching reaction rules). Here, all rules are persistent and reaction works by 'flipping' the activation state of its target arrow.

$$\begin{aligned} c_0 &: S_0 \to aS_1 \\ c_1 &: S_1 \to bS_2 \\ c_2 &: S_2 \to dS_2 \\ c_3 &: S_2 \to c \end{aligned}$$

The reaction rule is:

$$r : c_2 \to c_3$$

Let all the rules be 'on' initially. Then rule c_3 is 'on' only after an even number of firings of rule c_2. So the language generated is $\{abd^{2n}c \mid n \geq 0\}$.

This reactive grammar is directly equivalent to (i.e. has the same derivations as) the normal grammar:

$$\begin{aligned} c_0 &: S_0 \to aS_1 \\ c_1 &: S_1 \to bS_2 \\ c_2 &: S_2 \to dS'_2 \\ c_2 &: S'_2 \to dS_2 \\ c_3 &: S_2 \to c \end{aligned}$$

In fact, for this form of reactivity, even for higher-order reactivity (when reaction rules act on reaction rules), it appears that the grammar is equivalent, by an expansion and copying technique, to a normal (non-reactive) grammar.

Let us consider another form of switch reactive grammar. In this case, all rules are transient.

Example 9.16 (All rules transient, switching reaction rules). Let us consider again the language of Example 1, $\{a^n b^n \mid n \geq 1\}$. The base rules are the same:

$$\begin{aligned} c_0 &: S \to AB \\ c_1 &: A \to a \\ c_2 &: A \to aA \\ c_3 &: B \to b \\ c_4 &: B \to bB \end{aligned}$$

with c_0 initially 'on'. Let us see how to add transient reaction rules so that this grammar is restricted to generate the required language. Consider, for example, the rule c_2. Part of its reaction is to switch rule c_4 together with all its reaction rules. But the latter reaction rules include a rule to switch rule c_2 and all its reaction rules...

So far, we have labelled the reaction rules with *rule names*, but they have not been used. Now, in order to describe this dependency of reaction rules on others, the names are crucial and allow us to define this via mutual recursion, as follows:

$$\begin{aligned}
r_0 &: c_0 \rightarrow c_1, c_2, r_1, r_2 \\
r_1 &: c_1 \rightarrow c_2, c_3, r_2 \\
r_2 &: c_2 \rightarrow c_1, c_4, r_1, r_4 \\
r_3 &: \\
r_4 &: c_4 \rightarrow c_1, c_2, r_1, r_2
\end{aligned}$$

The rule name r_3 labels an empty rule. Notice that r_2 changes the activation of r_4 and vice versa. Initially r_0 is active. A derivation proceeds by firing c_0, which then activates c_1 and c_2 via reaction rule r_0. If c_1 then fires, c_2 is turned off and c_3 turned on via reaction rule r_1, etc.

Notice that *if no rule names are introduced for the reaction rules* and instead we use a hierarchical assembly of 'reactions' of the form $c_0 \rightarrow c_1, c_2$, then the mutual recursion can only be expressed through infinite rules or an infinite set of rules.

In fact, grammars of the form in Example 3 are equivalent to switching grammars, as we now show.

Example 9.17 (Example 3 continued). First notice that switching grammars specify only which rules are to be active at the next step, instead of the 'flipping' reactivity used in Example 3. However, we can convert Example 3 to this positive form of reaction, in which case the grammar is:

$$\begin{aligned}
c_0 &: S \rightarrow AB \\
c_1 &: A \rightarrow a \\
c_2 &: A \rightarrow aA \\
c_3 &: B \rightarrow b \\
c_4 &: B \rightarrow bB
\end{aligned}$$

$$\begin{aligned}
r_0 &: c_0 \rightarrow c_1, c_2, r_1, r_2 \\
r_1 &: c_1 \rightarrow c_3 \\
r_2 &: c_2 \rightarrow c_4, r_4 \\
r_3 &: \\
r_4 &: c_4 \rightarrow c_1, c_2, r_1, r_2
\end{aligned}$$

From this combination of base rules and the reaction rules, we construct a switching grammar whose labels are derived from those of the base and reaction rules as follows:

$$\begin{aligned}
c_{0r_0} &: S \rightarrow AB, \{c_{1r_1}, c_{2r_2}\} \\
c_{1r_1} &: A \rightarrow a, \{c_{3r_3}\} \\
c_{2r_2} &: A \rightarrow aA, \{c_{4r_4}\} \\
c_{3r_3} &: B \rightarrow b, \{\} \\
c_{4r_4} &: B \rightarrow bB, \{c_{1r_1}, c_{2r_2}\}
\end{aligned}$$

In summary, we have explored various types of switch reactive grammars and shown their equivalence to normal non-reactive grammars or to switching grammars. Some of these conversions are economical (do not lead to a great expansion of the grammar). Others (e.g. Example 2) are less so.

9.4 A reduction result

We present a result showing how, using switch reactive grammars, a linear grammar may be reduced, in terms of the number of non-terminal symbols, to a reactive linear grammar generating the same language. In fact, we establish a 'simulation' between a grammar and its reduction. The reduction in the number of non-terminal symbols is considerable: from k^n to kn. The form of switch reactive grammars used is that all rules are transient (or self-cancelling), i.e. as soon as a rule is fired its activation status is switched off. Reaction rules are switching, i.e. change the activation status of their targets.

Theorem 9.18. *A linear grammar G with k^n non-terminal symbols can be simulated by a linear switch reactive grammar $\mathbb{T}(G)$ with kn non-terminal symbols.*

Proof outline. Suppose G is a linear grammar with k^n non-terminal symbols – i.e. the non-terminals are in an n-dimensional cube of k entries in each dimension, so each non-terminal symbol may be represented as an n-tuple of symbols from k sets.

Let $[a^i_j]$ be a matrix of distinct elements for $i = 1, \ldots, n$ and $j = 1, \ldots, k$. Let $x = (x_1, \ldots, x_n)$ be a vector such that $x_i \in \{a^i_1, \ldots, a^i_k\}$. There are exactly k^n such vectors.

We construct a reactive grammar $\mathbb{T}(G)$ with non-terminals $[a^i_j]$ which simulates G.

Let us show how it is done for the case $k = 3$, $n = 5$ (these numbers are chosen simply to allow us to draw the appropriate figures).

The aim is to simulate the effect of the rule of G

$$x \longrightarrow w_1 y w_2$$

where w_1 and w_2 are terminal letters. (For regular grammars, consider rules of the form $x \rightarrow y w_2$ instead.)

Let $x = (x_1, \ldots, x_n)$ and $y = (y_1, \ldots, y_n)$. The construction is illustrated in Figure 9.8. The arrows in this diagram are grouped into various types:

Group 1. These are the (red) vertical arrows $(x_i, x_{i+1}) : x_i \rightarrow x_{i+1}$, $i = 1, 2, 3, 4$ and $(y_j, y_{j+1}) : y_j \rightarrow y_{j+1}$, $j = 1, 2, 3, 4$. They can be either on or off. Each chain of such arrows forms the basis of the encoding of a non-terminal from the original grammar.

Group 2. The possibly multi-targeted (blue) arrows are reactivity arrows. The source is an arrow and the target is a set of arrows, e.g. $(\alpha, \{\beta_1, \beta_2, \ldots, \beta_m\})$. This reads as follows: if you use the rule α then flip on or off the activation status of the target rules $\beta_1, \ldots, \beta_m$, e.g. if β_1 is on, flip to off, or if it is off, flip it to on.

Group 3. Other arrows connect arrows of any group to those of any group. For example the multi-targeted (green) bold-face arrow in the diagram marked (3) denotes the grammar rule $x_5 \rightarrow w_1 y_1 w_2$.

Note the following: The vector $x = (x_1, x_2, x_3, x_4, x_5)$ is a non-terminal of G. This is represented in the diagram by the left-hand side, with the vertical arrows, and

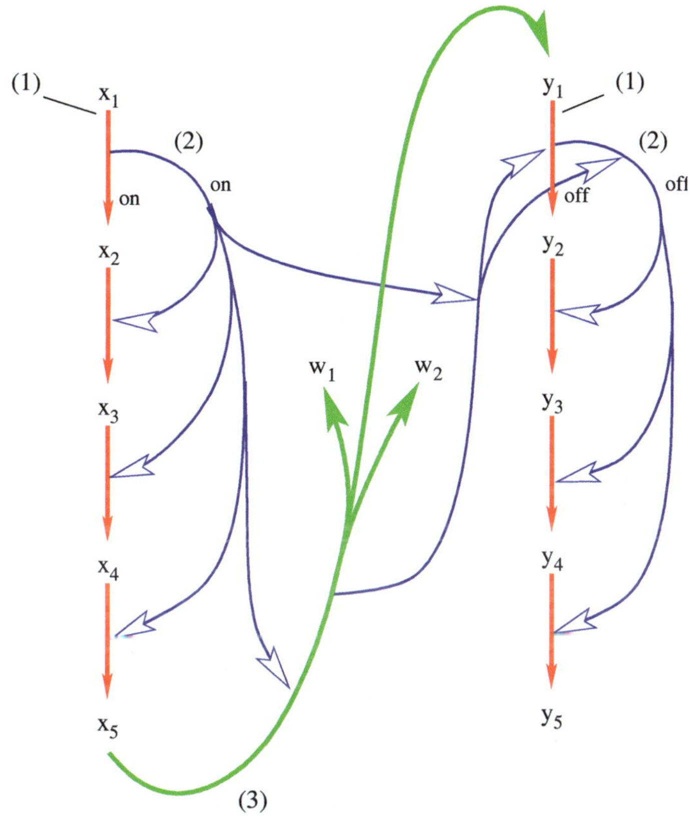

Fig. 9.8: An example of the construction for a rule of G.

the hierarchy of curved arrows between them. The last curved arrow, to $(x_4 \to x_5)$, contains all the information of the vector x. So from this arrow, we send arrows to activate y and all its arrows.

How do we simulate $x \to w_1 y w_2$? As an example, let the non-terminal x be encoded by the vector $(a_1^1, a_1^2, a_2^3, a_1^4, a_3^5)$ of non-terminals from the reduced grammar and y encoded by the vector $(a_1^1, a_2^2, a_1^3, a_1^4, a_1^5)$. Thus the original grammar rule $x \to w_1 y w_2$ is transformed into the following collection of grammar rules

$$
\begin{aligned}
a_1^1 &\to a_1^2 \\
a_1^2 &\to a_2^3 \\
a_2^3 &\to a_1^4 \\
a_1^4 &\to a_3^5 \\
a_3^5 &\to w_1 a_1^1 w_2 \\
a_1^1 &\to a_2^2 \\
a_2^2 &\to a_1^3 \\
a_1^3 &\to a_1^4 \\
a_1^4 &\to a_1^5
\end{aligned}
$$

with the associated reactivity rules

$$
\begin{aligned}
&((a_1^1, a_1^2), \{(a_1^2, a_2^3), (a_2^3, a_1^4), (a_1^4, a_3^5), \\
&\qquad (a_3^5, w_1 a_1^1 w_2), \\
&\qquad ((a_3^5, w_1 a_1^1 w_2), \{(a_1^1, a_2^2), ((a_1^1, a_2^2), \{(a_2^2, a_1^3), (a_1^3, a_1^4), (a_1^4, a_1^5)\})\})\}) \\
&((a_3^5, w_1 a_1^1 w_2), \{(a_1^1, a_2^2), ((a_1^1, a_2^2), \{(a_2^2, a_1^3), (a_1^3, a_1^4), (a_1^4, a_1^5)\})\}) \\
&((a_1^1, a_2^2), \{(a_2^2, a_1^3), (a_1^3, a_1^4), (a_1^4, a_1^5)\})
\end{aligned}
$$

We start with the grammar rule $a_1^1 \to a_1^2$ and its associated reactivity rule switched on. Then, for a word W containing the non-terminal a_1^1, the rule $a_1^1 \to a_1^2$ can be applied to produce a new word derived from W by replacing a_1^1 by a_1^2 in the usual way. However, at the same time, the reactivity rule is fired and the activation status of the remaining rules $a_1^2 \to a_2^3$, $a_2^3 \to a_1^4$ and $a_1^4 \to a_3^5$ are switched on, as is the grammar rule $a_3^5 \to w_1 a_1^1 w_2$ together with its associated reactivity rule. The derivation will then continue to "decode" the original non-terminal x. When the grammar rule $a_3^5 \to w_1 a_1^1 w_2$ is applied, its reactivity will switch on the header rule and its reactivity for the decoding of the (original) y nonterminal.

The system is set up with arrows for every rule $u \to w_1' v w_2'$ in the original grammar G. Figure 9.8 thus illustrates the set-up for the rule $x \to w_1 y w_2$.

Note that if the original grammar is regular, so is the reactive one.

Cycles and choice

Let us consider an encoding of the following two linear rules

$$X \to a \qquad\qquad X \to bX$$

for non-terminal symbol X and terminal symbols a and b, extracted from a grammar with 5^4 non-terminals. The reactive grammar will encode the non-terminals of the grammar as vectors of length 4.

For ease of initial presentation, we label each grammar and reactivity rule for the reactive linear (sub)grammar depicted in Figure 9.9.

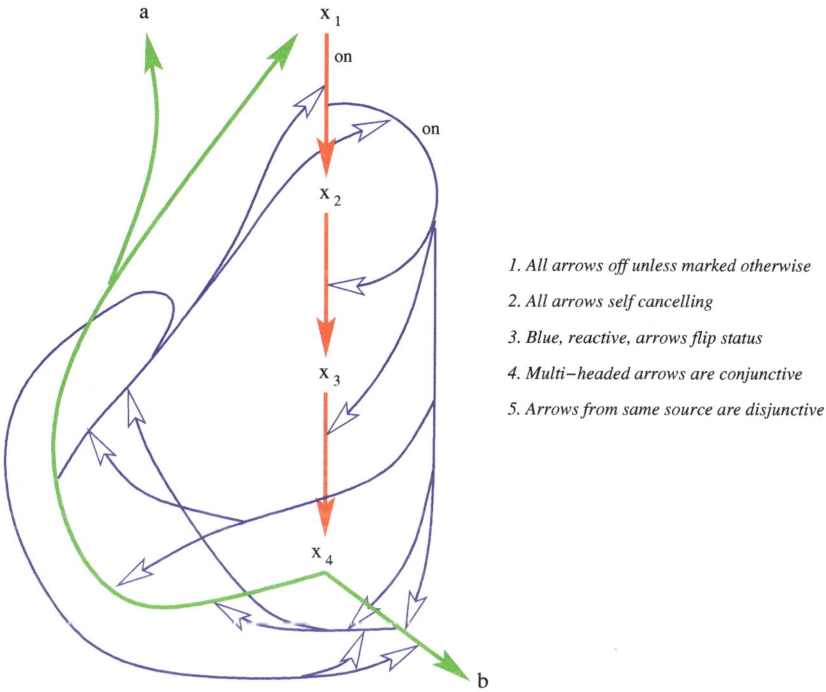

Fig. 9.9: Cycle and choice

Grammar rules	Reactivities
$g_1 : x_1 \longrightarrow x_2$	$r_1 : (g_1, \{g_2, g_3, g_4, r_4, g_5, r_5\})$
$g_2 : x_2 \longrightarrow x_3$	$r_4 : (g_4, \{g_5, r_5\})$
$g_3 : x_3 \longrightarrow x_4$	$r_5 : (g_5, \{g_4, r_4, g_1, r_1\})$
$g_4 : x_4 \longrightarrow b$	
$g_5 : x_4 \longrightarrow ax_1$	

A rule can be named by its source and target, e.g. g_1 by (x_1, x_2), hence, the above can be represented by labelling just one rule, as below.

Grammar rules

$$
\begin{aligned}
x_1 &\rightarrow x_2 \\
x_2 &\rightarrow x_3 \\
x_3 &\rightarrow x_4 \\
x_4 &\rightarrow b \\
x_4 &\rightarrow a x_1
\end{aligned}
$$

Reactivities

$$
\begin{aligned}
&((x_1, x_2), \{(x_2, x_3), (x_3, x_4), \\
&\qquad (x_4, b), ((x_4, b), \{(x_4, a x_1), r_5\}), \\
&\qquad (x_4, a x_1), r_5\}) \\
&r_5 : ((x_4, a x_1), \{(x_4, b), ((x_4, b), \{(x_4, a x_1), r_5\}), \\
&\qquad\qquad (x_1, x_2), ((x_1, x_2), \{(x_2, x_3), (x_3, x_4), \\
&\qquad\qquad\qquad (x_4, b), ((x_4, b), \{(x_4, a x_1), r_5\}), \\
&\qquad\qquad\qquad (x_4, a x_1), r_5\})\})
\end{aligned}
$$

Whilst the above example encodings indicate how a linear grammar with k^n non-terminals can be represented by a reactive linear grammar with kn non-terminals, it is clear there may be a very large number of reactivity rules. Furthermore, the reactivities will be mutually recursive as soon as there are non-terminal cycles in the original grammar, as illustrated in the above example.

9.4.1 Is there a reduction for non-linear context-free grammars?

It is natural to question whether a similar reduction result can be obtained for non-linear grammars. Here we briefly explore what happens when the above scheme is applied to non-linear context-free production rules. Consider a rule

$$X \rightarrow YZ$$

for non-terminals X, Y and Z. Let the vector of non-terminals $(x_1, x_2, \ldots, x_n)$ from the simulating grammar encode X, and then similarly for Y and Z. The context-free rule would then be translated into a set of grammar rules such as

$$
\begin{array}{llll}
x_1 \rightarrow x_2 & x_n \rightarrow y_1 z_2 & y_1 \rightarrow y_2 & z_1 \rightarrow z_2 \\
x_2 \rightarrow x_3 & & y_2 \rightarrow y_3 & z_2 \rightarrow z_3 \\
\vdots & & \vdots & \vdots \\
x_{n-1} \rightarrow x_n & & y_{n-1} \rightarrow y_n & z_{n-1} \rightarrow z_n
\end{array}
$$

together with reactivities

$$
\begin{aligned}
&((x_1, x_2), \{(x_2, x_3), \ldots (x_{n-1}, x_n), (x_n, y_1 z_1), ((x_n, y_1 z_1), \{\ldots\})\}) \\
&((x_n, y_1 z_1), \{\ldots\}) \\
&((y_1, y_2), \{(y_2, y_3), \ldots (y_{n-1}, y_n), \ldots\}) \\
&((z_1, z_2), \{(z_2, z_3), \ldots (z_{n-1}, z_n), \ldots\})
\end{aligned}
$$

Superficially this seems adequate. Unfortunately, there are problems. Firstly it is important to remember that, for any given i, the non-terminals denoted by x_i, y_i and z_i are not necessarily disjoint (of course, those from different levels in the encoding chain are necessarily disjoint). Since the non-terminals Y and Z are jointly active, there may be conflict between their respective encoding chains. For example, if $y_1 = z_1 = a_1^1$ and $y_2 = z_2 = a_3^2$ then the application of the rule $y_1 \to y_2$ to the derived word $y_1 z_1$ will deactivate itself (all rules are self-cancelling) with the consequence that the rule $z_1 \to z_2$ is deactivated. Thus the Z non-terminal will not get "decoded" and applied.

The above reduction result works for linear grammars because there is only one non-terminal present (i.e. active) in a derived word. This therefore suggests that perhaps the context-free case can be made to work by encoding, for example, a leftmost derivation strategy in which only the leftmost non-terminal in a derived word is allowed to be active. This would require a mechanism to ensure that the decoding rules and reactivities associated with non-leftmost non-terminals are disabled until the leftmost non-terminal has been fully expanded. Consider again the rule $X \to YZ$ in the original grammar. For the situation where the sets of non-terminals derived from Y and Z are always disjoint, this is possible. However, in general, there may be dependencies between the Y and Z non-terminals, as in Figure 9.10, which then require a stack in order to determine when a derivation from a non-terminal has been completed.

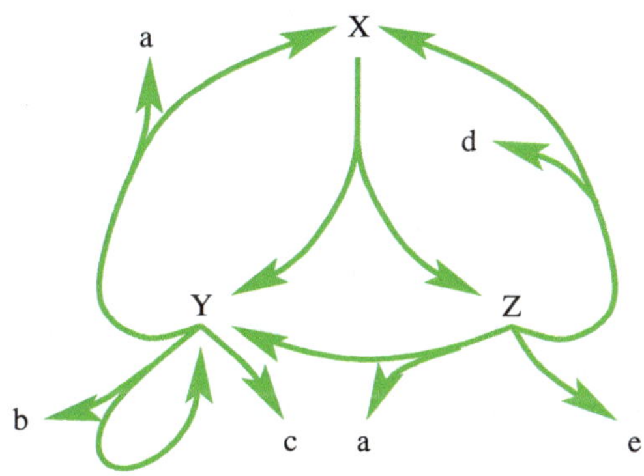

Fig. 9.10: Non-terminal Graph

9.5 String transformer grammars

We now consider a quite different example of reactivity in grammars in which each string produced by the application of a rule may undergo a transformation before further rules may be applied. Such transformations may be thought of as 'mutations' of a string in which case their occurrence is spontaneous and unpredictable, or as a form of 'corruption' caused by, say, noise or other forms of degradation, in which case transformations of each string may be 'small' but unavoidable.

In fact, we consider a general formulation of such grammars which covers various interpretations of the string transformations. To do so we introduce a set of string transformers and define *string transformer grammars* in terms of rules where each rule has an associated string transformer. After application of the rule, the string undergoes the associated transformation before application of further rules. Special cases are standard grammars where no transformation takes place (i.e. allowing only the identity transformation), the case when transformations are independent of the rules used, and the case where string transformations may apply optionally. We concentrate our attention on certain forms of string transformations – those induced by actions on the non-terminal symbols in a string, which therefore interact directly with the application of rules.

Introducing this form of reactivity into grammars may considerably extend the expressivity of grammars. We begin to explore the classification of string transformer grammars. Moreover, this reactivity introduces an additional 'degree of freedom' into grammars and we consider how varying the string transformations affects the language generated.

Definition 9.19. *A string transformer grammar is a tuple consisting of a finite set of non-terminal symbols, N, a finite set of terminal symbols, Σ, a set of string transformers, $T \subseteq (N \cup \Sigma)^* \to (N \cup \Sigma)^*$, a set of production rules P with each rule of the form*

$$\alpha A \beta \xrightarrow{t} \gamma$$

for $A \in N$, $\alpha, \beta, \gamma \in (N \cup \Sigma)^*$, $t \in T$, and an initial non-terminal symbol $S \in N$.

Definition 9.20. *Let $G = \langle N, \Sigma, T, P, S \rangle$ be a string transformer grammar. G derives the string s from $u\alpha A\beta v$ in one step (written as $u\alpha A\beta v \Rightarrow s$) for $u, v, s \in (N \cup \Sigma)^*$ iff $\alpha A \beta \xrightarrow{t} \gamma$ and $t(u\gamma v) = s$. A word $w \in \Sigma^*$ is derived in G iff there is a sequence of single step derivations $s_0 \Rightarrow \ldots s_k \Rightarrow \ldots \Rightarrow s_n$ such that $s_k \in (N \cup \Sigma)^*$ and $S = s_0$ and $w = s_n$. The language generated by G is the set of its derivable words w.*

Note that the strings resulting from the application of string transformers need not be (and will not be, in general) strings generated from the underlying grammar. The definition is very general in terms of the string transformers that are considered, allowing any function from the set of strings of terminals and non-terminal symbols to itself. As a consequence, if we wish to consider ϵ-freeness in this context, we need a modified definition. Restricted forms of string transformations are of particular interest. One such restriction is to limit string transformations to those that are induced by functions $f : N \to N$ from non-terminal symbols to non-terminal symbols. Such a function induces a string transformation by replacing each non-terminal symbol X in a string with the non-terminal symbol $f(X)$. Notice that interchanging non-terminal symbols in this way interacts with the application of production rules. It is in this interaction that we find new forms of expressivity of grammars. Often the form of the functions on non-terminal symbols that we consider may be succinctly expressed in terms of labelled symbols, that is the non-terminal symbols are of the form $\langle X, l \rangle$ with n a symbol and l a label from a label set Λ. A label transformer $f : \Lambda \to \Lambda$ defines a function f^* on these non-terminal symbols in an obvious way: $f^*\langle X, l \rangle = \langle X, f(l) \rangle$.

Let us look at a few examples of string transformer grammars to show how words are generated and illustrate the power of these grammars.

First consider a right regular grammar written with labelled non-terminal symbols:

Example 9.21 (Odds and Evens). Consider the following string transformer grammar with labelled non-terminal symbols

$$G = \langle \{S\} \times \{even, odd\}, \{a\}, \{t_1, t_2, t_3\}, P, \langle S, even \rangle \rangle$$

where the set of production rules P is

$$\begin{aligned} &\langle S, even \rangle \xrightarrow{t} a\langle S, odd \rangle \\ &\langle S, odd \rangle \xrightarrow{t} a \\ &\langle S, odd \rangle \xrightarrow{t} a\langle S, even \rangle \end{aligned}$$

Notice that, for this example, all the rules have the same associated string transformer. When t is the identity function it is easy to see that G generates the language $\{a^{2n} \mid n > 0\}$. What do we get for non-identity transformers? Defining the transformers as functions on the label set, there are three other transformations to consider:

1. When $t = \{even \mapsto even, odd \mapsto even\}$, the first rule application yields the string $a\langle S, even \rangle$ since the generated label of S is odd which is then transformed under t to $even$. No finite word can therefore be generated and the language generated is thus empty.
2. $t = \{even \mapsto odd, odd \mapsto even\}$ also generates the empty language for the same reason.
3. $t = \{even \mapsto odd, odd \mapsto odd\}$ generates the language $\{a^{n+1} \mid n > 0\}$.

We may associate different string transformers with each production rule, as in the next example. Consider the above production rules, but with differing string transformers:

$$\begin{aligned} &\langle S, even \rangle \xrightarrow{t_1} a\langle S, odd \rangle \\ &\langle S, odd \rangle \xrightarrow{t_2} a \\ &\langle S, odd \rangle \xrightarrow{t_3} a\langle S, even \rangle \end{aligned}$$

where the string transformers are defined by the following functions on labels:

$$\begin{aligned} t_1 : &\quad \{even \mapsto even, odd \mapsto even\} \\ t_2 : &\quad \{even \mapsto odd, odd \mapsto even\} \\ t_3 : &\quad \{even \mapsto odd, odd \mapsto odd\}. \end{aligned}$$

The grammar generates the set of words $\{a^n \mid n > 0\}$.

Notice that, in these examples, each right regular grammar generates a regular language. This is, in fact, a general result.

Theorem 9.22. *The class of languages generated by right regular string transformer grammars whose transformers are defined by functions on non-terminal symbols coincides with regular languages.*

Theorem 9.23. *Given two right regular grammars G_1 and G_2 that generate languages $\mathcal{L}_1$ and $\mathcal{L}_2$ respectively, there is a right regular string transformer grammar G with an identity string transformer which generates $\mathcal{L}_1$ and for which a different string transformer t makes G generate the language $\mathcal{L}_2$, moreover t is induced by a function on non-terminal symbols.*

Proof. Without loss of generality we assume that the regular grammars G_1 and G_2 have disjoint non-terminal sets apart from their initial non-terminal symbol S. Let $G_i = \langle N_i, \Sigma_i, P_i, S\rangle$ for $i = 1, 2$. Construct a string transformer grammar $G_l = \langle N_1 \cup N_2, \{1, 2\}, \Sigma_1 \cup \Sigma_2, \{\iota\}, S_l, 1\rangle$ where the production rules in P are given by the set

$$\begin{aligned}
&\{\langle S_l, 1\rangle \xrightarrow{\iota} \langle S, 1\rangle\} \cup \\
&\{\langle X, 1\rangle \xrightarrow{\iota} \alpha[\langle Y, 1\rangle] \mid X \to \alpha(Y) \in P_1,\ X, Y \in N_1\} \cup \\
&\{\langle X, 2\rangle \xrightarrow{\iota} \alpha[\langle Y, 2\rangle] \mid X \to \alpha(Y) \in P_2,\ X, Y \in N_2\}
\end{aligned}$$

and ι is the identity string transformer. It is easy to see that G_l will generate the language of G_1.

Now replace the transformation ι by that defined by the relabelling $\{1 \mapsto 2, 2 \mapsto 2\}$. This revised grammar now generates the language of G_2.

It is not the case that any standard grammar that generates a regular language also generates a regular language when extended with string transformers, even those defined by functions on non-terminal symbols.

We illustrate this with an example which also shows that a very simple grammar with simple string transformers can generate fairly complex languages. Indeed, our experience is that even in simple cases of string transformer grammars it is often very difficult to determine what the language generated is.

Example 9.24 (A simple grammar for a complex language). The context-free grammar

$$G = \langle\{S, A, B\}, \{a, b\}, P, S\rangle$$

where P is the set of production rules

$$\begin{aligned}
S &\to AB \\
A &\to aA \\
A &\to a \\
B &\to b
\end{aligned}$$

generates the regular language $\{a^n b \mid n > 0\}$. A regular grammar for the same language is straightforward to define, for example:

$$S \to aS$$
$$S \to ab$$

with S as the only non-terminal symbol and hence initial symbol. As an example of a string transformer grammar consider associating with each rule the string transformer induced by the function $\{A \mapsto B, B \mapsto A\}$ on non-terminal symbols.

Different derivation strategies give rise to different languages. The language generated by just leftmost derivations is $\{bb\}$ for we have only one derivation, namely:

$$S \Rightarrow BA \Rightarrow bB \Rightarrow bb$$

Note that $S \Rightarrow BA$ since $S \to AB$ and the transformation then yields the string BA. Similarly, $BA \Rightarrow bB$ as $B \to b$ and hence BA produces the string bA which is then transformed to the string bB. The language generated by just rightmost derivations is $\{aa, aba, bab\}$, for we have

$$\begin{aligned}
&S \Rightarrow BA \Rightarrow Aa \Rightarrow aa\\
&S \Rightarrow BA \Rightarrow Aa \Rightarrow aBa \Rightarrow aba\\
&S \Rightarrow BA \Rightarrow AaB \Rightarrow Bab \Rightarrow bab
\end{aligned}$$

Note that both the above are regular languages. Consider now, however, the language generated through the unrestricted use of production rules. Here is an exhaustive list of derivations beginning with the initial symbol S.

$$\begin{aligned}
&S \Rightarrow BA \Rightarrow bB \Rightarrow bb\\
&S \Rightarrow BA \Rightarrow Aa \Rightarrow aa\\
&S \Rightarrow BA \Rightarrow Aa \Rightarrow aBa \Rightarrow aba\\
&S \Rightarrow BA \Rightarrow AaB \Rightarrow aaA \Rightarrow aaa\\
&S \Rightarrow BA \Rightarrow AaB \Rightarrow aaA \Rightarrow aaaB \Rightarrow aaab\\
&S \Rightarrow BA \Rightarrow AaB \Rightarrow aBaA \ldots\\
&S \Rightarrow BA \Rightarrow AaB \Rightarrow Bab \Rightarrow bab
\end{aligned}$$

All bar the sixth branch are complete derivations. The partial derivation given by the sixth branch has been stopped at a point at which it recurses. The above set is sufficient for us to obtain a description of the generated language in a closed form. Before doing so, let us continue the derivation from $aBaA$ for one more unfolding of the derivation tree to $aaBaaA$.

$$\begin{aligned}
&\ldots aBaA \Rightarrow abaB \Rightarrow abab\\
&\ldots aBaA \Rightarrow aAaa \Rightarrow aaaa\\
&\ldots aBaA \Rightarrow aAaa \Rightarrow aaBaa \Rightarrow aabaa\\
&\ldots aBaA \Rightarrow aAaaB \Rightarrow aaaaA \Rightarrow aaaaa\\
&\ldots aBaA \Rightarrow aAaaB \Rightarrow aaaaA \Rightarrow aaaaaB \Rightarrow aaaaab\\
&\ldots aBaA \Rightarrow aAaaB \Rightarrow aaBaaA \ldots\\
&\ldots aBaA \Rightarrow aAaaB \Rightarrow aBaab \Rightarrow abaab
\end{aligned}$$

If we examine the form of this recursion, we find that the language (of finite words) generated may be expressed in the following closed form:

$$\{a^n ba^n b \cup aaa^n \cup a^{n+1}ba^{n+1} \cup aaa(aa)^n b \cup a^n ba^{n+1}b \mid n \geq 0\}$$

The first term specifies the words derivable via the the first branch; the second term covers words derivable via the second and fourth branches; the third term covers those words derivable via the third branch; the fourth term for the fifth branch; and finally the fifth term covers those words derivable via the seventh branch. Note that the terms denote disjoint sets of finite words. The pumping lemma for regular languages can be used to show that the language is not regular.

In Example 9.12, we saw that switching context-free grammars can define languages beyond context-free, using the example of $a^n b^n c^n$. We now revisit this language showing how it may be defined using a context-free string transformer grammar. In fact, there are many ways of generating this language using simple string transformer languages, each corresponding to a restricted derivation strategy. This suggests a general result which we present later and which relates derivation strategies to string transformer grammars.

Example 9.25 (The language $\{a^n b^n c^n \mid n \geq 1\}$ again). In this example, we show how the labelling and associated transformers can be used to control the derivation and therefore restrict the language associated with a grammar. The context-free grammar $G = \langle\{A, B, C, S\}, \{a, b, c\}, P, S\rangle$ where P is the set of production rules

$$\begin{array}{ll} 1: & S \rightarrow ABC \\ 2: & A \rightarrow a \\ 3: & A \rightarrow aA \\ 4: & B \rightarrow b \\ 5: & B \rightarrow bB \\ 6: & C \rightarrow c \\ 7: & C \rightarrow cC \end{array}$$

generates the language $a^n b^m c^l$ for $n, m, l \in \mathbb{N}$. Let us now consider a labelled version of this in which functions on the labels are used to ensure that the production rules numbered 3, 5 and 7 are always applied in that sequential order and, similarly, the rules numbered 2, 4 and 6 are applied sequentially. These constraints ensure that the only words generated are those for which $n = m = l$ holds.

More formally, define the string transformer grammar

$$G = \langle\{A, B, C, S\} \times \{0, 1, 2, 3, 4\}, \{a, b, c\}, \{t_1, t_2, t_3\}, P, S, 0\rangle$$

where P is the set of production rules

$$\begin{array}{ll} 1: & \langle S, 0\rangle \xrightarrow{t_1} \langle A, 1\rangle\langle B, 0\rangle\langle C, 2\rangle \\ 2: & \langle A, 2\rangle \xrightarrow{t_2} a \\ 3: & \langle A, 2\rangle \xrightarrow{t_1} a\langle A, 2\rangle \\ 4: & \langle B, 3\rangle \xrightarrow{t_3} b \\ 5: & \langle B, 2\rangle \xrightarrow{t_1} b\langle B, 2\rangle \\ 6: & \langle C, 4\rangle \xrightarrow{t_3} c \\ 7: & \langle C, 2\rangle \xrightarrow{t_1} c\langle C, 2\rangle \end{array}$$

and the functions on non-terminal symbols are defined by the following label transformers:

$$t_1 \text{ is } +1 \text{ modulo } 3$$
$$t_2 \text{ is the constant function } 3$$
$$t_3 \text{ is } +1$$

This grammar has derivations

$$\begin{aligned}\langle S,0\rangle &\Rightarrow \langle A,2\rangle\langle B,1\rangle\langle C,0\rangle\\ &\Rightarrow a\langle A,0\rangle\langle B,2\rangle\langle C,1\rangle \Rightarrow a\langle A,1\rangle b\langle B,0\rangle\langle C,2\rangle \Rightarrow a\langle A,2\rangle b\langle B,1\rangle c\langle C,0\rangle\\ &\ \vdots\\ &\Rightarrow a^{n-1}\langle A,2\rangle b^{n-1}\langle B,1\rangle c^{n-1}\langle C,0\rangle\\ &\Rightarrow a^n b^{n-1}\langle B,3\rangle c^{n-1}\langle C,3\rangle \Rightarrow a^n b^n c^{n-1}\langle C,4\rangle \Rightarrow a^n b^n c^n\end{aligned}$$

and thus generates the language $\{a^n b^n c^n \mid n \geq 1\}$.

As a final example, illustrating again the complexity of derivation in string transformer grammars even when the rules are simple and so are the string transformers, consider a variant of Example 9.24.

Example 9.26 (Another simple grammar for a complex language). Consider the grammar $G = \langle\{S, A, B\}, \{a, b\}, P, S\rangle$ where the production rule set P is

$$\begin{aligned}S &\to AB\\ A &\to aA\\ A &\to a\\ B &\to bB\\ B &\to b\end{aligned}$$

This generates the regular language $\{a^n b^m \mid n, m > 0\}$. Now associate with each rule the same string transformer defined by the function interchanging non-terminal symbols $\{A \mapsto B, B \mapsto A\}$.

What language is now generated?

For leftmost derivations, we obtain the following prefixes of derivation paths.

$$\begin{aligned}S &\Rightarrow BA \Rightarrow bB\ldots\\ S &\Rightarrow BA \Rightarrow bAB \Rightarrow baA\ldots\\ S &\Rightarrow BA \Rightarrow bAB \Rightarrow baBA\ldots\end{aligned}$$

Given that B generates the language $(ba)^*b \cup (ba)^+$ and A generates the language $(ab)^*a \cup (ab)^+$, the leftmost derivation language will be

$$(ba)^*(b((ba)^*b \cup (ba)^+) \cup ba((ab)^*a \cup (ab)^+)).$$

For rightmost derivations, we obtain the prefixes

$$\begin{aligned}S &\Rightarrow BA \Rightarrow Aa\ldots\\ S &\Rightarrow BA \Rightarrow AaB \Rightarrow Bab\ldots\\ S &\Rightarrow BA \Rightarrow AaB \Rightarrow BabA\ldots\end{aligned}$$

Because of the third sequence, it is slightly trickier to derive the closed form. We have the recursion

$$Z(\mu) \Rightarrow A\mu a$$
$$Z(\mu) \Rightarrow B\mu ab$$
$$Z(\mu) \Rightarrow Z(\mu ab)$$

where $S \Rightarrow Z(\epsilon)$. The general form for μ is therefore $(ab)^*$ and hence the rightmost derivations generate the language

$$((ab)^*a \cup (ab)^+)(ab)^*a \cup ((ba)^*b \cup (ba)^+)(ab)^+$$

or more simply

$$(ab)^*a(ab)^*a \cup (ab)^+a \cup (ba)^+b \cup (ba)^+(ab)^+.$$

Now consider the case of unrestricted rule application. Let us again give some of the prefixes of derivation paths.

$$1\ S \Rightarrow BA \Rightarrow bB \ldots$$
$$2\ S \Rightarrow BA \Rightarrow bAB \ldots$$
$$3\ S \Rightarrow BA \Rightarrow AaB \ldots$$
$$4\ S \Rightarrow BA \Rightarrow Aa \ldots$$

Branches 1 and 4 are easy to close off. The first will yield words $b(ba)^*b \cup b(ba)^+$. Similarly, the fourth generates finite words $(ab)^*aa \cup (ab)^+a$.

What happens with the second and third branches? Let us expand the second branch.

$$2.1\ bAB \Rightarrow baA \ldots$$
$$2.2\ bAB \Rightarrow baBA \ldots$$
$$2.3\ bAB \Rightarrow bBbA \ldots$$
$$2.4\ bAB \Rightarrow bBb \ldots$$

Again, we know what will happen with the first and fourth branches — as we had above apart from being prefixed by the letter b. The branch (2.2) now recurses having generated prefix ba. Thus it will generate a set of finite words that are prefixed by $(ba)^*$. But what follows the prefix? More interestingly, consider branch 2.3. There is recursion again, however, the letter b is both prefixed and placed between the non-terminal symbol B and A — if one iterated just using this branch, then words of the form b^nBb^nA are derived. Let us expand the third branch (3) of the derivation.

$$3.1\ AaB \Rightarrow aaA \ldots$$
$$3.2\ AaB \Rightarrow aBaA \ldots$$
$$3.3\ AaB \Rightarrow BabA \ldots$$
$$3.4\ AaB \Rightarrow Bab \ldots$$

The first and fourth branches are straightforward. The second and third branches maintain the presence of non-terminal symbols A and B but switched in order, as in the case above.

It is instructive to see what happens with this string transformer grammar if we re-present it as a standard grammar using parameterised production rules. The new grammar will be directly constructed from the derivation tree, the beginnings of which we've written out above.

$$
\begin{aligned}
&S \to Z(\epsilon)\\[4pt]
&Z(\mu) \to b\mu B\\
&Z(\mu) \to ba\mu A\\
&Z(\mu) \to baZ(\mu)\\
&Z(\mu) \to bZ(\mu b)\\
&Z(\mu) \to bB\mu b\\
&Z(\mu) \to a\mu aA\\
&Z(\mu) \to aZ(\mu a)\\
&Z(\mu) \to Z(\mu ab)\\
&Z(\mu) \to B\mu ab\\
&Z(\mu) \to A\mu a\\[4pt]
&B \to b\\
&B \to bA\\
&A \to a\\
&A \to uB
\end{aligned}
$$

The non-terminal symbol Z is parameterised by a terminal string μ; $Z(\mu)$ represents derived words of the form $B\mu A$. The initial production S fires Z with an empty string, i.e. denoting the non-terminal pairing BA, the result of the first production application (after transformation) in the original grammar.

First of all, note the non-terminal symbol B yields the set of words $(ba)^*b \cup (ba)^+$. Similarly, A yields the set of words $(ab)^+ \cup (ab)^*a$. Thus, we can easily compute the words generated from the non-recursive rules for $Z(\mu)$. The other recursive cases are more interesting! A general pattern for words after a number of iterations from $Z(\mu)$ is as follows

$$((ba)^* b^{n_p} a^{m_p})^p Z(\mu\ ((ab)^* b^{n_p} a^{m_p})^p)$$

Note that we have the same indices occurring in the prefix substring as in the μ extension. We can thus substitute in the words generated through no further recursion of Z, obtaining the set of words

$$
\begin{aligned}
&\{((ba)^* b^{n_p} a^{m_p})^p\ b\ ((ab)^* b^{n_p} a^{m_p})^p\ ((ba)^* b \cup (ba)^+) \mid p, n_1, \ldots, n_p, m_1, \ldots, m_p \in \mathbb{N}\}\\
&\quad\cup\\
&\{((ba)^* b^{n_p} a^{m_p})^p\ ba\ ((ab)^* b^{n_p} a^{m_p})^p\ ((ab)^+ \cup (ab)^* a) \mid p, n_1, \ldots, n_p, m_1, \ldots, m_p \in \mathbb{N}\}\\
&\quad\cup\\
&\{((ba)^* b^{n_p} a^{m_p})^p\ b\ ((ba)^* b \cup (ba)^+)\ ((ab)^* b^{n_p} a^{m_p})^p\ b \mid p, n_1, \ldots, n_p, m_1, \ldots, m_p \in \mathbb{N}\}\\
&\quad\cup\\
&\{((ba)^* b^{n_p} a^{m_p})^p\ a\ ((ab)^* b^{n_p} a^{m_p})^p\ a\ ((ab)^+ \cup (ab)^* a) \mid p, n_1, \ldots, n_p, m_1, \ldots, m_p \in \mathbb{N}\}\\
&\quad\cup\\
&\{((ba)^* b^{n_p} a^{m_p})^p\ ((ba)^* b \cup (ba)^+)\ ((ab)^* b^{n_p} a^{m_p})^p\ ab \mid p, n_1, \ldots, n_p, m_1, \ldots, m_p \in \mathbb{N}\}\\
&\quad\cup\\
&\{((ba)^* b^{n_p} a^{m_p})^p\ ((ab)^+ \cup (ab)^* a)\ ((ab)^* b^{n_p} a^{m_p})^p\ a \mid p, n_1, \ldots, n_p, m_1, \ldots, m_p \in \mathbb{N}\}
\end{aligned}
$$

Thus we see that a simple context-free grammar with a simple string transformer has given rise to a very complex language indeed.

We have already shown that context-free string transformer grammars may generate non-context-free languages. Thus, we are led to ask: what is the expressivity of context-free string transformer grammars?

Theorem 9.27. *A deterministic Turing machine (TM) can be simulated by a context-free string transformer grammar.*

Proof. The essence of the simulation is to represent the tape of the TM by the currently derived word and the finite control of the TM is represented by the context-free grammar. Tape head movements are simulated by the string transformers. The non-terminal symbols of the regular grammar are triples formed from the state of the TM with the content of a tape cell and a marker 0 or 1 denoting whether the cell is under the head. The string transformers representing tape head movement may change the state component of a non-terminal symbol that is to come under the tape head following a left or right move.

The form of the string transformers here is, in fact, replacement of non-terminal symbols with non-terminal symbols but is not defined in terms of a function on the set of non-terminal symbols, as the determination of which symbol will replace a given one depends on its context in the string. This is the mechanism by which tape head movement is encoded.

In more detail, given a Turing Machine $TM = \langle Q, \Sigma, V, R, q_0, F\rangle$ where Q are the states of the finite control, Σ is the input alphabet, V is an alphabet, $R \subseteq (Q \times ((\Sigma \cup V) \times (\Sigma \cup V) \times \{L, R\}) \times Q)$ is the transition relation, and $F \in Q$ is the distinguished halt state, and q_0 is a starting state of the finite control. We assume an initial finitely populated tape $\sigma = \#_b\sigma_1 \ldots \sigma_n\#_e$ where $\#_b$ and $\#_e$ are beginning and end markers on the tape and $\sigma_i \in \Sigma$. We further assume that the tape head is positioned under σ_n.

We construct a context-free string transformer grammar with labelled non-terminal symbols

$$G = \langle N_{(\Sigma\cup V\cup\{\#_b,\#_e\})\cup\{S\},Q\times\{0,1\}}, \Sigma, T(\{L,R\} \times Q), P, S\rangle$$

where $N(\Sigma \cup V \cup \{\#_b, \#_e\})$ is a set of non-terminal symbols corresponding to the tape symbols of the Turing machine, i.e. the set $\{N_x \mid x \in (\Sigma \cup V \cup \{\#_b, \#_e\})\}$; S is a unique initial non-terminal symbol; the non-terminal labelling is drawn from $Q \times \{0, 1\}$, i.e. the states of the Turing machine paired with a 0 or 1 marker to denote whether the tape symbol corresponding to the non-terminal symbol is under the tape head; $T_{(\{L,R\}\times Q)}$ is a set of string transformers defined in terms of functions on labels, each one corresponding to either a left or right tape head movement into state $q \in Q$. Finally P is a set of context-free production rules related to the transitions of the Turing machine and is defined as follows.

1. The initial production for S generates the initial tape, i.e.

$$S \to \langle\sigma_1, q_0, 0\rangle \ldots \langle\sigma_n, q_0, 1\rangle$$

For ease of notation, in the labelled non-terminal symbol, we let σ_1 stand for non-terminal symbol N_{σ_1}.

2. For each TM rule of the form $q \overset{x,y,L}{\rightarrow} q'$ for $q \neq F$ and $x \neq \#_b$, we have the production rule
$$\langle x, q, 1\rangle \overset{t_{Lq'}}{\rightarrow} \langle y, q, 1\rangle$$
3. For each TM rule of the form $q \overset{x,y,R}{\rightarrow} q'$ for $q \neq F$ and $x \neq \#_e$, we have the production rule
$$\langle x, q, 1\rangle \overset{t_{Rq'}}{\rightarrow} \langle y, q, 1\rangle$$
4. For each TM rule of the form $q \overset{\#_b,y,L}{\rightarrow} q'$ for $q \neq F$, we have the production rule
$$\langle \#_b, q, 1\rangle \overset{t_{Lq'}}{\rightarrow} \langle \#_b, q, 0\rangle\langle y, q, 1\rangle$$
5. For each TM rule of the form $q \overset{\#_e,y,R}{\rightarrow} q'$ for $q \neq F$, we have the production rule
$$\langle \#_e, q, 1\rangle \overset{t_{Rq'}}{\rightarrow} \langle y, q, 1\rangle\langle \#_e, q, 0\rangle$$
6. Three termination rules that rewrite the non-terminal symbols to their corresponding terminal letter (tape symbol).
$$\begin{array}{ll} \langle x, F, p\rangle \rightarrow x & \text{for } x \neq \#_b, \#_e \\ \langle \#_b, F, p\rangle \rightarrow \epsilon & \\ \langle \#_e, F, p\rangle \rightarrow \epsilon & \end{array}$$

The label sequence transformation functions are defined as below.

$$\begin{array}{ll} t_{Lq} : (Q \times \{0,1\})^* \rightarrow (Q \times \{0,1\})^* & t_{Rq} : (Q \times \{0,1\})^* \rightarrow (Q \times \{0,1\})^* \\ t^1_{Lq} : (Q \times \{0,1\})^+ \rightarrow (Q \times \{0,1\})^+ & t^1_{Rq} : (Q \times \{0,1\})^+ \rightarrow (Q \times \{0,1\})^+ \end{array}$$

$$\begin{array}{ll} t_{Lq}(\langle\rangle) = \langle\rangle & t_{Rq}(\langle\rangle) = \langle\rangle \\ t_{Lq}(\sigma :: \langle s, 0\rangle) = t_{Lq}(\sigma) :: \langle q, 0\rangle & t_{Rq}(\langle s, 0\rangle :: \sigma) = \langle q, 0\rangle :: t_{Rq}(\sigma) \\ t_{Lq}(\sigma :: \langle s, 1\rangle) = t^1_{Lq}(\sigma) :: \langle q, 0\rangle & t_{Rq}(\langle s, 1\rangle :: \sigma) = \langle q, 0\rangle :: t^1_{Rq}(\sigma) \\ t^1_{Lq}(\langle\langle s, 0\rangle\rangle) = \langle\langle q, 1\rangle\rangle & t^1_{Rq}(\langle\langle s, 0\rangle\rangle) = \langle\langle q, 1\rangle\rangle \\ t^1_{Lq}(\sigma :: \langle s, 0\rangle) = t_{Lq}(\sigma) :: \langle q, 1\rangle & t^1_{Rq}(\langle s, 0\rangle :: \sigma) = \langle q, 1\rangle :: t_{Rq}(\sigma) \end{array}$$

We now claim that the string transformer grammar generates a word w if and only the Turing machine halts with tape contents $\#_b w \#_e$. The proof can be given by showing a bisimulation between the Turing machine and the string transformer grammar. We leave the details as an exercise for the interested reader.

9.6 Reactivity and derivation strategies

We note from some of the previous examples that we may use the labelling and associated transformations of string transformer grammars to mimic derivation strategies.

Can all derivation strategies be encoded this way? We answer this positively by first defining a suitable general notion of derivation strategy, then encoding through labels and relabelling functions a general scheme for constructing string transformer grammars.

Given a context-free grammar, let us uniquely name each of its production rules by, say, R_i, for $i \in \mathbb{N}$. For a given derivation sequence

$$S \Rightarrow \alpha_1 \Rightarrow \ldots \alpha_i \Rightarrow \ldots s$$

each derivation step can be labelled by the rule name applied at that step and the non-terminal position (amongst the non-terminal symbols of the currently derived string) at which the application took place. For example, given $G = \langle\{S, A, B\}, \{a, b\}, P, S\rangle$ with P the set of named production rules:

$$\begin{aligned} R_0 &: S \rightarrow AB \\ R_1 &: A \rightarrow a \\ R_2 &: A \rightarrow aA \\ R_3 &: B \rightarrow b \\ R_4 &: B \rightarrow bB \end{aligned}$$

This generates $\{a^m b^n \mid m, n \geq 1\}$. Here is an example of a (labelled) derivation sequence:

$$S \stackrel{R_0,1}{\Rightarrow} AB \stackrel{R_1,1}{\Rightarrow} aB \stackrel{R_3,1}{\Rightarrow} ab.$$

A leftmost derivation sequence is clearly a derivation sequence for which the non-terminal position is always 1, as above. Indeed, we can define a leftmost derivation strategy as one which only allows derivation sequences where the non-terminal position is always 1.

Of course, there are many other derivation strategies of interest. To move towards the more general situation let us consider a strategy for generating words of the form $a^{2n}b^n$ for $n \geq 1$ using the above production rules. An example of such a strategy is alternating the generation of 2 a's with 1 b, and so on. We can write this strategy explicitly in terms of derivation sequences, using only the last two rules used (at most). We can write this strategy as a function σ which takes a current derivation sequence and yields the set of rules which may be used next together with the position of the occurrence in the current string of the non-terminal symbol at which the rule is used. In the following definition of the function σ, let δ be an arbitrary derivation sequence, a dot denotes the concatenation of an item onto a derivation sequence, and ϵ is the empty sequence.

$$\begin{aligned} &\sigma(\epsilon) = \{(R_0, 1)\} \\ &\sigma((R_0, 1)) = \{(R_2, 2)\} \\ &\sigma(\delta.(R_2, 1).(R_2, 1)) = \{(R_4, 2)\} \\ &\sigma(\delta.(R, j).(R_2, 1)) = \{(R_1, 1), (R_2, 1)\} \qquad \text{for } R \neq R_2 \\ &\sigma(\delta.(R_1, 1)) = \{(R_3, 2)\} \\ &\sigma(\delta.(R_4, 2)) = \{(R_2, 1)\} \end{aligned}$$

For all other sequences the result is the empty set.

We are now in a position to introduce the general notion of a *derivation strategy* in a derivation tree of a context-free grammar.

Definition 9.28. *Consider a context-free grammar G with uniquely named production rules (with Rule the set of rule names in G). A* derivation strategy *(or simply a* strategy*) of G is a function*

$$\sigma : (Rule \times \mathbb{N}_1)^* \rightarrow \mathcal{P}_f(Rule \times \mathbb{N}_1).$$

Here $\mathcal{P}_f$ is the finite powerset function and $\mathbb{N}_1$ the positive integers. The argument of the strategy σ is a derivation sequence — the history of the derivation so far – recording the (possibly empty) sequence of named rules and the non-terminal positions in the strings at which the rule is applied. Given a derivation sequence, a strategy yields a (possibly empty) finite set of pairs (R, i) consisting of a rule name R and a non-terminal position i in the current string. For such a pair (R, i), a possible next step in the derivation is to apply the rule named R to the non-terminal at position i.

A derivation sequence δ is said to be a derivation in strategy σ if, at any position j in δ, $(0 \leq j \leq n)$, we have $\delta_j \in \sigma(\delta|_{j-1})$, where δ_j is the j-th step in δ and $\delta|_{j-1}$ is the initial segment of δ up to the $(j-1)$-th entry.

A strategy σ of grammar G is said to generate a word w if there is a derivation of w in σ. The language generated by a strategy is the set of all its generated words.

As in games, strategies may depend on only certain aspects of the history of a derivation (e.g. on the number of rule applications, as in the examples above) or, in fact, may be *history-free*, in which case they do not depend on the history of a derivation at all but only on the current string. In these cases, the strategy, as a function, is constant over certain collections of inputs.

Example 9.29 (History-free strategies). A *leftmost* derivation strategy σ for a context-free grammar G (with uniquely named rules) is history-free, and $(R, i) \in \sigma(s)$ for derived string s iff $i = 1$ and R is the name of a rule of G for rewriting the non-terminal at position 1 in s.

A *leftish* derivation strategy σ for a context-free grammar G (with uniquely named rules) is history-free, and $(R, i) \in \sigma(s)$ for derived string s iff for all j, $j < i$, there are no rules in G for rewriting the non-terminal at position j, and R is the name of a rule for rewriting the i-th non-terminal symbol.

This notion of a derivation strategy is *general* in the following sense.

Theorem 9.30. *For a context-free grammar, any set Δ of (finite or infinite) derivation sequences defines a strategy σ such that $\delta \in \Delta$ iff δ is a derivation sequence in σ.*

We now turn to the main result of this section, showing that any strategy for deriving words in a context-free grammar corresponds exactly to a string transformer grammar with unrestricted use of production rules. This result both shows the expressivity of string transformer grammars and also provides a general construction of such grammars for various languages.

Theorem 9.31. *Let G be a context-free grammar with a finite set of uniquely named production rules and σ a derivation strategy of G. Then there is a context-free string transformer grammar (possibly with an infinite set of rules) whose unrestricted rule application generates the language of σ. In fact, we may construct the context-free string transformer grammar so as to exactly simulate the strategy σ (in the sense of derivation steps being in 1-1 correspondence).*

Proof. The key to the proof is the construction of a context-free string transformer grammar G' using labelled non-terminal symbols, where the labels code the current history together with a non-terminal position. We then ensure that G' has a rule for a particular labelled non-terminal just when the strategy σ allows that rule to be applied. This, together with suitable label transformations, ensures that the derivations of G' correspond exactly to those of the strategy σ of grammar G.

Thus, the non-terminal symbols of G' consist of the non-terminal symbols of G labelled with a pair (δ, i) where δ is a derivation sequence of σ and i is a position of a non-terminal in the string derived by δ. This may produce a finite or an infinite set of labelled non-terminal symbols depending on whether there is a finite or infinite number of derivation sequences of σ.

For the rules of G', consider a derivation sequence δ in strategy σ. For each $(R, i) \in \sigma(\delta)$ with rule $X \to s$ in G named by R, we create the rule of G',

$$\langle X, (\delta, i)\rangle \xrightarrow{t} \tilde{s}$$

where $\tilde{s}$ consists of the string s with each non-terminal labelled by the derivation δ and its non-terminal position in s. The transformation t acts on labels by 'advancing the history' and adding the current position of the non-terminal, that is, for the non-terminal at position j in the current string,

$$t(\delta, \_) = (\delta.(R, i), j).$$

The initial symbol of G' is $\langle S, (\epsilon, 1)\rangle$ where S is the initial symbol of G and ϵ is the empty derivation sequence.

It is straightforward to verify that, for each word w, there is a bijection between derivation sequences in strategy σ of word w and unrestricted derivations in G' of word w.

Whilst, in general, this construction of a string transformer grammar may yield a grammar with an infinite set of rules, we shall see that for many natural strategies for generating languages, not only is the resulting grammar finite, but it is a succinct and useful presentation of the language. In other cases, the infinite set of rules is presented as a finite set of rule schema (i.e. parameterised rules).

We give examples of both these cases, constructing string transformer grammars from strategies in context-free grammars.

Example 9.32 (An alternation strategy as a string transformer grammar). We revisit the strategy described above for the language $\{a^{2n}b^n \mid n \geq 1\}$. The strategy is one of strict alternation between generating pairs of as and single bs. This strategy depends only on the last two (at most) rules used, and so we can label the non-terminal symbols in the constructed string transformer grammar with the necessary one or two previous rules:

$$\begin{array}{ll}
\langle S, (\epsilon, 1)\rangle \xrightarrow{t_0} \langle A, (\epsilon, 1)\rangle\langle B, (\epsilon, 2)\rangle & \\
\langle A, ((R_0, 1), 1)\rangle \xrightarrow{t_1} a\langle A, ((R_0, 1), 1)\rangle & \\
\langle B, ((R_2, 1).(R_2, 1), 2)\rangle \xrightarrow{t_2} b\langle B, ((R_2, 1).(R_2, 1), 2)\rangle & \\
\langle A, ((R, j).(R_2, 1), 1)\rangle \xrightarrow{t_3} a \qquad R \neq R_2 & \\
\langle A, ((R, j).(R_2, 1), 1)\rangle \xrightarrow{t_4} a\langle A, ((R, j).(R_2, 1), 1)\rangle & R \neq R_2 \\
\langle B, ((R_1, 1), 2)\rangle \xrightarrow{t_5} b & \\
\langle A, ((R_4, 2), 1)\rangle \xrightarrow{t_6} a\langle A, ((R_4, 2), 1)\rangle &
\end{array}$$

with label transformations:

$$\begin{array}{l}
t_0(\epsilon, _) = ((R_0, 1), j) \\
t_1((R_0, 1), _) = ((R_0, 1).(R_2, 1), j) \\
t_2((R_2, 1).(R_2, 1), _) = ((R_4, 2), j) \\
t_3((R, j)(R_2, 1), _) = ((R_1, 1), j) \\
t_4((R, j)(R_2, 1), _) = ((R_2, 1).(R_2, 1), j) \\
t_5((R_1, 1), _) = ((R_3, 2), j) \\
t_6((R_4, 2), _) = ((R_4, 2).(R_2, 1), j)
\end{array}$$

with j =1 for non-terminal symbol A and $j = 2$ for B.

Simplifying the labels reduces this grammar to the set of rules:

$$\begin{array}{l}
S \rightarrow \langle A, 0\rangle\langle B, 0\rangle \\
\langle A, 0\rangle \xrightarrow{t_0} a\langle A, 0\rangle \\
\langle B, 22\rangle \xrightarrow{t_1} b\langle B, 22\rangle \\
\langle A, 2\rangle \xrightarrow{t_2} a \\
\langle A, 2\rangle \xrightarrow{t_3} a\langle A, 2\rangle \\
\langle B, 1\rangle \xrightarrow{t_4} b \\
\langle A, 4\rangle \xrightarrow{t_5} a\langle A, 4\rangle
\end{array}$$

with label transformations: $t_0(0) = 2, t_1(22) = 4, t_2(2) = 1, t_3(2) = 22, t_4(1) = 3, t_5(4) = 2$.

What about other strategies for the same language $\{a^{2n}b^n \mid n \geq 1\}$ using the same grammar?

Example 9.33 (Another string transformer construction). As another strategy for the language $\{a^{2n}b^n \mid n \geq 1\}$, we could allow arbitrary interleaved generation of as and bs but allow termination only when the correct number of as and bs are generated. This again provides a derivation strategy which, using the above construction, produces the following equivalent string transformer grammar.

$$\begin{array}{ll}
\langle S, ([0,0,0,0,0],1)\rangle \xrightarrow{t_0} \langle A, ([0,0,0,0,0],1)\rangle\langle B, ([0,0,0,0,0],1)\rangle & \\
\langle A, ([1,0,m,0,n],1)\rangle \xrightarrow{t_1} a\langle A, ([1,0,m,0,n],1)\rangle & m,n \geq 0 \\
\langle B, ([1,0,m,0,n],2)\rangle \xrightarrow{t_2} b\langle B, ([1,0,m,0,n],2)\rangle & m,n \geq 0 \\
\langle A, ([1,0,2n-1,0,n-1],1)\rangle \xrightarrow{t_3} a \quad n \geq 1 & \\
\langle B, ([1,0,2n-1,0,n-1],2)\rangle \xrightarrow{t_4} b \quad n \geq 1 & \\
\langle B, ([1,1,m,0,n],1)\rangle \xrightarrow{t_5} b & \\
\langle A, ([1,0,m,1,n],1)\rangle \xrightarrow{t_6} a &
\end{array}$$

Here the arrays encode information about the derivation so far: The ith entry records the number of applications of rule R_i of the grammar at the beginning of this section. The transformations of labels extend the count according to which rule has just been used:

$$\begin{array}{l}
t_0([0,0,0,0,0],_) = ([1,0,0,0,0],j) \\
t_1([1,0,m,0,n],_) = ([1,0,m+1,0,n],j) \\
t_2([1,0,m,0,n],_) = ([1,0,m,0,n+1],j) \\
t_3([1,0,2n-1,0,n-1],_) = ([1,1,2n-1,0,n-1],j) \\
t_4([1,0,2n-1,0,n-1],_) = ([1,0,2n-1,1,n-1],j) \\
t_5([1,1,2n-1,0,n-1],_) = ([1,1,2n-1,1,n-1],j) \\
t_6([1,0,2n-1,1,n-1],_) = ([1,1,2n-1,1,n-1],j)
\end{array}$$

with j =1 for non-terminal symbol A and $j = 2$ for B.

Note that this is a context-free string transformer grammar with a set of *parameterised* rules, i.e. a finite schema for an infinite set of rules, one for each valid m and n.

We now turn to another example – the running example of a non-context-free language: $\{a^n b^n c^n \mid n \geq 1\}$

Example 9.34 ($a^n b^n c^n$ yet again). Consider the context-free grammar G with named rules:

$$\begin{array}{l}
R_0 : S \to AC \\
R_1 : A \to aAb \\
R_2 : A \to ab \\
R_3 : C \to cC \\
R_4 : C \to c
\end{array}$$

A strategy for generating the language $\{a^n b^n c^n \mid n \geq 1\}$ is to alternate use of rules R_1 and R_3. This derivation strategy σ of G is defined as follows. For any derivation sequence δ:

$$\begin{array}{l}
\sigma(\epsilon) = \{(R_0,1)\} \\
\sigma((R_0,1)) = \{(R_1,1),(R_2,1)\} \\
\sigma(\delta.(R_1,1)) = \{(R_3,2)\} \\
\sigma(\delta.(R_2,1)) = \{(R_4,2)\} \\
\sigma(\delta.(R_3,2)) = \{(R_1,1),(R_2,1)\}
\end{array}$$

This strategy generates the language $\{a^n b^n c^n \mid n > 0\}$.

Notice that this strategy depends only on the previous rule applied, thus we may simplify the labelling of non-terminal symbols using the previously applied rule in place of the entire derivation history. Thus, labels consist of a rule name and the non-terminal position of its application, together with the current position of the non-terminal symbol.

Following the construction in the proof above, we construct a string transformer grammar G' with labelled non-terminal symbols which mimics this strategy to generate the same language. The rules of G' are:

$$
\begin{aligned}
&\langle S, (\epsilon, 1)\rangle \xrightarrow{t_0} \langle A, (\epsilon, 1)\rangle\langle C, (\epsilon, 1)\rangle\\
&\langle A, ((R_0, 1), 1)\rangle \xrightarrow{t_1} a\langle A, ((R_0, 1), 1)\rangle b\\
&\langle A, ((R_0, 1), 1)\rangle \xrightarrow{t_2} ab\\
&\langle C, ((R_1, 1), 2)\rangle \xrightarrow{t_3} c\langle C, ((R_1, 1), 1)\rangle\\
&\langle C, ((R_2, 1), 2)\rangle \xrightarrow{t_4} c\\
&\langle A, ((R_3, 2), 1)\rangle \xrightarrow{t_5} a\langle A, ((R_3, 2), 1)\rangle b\\
&\langle A, ((R_3, 2), 1)\rangle \xrightarrow{t_6} ab
\end{aligned}
$$

The string transformations t_m act on the labels of non-terminal symbols as follows, for the j-th non-terminal symbol in a string:

$$
\begin{aligned}
&t_0(\epsilon, \_) = ((R_0, 1), j)\\
&t_1((R_0, 1), \_) = ((R_1, 1), j)\\
&t_2((R_0, 1), \_) = ((R_2, 1), j)\\
&t_3((R_1, 1), \_) = ((R_3, 2), j)\\
&t_4((R_1, 1), \_) = ((R_4, 2), j)\\
&t_5((R_3, 2), \_) = ((R_1, 1), j)\\
&t_6((R_3, 2), \_) = ((R_2, 1), j)
\end{aligned}
$$

By observing the pattern of the non-terminal labels in the rules of this grammar, we may simplify it considerably, using rule numbers as labels:

$$
\begin{aligned}
&S \rightarrow \langle A, 0\rangle\langle C, 0\rangle\\
&\langle A, 0\rangle \xrightarrow{t_1} a\langle A, 0\rangle b\\
&\langle A, 0\rangle \xrightarrow{t_2} ab\\
&\langle C, 1\rangle \xrightarrow{t_3} c\langle C, 1\rangle\\
&\langle C, 2\rangle \xrightarrow{t_4} c\\
&\langle A, 3\rangle \xrightarrow{t_5} a\langle A, 3\rangle b\\
&\langle A, 3\rangle \xrightarrow{t_6} ab
\end{aligned}
$$

with transformations $t_1(0) = 1, t_2(0) = 2, t_3(1) = 3, t_4(1) = 4, t_5(3) = 1$, $t_6(3) = 2$.

Example 9.35 (A context-sensitive grammar). In [142], Rosenkrantz uses the following context-sensitive language to demonstrate a programmed context-free grammar with both success and failure continuation rule sets:

$$\{\text{bin}(n)ha^n \mid n \in \mathbb{N}\}$$

where bin(n) is a binary representation (with, say, leftmost digit as the most significant) of the positive integer n. Let us first give an unrestricted grammar for this language. Take $G_u = \langle\{S, T, T', D, Z, F\}, \{1, 0, h, a\}, P, S\rangle$ with P the set of rules

$$\begin{aligned}
S &\rightarrow TZ\\
T &\rightarrow 1T'aD\\
T &\rightarrow 0T'D\\
T' &\rightarrow 1T'aD\\
T' &\rightarrow 0T'D\\
T' &\rightarrow hF\\
Da &\rightarrow aaD\\
DZ &\rightarrow Z\\
Fa &\rightarrow aF\\
FZ &\rightarrow \epsilon
\end{aligned}$$

G_u thus has derivations such as:

$$\begin{aligned}
S &\Rightarrow TZ \Rightarrow 1T'aDZ \Rightarrow 11T'aDaDZ \Rightarrow 11T'aaaDDZ \Rightarrow 11hFaaaDDZ\\
&\Rightarrow 11haFaaDDZ \Rightarrow 11haaFaDDZ \Rightarrow 11haaaFDDZ \Rightarrow 11haaaFDZ\\
&\Rightarrow 11haaaFZ \Rightarrow 11haaa
\end{aligned}$$

Consider now the following context-free grammar with named rules

$$G_{cf} = \langle\{S, A\}, \{1, 0, h, a\}, \{r_0, r_1, r_2, r_3, r_4\}, P, S\rangle$$

with rules in P as

$$\begin{aligned}
r_0 &: S \rightarrow 1SA\\
r_1 &: S \rightarrow 0S\\
r_2 &: S \rightarrow h\\
r_3 &: A \rightarrow AA\\
r_4 &: A \rightarrow a
\end{aligned}$$

The language generated by this context-free grammar clearly contains the language of G_u. So by defining an appropriate strategy for G_{cf}, we can obtain the language of G_u.

9.6.1 Switching grammars and string transformers

We now consider the relationship between switching grammars, introduced earlier in the chapter, and string transformer grammars.

Theorem 9.36. *Every context-free switching grammar can be exactly simulated by a context-free string transformer grammar, i.e. there is a context-free string transformer grammar whose derivation steps correspond exactly to those of the switching grammar. In particular, the class of languages generated by context-free switching grammars is contained in that of context-free string transformer grammars.*

Proof. Let $G = \langle N, \Sigma, I, P, S, i_0 \rangle$ be a context-free switching grammar. For each $i \in I$, we form a set of string transformer rules $\tilde{P}_i$ as follows: For each rule $X \rightarrow \gamma, j$ in P_i form the rule $\langle X, i \rangle \xrightarrow{t} \tilde{\gamma}$ in $\tilde{P}_i$ where $\tilde{\gamma}$ is γ with each non-terminal symbol Y replaced by $\langle Y, i \rangle$, and the string transformer t is defined by the function on non-terminal symbols $t(\langle Y, k \rangle) = \langle Y, j \rangle$.

Now form the set of production rules of a string transformer grammar $\tilde{P} = \bigcup_{i \in I} \tilde{P}_i$. It is easy to verify that this string transformer grammar exactly simulates the original switching grammar.

We have shown that any derivation strategy may be represented (by an exact simulation) by a context-free string transformer grammar with unrestricted rule application. Context-free switching grammars have the same degree of expressivity for arbitrary derivation strategies but the encoding of strategies is different and the encoding in string transformer grammars does not factor through the construction for switching grammars. A difficulty in both cases is to identify the n-th non-terminal symbol (which is used in the definition of strategies) in a string after a substitution has taken place. With string transformers, we relabel the non-terminal symbols after a rule applications, for switching grammars this option is not available and we have to code, within the labels, information about relative positions of the non-terminal symbols. Switching grammars as representations of strategies tend to be less succinct than string transformer grammars, partly because of this more elaborate labelling of non-terminal symbols.

Theorem 9.37. *Let G be a context-free grammar with uniquely named production rules and σ a derivation strategy of G. Then there is a context-free switching grammar with unrestricted rule application which exactly simulates σ.*

Proof. Consider a valid derivation δ of string s in strategy σ with non-empty $\sigma(\delta)$. We proceed by induction on the length of δ. Assume that there is a switching grammar with rule sets indexed by initial substrings of δ (not including δ) which exactly simulates the steps of δ to produce a string $\tilde{s}$. Form a set of rules P_δ as follows. For each $\langle l, n \rangle \in \sigma(\delta)$, let $X \rightarrow \gamma$ be the rule labelled l. Form switching rule $\langle X, \lambda \rangle \rightarrow \tilde{\gamma}, \delta'$ in P_δ where λ is the label of the n-th non-terminal symbol in $\tilde{s}$, $\tilde{\gamma}$ is γ with non-terminal symbol Y at the j position in γ replaced by $\langle Y, \lambda.j \rangle$ and $\delta' = \delta.\langle l, n \rangle$.

Then for all valid derivation sequences δ of σ, the indexed set of rule sets P_δ, together with initial symbol $\langle S, 0 \rangle$ where S is the initial symbol of G, and initial set of rules $\sigma([])$, forms a switching grammar that exactly simulates the strategy σ of G.

Example 9.38 (The language $\{a^n b^n c^n \mid n \geq 1\}$). Consider the context-free grammar G with rules

$$\begin{array}{ll} 1: & S \to AX \\ 2: & X \to bXc \\ 3: & X \to bc \\ 4: & A \to AA \\ 5: & A \to a \end{array}$$

As a strategy σ for $a^n b^n c^n$, for $n \geq 1$, consider alternating between extending the As and generating Xs, then, when there are no Xs, finally reducing As to as in an arbitrary order. Thus:

$\sigma([]) = \{\langle 1, 1\rangle\}$
$\sigma([\langle 1, 1\rangle]) = \{\langle 4, 1\rangle, \langle 5, 1\rangle\}$
$\sigma(\delta.\langle 4, 1\rangle) = \{\langle 5, |\delta| + 1\rangle\}$
$\sigma(\delta.\langle 5, i\rangle) = \{\langle 4, 1\rangle, \langle 3, i\rangle\}$
$\sigma(\delta.\langle 3, i\rangle) = \{\langle 5, j\rangle \mid 1 \leq j \leq |\delta|\}$
$\sigma(\delta.\langle 5, i\rangle) = \{\langle 5, j\rangle \mid 1 \leq j \leq k\}$ where k is the number of A symbols in the string generated by $\delta.\langle 5, i\rangle$

We consider the first few steps of the construction using the derivation sequence $\delta = S \Rightarrow AX \Rightarrow AAX$. The initial set is

$$P_{[]} = \{\langle S, 0\rangle \to \langle A, 0.1\rangle\langle X, 0.2\rangle, [\langle 1, 1\rangle]\}.$$

The corresponding step in the switching grammar for $S \Rightarrow AX$ is therefore $\langle S, 0\rangle \Rightarrow \langle A, 0.1\rangle\langle X, 0.2\rangle$. For the next step, the strategy gives two rules, $A \to AA$ or $A \to a$. Thus, the ruleset for the switching grammar is

$$\begin{aligned} P_{[\langle 1,1\rangle]} = \{\ & \langle A, 0.1\rangle \to \langle A, 0.1.1\rangle\langle A, 0.1.2\rangle, [\langle 1, 1\rangle, \langle 4, 1\rangle], \\ & \langle A, 0.1\rangle \to a, [\langle 1, 1\rangle, \langle 5, 1\rangle]\}. \end{aligned}$$

Then the derivation δ, represented in the switching grammar, yields the string $\langle A, 0.1.1\rangle\langle A, 0.1.2\rangle\langle X, 0.2\rangle$. Notice how the labels are used to code the tree of substitution positions so that the correct rules may be applied according to the form of the labelled non-terminal symbol.

9.7 Embedded reactivity

We now turn to examples of reactive grammars where the reactivity, instead of being associated with rules of the grammar, is determined by the generated string or parts of the string. The presentation and formalisation of such grammars is not straightforward because what is meant by 'parts of a string' depends, in general, on the parsing of the string, i.e. on the derivation sequence used to generate it. Moreover, the effect of reactivity will depend again on the derivation sequence and how subtrees in a parse tree are related to each other.

Let us begin by looking at some examples of embedded reactivity.

Example 9.39 (Correction grammars). Consider the following example text:

'*Tom likes playing basketball, sorry I mean football.*'

Here, the phrase '*sorry I mean X*' is a reactive component of the text, that is, it describes a transformation to be performed on the text. In this case, the transformation is to replace the word '*basketball*' with '*football*' and remove the reactive component, resulting in the string:

'*Tom likes playing football.*'

Let us look at the process in more detail. How is the '*X*' in '*sorry I mean X*' matched to part of the remaining text? Clearly, it ought to have the same 'phrase type' which in this case is a noun phrase (indeed it is a noun). But there are two noun phrases in the text '*Tom likes playing basketball.*' Which to choose? Both semantics and proximity may have a role in determining the action to be associated with a reactive component.

Reactive components may interact. Consider the example text:

'*Jack loves Mary. Mary rejected Jack, sorry I mean the man in blue, sorry I mean red, sorry I mean accepted.*'

Of course, as a natural language text, this is rather contrived, but as an example of an embedded reactivity, it illustrates the complex phenomena that may be involved. One possible resolution of this (i.e. determination of the actions associated with the reactive components) is the text:

'*Jack loves Mary. Mary accepted the man in red.*'

Other resolutions are possible, for example:

'*The man in red loves Mary. Mary accepted the man in red.*'

When a string has multiple derivations, i.e. ambiguity in parsing, then the reactivity may depend upon the derivation. For example:

'*British left waffles on Falkland Islands, sorry I mean toast!*'

We are not proposing embedded reactive grammars as a technique for analysing such natural language phrases as in the example above. Rather, we are using the example to illustrate embedded reactivity and its analysis.

Example 9.40 (Spelling reform). A widely quoted text, purporting to be an incremental spelling reform, is:

> For example, in Year 1 that useless letter 'c' would be dropped to be replased either by 'k' or 's,' and likewise 'x' would no longer be part of the alphabet. The only kase in which 'c' would be retained would be the 'ch' formation, which will be dealt with later. Year 2 might reform 'w' spelling, so that 'which' and 'one' would take the same konsonant, wile Year 3 might well abolish 'y' replasing it with 'i' and Iear 4 might fiks the 'g/j' anomali wonse and for all. Jenerally, then, the improvement would kontinue iear bai iear with Iear 5 doing awai with useless double konsonants, and Iears 6-12 or so modifaiing vowlz and the rimeining voist and unvoist konsonants. Bai Iear 15 or sou, it wud fainali bi posibl tu meik ius ov thi ridandant letez 'c,' 'y'

> and 'x"–bai now jast a memori in the maindz ov ould doderez–tu riplais 'ch,' 'sh,' and 'th' rispektivli. Fainali, xen, aafte sam 20 iers ov orxogrefkl riform, wi wud hev a lojikl, kohirnt speling in ius xrewawt xe Ingliy-spiking werld.

The reactivity here is the replacement of letters, or letter combinations. Notice how the action associated with a reactive component may depend upon the phonetics as well as on letters and letter combinations. Notice also that, unlike the previous example where the action was retrospective and we are given the phrase before the action, here it is applied to subsequent text and we are given the result of the application. Again, reactive components are acted on by other reactive components.

Example 9.41 (Grammar and word order). Consider the example text:

> Tom loves Mary. From now on, subjects switch with objects. Pizza likes Jack. From now on, verbs switch with objects. Cat Mary loves.

Here again there are reactive components, now acting on word order according to phrase classes occupied by words. Like the previous example, this is forward action. Notice also that the reactive component '*verbs switch with objects*' is subject to the component '*subjects switch with objects*' and so the actual action is '*objects switch with verbs*', which, in this case, is the same action but, in general, would be different. Again, this brings out the complexity of embedded reactivity, where the reactivity can act on other reactive components, or indeed allows the possibility of two reactive components acting on each other.

9.7.1 Formalising grammars with embedded reactivity

We begin the process of formalising a general notion of an *embedded reactive grammar*. To do so, we first formalise parse trees in terms of derivations.

We introduce a labelling of non-terminals in a derivation sequence using lists of numbers. For example, consider a derivation step

$$aAcB \to aaCAcB$$

which is the result of replacing A using a rule

$$r : A \to aCA.$$

If the original non-terminal symbol A is labelled p then the C in the result string is labelled $p.1$ and the A in the resultant string is labelled $p.2$; the label of B remains unchanged. The initial symbol is labelled with the empty sequence. Each derivation step is denoted by a pair (p, r) consisting of a non-terminal label p (the label of the non-terminal to be replaced) and a rule name r (the rule used).

Definition 9.42. *For derivation steps $\pi = (p, r)$ and $\pi' = (p', r')$, we say that (p', r')* extends *(p, r) iff sequence p is a prefix of p'.*

For a derivation sequence, $\delta = \pi_0, \ldots, \pi_n$ a subsequence $\pi_{i_0}, \ldots \pi_{i_k}$ is compatible *iff (1) $\pi_{i_{j+1}}$ extends π_{i_j} for $0 \le j \le k-1$, and (2) no intermediate derivations extend any of the elements of the subsequence.*

In a derivation sequence δ of string s, the collection of compatible subsequences which are left closed (i.e. may not be extended leftwards) forms a tree (i.e. is prefix closed). This is the parse tree parsetree(δ, s) *determined by the derivation sequence. A* path *p in a parse tree (as a list of numbers) determines a subtree* subtree(δ, s, p). *The set of paths in a parse tree, we denote by* paths(δ, s).

Definition 9.43 (Embedded reactive grammar). *An* embedded reactive grammar *consists of a set of non-terminal symbols N, an initial non-terminal $S \in N$, a set of terminal symbols Σ, a set of (production) rules P as in Definition 9.7, together with for each derivation sequence δ of string s a set of reactive components $R(\delta, s) \subseteq$* paths(δ, s). *In addition, there is a transformation of strings T, which associates with each reactive component $p \in R(\delta, s)$ a transformed string $s' = T(p, \delta, s)$* together with *a derivation sequence δ' of s'.*

The definition is fairly general. It allows reactive components to contain non-terminals (i.e. allows some forms of 'reactive patterns'). The transformations associated with a reactive component are unrestricted – any string may be the result of such a transformation. Note that the result of a transformation is not simply a string but a derivation sequence of a string. Thus, transformations associated with reactive components map derivation sequences to derivation sequences. However, the transformation is functional – a unique string and its derivation is determined by a reactive component in a derivation of a string. The recognition of reactive components is, in general, dependent upon the context, i.e. on the string in which it is embedded. For example, in the case of text correction, a phrase '*sorry I mean football*' may be considered not to be a reactive component if, for example, it appears in a direct quote.

In defining a grammar, it is usual to define what is meant by a derivation step. However, this is no longer a simple matter because embedded reactivities move us between derivation sequences. Several options are available which correspond to strategies for applying the actions associated with reactive components. One is to first generate a terminal string using the underlying grammar, then apply the actions of any reactive components (in any order, each time producing a new derivation sequence), to end possibly with a word with no reactive components (this will not always occur - it depends on the transformations). Another possibility is to interleave applications of production rules with reactive transformations. In either case, we need a new term to denote the combined processes of applying production rules and reactive transformations.

Definition 9.44. *A string s' with derivation δ' is a* single step result *from string s with derivation δ, written $(\delta, s) \to (\delta', s')$ in an embedded reactive grammar iff either (1) this is a single step derivation in the underlying grammar, or (2) (δ', s') is the result of a transformation associated with a reactive component of (δ, s).*

Chains of such single steps arise from various strategies for applying grammar productions and reactive transformations.

Let us now consider how one of the examples above is an embedded reactive grammar in this sense.

Example 9.45 (Correction grammars revisited). We revisit the example of correction grammars above (Example 9.39) and show how it is an example of an embedded reactive grammar.

The reactive components are of the form ρ = '*sorry I mean w*' where w is a terminal word generated from a non-terminal X. In the tree $t = \mathsf{parsetree}(\Delta, s)$, the action of a subtree determining the reactive component ρ is the replacement of a suitably chosen subtree of t and the removal of subtree ρ from t. The resultant tree must itself be a parse tree in the underlying grammar, i.e. have an associated derivation Δ'. Which subtree is replaced is a choice determined by our understanding of the notion of text correction.

This is one form of an embedded reactive grammar that describes the correction process. Notice that we only replace with terminal words. This is not always necessary (depending upon how trees are matched in the correction phase) and an alternative formulation would allow reactive components to be of the form '*sorry I mean s*' where $s \in (N \cup \Sigma)^*$.

Thus, in the case of correction grammars, we present the string before any actions of reactive components are undertaken, and then consider the result of these actions. For the other two examples above, the text presented is the result of applying the actions. An analysis of the text requires us to find a chain of single steps (consisting of applying grammar rules and reactive transformations) which result in the given text. Notice also that, in these cases, the reactive transformation leaves in place the reactive component (i.e. the instructions for modifying the text) instead of deleting it.

9.8 Reactivity and contrary-to-duty

The previous sections entertained the idea of reactive rules, namely when a certain rewrite rule r_1 is used then another rule r_2 gets activated (or de-activated) and is available (resp. no longer available) to use. This concept allows us to achieve two more goals:

1. Introduce the idea of contrary-to-duty rules into grammars.
2. Use the expanded types of grammars to provide proof theory to contrary-to-duty obligations in deontic logic.

Let us begin with an example.

Example 9.46 (A simple grammar). We introduce a simple grammar. Let the non-terminal letters be $\infty, 0, x_1, \neg x_1, x_2, \neg x_2$.

The letters "$\neg x$" are considered atomic. The significance of the prefix "$\neg$" will become apparent when we make a connection with deontic logic, see [38, 81].

Consider the following rules:

$$\begin{aligned} r_{\infty,0} &: \quad \infty \to 0 \\ r_{0,x_1} &: \quad 0 \to x_1 \\ r_{0,\neg x_1} &: \quad 0 \to \neg x_1 \\ r_{\neg x_1 x_2} &: \quad \neg x_1 \to x_2 \\ r_{\neg x_1,\neg x_2} &: \neg x_1 \to \neg x_2. \end{aligned}$$

Assume our grammar has the starting letter as ∞ and has the rules $r_{0,\infty}$ and r_{∞,x_1} active and all the other rules not active and that this does not change during derivation. Our starting point is always ∞. We can derive in this grammar

$$\infty \to 0 \to x_1$$

and no more.

So the above reactive grammar is equivalent to the traditional context-free grammar with the rules

$$\infty \to 0 \text{ and } 0 \to x_1.$$

We can ask ourselves what if we are naughty and decide to use rules which are not active and derive words using them?

Suppose we use the following derivation:

$$\infty \to 0 \to \neg x_1.$$

Having done that we can continue to illegitimately derive either $\neg x_1 \to x_2$ or $\neg x_1 \to \neg x_2$.

A contrary-to-duty approach might say that if the inactive rule $0 \to \neg x_1$ is used then activate the rule $\neg x_1 \to x_2$. So the way to understand this intuitively is as follows:

(a) From 0 you should go to x_1 not to $\neg x_1$.
(b) If however you do illegitimately go to $\neg x_1$ then at least be good and continue to x_2 and not to $\neg x_2$.

Let us write the reactive grammar for this:

1. ∞ starting point.
   $r_{\infty,0}$, r_{∞,x_1} active. All other rules inactive.
2. If $r_{0,\neg x_1}$ is used then activate rule $r_{\neg x_1,x_2}$.
   Note that (2) leaves rule $0 \to x_1$ still active and so it is possible also to derive x_1.
   We could have used (2*) instead of (2).

2*. If rule $r_{0\neg x_1}$ is used then deactivate rule r_{0,x_1} and activate rule $r_{\neg x_1,x_2}$.

To see the connection with deontic logic, consider the next example.

Example 9.47 (Local housing estate). The following is a set of rules for a local housing estate.

1. It is obligatory to have a fence around your house. Formally

$$\bigcirc\text{Fence}$$

2. If you do not have a fence, you should have a dog. Formally

$$\neg\text{Fence} \rightarrow \bigcirc\text{Dog}$$

Deontic logic traditionally uses **SDL** (standard deontic logic) to model obligations. This is the modal logic **K** with the modal operator $\bigcirc$ and the axiom $\neg \bigcirc \bot$.

Obligation statements are written in this logic. Statement 2 above is a contrary-to-duty statement. You should have a fence but failing that (i.e. if you violate the fence obligation), then at least have a dog. It is not clear (open) whether once you have a dog, the obligation to have a fence is still in force. **SDL** has difficulties modelling contrary to duty obligations and one such famous difficulty is the Chisholm paradox set, see [38, 81].

C1: You should go and help your neighbour:

$$\bigcirc\text{go}$$

C2: If you go you ought to tell him you are coming

$$\text{go} \rightarrow \bigcirc\text{tell}$$

C3: If you do not go you ought not to tell him you are coming:

$$\neg\text{go} \rightarrow \bigcirc\neg\text{tell}$$

C4: Fact: you do not go.

$$\neg\text{go}$$

This is a set of logically independent clauses. It is difficult to model in **SDL** (i.e. find a uniform translation into **SDL** such that the four clauses are consistent, independent and closed under substitution) and many logic papers have been written about it. Gabbay [81] proposed a reactive solution. This is our point of contact with reactive grammars.

Remark 9.48. To show the connection of deontic logic with reactive grammars we need an intermediate language which can be read by both disciplines. We explain this by an example.

Consider Figure 9.11.

This figure has a set of nodes $S = \{\infty, 0, x_1, \neg x_1, x_2, \neg x_2\}$ and a binary relation $R \subseteq S \times S$ denoted by the arrows and another relation $\mathbb{R} \subseteq S^2 \times S^2$ denoted by the double arrows.

The figure can be read in two ways

(i) As a carrier figure for a denotic contrary to duty system.
(ii) As a carrier figure for some specialised reactive grammar.

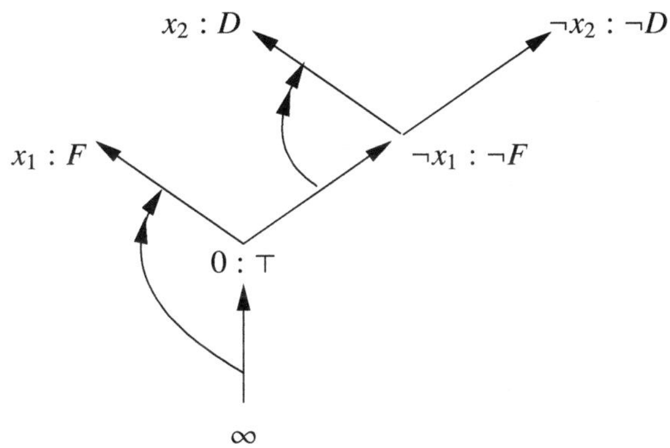

Fig. 9.11

Let us address (i) first. We read ∞ as home and 0 as office. $\infty \to 0$ means you go from home to office and thus your obligations begin. The double arrow

$$(\infty \to 0) \twoheadrightarrow (0 \to x_1)$$

says your obligation is to continue to x_1 (i.e. $\bigcirc$Fence). If you go there then you are done. If you go to $\neg x_1 : \neg F$ you trigger a new obligation

$$(0 \to \neg x_1) \twoheadrightarrow (\neg x_1 \to x_2)$$

namely to go to $x_2 : D$

$$(\neg\text{Fence } \to \bigcirc\text{Dog}).$$

Formally our graph figure is a tree with root ∞, with 0 the only successor to ∞ and from each $y_1 \to y_2$ there exists at most one double arrow and this double arrow is of the form

$$(y_1 \to y_2) \twoheadrightarrow (y_2 \to z).$$

in which case we write $y_2 \to \bigcirc z$.

Let us address reading (ii) now. The graph nodes are the grammar's non-terminals.

- ∞ is the starting point of the grammar.
- $\infty \to 0$ is active at the start. Everything else is nonactive.
- Whenever we have $x \to y$ in the graph this indicates a rule of grammar $x \to y$.
- Whenever we use a rule $x \to y$ and in the graph we have

$$(x \to y) \twoheadrightarrow (y \to z)$$

then $(y \to z)$ gets activated and all other rules get deactivated.

We allow naughty illegitimate derivations as well (naughty meaning one uses rules even if they are not active) but record violations (a violation instance is the use of a rule even though it is not active).

The following are the only possible derivations in the grammar of Figure 9.11. 1 means legitimate move, 0 means illegitimate move:

$$\infty \to_1 0 \to_1 x_1$$
$$\infty \to_1 0 \to_0 \neg x_1 \to_1 x_2$$
$$\infty \to_1 0 \to_0 \neg x_1 \to_0 \neg x_2$$

Example 9.49 (General Chisholm set). Let us give the graph for the general Chisholm set, see [81].

1. The Chisholm set written in semi-logical language:
   C1: $\top \to \bigcirc x_1$
   Ck: $x_k \to \bigcirc x_{k+1}$
   $\neg x_k \to \bigcirc \neg x_{k+1}$
   $k = 1, \ldots, n-1.$

   Facts: $\neg x_1, \pm x_2, \ldots, \pm x_n.$
2. The Chisholm set as a graph. See Figure 9.12

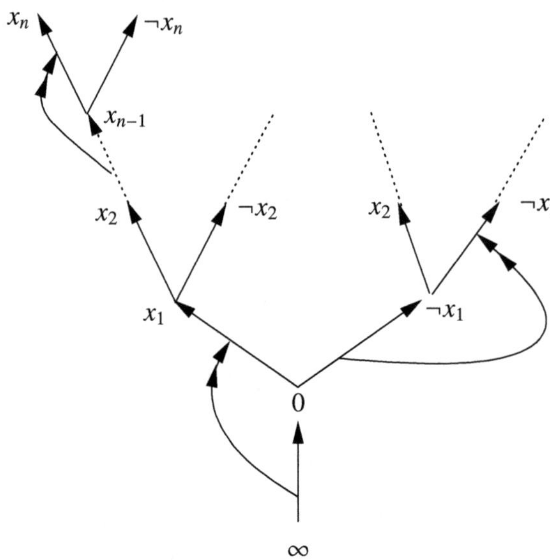

Fig. 9.12

3. The Chisholm set as a grammar

$$\infty \to 0$$
$$0 \to x_1$$
$$0 \to \neg x_1$$
$$\vdots$$
$$x_k \to x_{k+1}$$
$$\neg x_k \to \neg x_{k+1}$$

- ∞ the starting point
- $\infty \to 0, 0 \to x_1$ active
- All other rules not active
- If we use $z \to \pm x_k$ then this makes $\pm x_k \to \pm x_{k+1}$ active and all other rules inactive.

Definition 9.50. *1. A reactive grammar is said to be a linear-resource grammar (the word "linear" here comes from the resource logic known as "linear logic") if whenever a rule is used in a derivation it is permanently deactivated. So for example switch grammars are not a linear resource.*

2. A linear resource grammar is said to be Chisholm if it has only non terminal letters and rules of the form $x \to y$, where x, y are single non terminals. We further have

a) Let ∞ be the starting point, then each point $y \neq \infty$ has a unique $x \neq y$ such that $x \to y$ is a rule.

b) Whenever rule $x \to y$ is used then maybe some rules $y \to z$ (same y) are activated and all other rules are deactivated.

c) Let $(\infty, x_1, x_2 \dots)$ be a maximal path in the tree. (Such a path is called a complete contrary-to-duty run.) We annotate each connection $x_i \to x_{i+1}$ by 1 (legitimate) or 0 (not legitimate) as follows:

c1. $\infty \to x_1$ is annotated 1 if active at start otherwise annotated 0.

c2. Assume at x_i it is clear which rules are active or not. Then annotate $x_i \to x_{i+1}$ by 1 if active at x_i and by 0 if not active at x_i. Use clause (b) above to decide which rules $u \to v$ are active at x_{i+1}.

c3. The annotated contrary-to-duty run (path) tells us about the "violations" of the derivation constructing the path.

3. A Chisholm linear resource grammar is said to be with a meta-level policy if it has an associated algorithm for deciding, for each complete annotated contrary-to-duty run, whether it is acceptable or not (for example the meta-level policy not to accept any run with more than one violation).

The ideas behind the above definitions are very promising. They invite the systematic study of how to logically add "compensation/counter/retraction" to any evolving system with choice. This is a subject for further papers.

9.9 Conclusions

Reaction to movement is a pretty natural concept. If one considers the dynamic operation of an automaton, or state machine, there is movement as the automaton transits from one state to another as a symbol is accepted (or generated). Thus, it is natural to explore how automata themselves might be extended to react to such movement. Guided by prior work applying ideas of reactivity in automata and Kripke structures, we have made an initial exploration into how different forms of reactivity can be applied in the context of grammars. Given the considerable research that has been undertaken in the area of grammars over the past 40 years, it comes as little surprise that some of the reactivities we've considered coincide with various, well-known, extensions to grammar systems. Our interests, however, have been driven primarily by an investigation of principles, rather than finding ways to extend the expressiveness of particular forms of grammar rules. Our work in this area is preliminary and, although results are few, we believe that the range of reactivities explored is itself of considerable interest. One particular area that we have not explored in this chapter and which we believe warrants further work is the use of reactive grammars in language processing, for example, parsing, with the intention of generating efficient algorithms for fairly complex languages which can, nevertheless, be expressed using simple reactive grammars.

10

Reactive Flow Products

10.1 Introduction and orientation

Products of modal logics have been widely studied in the literature, see [70]. Flat products, see [131] all have the following type of models,

$$\mathbf{M} = \mathbf{M}_1 \times \mathbf{M}_2 \times \ldots \times \mathbf{M}_n.$$

where $\mathbf{M}_i = (S_i, R_i, h_i)$ are Kripke models where S_i is the set of possible worlds, $R_i \subseteq S_i \times S_i$ and h_i is the assignment.

Evaluation is done at a vector point $\mathbf{x} = (x_1, \ldots, x_n)$, with x_i from $\mathbf{M}_i$, and we have: $\mathbf{x} \vDash q$ for q atomic, if some Boolean combination of $\{x_i \vDash q_i\}$ holds. The Boolean combination is fixed and is part of the definition of the product.

$\mathbf{x} \vDash \Diamond_i A$ iff for some y such that $x_i R_i y$ we have $(x_1, \ldots, x_{i-1}, y, x_{i+1}, \ldots, x_n) \vDash A$.

The connective $\Diamond_i$ "moves" the coordinate x_i only, to the new y while the coordinates x_j, for $j \neq i$, remain fixed.

In our paper [131] we observed that when we have a product of two coordinates, space and time, it is not possible to move in space, without also moving in time. This gave rise to the concept of flow products. Let $\mathbf{M} = \mathbf{M}_1 \times \ldots \mathbf{M}_n$. For each i, j we have functions $f_{i,j}(\mathbf{x}, y)$ such that if coordinate x_i moves to y (such that $x_i R_i y$), then coordinate x_j moves to $z = f_{i,j}(\mathbf{x}, y)$ such that $x_j R_j z$.

Figure 10.1 shows how this works for the case of space-time.

If we move from x_1 to y in space, then we move from x_2 to z in time. Suppose the relative speed of the space movement is $\frac{1}{u}$. Then $z = x_2 + u(y - x_1)$. Thus $f_{1,2}(x_1, x_2, y) = x_2 + u(y - x_1)$.

This chapter offers a further generalisation of flow products, namely reactive flow products. We now shift our point of view step by step to turn the situation in Figure 10.1 into a reactive flow product. We do this through a series of observations.

D.M. Gabbay, *Reactive Kripke Semantics*, Cognitive Technologies,
DOI 10.1007/978-3-642-41389-6_10, © Springer-Verlag Berlin Heidelberg 2013

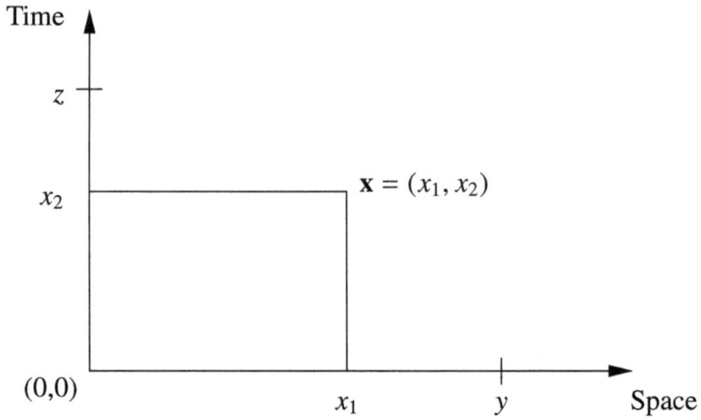

Fig. 10.1

Observation 1

Figure 10.1 is a product of the same Kripke model. It is a square with the possible worlds being the real numbers $(S, R) = (\mathbb{R}, <)$. So our reactive flow products will be defined on models of the form $\mathbf{M} = \mathbf{M}_1^n$.

Observation 2

Assume we start our movement at point (0,0). then if we move and have x_1 coordinate a, then the x_2 coordinate must be ua. Thus the coordinate pair (a, a) is possible only if $a = 0$ or $u = 1$.

Let G be a function telling us whether a coordinate pair (x_1, x_2) is allowed. So $G(x_1, x_2) \in \{0, 1\}$.

Observation 3

Imagine two agents, Agent x_1 and Agent x_2, with a Kripke frame $\mathbf{M} = (S, R)$. Let us say that x_1 is located at world $a_1 \in S$ and x_2 is located at world $a_2 \in S$. We write $\pi(x_1) = a_1, \pi(x_2) = a_2$.

We can view this situation as giving us (x_1, x_2) coordinates in the product $M \times M$. Viewed this way, if agent x_1 moves to a new node b_1 such that $a_1 R b_1$, then the system being a flow product, Agent x_2 will have to move to node b_2 such that $a_2 R b_2$. When we use the real numbers axis of Figure 10.1 we have $b_2 = b_1 + u(a_2 - a_1)$.

This point of view can be generalised using reactive arrows. See Figure 10.2.

Agent x_1 moves from position a_1 to position a_2. This is symbolised by the accessibility arrow $a_1 \to a_2$. Going though this arrow, a double arrow $(a_1 \to a_2) \twoheadrightarrow b_1$ is sent to node b_1. The double arrow says to agent x_2 to move out of b_1. Agent x_2 moves to $b_1 + 1$. However $G(b_1 + 1, a_2) = 0$, because the shift in time is $u(a_2 - a_1)$. So Agent x_2 will continue to move until it gets to node $b_2 = b_1 + u(a_2 - a_1)$. The function $G(b_2, a_2)$ will equal 1. Agent x_2 will stop at b_2.

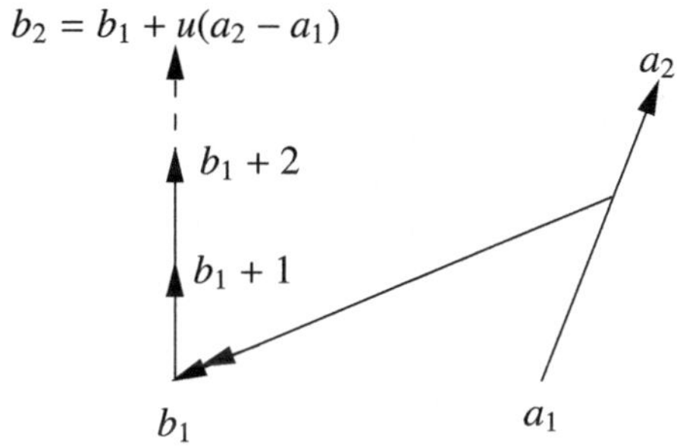

Fig. 10.2

Observation 4

We allow also for agents not being able to move. To do this we annotate arrows with + or −, saying the connection is on or off. When an agent moves along an arc it can also change the status of another arc (by sending a double arrow to that arc) from + to − or from − to +.

Let us summarise the main points of our observation.

1. We deal with products of the same model $\mathbf{M} = (S, R)$ of the form $\mathbf{M}\times\mathbf{M}\times\ldots\times\mathbf{M}$.
2. Such products are represented inside $\mathbf{M}$ itself by allowing for n agents to hold positions in M. The positions of the agents are the coordinates in the product.
3. The "flow" part is the idea that when an agent moves, he forces other agents also to move.
4. The shift is executed by double arrows.

Example 10.1. We illustrate our ideas giving a practical example. Consider a partnership with two partners, Agent x and Agent y. The question is which of the partners is to be the treasurer, and sign all cheques. There are 3 possibilities.

1. There is no treasurer
2. Agent x is the treasurer
3. Agent y is the treasurer
4. We cannot have two treasurers
   a) When there is no treasurer each agent can make a move and declare herself treasurer
   b) When Agent x is a treasurer, Agent y cannot make any move and vice versa.
   c) When Agent x is treasurer, she can resign. Similarly for Agent y.

It is convenient to represent this model as a blocks world model with two blocks x and y. We have three states for the blocks.

1. x and y are on the table
2. x is in top of y and y is on the table
3. y is on top of x and x is on the table

4. We cannot have both x on y and y on x.

We now describe this situation as a flat product of two models $\mathbf{M}_x$ and $\mathbf{M}_y$. $\mathbf{M}_x$ has states $S_x = \{x1, x2, x3\}$ and $\mathbf{M}_y$ has states $S_y = \{y1, y2, y3\}$. See Figure 10.3.

We consider the product $M_x \times M_y$ and we can put the models side by side as in Figure 10.5 (this is a flat product as defined in [131]). This needs to be done systematically so to continue let us talk first about the general blocks world with n elements.

Example 10.2 (Global view of the blocks world). We begin with a set S of blocks and a binary relation $\mathrm{On}(x, y)$ on S; $\mathrm{On} \subseteq S \times S$. Without additional axioms this is just a set with a binary relation. We assume a distinguished element $\mathbf{t} \in S$ (called table) and we ask for the following axioms to hold:

$\mathbb{B}$ axioms:

$\mathbb{B}1.$ $\forall x \exists y \mathrm{On}(x, y)$
$\mathbb{B}2.$ $\mathrm{On}(x, z) \wedge \mathrm{On}(y, z) \to (x = y) \vee z = \mathbf{t}$
$\mathbb{B}3.$ $\mathrm{On}(x, y) \wedge \mathrm{On}(x, z) \to y = z$
$\mathbb{B}4.$ $\neg\mathrm{On}(x, x)$
$\neg\mathrm{On}(\mathbf{t}, x)$
$\mathbb{B}5.$ $\neg\exists x_1, \ldots, x_k(\mathrm{On}(x_1, x_2) \wedge \ldots \wedge \mathrm{On}(x_{k-1}, x_k) \wedge \mathrm{On}(x_k, x_1)).$

Let us assume a finite model of $n + 1$ elements, (i.e. n blocks plus one table). Given the axioms we can represent the model as a family of towers

$$\tau_i = (x^i_1, \ldots, x^i_{k_i}), i = 1, \ldots, r$$

where for each i,

$$\mathrm{On}(x^i_j, x^i_{j+1}), 1 \le j < k_i$$

and

$$\mathrm{On}(x^i_{k_i}, \mathbf{t})$$

and

$$\sum_{i=1}^{r} k_i = n.$$

Let us fix $n + 1$ and allow in the model unique names for each block. We now consider different models based on this fixed set of blocks.

Denote such models by $\mathbf{M}, \mathbf{N}$, etc.

We say we are working in the blocks worlds with n blocks.

We can also allow for actions on such finite models of $\mathbb{B}$.

The language of the set of actions, which we denote by $\mathbb{A}$, uses actions having the form $\mathbf{move}(x, y)$ where $x \neq \mathbf{t}$ and $x \neq y$. The action means that block x moves from where it is and sits on top of block y. The action moves us from any model $\mathbf{M}_1$ of the blocks to a new model $\mathbf{M}_2$. We read the action as a move from $\mathbf{M}_1$ to a new model $\mathbf{M}_2$ where $\mathrm{On}(x, y)$ holds, and if $\mathrm{On}(x, z)$ holds in $\mathbf{M}_1$ then cancel it. Otherwise $\mathbf{M}_2$ is like $\mathbf{M}_1$.

This action, $\mathbf{move}(x, y)$ has the following preconditions in $\mathbf{M}_1$:

- $\text{On}(x, u)$, for some u
- $\neg\exists z(\text{On}(z, x) \vee (y \text{ is not } \mathbf{t} \wedge \text{On}(z, y)))$

and the post conditions in $\mathbf{M}_2$:

- $\neg\exists x\text{On}(x, u)$, when u is not $\mathbf{t}$
- $\text{On}(x, y)$

and otherwise $\mathbf{M}_2$ is like $\mathbf{M}_1$.

The above view of a model of n blocks is really an n-dimensional flat product to be explained.

Note that the system is characterised by the number of blocks, the axioms $\mathbb{B}$ and the set of allowed actions $\mathbb{A}$. If we change any of these we get a new system. For example if we have a block called s and we include in the axioms $\text{On}(s, Bt)$ and we do NOT list among the actions any action of the form $\text{move}(s, x)$, then this means that we want the block s to remain on the table and it cannot be moved.

Example 10.3 (Example 10.1 continued. Product view of the two blocks world). We now offer a product view of the blocks world.

To simplify matters, let us now assume that we have only two blocks x, y and a table.

The local states for x are:

$x1$. $\text{On}(x, \mathbf{t}) \wedge \neg\exists z\text{On}(z, x)$
$x2$. $\text{On}(y, x) \wedge \text{On}(x, \mathbf{t})$
$x3$. $\text{On}(x, y) \wedge \neg\exists z\text{On}(z, x)$.

Note that in $x3$, x does not know or care where y is. All x can see is who is directly on top of or beneath him (if any).

Similarly the local states for y are:

$y1$. $\text{On}(y, \mathbf{t}) \wedge \neg\exists z\text{On}(z, y)$
$y2$. $\text{On}(x, y) \wedge \text{On}(y, \mathbf{t})$
$y3$. $\text{On}(y, x) \wedge \neg\exists z\text{On}(z, y)$

Figure 10.3 shows the configurations. x, y are the blocks and "—" is the table.

Remark 10.4. Note that $x3$ is different from $y2$ because in case there are more than two blocks, $x3$ does not say where y is. y may not be on the table but on another block.

Similarly $x2$ is different from $y3$. Further note that if we want to simplify the system in this particular case (of two blocks) and identify $x2$ and $y3$ as the same state and identify $x3$ and $y2$ as the same state we get that $x3 = y2 = \mathbf{M}_1$ and $x2 = y3 = \mathbf{M}_2$. Here $\mathbf{M}_1$ and $\mathbf{M}_2$ are the global states of Figure 10.4.

If we compare Figure 10.4 with the possible global states of the world of $\{x, y, \mathbf{t}\}$ we get globally

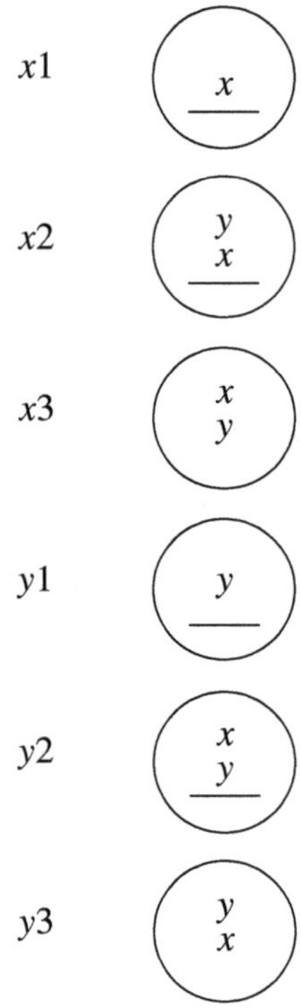

Fig. 10.3

$$
\begin{aligned}
&\mathbf{M}_1 : \tau = (x, y) \text{ i.e. } \mathrm{On}(x, y) \wedge \mathrm{On}(y, \mathbf{t})\\
&\mathbf{M}_2 : \tau = (y, x) \text{ i.e. } \mathrm{On}(y, x) \wedge \mathrm{On}(x, \mathbf{t})\\
&\mathbf{M}_3 : \tau_1(x) \qquad \text{i.e. } \mathrm{On}(x, t)\\
&\qquad \text{and}\\
&\qquad \tau_2 = (y) \quad \text{i.e. } \mathrm{On}(y, t).
\end{aligned}
$$

Figure 10.4 illustrates the global states.
The global states can be obtained as a combination of local states

$$
\begin{aligned}
\mathbf{M}_1 &= x3 \text{ and } y1\\
\mathbf{M}_2 &= y3 \text{ and } x1\\
\mathbf{M}_3 &= x1 \text{ and } y1
\end{aligned}
$$

Note that the combination of, for example, $x2$ and $y1$ is not allowed.
Table 10.1 shows which combinations are allowed and which are r not.

Example 10.5 (The two blocks world local actions). We now describe the two blocks world from a local point of view. Let S be a set of states

$$S = \{x2, x2, x3, y1, y2, y3\}.$$

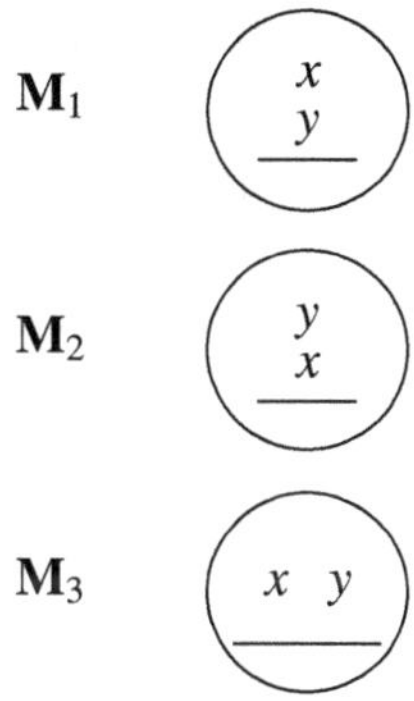

Fig. 10.4

Table 10.1

	$y1$	$y2$	$y3$
$x1$	✔	✘	✔
$x2$	✘	✘	✔
$x3$	✔	✔	✘

We have two agents x and y. x is in state x_i and y is in state y_j. The pair (x_i, y_j) must be allowed by Table 10.1.

We now want to figure out the allowable single moves of the system. Consider the following possibilities.

P1

x is on the table and y is on the table. x cannot move to the state where x is under y. So the move $x1 \to x2$ is not allowed.

We write

$$x1 \longrightarrow^{-} x2$$

Similarly

$$y1 \longrightarrow^{-} y2.$$

P2

x is on the table and y is on the table. I.e. x is at $x1$ and y is at $y1$. Then if x makes the first move then x can move on top of y, i.e. $x1 \longrightarrow^{+} x3$ (remember one action at a time!) and so y does not move. We also have that if y makes the first move then y can move on top of x and x does not move. In symbols $y1 \longrightarrow^{+} y3$.

If x moves on top of y, two things happen.

1. x traverses the connection $x1 \longrightarrow^{+} x3$

2. at the same time x forces y to move from $y1$ to $y2$.

So, Figure 10.5 describes the initial configuration, and becomes Figure 10.6 describing the resulting configuration after the move by x.[1]

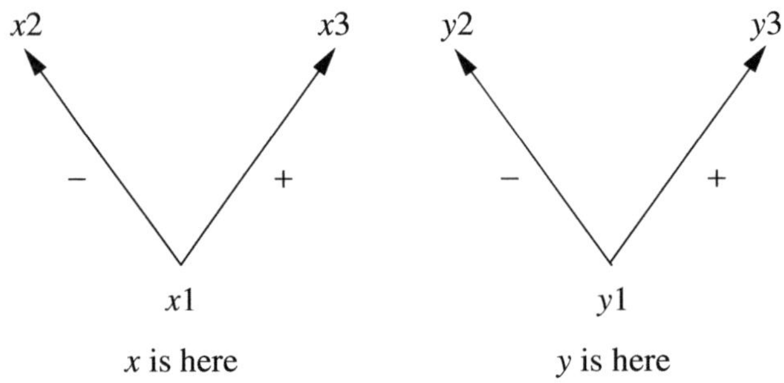

Fig. 10.5

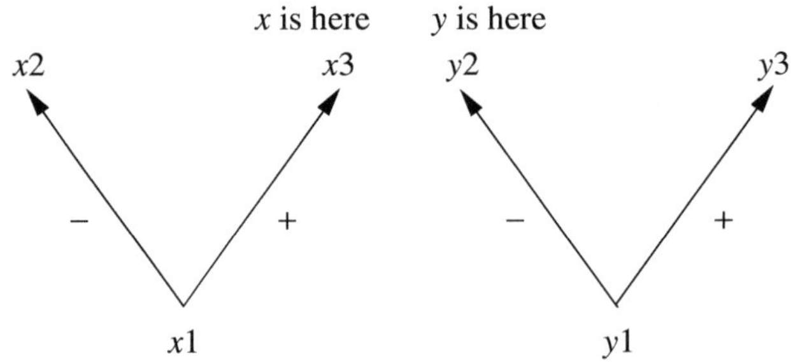

Fig. 10.6

The way we "implement" this transition is by introducing reactivity into Figure 10.5. We use Figure 10.7 instead of Figure 10.5

The meaning of the lightning double arrow $(a \to b) \rightsquigarrow (c \to d)$ is that if an agent x traverses the connection $a \to b$ from which this double arrow emanates, then its target arrow $c \to d$ forces any agent y at c to pass through and go to d.

Note that it does not matter whether the arrow $c \to d$ is on or off. The agent y is forced to go through it.

So when x moves from $x1$ to $x3$, a signal is sent to $y1 \longrightarrow^{-} y2$ and this forces any agent at $y1$ (in our case y is at $y1$) to move across this connection. We thus end up at the configuration of Figure 10.8.

[1] This makes our model a two-dimensional flow product.

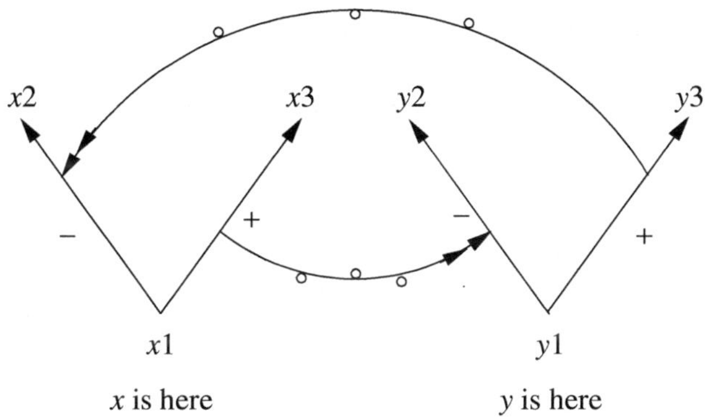

Fig. 10.7

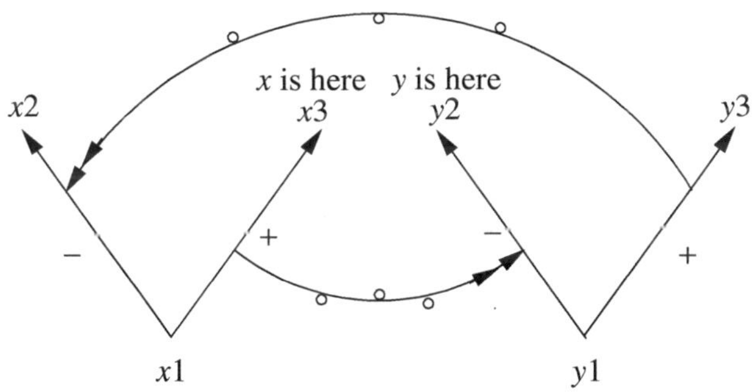

Fig. 10.8

The following Figure 10.9 describes the entire map of the two blocks world local system.

The allowed positions for the agents (x, y) are

$$(x1, y1)$$
$$(x2, y3)$$
$$(x3, y2)$$

We now consider possible concurrent double moves of the system, where both x and y make their move at the same time.[2] This is not a problem to define given that we have Figure 10.9, we simply let x and y move simultaneously in Figure 10.9 and activate all double arrows emanating from the arcs they traverse. So assume x is at $x1$ and y is at $y1$. x is allowed to move to $x3$ in Figure 10.9 and y is allowed to move to $y3$. Let them do it simultaneously. Now x is at $x3$ and y at $y3$. This is not an allowable position. What do we do? We need a policy. Before we examine our

[2] This corresponds to a diagonal modal operator which in the space-time case moves in both the space and time coordinates.

options, note that for example the double arrow $(x1 \rightarrow x3) \twoheadrightarrow (y1 \rightarrow y2)$, which has been activated by x moving from $x1$ to $x3$, is not effective because y simultaneously has already left $y1$, and so cannot be forced to go to $y3$. (Note that this is a design decision and could be made differently.) Similarly y going to $y3$ cannot force x to go to $x2$ because x has already left $x1$. We thus end up with the disallowed position of x at $x3$ and y at $y3$.

We can adopt the policy of asking x and y to move on until they get into an allowable position or we can adopt the policy of asking x and y to go back (i.e. cancel the move). In this case both policies give the same result, with x at $x1$ and y at $y1$. There is a third possibility, allow x and y to remain at a position which is not allowed and adjust the system to behave differently in case the position is allowed and in case the position is not allowed.We leave the detailed discussion and options to the formal Section 2. Note also that we should allow the system to react in case x or y make a single move which is not allowed, i.e. traverse an arc marked with "–", (i.e. for which **f** gives value 0).

Another problem which may arise in simultaneous parallel actions is that another agent may be asked to move forcibly by two different lightning arrows in two different directions.[3] In this case we can agree that the agent can choose which one to use. Perhaps we can read the instruction as "get out of your state and go along one of these arrows as indicated."

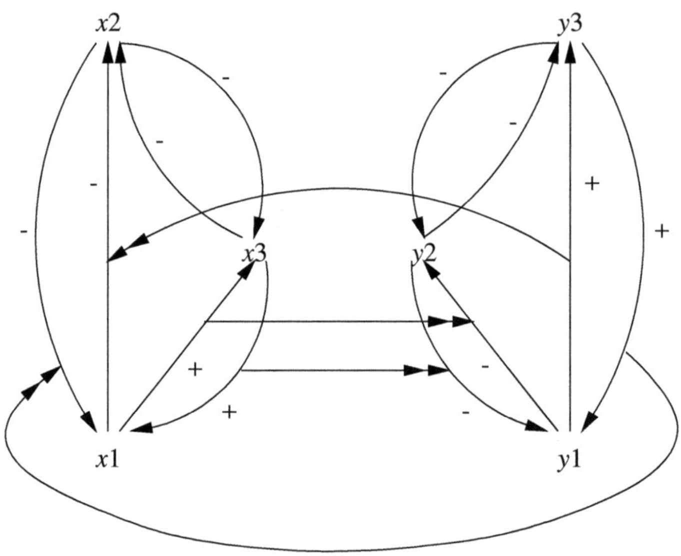

Fig. 10.9

[3] In a three-dimensional product.

Remark 10.6. Note that following Remark 10.4, we can identify in Figure 10.9 the nodes $x3 = y2 = \mathbf{M}_1$ and $x2 = y3 = \mathbf{M}_2$. Note the symmetry in Figure 10.9 which supports this identification! As a result we get the simpler Figure 10.10, with only four nodes. In this figure the two agents can occupy the same state.

Remark 10.7 (Agent Norms). The perceptive reader, looking at the product view of the blocks world, as presented for example in Figure 10.9 for the case of two blocks, might wonder whatever happened to the list of $\mathbb{B}$ axioms introduced in Example 10.2, for the global view of the blocks world. These global axioms become Agent Norms in the local view or in the product language, product rules. These norms manifest themselves in three ways.

1. In the list of which combinations of local states can be accepted as generating a global state, see for example Table 10.1 for the case of two blocks.
2. In the locally allowable transitions for each of the agents from each of the states. In Figure 10.9 these are the plus–minus annotations of the arrows.
3. In the nature of the double arrows in the system.

The reader should note that the above items (1)–(3) are typical to how we implement General Norms for multi-agent systems and can be used for the global view as well.

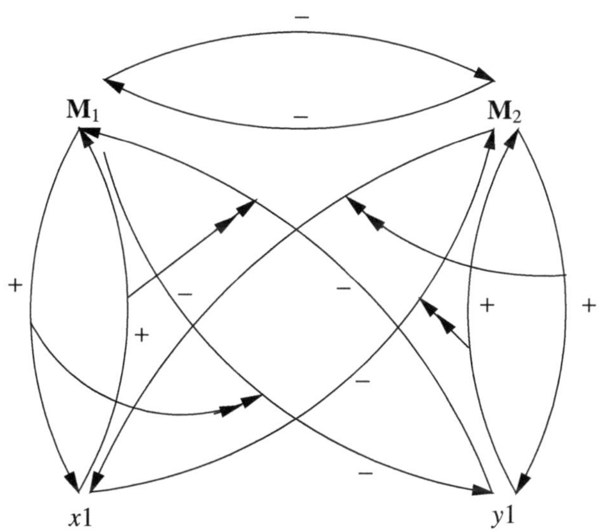

Fig. 10.10

Remark 10.8 (A reactive flow product model). The previous two blocks example gives us an idea of how to define a reactive flow product model. We need the following:

1. A set S of states (these are the coordinates).
2. A set $\mathbb{A}$ of agents (this is the dimension of the product).
3. A function $\pi : \mathbb{A} \mapsto S$ telling us for each agent at which local state he is.
4. A function G telling us the legitimacy of any π. G is $\{0, 1\}$. If $G(\pi)$ is not 1 then π is a disallowed distribution of coordinate states. This means the coordinates indicated by π are not legitimate.
5. A relation $R_1 \subseteq S^2$ giving us the connections between states. This defines the Kripke model participating in the product.
6. A reactive relation $R_5 \subseteq R_1^2$ giving us the lightning double arrows.
7. A function $\mathbf{f} : R_1 \mapsto \{0, 1\}$ giving us which R_1 connections are allowed. ($\mathbf{f}$ and R_5 together give us the flow part of the product.)

Note that in item 6 we use R_5 and not say R_2 because in other papers and other applications of reactivity, R_2 is used for a switch double arrow. So a switch double arrow

$$(a \to b) \twoheadrightarrow (c \to d)$$

means that if you traverse $a \to b$ then a signal goes to $c \to d$ and changes it. If $c \to d$ is off, it becomes on and if it is on it becomes off.

Thus a model

$$\mathbf{m} = (S, \mathbb{A}, \pi, G, R_1, R_2, R_5, \mathbf{f})$$

will have, once an agent moves along an arc $(c \to d) \in R_1$, both π changing to $\pi_{(c,d)}$ and $\mathbf{f}$ changing to $\mathbf{f}_{(c,d)}$. The next example, of the three blocks world, will require both types of double arrows, the lightning double arrows and the switch double arrows.

Example 10.9 (Three blocks world). Let us see what happens when we have three blocks x, y, z and a table. This is a three-dimensional product.

The possible global models are

$\mathbf{M}_1$ x
y
z
______

$\mathbf{M}_2$ x
z
y
______

$\mathbf{M}_3$ y
x
z
______

$\mathbf{M}_4$ y
z
x

$\mathbf{M}_5$ z
x
y

$\mathbf{M}_6$ z
y
x

$\mathbf{M}_7$ x
y z

$\mathbf{M}_8$ y
x z

$\mathbf{M}_9$ x
z y

$\mathbf{M}_{10}$ z
x y

$\mathbf{M}_{11}$ y
z x

$\mathbf{M}_{12}$ z
y x

$\mathbf{M}_{13}$ x y z

The coordinate states are

$x1$ x

$x2$ y
x

$x3$ z
x

$x4$ x
y
$x5$ x
z

$x6$ z
x
y

$x7$ y
x
z

$y1$ y

$y2$ z
y

$y3$ x
y

$y4$ y
z

$y5$ y
x

*y*6 x
y
z

*y*7 z
y
x

*z*1 z ______

*z*2 x
z ______

*z*3 y
z ______

*z*4 z
x

*z*5 z
y

*z*6 y
z
x

*z*7 x
z
y

The allowable combined coordinate states are the ones that combine to give full global states:

$$
\begin{aligned}
\mathbf{M}_1 &= x4 + y6 + z3\\
\mathbf{M}_2 &= x5 + y2 + z7\\
\mathbf{M}_3 &= x7 + y5 + z2\\
\mathbf{M}_4 &= x3 + y4 + z6\\
\mathbf{M}_5 &= x6 + y3 + z4\\
\mathbf{M}_6 &= x2 + y7 + z5\\
\mathbf{M}_7 &= x4 + y3 + z1\\
\mathbf{M}_8 &= x2 + y5 + z1\\
\mathbf{M}_9 &= x5 + y1 + z2\\
\mathbf{M}_{10} &= x3 + y1 + z4\\
\mathbf{M}_{11} &= x1 + y4 + z3\\
\mathbf{M}_{12} &= x1 + y2 + z5\\
\mathbf{M}_{13} &= x1 + y1 + z1
\end{aligned}
$$

Note that there cannot be two different blocks with the same coordinate state, not if we include the block itself in the description of the state. A local state of a block u has the form

$$\begin{pmatrix}\alpha\\ u\\ \beta\end{pmatrix}$$

where α = some other block or nothing $\varnothing$ and β is some block or the table.

So if $u_1 \neq u_2$ then

$$\begin{pmatrix}\alpha_1\\ u_1\\ \beta_1\end{pmatrix} \neq \begin{pmatrix}\alpha_1\\ u_2\\ \beta_2\end{pmatrix}$$

If we do not include the block itself in the state then we can have $\alpha_1 = \alpha_2, \beta_1 = \beta_2$.

However, we need the block to be part of the state to decide where it can go and what happens when it moves. This makes the system a product.

Let us check what actions are allowed in some cases, to get an idea of what can happen.

Let $S = \{x1, \ldots, x7, y1, \ldots, y7, z1, \ldots, z7\}$. Let $\pi(x) = x1, \pi(y) = y1, \pi(z) = z1$. Thus the global state corresponding to these local states is $\mathbf{M}_{13}$.

Figure 10.11 is a partial figure showing what transitions x, y, z can do from where they are.

Since y can move from $y1$ to $y5$, it forces x to move from $x1$ to $x2$ and at the same time prevents z from moving from $z1$ to $z4$. This is why in the partial figure 10.11 we have two types of double arrows:

$$(y1 \to y5) \rightsquigarrow (x1 \to x2)$$

and

$$(y1 \to y5) \twoheadrightarrow (z1 \to z4)$$

The first double arrow is a lightning arrow forcing x to move from $x1$ to $x2$ and the second double arrow is a switch arrow switching $z1 \to z4$ from allowed ("+") to not allowed ("–"). z is not forced to move from $z1$.

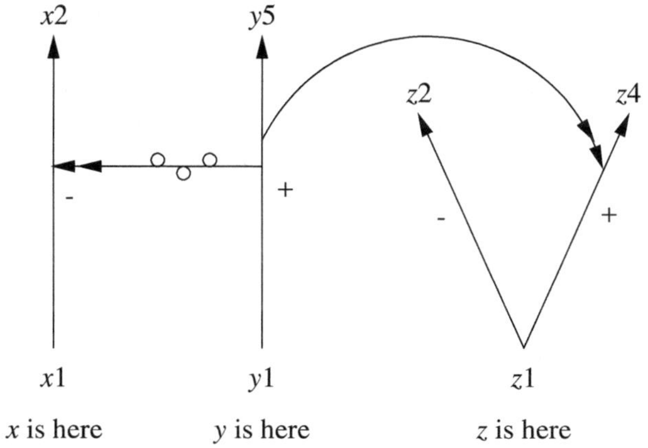

Fig. 10.11

The new global model is $x2 + y5 + z1 = \mathbf{M}_8$

Example 10.10. It is instructive to give a non-reactive blocks example for comparison.

Consider the two blocks x and y, as well as two fixed blocks on the table, called **s** and **b**. We do not allow **s** and **b** to be moved. The only allowed actions in this system are

$$\begin{array}{l} \text{move}(x, \mathbf{s}) \\ \text{move}(x, \mathbf{b}) \\ \text{move}(y, \mathbf{s}) \\ \text{move}(y, \mathbf{b}) \end{array}$$

Now change the axiom set $\mathbb{B}$ by allowing more than one block to be on **b** or on **s**, (i.e. we delete axioms $\mathbb{B}2$).

We now have a system which is not reactive. If the initial position is that x and y are on the table, we have four possible moves, two for block x and two for block y and these are the moves decided by the choice for each block of whether to move it onto **b** or onto **s**.

Let

b_1 = block x is on **b**
b_2 = block y is on **b**
s_1 = block x is on **s**
s_2 = block y is on **s**

Then every assignment of truth values to these variables (i.e. to $\{b_1, b_2, s_2, s_2\}$) gives us a state of the system. There is the restriction that a block cannot move in one go both to **s** and to **b**, and so the assignment must respect the equivalences.

$$\begin{array}{l} s_1 \text{ iff } \neg b_1 \\ s_2 \text{ iff } \neg b_2 \end{array}$$

The movements of x and y are local but independent and they are not reactive (do not interfere) with each other.

The system is a product of two single-agent global state systems.

We can add a cost function $c(b_1), c(b_2), c(s_1), c(s_2)$ for moving onto **b** and **s** respectively for x and y respectively. We can also give each agent a goal, e.g. $b_1 \wedge \neg s_2$ for x and $s_1 \wedge b_2$ for y and add preferences, payoffs, coalition formations, etc, etc.

We can calculate the complexity of various scenarios, etc, etc.

This is standard. Compare with Example 1 of [46, p. 1016].

Remark 10.11 (Normative positions). Figure 10.11 gives us a clue about how to apply our models to a completely different area, the area of normative positions [117]. We consider expressions like OE_aA reading "It is obligatory that agent a sees to it that A becomes true".

The current modelling of normative positions is to view O and E_a as modal operators. The semantics for these operators does not explain much about how a is going to "see to it" that A holds. This is especially noticeable when we iterate

$$OE_aE_bA$$

saying that it is obligatory for a to "see to it" that b "sees to it" that A becomes true. The modal semantics will say:

> In all a-accessible worlds y (to our current world x, i.e. xR_ay) it is true that in all b-accessible worlds z to y (yR_bz) we have $z \vDash A$.

To see the problem, let us try the following:

Dov Gabbay "sees to it" that President Obama "sees to it" that Bar-Ilan University receives a grant of $ 10 million.

So the semantics for this is:

> In all worlds accessible according to Dov, it is true that in all worlds accessible to Obama it is true that Bar-Ilan University gets $ 10 million.

Are you, the reader, any wiser about how a humble Professor like me is going to influence Obama?

Now consider Figure 10.12, which is a variation of Figure 10.11. This figure gives more details about how Dov Gabbay can bring about that Obama gives money.

Our models interpret

> x_1 "sees to it" that x_2 "sees to it" ... that x_n "sees to it" that A

x_1 makes a move that forces x_2 to make a move which forces x_n in a "chain reaction" to move to a position where A holds in the global model. Formally in the model $x_1, \dots, x_n$ are in certain states. x_1 makes a move which forces all the other xs to new states and in the corresponding new global state A holds.

The details need to be worked out and this is done in [28].

So the theory of normative positions of agents uses a local multi-agents point of view.

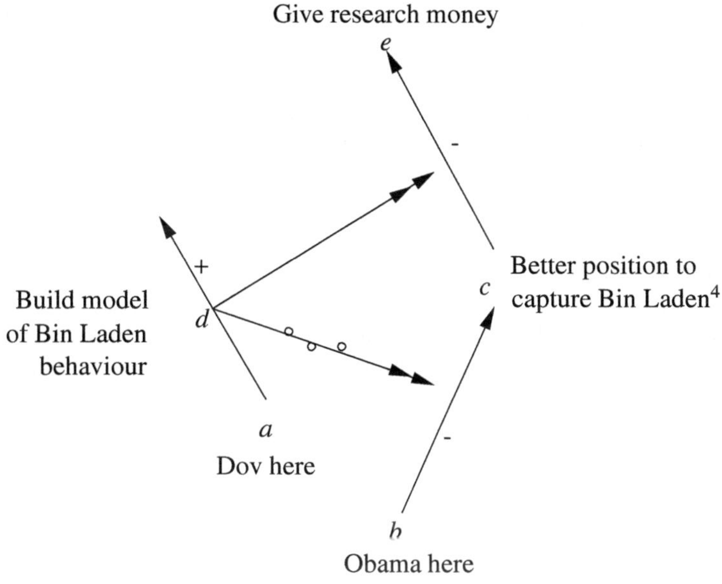

Fig. 10.12

Remark 10.12 (Parallel computation). As computer chips run faster they generate more heat. This physical fact places limits on how quickly computer processors can run. This is why recently computer clock speeds have levelled off at approximately 4GHz — if they ran any quicker they would melt.

Transistors are still getting smaller though, so manufacturers are producing chips with *multiple cores* — two, four, six, eight, or more processing units on every chip.

Examples include any modern computing chip, including for example Intel's Core 2 Duo series.

It takes time to propagate a signal across a computer chip and in any case bandwidth — the information that can be carried — is limited.

What this means is that notions of *geometry* are necessarily introduced into the design of computer algorithms. Efficient computation with modern processors occurs on a grid of processing units.

There is a global memory but the cost of communication with it is relatively high so each processing unit is equipped with a small amount of fast local memory, called its *cache*. Processing units no longer have direct access to the global memory of the computer. Instead, the processing unit communicates with its cache, and its cache communicates with the other caches (more slowly) and with the global memory (much more slowly).

In effect, computation ceases being about processes updating global memory and becomes instead a geometric process which propagates itself on the processing units.

This is not a completely sharp distinction. There is still a global memory, but it is slow and far away. Thus *for shorter time scales*, processing becomes 'geometric' and 'local'.

What this means, furthermore, is that there can be no guarantee that *local processors know about other processors' updates to global memory*. There is no longer a global state: instead there is a collection of local states which may or may not all be competing to update the same address in global memory.

Put another way, local caches may not be coherent. They may disagree about what should be written to global memory.

Notions of linear time are lost; indeed, the local processors may be running at different clock speeds. Therefore situations may arise where computations on one local processor may affect the admissibility of computations that have already taken place, or are taking place, or are about to take place, on another processor.

In summary: the old computational model of a Turing machine accessing a single global state is slowly but unavoidably breaking in the face of the constraints of modern hardware. The new model is based on computation as a local phenomenon which takes place on a geometric array of processing sites. Notions of both memory and time are local. At some point, these local processes have to communicate and coordinate their actions. Thus this new computational model provides rich pickings (and concrete motivations) for notions of agency based on locality and indeed on backwards causation. For notions of agency we need to look at multi-agent systems whose agents want to cooperate without being able to communicate or synchronise. For backward causation we need to study how to undo or at least salvage past actions (backward causation appears in law, see [1]).

Example 10.13 (Battle of Thermopylae). This is the historically famous battle of the Greeks against the Persians held in the summer of 480 B.C.

It is also known through the Hollywood film *300*. It is an example of a coalition of local agents (the 300 Spartans) intending to change a global state, even though it is detrimental to their local state (they died).

According to historians, this heroic battle was not important for the overall war. It is famous for the example it set for heroism and for sacrifice for one's freedom and one's country.

This very point of view of the battle throughout history proves that the local view of a multi-agent system is the primary one and when a group of agents take a global view, and die for it, it is to be admired!

We now conclude Section 1, which introduced our ideas and we move to Section 2 which introduces our formal machinery.

10.2 Formal models of reactive flow product systems

This section formally develops our view of local flow product systems. First we need some changes in our point of view regarding semantic evaluation at possible worlds.

Change in point of view ♯1:
Consider a Kripke model for some modal language $\mathbb{L}$. The basic semantical evaluation unit is of the form

$$t \vDash A$$

where t is a possible world and A a wff of $\mathbb{L}$. We get a truth value as the result of the evaluation.

Let us write this basic situation as

$$t?A = x$$

where $x \in \{0, 1\}$. We first note that we can let

$$\Delta_t = \{X | t \vDash X\}$$

and so write

$$\Delta_t \vdash A$$

instead of

$$t \vDash A$$

and so we can write

$$\Delta_t?A = x.$$

This is not new, this is the situation we have when we construct a canonical model.

Now let us make the step that with each t we associate a Δ_t and a query algorithm "?" which can give values x from some set.

Now that we look at the situation in this way, we can let Δ_t be something completely different, e.g. some network, let A be a wff in some query language from this network and let ? be some algorithm giving value x in some range, e.g. the real numbers. See, for example, [14].

Change in point of view ♯2:
This change is a shift from a global to local point of view. When we write

$$t \vDash A \text{ or } \Delta_t?A = x$$

we take a global view of t. All the information is there at t. We need only ask. However, if we take a local view of t, we cannot ask from t, we may need to combine several worlds. So we write

$$(t_1, \ldots, t_k) \vDash A$$

or

$$\Delta_{(t_1,\ldots,t_k)}?A = x.$$

Here k is fixed. This is reminiscent of many-dimensional modal logic. The difference is in how we manipulate it. Some sequences $(t_1, \ldots, t_k)$ will be allowed and some will not be allowed. We saw this happening in Example 10.3.

Definition 10.14 ($\mathbb{QM}$ systems). *A query (semantic or syntactic) system* $\mathbb{QM}$ *has the following components:*

1. *A language $\mathbb{Q}$ for queries $Q \in \mathbb{Q}$.*
2. *A language $\mathbb{M}$ for databases or models Δ.*
3. *An algorithm "?" capable of giving a value $x \in \mathbb{V}$ for any pair (Δ, Q), where $\mathbb{V}$ is a set of values. We write*

$$\Delta ? Q = x.$$

Example 10.15 ($\mathbb{QM}$ = classical propositional calculus). To give an example of such a language $\mathbb{QM}$ let us take the classical propositional calculus as our language. Thus the language $\mathbb{M}$ will be the language of assignments to the classical propositional atoms, yielding a classical propositional model, and the language $\mathbb{Q}$ will be the full set of well-formed formulas of classical propositional logic. The query algorithm "?" is just evaluation of a formula in a model according to the classical truth tables.

Definition 10.16 (Reactive flow product). *A reactive flow product model has the form*

$$\mu = (S, \mathbb{A}, \pi, G, R_1, R_2, R_5, \mathbf{f}, \mathbb{QM})$$

where

1. *$\mathbb{QM}$ is a query system as in Definition 10.14.*
2. *S is a set of local states $S \neq \varnothing$.*
3. *$\mathbb{A}$ is a finite set of agents.*
4. *π is a function from $\mathbb{A}$ into S (telling us the local state of each agent).*
5. *G is a partial function giving us, when defined for a function π, a database $G(\pi)$ in the query system $\mathbb{QM}$.*
6. *$R_1 \subseteq S^2 \times \mathbb{A}$ giving us the transitions possible for each agent.*
7. *$R_2 \subseteq R_1^2$ is a set of switch double arrows, switching on and off the elements of R_1.*
8. *$R_5 \subseteq R_1^2$ is a set of lightning double arrows forcing agents to move.*
9. *A function $\mathbf{f} : R_1 \mapsto \{0, 1\}$ telling us which elements in R_1 are on or off.*

Note that in (8), we are not excluding the possibility that several arcs of the form $x \xrightarrow{a} y_i \in R_1$ can be the target of a lightning double arrow in R_5 emanating from the same or several $u \xrightarrow{b} w$. In this case our semantic interpretation is that agent a is forced to move along one of the arcs $x \xrightarrow{a} y_i$. We can use the notation, for $x \in S$,

$$(u \xrightarrow{b} w) \overset{a}{\twoheadrightarrow} x$$

to mean that a should move out of x in any direction.

In other words, this is shorthand for

$$\{(u \xrightarrow{b} w) \twoheadrightarrow (x \xrightarrow{a} y) \in R_5 | y \text{ is such that } (x \xrightarrow{a} y) \in R_1\}.$$

Similarly note that we can use a shorthand for case (7): when a double arrow in R_2 leads to a point instead of an arc we can read it as leading to every arc emanating from the point, or to every arc leading into the point or both, depending on how many abbreviations we want to use.

Example 10.17. Consider Figure 10.11. This figure is almost a model in the sense of Definition 10.16. The set of agents is the set $\mathbb{A} = \{x, y, z\}$. π is defined, so are R_1, R_2 and R_5. If $\pi' : \{x, y, z\} \mapsto S$ is given, then we can let

$$G(\pi') = \pi'(x) \cup \pi'(y) \cup \pi'(z)$$

and say that it is well defined if it is consistent, i.e. if we get a global model of the three blocks. See the list in Example 10.9.

Remark 10.18. The query system $\mathbb{QM}$ replaces the traditional notion of assignment. It allows us to ask queries from the global state.

So we can have

$$\mu \vDash Q \text{ iff } G(\pi)?Q = 1.$$

Example 10.19. We give a completely different example now. Consider 11 members of a football team and consider the numbers 11 to 99. The members can freely choose a number for their T-shirt. The restrictions are that different members wear different numbers, and that the numbers be all primes (there are 21 prime numbers in the range [11–99], say $p_1, \ldots, p_{21}$).

So each agent $a_1, \ldots, a_{11}$ can be in 21 local states, and can move locally from state to state by choosing another number. The allowable distributions π are as we described, namely 11 different primes, one for each player.

Assume that each player has his preference relation on the prime numbers. For example, player 1 may prefer bigger numbers. So his preference relation is

$$11 < 13 < 17 < 19 < 23 < 29 < 31 < 37 < 41 < 43 < 47 < 53 < 57 <$$
$$61 < 67 < 71 < 73 < 79 < 83 < 89 < 97.$$

Player 2 may have a preference for smaller numbers so his ordering is the opposite, while the other players may have orderings dictated by birth dates, marriages and superstition.

So let us now describe the model.

Let $\mathbb{QM}$ be a query system looking at 11 prime numbers and giving a "beauty value" $\in \{0, 1, \frac{1}{2} = \text{don't care}\}$.

So $\mathbb{M}$ = 11-tuples of 11 different primes. $\mathbb{Q}$ = {do you like this set} and $\mathbb{V} = \{0, 1, \frac{1}{2}\}$.

S = the set of all primes from 11 to 97.

$\mathbb{A}$ = set of 11 players.

R_1 = all arrows of the form $x \xrightarrow{a} y$

where $x \neq y$ and x, y are prime numbers and a is a player.

$R_2 = \varnothing$ (we still have no switch arrows).

R_5 is $\{(x \xrightarrow{a} y) \overset{b}{\twoheadrightarrow} y\}$ for $a \neq b$.

i.e. if a chooses the new prime number y then any player b with y must move to another number.

Of course if we allow one move at a time then if a moves first and b is next, he will have to move and a chain reaction of forced moves may start even possibly looping back to a.

We can make the model more interesting by forcing our players to be rational and let $\mathbf{f}^{\pm}(x \xrightarrow{a} y) = 1$ if y is more (resp. less) preferable than x in the preferences of a. Otherwise the value is 0.

We can start with some distribution π of numbers to the players and see what happens!

In fact we can let the system run until it stops and view this as an algorithm for negotiations between the players concerning their choice of numbers. In this case we start with π giving each player his highest preference and use $\mathbf{f}^{-}$.

Remark 10.20 (Point-to-point one-step transition). Consider a model μ and an allowable position π for the agents of $\mathbb{A}$.

Take a subset of agents $\mathbb{C} \subseteq \mathbb{A}$. Regard $\mathbb{C}$ as a coalition. What can they do?

1. Each agent $a \in \mathbb{C}$ can check whether he has an allowable move. This means that if $\pi(a) = x \in S$, then for some $y \in S$ we have $x \xrightarrow{a} y \in R_1$, and $\mathbf{f}(x \xrightarrow{a} y) = 1$.

If no such move is available we say that the agent a is *immediately* (level 0) *stuck at* x.

If all agents $a \in \mathbb{C}$ are immediately (level 0) stuck at their respective states, we say that $\mathbb{C}$ is *immediately* (level 0) *stuck* at π.

Suppose some of agents $a_1, \ldots, a_k$ can move. Let them simultaneously move to their new positions. So since a_i can move we have that some $x_i \xrightarrow{a_i} y_i$ is live and a_i is at x_i and a_i moves to y_i. Several things happen when all or some a_i move simultaneously to their respective y_i.

1. All R_2 switch double arrows of the form

$$(x_i \xrightarrow{a_i} y_i) \twoheadrightarrow (u \xrightarrow{b} w)$$

   fire (i.e. send a signal), where $u \xrightarrow{b} w$ can be any connection in R_1 which happens to be the target of at least one such $(x_i \xrightarrow{a_i} y_i) \twoheadrightarrow (u \xrightarrow{b} w)$.

Now if the number of "hits" on any specific $u \xrightarrow{b} w$ is odd then if under $\mathbf{f}$, $u \xrightarrow{b} w$ is on (off) then it changes value to off (resp. on). If the number of hits is even then there is no change.

Let us figure out what happens when each a_i moves from x_i to y_i, resp.

We now get new positions for agents a_i (they are now at y_i instead of x_i) and we have a new function $\mathbf{f}' = \mathbf{f}_{\{x_i \xrightarrow{a_i} y_i\}}$ replacing the original function $\mathbf{f}$.

This also means that we have a new function $\pi' = \pi_{\{x_i \xrightarrow{a_i} y_i\}}$.

Two things can happen at the new position π':

1. The new position π' is not acceptable by G.
2. While moving along $x_i \xrightarrow{a_i} y_i$ several R_5 lightning arrows were triggered, e.g. for at least one $u \in S$ and $w_j \in S$

$$x_i \xrightarrow{a_i} y_i \rightsquigarrow (u \xrightarrow{b} w_j),\ j = 1, 2, \ldots$$

were triggered.
If b is at u (i.e. $\pi(b) = u$, *not* $\pi'(b)$!) then b has to move to one of the w_j. Note that if $b = a_i$ and $u = x$ then a_i has already moved. If $b = a_i$ and $u = y_i$, we need not worry because a_i has 'not yet' been at u.

So if (1) is true but (2) is also true we simply follow the chain reaction and let b move. If (2) is not true, i.e. no b is forced to move and (1) is not true (i.e. π' is acceptable) then we say that we have completed an immediate (level 1) point-to-point move from μ to μ' and the new model μ' has

$$\pi' = \pi_{\{x_i \overset{a_i}{\to} y_i\}} \text{ and } \mathbf{f}' = \mathbf{f}_{\{x_i \overset{a_i}{\to} y_i\}}.$$

If, however, π' is not a G-acceptable position, then we distinguish two further cases:

3. All a_i are stuck at y_i.
   In this case we go back to π and $\mathbf{f}$. The group $\mathbb{C}$ cannot move along $x_i \overset{a_i}{\to} y_i$. Their possible move along $x_i \overset{a_i}{\to} y_i$ is level 1 stuck.
4. Some members, say $\mathbb{C}' \subseteq \mathbb{C}$ can go on moving.
   We now start moving again with $\mathbb{C}'$ and $y_k \overset{a'_k}{\to} z_k$, with $a'_k \in \mathbb{C}', k = 1, 2 \ldots$. This chain reaction continues until either of the following holds:
5. We first get to a position $\pi^{(r)}$ which is stable and acceptable, in which case we say we completed a level r, single point-to-point transition from $(\pi, \mathbf{f})$ to $(\pi^{(r)}, \mathbf{f}^{(r)})$, along the path $\mathbb{C}, \mathbb{C}^1, \ldots, \mathbb{C}^{(r)}$ and the respective arcs used.
6. We either get stuck (i.e. case 3 above) in which case we go back to $(\pi, \mathbf{f})$ and say we are level r stuck along the path or loop and again we go back to $(\pi, \mathbf{f})$ and say we loop along the path.
   If we get stuck or loop along every possible path we say we cannot make any single point-to-point transition from $(\pi, \mathbf{f})$.

All the above described how to accomplish a single point-to-point transition from a μ with $(\pi, \mathbf{f})$ to a μ^* with $(\pi^*, \mathbf{f}^*)$.

Example 10.21. Consider Figure 10.13.
In this model we have

$$\begin{array}{lcl} S & = & \{x, y, z, w\} \\ \mathbb{A} & = & \{a, b\} \\ R_1 & = & \{x \overset{a}{\to} y, x \overset{b}{\to} z, z \overset{a}{\to} w, z \overset{b}{\to} w\} \\ R_2 & = & \varnothing \\ R_5 & = & \{(x \overset{a}{\to} y) \twoheadrightarrow (x \overset{b}{\to} z), (x \overset{b}{\to} z) \twoheadrightarrow (z \overset{a}{\to} w), (x \overset{b}{\to} z) \twoheadrightarrow (z \overset{b}{\to} w)\} \\ & \mathbf{f} \text{ is always } 1 & \\ \pi(a) & = & \pi(b) = x \end{array}$$

We leave the definition of G open for now, except that we agree that π is acceptable to G.

Let us examine several scenarios:

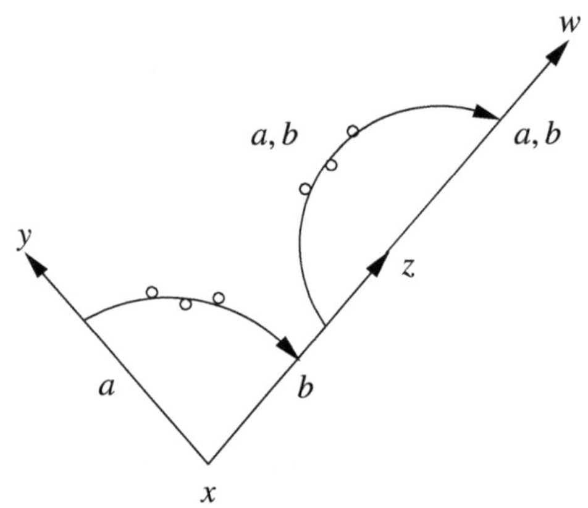

Fig. 10.13

1. $\{a\}$ moves first alone

a can move only to y. As he traverses $x \stackrel{a}{\rightarrow} y$, the lightning double arrow sends a signal to $x \stackrel{b}{\rightarrow} z$ which forces b to move to z. As b moves to z a signal is sent to $z \stackrel{a}{\rightarrow} w$ and to $z \stackrel{b}{\rightarrow} w$, which forces a and b, if they are at z, to move to w. a is not at z and b has just arrived at z but was not at z at the time the signal was sent. So b can remain at z, and the new position is that a is at y and b is at z. (Note that this policy is a design decision!) The question is, is this an acceptable position to the function G?

If it is acceptable, we say that a performed a point-to0-point move from π to π' where $\pi(a) = \pi(b) = x$, $\pi'(a) = y$ and $\pi'(b) = z$.

If π' is not acceptable to G, we ask a to continue to move on. a cannot move so we say a is stuck and the move cannot be made. We are back at π.

There is the option to ask b to move on to w. However, since the original move started as an $\{a\}$ move, we have not allowed for this option. b is not part of the original 'team'. (Note again that this is a design decision.)

2. $\{a, b\}$ move together

In this case both a and b move in parallel. As before a moves to y and this sends a signal to b to move out of x. Well, b is moving to z anyway. So again if π' (with a at y and b at z) is an acceptable position then we say $\{a, b\}$ performed a single point-to-point move from π to π'.

If π' is not acceptable, then we can ask b to move further on to w along $z \stackrel{b}{\rightarrow} w$. We now get π'' with $\pi''(a) = y$ and $\pi''(b) = w$. If π'' is acceptable to G we say the team $\{a, b\}$ performed a single point-to-point move from π to π''. (This move was of level 2 as b had to move twice to get to an acceptable position, but overall it executed a single point-to-point move.)

If π' is not acceptable to G then $\{a, b\}$ cannot execute a single point-to-point transition/move.

3. $\{b\}$ moves first alone

In this case b moves to z along the arc $x \xrightarrow{b} z$. If π' (where $\pi'(a) = x$ and $\pi'(b) = z$, is acceptable then we say that $\{b\}$ executed a single point-to-point move from π to π'. Otherwise b can move on to w, along the arc $z \xrightarrow{b} w$. We get π'' where $\pi''(a) = x$ and $\pi''(b) = w$.

If π'' is acceptable then we say that $\{b\}$ executed a single point-to-point move from π to π''. Otherwise b remains at x and we say that b is stuck at x ("stuck" of level 2 as b makes 2 moves to discover he is stuck) and cannot execute any single point-to-point move.

Example 10.22 (Looping). Figure 10.14 is an example of a loop.

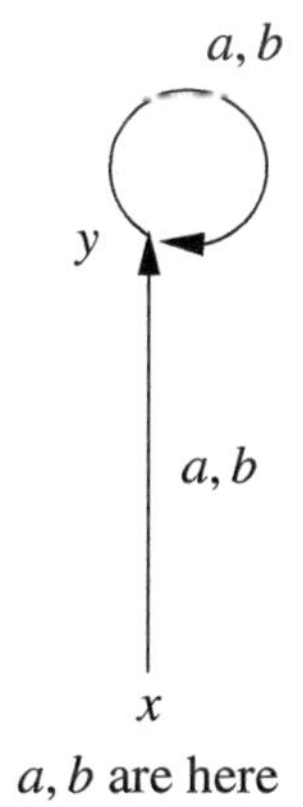

Fig. 10.14

a, b are both at x. This is acceptable. They want to make a point-to-point joint move to y. Moving to y they discover they are at an unacceptable state. So they must move on. They keep on moving to y. Here we need a loop checker to realise there is a loop and declare that $\{a, b\}$ cannot make a point-to-point move from x to y.

Definition 10.23 (A model in transition).

1. *Let μ be a model as in Definition 10.16. Let $\mathbb{C} \subseteq \mathbb{A}$ be a coalition. If the members of $\mathbb{C}$ decide to move along some arcs, we get a new model μ_1 which may be temporary as it may need a further chain of moves until an acceptable model is reached (see Example 10.21). To describe the chain of models along the way we*

need the notion of a 'model $\tilde{\mu}$ in transition' (the wave above μ shows it may be in transition). $\tilde{\mu}$ has the following components:

1.1. $S, \mathbb{A}, \pi, G, R_1, R_2, R_5, \mathbf{f}$ and $\mathbb{QM}$ are as in Definition 10.16.
In addition we have the following:

1.2. A non-empty set $\mathbb{C} \subseteq \mathbb{A}$, this is the coalition which is making the moves.

1.3. We allow $G(\pi)$ to be undefined, i.e. π is not necessarily acceptable.

1.4. We have another set $\mathbb{E} \subseteq \mathbb{A}$, of agents who are forced to move. For each $a \in \mathbb{E}$, we have a set of R_1 connections ρ_a, of the form $\pi(a) \xrightarrow{a} y$ such that a has to move along one of the members of ρ_a, irrespective of whether $\mathbf{f}(\pi(a) \xrightarrow{a} y) = 1$ or $\mathbf{f}(\pi(a) \xrightarrow{a} y) = 0$.
Note that if π is G-acceptable and $\mathbb{E}$ is empty (nobody needs to move) we have an ordinary model μ with a coalition $\mathbb{C}$. This model is stable.

2. Let $\tilde{\mu}$ be a model in transition. Let us execute a transition for this model.

2.1. If $\mathbb{E} = \varnothing$ and π is not acceptable by G then

- *If none of $\mathbb{C}$ can move (i.e. there is no $c \in \mathbb{C}$ and a y such that $\mathbf{f}(\pi(c) \xrightarrow{c} y) = 1$) then we say that $\tilde{\mu}$ is stuck.*
- *If $\mathbb{E} = \varnothing$ and $\mathbb{C}' \subseteq \mathbb{C}$ is a maximal subset of the elements which can move then let us move with this set $\mathbb{C}'$.*
- *If $\mathbb{E} \neq \varnothing$ then we move with $\mathbb{E}$ (disregarding $\mathbb{C}'$).*

2.2. Let $\mathbb{E}^1$ be the set of elements with which we need to move, as discussed above. Choose for each $a \in \mathbb{E}^1$ a connection $\pi(a) \xrightarrow{a} y_a$ along which a can move. Call this choice a move option for *$\tilde{\mu}$ and denote it by* $\mathbf{mo}(\tilde{\mu})$.

Remember that if $a \in \mathbb{C}$ then we must have $\mathbf{f}(\pi(a) \xrightarrow{a} y_a) = 1$ and if $(\pi(a) \xrightarrow{a} y_a) \in \rho_a$ then we don't care what $\mathbf{f}$ says, and we may have that $\mathbf{f}(\pi(a) \xrightarrow{a} y_a) = 0$.
Let the agents in $\mathbb{E}^1$ simultaneously move from $\pi(a)$ to y_a respectively.
We now get new $\pi', \mathbf{f}'$ as follows:

$$\begin{aligned} \pi'(a) &= \pi(a) \text{ if } a \notin \mathbb{E}^1 \\ \pi'(a) &= y_a \quad \text{ if } a \in \mathbb{E}^1 \end{aligned}$$

To define $\mathbf{f}'(u \xrightarrow{d} w)$ proceed as follows:
Consider $\{a \in \mathbb{E}^1 | \pi(a) \xrightarrow{a} y_a) \twoheadrightarrow (u \xrightarrow{d} w)$ is in $R_2\}$.
If the size of this set is odd then let $\mathbf{f}'(u \xrightarrow{d} w) = 1 - \mathbf{f}(u \xrightarrow{d} w)$.
If the size of this set is even then let $\mathbf{f}'(u \xrightarrow{d} w) = \mathbf{f}(u \xrightarrow{d} w)$.
To define ρ'_d for $d \in (\mathbb{A} - \mathbb{E}^1)$ proceed as follows:
let $(\pi(d) \xrightarrow{d} y)$ be in ρ'_d if for some $a \in \mathbb{E}^1$ we have that

$$(\pi(a) \xrightarrow{a} y_a) \twoheadrightarrow (\pi(d) \xrightarrow{d} y)$$

is in R_5.
Otherwise, ρ'_d is empty.
Let $\mathbb{E}' = \{d | \rho'_d \neq \varnothing\}$.

Design note
Note that we said $d \in (\mathbb{A} - \mathbb{E}^1)$. If $d \in \mathbb{C}$ and d is stuck (so $d \notin \mathbb{E}^1$) then d may be forced to move. If $d \in \mathbb{E}^1$ then it is already moving.
We now have a new model $\tilde{\mu}'$ with $\pi', \mathbf{f}', \mathbb{E}'$ and ρ'. We write $\tilde{\mu}' = \tilde{\mu}'(\mathbf{mo}(\tilde{\mu}))$ to stress that $\tilde{\mu}'$ depends on our choice of move options for $\tilde{\mu}$.
We say that $\tilde{\mu}'$ is a result of executing the required moves dictated by $\tilde{\mu}$, using the move option $\mathbf{mo}(\tilde{\mu})$.
This means that we choose a move option for $\tilde{\mu}$ and execute the move as defined.
Note that if we start with a μ which is stable and choose a coalition $\mathbb{D} \subseteq \mathbb{A}$ in μ which is not stuck we can pretend it is in transition and choose a move option for μ.
In this way any ordinary model can be considered in transition if we pretend that π is not acceptable and start moving.

Definition 10.24 (Single point-to-point move). *Let μ be a model $\mu = (S, \mathbb{A}, \pi, G, R_1, R_2, R_5, \mathbf{f}, \mathbb{QM})$, as in Definition 10.16.*

Let $\mathbb{C} \subseteq \mathbb{A}$ be a coalition. We want to define the concept of:

- *$\mathbb{C}$ makes a single point-to-point move from μ to μ^* along the path*

$$\mathbf{Path} = (\mu, \tilde{\mu}_1, \ldots, \tilde{\mu}_n, \mu^*)$$

where the $\tilde{\mu}_i$ are models in transition.

Let $\tilde{\mu}_0 = \mu$ and $\tilde{\mu}_{n+1} = \mu^*$. Then each $\tilde{\mu}_{i+1}$ is obtained from $\tilde{\mu}_i$ by executing a move option for $\tilde{\mu}_i$ (as defined in Definition 10.23).

Case 0
There are no members $a \in \mathbb{C}$ such that for some $\pi(a) \xrightarrow{a} y \in R_1$ we have $\mathbf{f}(\pi(a) \xrightarrow{a} y) = 1$.

In this case we say that $\mathbb{C}$ is level 0 stuck and cannot make a move.

Case 1
μ is not stuck. So $\mathbb{C}$ can move. We choose a move option for *$\mathbb{C}$ and move from $\tilde{\mu}_0 = \mu$ to a new $\tilde{\mu}_1$. If we can find a move option for which $\mathbb{E}^{(1)}$ of $\tilde{\mu}_1$ is empty and $G(\pi^{(1)})$ is acceptable, we stop here and $\tilde{\mu}_1$, which is an ordinary model not in transit, is the result of our point-to-point move, through the chosen move option.*

If for every choice of move option leading to a $\tilde{\mu}_1$ we find that $\tilde{\mu}_1$ is not an ordinary model but $\tilde{\mu}_1$ is stuck, we say that μ is level 1 stuck.

The remaining case is that for some choices of move options $\tilde{\mu}_1$ is not stuck but there is no choice where $\tilde{\mu}_1$ is an ordinary model. So we have to move for each such $\tilde{\mu}_1$ to a $\tilde{\mu}_2$ by making another choice of move option for $\tilde{\mu}_1$.

Case $k + 1$
Similar analysis as in Case 1.

If $\tilde{\mu}_{k+1}$ is an ordinary model then we say that $\tilde{\mu}_{k+1}$ is the result of a single level $k+1$ point to point move, through the path $(\tilde{\mu}_0, \ldots, \tilde{\mu}_{k+1})$.

If $\tilde{\mu}_{k+1}$ is stuck and is not an ordinary model, then we go back to $\tilde{\mu}_k$ and seek other move options to continue to a new $\tilde{\mu}_{k+1}$ which is not stuck or is an ordinary model.

If all other choices of new move options in $\tilde{\mu}_k$ result in stuck $\tilde{\mu}_{k+1}$, we move back to $\tilde{\mu}_{k-1}$ and seek a new move option for it, etc.

It may be that there is no sequence of $\tilde{\mu}_0 = \tilde{\mu}, \ldots, \tilde{\mu}_{k+1}$ for which $\tilde{\mu}_{k+1}$ is not stuck or is an ordinary model. In this case we say that μ is level $k+1$ stuck.

We continue this construction until one of the following is obtained:

1. *We find a sequence of move options and models $(\tilde{\mu}_0 = \mu, \ldots, \tilde{\mu}_{k+1} = \mu^*)$ which is minimal in k. Then we say that μ can make a level $k+1$ single point-to-point move to μ^*.*
2. *Every possible move choice and every sequence $(\tilde{\mu}_0 = \mu, \ldots, \tilde{\mu}_{k+1})$ ends up with $\tilde{\mu}_{k+1}$ stuck, and not an ordinary model, and $k+1$ is minimal with this property. Then we say μ is level $k+1$ stuck.*
3. *Every sequence of choices and every $(\tilde{\mu}_0, \ldots, \tilde{\mu}_{k+1}, \ldots, \tilde{\mu}_{k+r} = \tilde{\mu}_k)$ is either stuck or loops (i.e. $\tilde{\mu}_{k+r} = \mu_k$).*
   In this case (where there are some loops) we say that μ cannot make any single point-to-point move.
4. *None of 1–3 obtains in which case we move on to case $k+2$ by making $k+1$ choices of move options along the way.*

Definition 10.25. *We can now define a sequence of point-to-point moves in the obvious way as follows. Let μ be a fixed model. Let $\mathbb{C} \subseteq \mathbb{A}, \mathbb{C} \neq \varnothing$.*

Define $\mu \mathbb{R}_{\mathbb{C}} \mu'$ iff there is a single point-to-point move for $\mathbb{C}$ from μ to μ'.

Let $\mathbb{R}$ be the transitive closure of the relation $\bigcup_{\mathbb{C} \subseteq \mathbb{A}} \mathbb{R}_{\mathbb{C}}$.

Definition 10.26 (Turning a flow product into a modal coalition logic). *Let μ be a model as defined in Definition 10.16. We turn it into a coalition multi-modal logic as follows.*

1. *The set of states $\mathbf{S}$ of our logic are all models μ' such that $\mu \mathbb{R} \mu'$ as well as μ itself.*
   Let $\mathbb{Q}$ of $\mathbb{QM}$ be the modal language with modalities $\Box_{\mathbb{C}}$ for each $\mathbb{C} \subseteq \mathbb{A}, \mathbb{C} \neq \varnothing$.
   Let $\mathbb{M}$ of $\mathbb{QM}$ be classical models.
   Define
   - *$\mu \vDash \Box_{\mathbb{C}} A$ iff for all single point to point moves of $\mathbb{C}$ from μ to μ' we have $\mu' \vDash A$.*

   Define $\mu \vDash A$ for A without modality iff $G(\pi) \vDash A$, where G and π are those functions of the model μ.
   If we look at $(\mathbf{S}, \mathbb{R})$ as a model we see we get a coalition logic in the sense of [135].
   Our advantage is that we get into the guts of the global state μ and can tell exactly how the coalition achieves its move to μ'.

10.3 Comparison with multi-modal logics for multi-agent systems

This section compares our local point of view with the traditional global point of view. We use as a starting point the modern papers of S. Borgo [29, 30]. We begin with a typical definition.

Definition 10.27 (Typical modal model for multi-agent systems). *Let $\mathbb{A}$ be a finite set of agents. With each $x \in \mathbb{A}$, let $\mathbf{Act}(x)$ be a set of actions which agent x can execute.*

Assume $\mathbb{A}, \mathbf{Act}(x), x \in \mathbb{A}$ are all pairwise disjoint. Let S be a non-empty set of states. Our modelling scenario is that the agents of $\mathbb{A}$ may be in the global state $s \in S$ and some of them decide to act. Let $\mathbb{C} \subseteq \mathbb{A}$ be the non-empty set of those who decide to act and let $\mathbf{C}$ be a finite set of actions such that

1. *For each $x \in \mathbb{C}$ there exists a unique $y \in \mathbf{Act}(x)$ such that $y \in \mathbf{C}$.*
2. *For each $y \in \mathbf{C}$ there exists a unique $x \in \mathbb{C}$ such that $y \in \mathbf{Act}(x)$.*

We write $\mathbf{C} \in \mathbf{Act}(\mathbb{C})$ for such $\mathbf{C}$s.

Once the actions are executed, the agents (all of them, i.e. all of $\mathbb{A}$) move to a new state s'. We use an indexed unary function and write $s' = \mathbb{F}_{\mathbf{C}}(s)$. We can also use a binary relation $R_{\mathbb{C}}$ and write $sR_{\mathbb{C}}s'$ iff for some $\mathbf{C} \in \mathbf{Act}(\mathbb{C})$, $s' = \mathbb{F}_{\mathbf{C}}(s)$.

Our Kripke models have the form $(S, R_{\mathbb{C}}, \mathbb{F})$, where $\emptyset \neq \mathbb{C} \subseteq \mathbb{A}$.

We also use a modal syntax for the above model, having a modality $\Box_{\mathbb{C}}$ for each $\emptyset \neq \mathbb{C} \subseteq \mathbb{A}$ as above. We adopt the semantic clause $(*)$ below

$$(*) \qquad t \vDash \Box_{\mathbb{C}} A \text{ iff for all } s \text{ such that } tR_{\mathbb{C}}s \text{ we have } s \vDash A.$$

Here A is a propositional wff built up from atoms, the classical connectives $\wedge, \vee, \neg, \to$ and all the modalities $\Box_{\mathbb{C}}$.

Remark 10.28. Note that using the semantics of Definition 10.27 we can define the Marc Pauly's concept [135] of

- the set of agents $\mathbb{C}$ being at the state $s \in S$, can force the next state to be in the subset $T \subseteq S$.

We write

- There exists a $\mathbf{C} \in \mathbf{Act}(\mathbb{C})$ such that $\mathbb{F}_{\mathbf{C}}(s) \in T$.

  We can also write

$$s \vDash \Diamond_{\mathbb{C}} \chi_T$$

where χ_T is the characteristic function of T, i.e.

$$t \vDash \chi_T \text{ iff } t \in T$$

Example 10.29. Note that in Definition 10.24 we have defined the notion of a single point-to-point move of a local multi-agent system. This notion can give rise to a model in the sense of Definition 10.27, where $\Box_{\mathbb{C}} A$ holds at μ means that A holds in all the results μ^* of all single point-to-point moves of the system as defined in Definition 10.24.

- $\mu \vDash \Box_{\mathbb{C}} A$ iff A holds at all μ^* such that $\mathbb{C}$ can make a single point to point move from μ to μ^*.

The models arising from Definition 10.24 are more restricted than the models defined in Definition 10.27. The reason is that for a coalition $\mathbb{C}$ to be able to make a move together, at least one of its members should be able to move alone. Thus although two agents may not be able to shift a heavy table from one room to another on their own, they may be able to do it together. This situation cannot be modelled in the reactive model because it satisfies

- If $\{x\}$ cannot move out of state s and $\{y\}$ cannot move out of state s then $\{x,y\}$ cannot move out of state s.

While in the model of Definition 10.27 we can have that the following is consistent:

- $\neg\exists t(sR_{\{x\}}t) \wedge \neg\exists t(sR_{\{y\}}t) \wedge \exists t(sR_{\{x,y\}}t)$.

We need to remedy this situation. For this we add the option of joint moves. See Figure 10.15

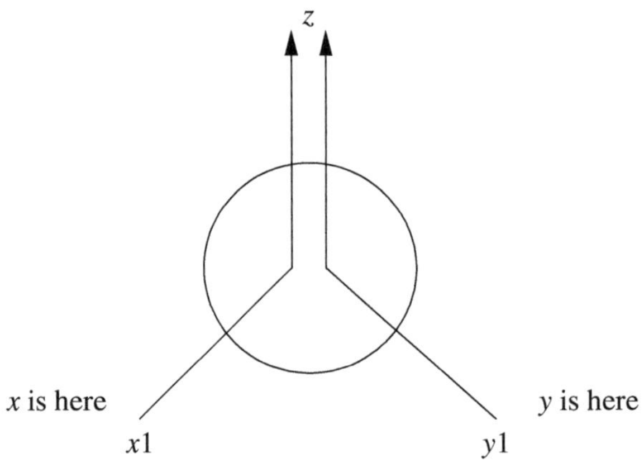

Fig. 10.15

Agent x is at local state $x1$ and agent y is at local state $y1$. They cooperate and make a joint move and end up in state z which is local and common to each of them, see Figure 10.16. This type of action is also used in argumentation networks where joint attacks are discussed. See [83, 131]. In [83] disjunctive actions are also introduced. These have the form of Figure 10.17.

Agent x takes an action at $x1$ and is happy if he non-deterministically ends up in either $z1$ or $z2$. For example, I apply for a prize competition and I am happy to get either first or second prize.

Example 10.30. Let us modify the case of the two blocks world and allow the two blocks to cooperate in a joint action and together put the table on top of themselves. Let us say that should they not like being under the table then their only action is to cooperate again and put themselves on top of the table.

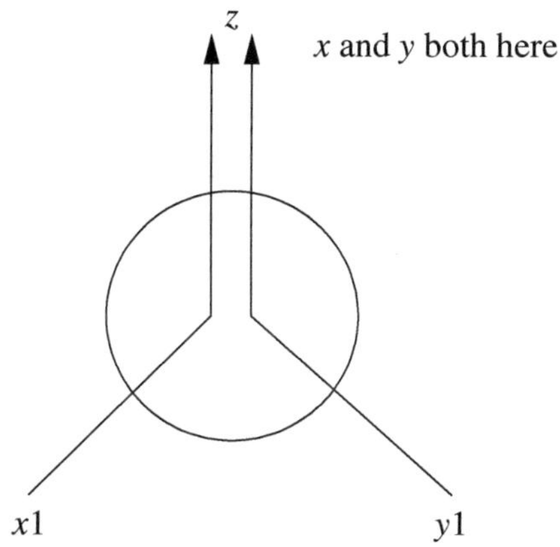

Fig. 10.16

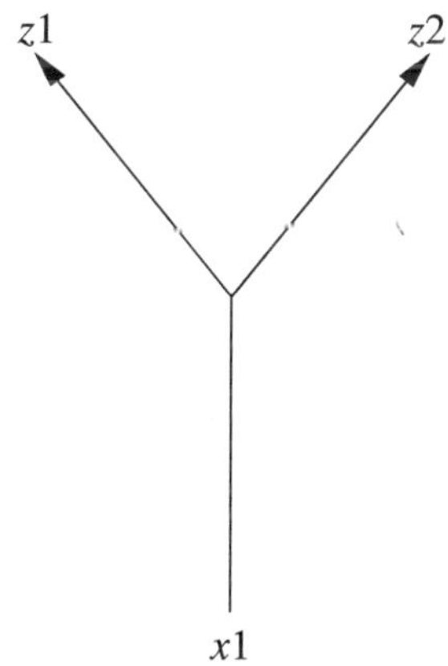

Fig. 10.17

Consider Figure 10.10. Add to it two more states $x0$ and $y0$

$x0$: x is under the table
$y0$: y is under the table

We add Figure 10.18 to Figure 10.10.

We can allow a different scenario where they need each other to put the table on top of them but each one of them can independently go back on his own on top of the table. However, in such a case the other agent is also forced to move back on top of the table (to avoid the heavy table resting on top of him alone). This scenario is described in Figure 10.19.

Example 10.31. Consider the utility agents, $\{x,y\}$ providing electric power to a large number of consumers. Suppose they are in states $x1$ and $y1$ respectively with respect

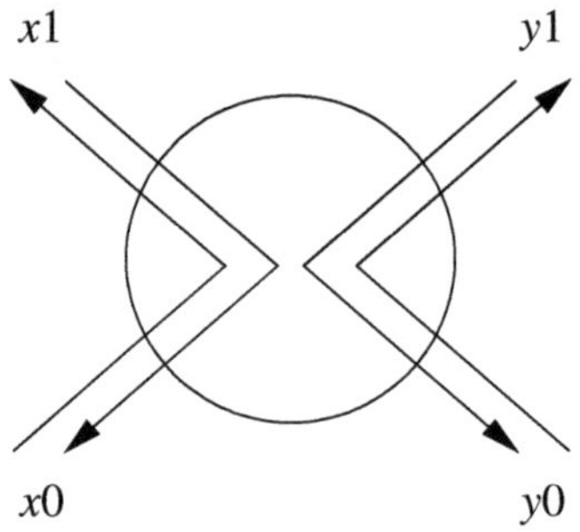

Fig. 10.18

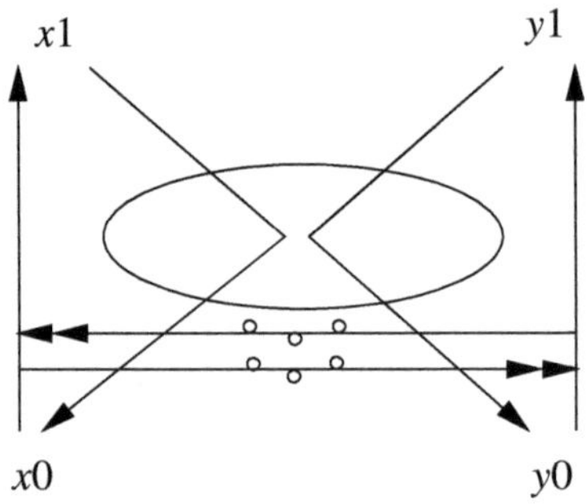

Fig. 10.19

to the price they charge. Because of market changes they each have the following possibilities:

1. raise prices by 10%
2. keep price as it is
3. form a coalition and raise price together (by 10% each).

The best scenario for them is to raise prices (both of them) but without forming a coalition (this way they cannot be accused of fixing the price). The worst scenario is that one raises the price and the other does not. The one who keeps the price down gains more, the other loses.

Figure 10.20 describes the options

The model of Definition 10.27 cannot make the distinction between simultaneous moves with cooperation and simultaneous moves without cooperation.

Example 10.32 (Accidents and partial knowledge). This example discusses the differences between the local view of agents taking actions based on their local states and the global view of agents having only partial knowledge of the global state but nevertheless taking actions based on this partial knowledge.The difference is illustrated by means of a simple example. Agent x backs his car out of his driveway into

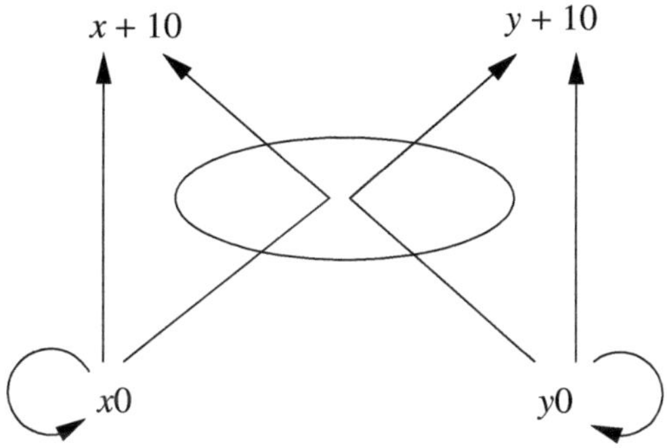

Fig. 10.20

the street. He can see Agent z coming towards him in his car but assumes z has ample space to go round the back end of x's car. At the same time Agent y turns left into the same street. Agent z is forced to veer and kills a pedestrian. Now if we adopt the local view, each agent changed state more or less reasonably and it is possible for each one of them to say that it was an accident. In our language we say that the three new local states of x, y and z combined into an unfortunate global state. If we adopt the view of one initial global state and each agent having a partial knowledge of it and taking action based on his respective partial knowledge, then it is very difficult to claim that there was an accident. Why on earth did the agents take action before they knew the full initial state! There cannot be accidents any more — only negligence! We could introduce the notion of having sufficient partial knowledge of a global state to safely execute an action. Yes indeed this is possible. But then are we not introducing local states through the back door? We can of course introduce this concept to the local state model as well and have the best of both worlds.

Example 10.33 (Eliminating an agent). The device of Example 10.30, the joint action arrow of Figures 10.16 and 10.20, allows us to have multi-agent models in which several agents can get together and eliminate (murder) another agent from the game. They could move jointly to local states which sends the victim to a dead end from which he cannot move. This cannot be done with such clarity in a global state multi-agent model.

11

Reactive Standard Deontic Logic

11.1 Standard Deontic Logic and its Problems

This chapter shows the power of the reactive idea in solving the Chisholm paradox, a difficult problem of **SDL**(Standard Deontic Logic). We introduce a reactive variant of **SDL**: **SDLR1** (reactive standard deontic logic). Given a Kripkean view of the semantics of **SDL** in terms of directed graphs where arrows $\to$ represent the accessibility relation between worlds, reactive models add two elements: arrows $\to$ are labelled as "active" or "inactive", and double arrows $\twoheadrightarrow$ connect arrows, e.g., $(x_1 \to x_2) \twoheadrightarrow (x_3 \to x_4)$. The idea is that passing through $x_1 \to x_2$ activates a switch represented by $\twoheadrightarrow$ which inverts the label of $x_3 \to x_4$ and hence activates resp. deactivates this arrow. This allows us to introduce two modalities: $\Box$ is the usual **KD**-modality of **SDL** and operates on the Kripkean graph where all labels and double arrows are ignored, while $\oslash$ takes them into account.

We demonstrate that **SDLR1** allows for an intuitive interpretation of 'ought'. The logic can handle contrary-to-duty cases such as several instantiations of the Chisholm set in a paradox-free way by means of using double arrows and annotations to block and give access to ideal worlds.

A logic with modality $\Box$ is a **KD** modality if we have the axioms

K0 All substitution instances of classical tautologies

K1 $\Box(p \wedge q) \equiv (\Box p \wedge \Box q)$

K2 $\vdash A \Rightarrow \vdash \Box A$

D $\neg\Box\bot$

It is complete for frames of the form (S, R, a) where $S \neq \varnothing$ is a set of possible worlds, $a \in S, andR \subseteq S \times S$ is serial (i.e., for all x there exists a y such that xRy).

Standard Deontic Logic **SDL** is a **KD** modality O. We read $u \vDash Op$ as saying p holds in all ideal worlds relative to u, i.e. for all t we have: if uRt then $t \vDash p$. So the set of ideal worlds relative to u is the set

$$I(u) = \{t \mid uRt\}.$$

D.M. Gabbay, *Reactive Kripke Semantics*, Cognitive Technologies,
DOI 10.1007/978-3-642-41389-6_11, © Springer-Verlag Berlin Heidelberg 2013

The **D** condition says

$$I(x) \neq \varnothing \text{ for } x \in S.$$

Following [115], let us quickly review some of the difficulties facing **SDL** in formalizing certain examples. These paradoxes are as follows, see [38, 90, 91, 62].

The Chisholm paradox

Consider the following statements:

1. It ought to be that a certain man goes to the assistance of his neighbour.
2. It ought to be that if he does go he tells them he is coming.
3. If he does not go then he ought not to tell them he is coming.
4. He does not go.

It is agreed that intuitively (1)–(4) of Chisholm are consistent and totally independent of each other. Therefore it is expected that their formal translation into logic **SDL** should retain these properties.

Let us semantically write the Chisholm set in semiformal English, with p and q as follows, p means HELP and q means TELL.

1. Obligatory p.
2. $p \to$ Obligatory q.
3. $\neg p \to$ Obligatory $\neg q$.
4. $\neg p$.

Consider also the following:

5. p.
6. Obligatory q.
7. Obligatory $\neg q$.

We intuitively accept that (1)–(4) are consistent and logically independent of each-other. Also we accept that (3) and (4) imply (7), and that (2) and (5) imply (6). Note that some authors would also intuitively expect to conclude (6) from (1) and (2).[1]

Now suppose we offer a logical system **L** and a translation τ of (1), (2), (3), (4) of the Chisholm set into **L**.

For example **L** could be Standard Deontic Logic or **L** could be a modal logic with a dyadic modality $\bigcirc_Y X$ (X is obligatory in the context of Y).

We now list coherence conditions for the translation τ and for **L**.

We expect the following to hold.

(a) (2) and (3) are translated the same way, i.e., we translate the form:
(23): $X \to$ Obligatory Y
to be $\psi(X, Y)$ and the translation does not depend on the fact that we have (4) $\neg p$ as opposed to (5) p.

[1] The principle that allows for such inferences is called *deontic detachment* (see [123]). See also [150] for a discussion of its role in the context of the Chisholm set.

(b) The translations of (1)–(4) remain logically independent in **L**.

(c) The translated system maintains its properties under reasonable substitution in **L**.

The notion of reasonable substitution is a tricky one. Let us say for the time being that if we offer a solution for one paradox, say $\Pi_1(p, q, r, \ldots)$ and by substitution for $p, q, r, \ldots$ we can get another well-known paradox Π_2, then we would like to have a solution for Π_2. This is a reasonable expectation from mathematical reasoning. We give a general solution to a general problem which yields specific solutions to specific problems which can be obtained from the general problem.

(d) The translation is essentially linguistically uniform and can be done item by item in a uniform way depending on parameters derived from the entire database. To explain what we mean consider in classical logic the set

(1) p

(2) $p \to q$.

To translate it into disjunctive normal form we need to know the number of atoms to be used. Item (1) is already in normal form in the language of $\{p\}$ but in the language of $\{p, q\}$ its normal form is $(p \wedge q) \vee (p \wedge \neg q)$. If we had another item

(3) r

then the normal form of p in the language of $\{p, q, r\}$ would be

$(p \wedge q \wedge r) \vee (p \wedge q \wedge \neg r) \vee (p \wedge \neg q \wedge r) \vee (p \wedge \neg q \wedge \neg r)$.

The moral of the story is that although the translation of (1) is uniform algorithmically, we need to know what other items are in the database to set some parameters for the algorithm.

Jones and Pörn, for example, examine in [115] possible translations of Chisholm (1)–(4) into **SDL**. They make the following points:

(1) If we translate according to, what they call, option *a*:

(1a) Op

(2a) $O(p \to q)$

(3a) $\neg p \to O\neg q$

(4a) $\neg p$

then we do not have consistency, although we do have independence.[2]

(2) If we translate Chisholm item (2) according to what they call option *b*:

(2b) $p \to Oq$

then we have consistency but not independence, since (4a) logically implies (2b).

(3) If (3a) is replaced by

(3b) $O(\neg p \to \neg q)$

then we get back consistency but lose independence, since (1a) implies (3b).

(4) Further, if we want (2) and (5) to imply (6), and (3) and (4) to imply (7) then we cannot use (3b) and (2a).

---

[2] Note that this translation is also not uniform in its treatment of conditional obligations (see (2a) and (3a)) and hence violates our coherence condition (a).

The translation of the Chisholm set is a "paradox" because known translations into Standard Deontic Logic (the logic with O only) are either inconsistent or dependent.

The fence paradox

The following example is formulated in [135]. Compare with items (1)–(4) of the Chisholm paradox.

1. There ought to be no fence.
2. [We are not dealing with this item here.]
3. If there is a fence it should be white.
4. There is a fence.

The fence paradox is a variation of the original gentle murderer paradox of Forrester [49].

1. It is obligatory that Smith does not murder Jones.
2. [we are not dealing with this item here.]
3. It is obligatory that, if Smith murders Jones, Smith murders Jones gently.
4. Smith murders Jones.

Of course the fence can be taken down, but the murder cannot be undone. This difference can be significant when we consider our options in a long chain of violations and contrary-to-duties.

The problems with **SDL** are so serious that nowadays researchers in the field call it 'silly deontic logic' instead of 'standard deontic logic'.

There have been many proposals for better systems to accommodate Contrary-to-Duties ranging from slight variations on **SDL** (see, for example, [115] and [28]) to considerably different systems (see, for example, [151, 135]), mainly the dyadic preferential approach and the temporal approach. See also [35, 94]. Our own reactive approach is new, see [64, 102, 64], and bridges the gap between the two extremes, as this chapter will show.

The community seems quite happy with a dyadic modal operator $\bigcirc_A B$, reading B is obligatory in the context of A. Thus we have for the fence example

1. $\bigcirc\neg$ fence
2. $\bigcirc_{\text{fence}}$ white-fence.

The dyadic deontic logic satisfies some axioms which can basically turn it into a preferential system. We need not go into details here. See [136, 151, 135].

Our aim is to show that **SDL** is not so silly; by turning it *reactive* and by looking at reactive standard deontic logic, **SDLR1**, we can save the situation. See [102, 64] for a more general reactive approach.

Consider the situation in Figure 11.1. Its graph describes a possible world model, of a reactive kind, for a special reactive modality $\oslash$.

The nodes $\{t, s, x, y\}$ are possible worlds. The syntactical word 'fence' represents an atomic proposition and by writing 'fence' next to x we inform the reader that

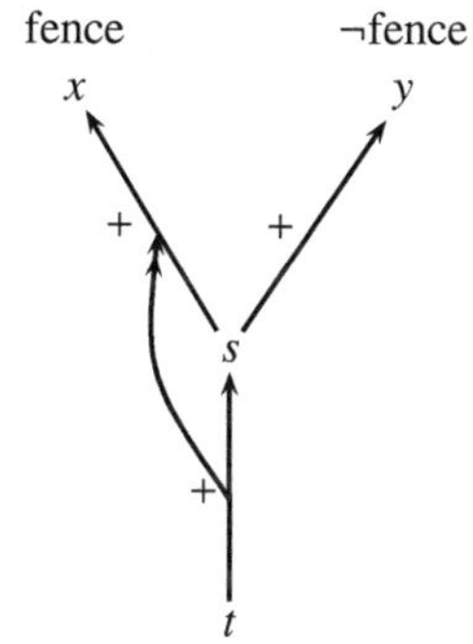

Fig. 11.1: A reactive graph

in some given assignment h (fixed for our discussion) we have $x \vDash$ fence under h. Similarly we have $y \vDash \neg$ fence under h. The arrows $s \to y, t \to s$ and $s \to x$ are annotated with either '+' or '−'. The '+' annotation means the connection is active (or connected) and the '−' annotation means the connection is not active or not connected. The double arrow from $t \to s$ to $s \to x$ is a switch. It says that if we pass from t to s along the arrow $t \to s$, then the double arrow gets activated and disconnects the arrow from s to x, if it is active, and connects it, if it is not active.

Suppose we want to evaluate $t \vDash \oslash\oslash\neg$ fence.

We have $t \vDash \oslash\oslash\neg$ fence iff (since $t \to s$ is active) $s \vDash \oslash\neg$ fence. Now since we passed through $t \to s$ to get to s, the connection $s \to x$ is switched off and we continue (since $s \to y$ is active):

$s \vDash \oslash\neg$ fence iff $y \vDash \neg$ fence

which indeed holds.

Thus, in the model of Figure 11.1 we indeed have that $t \vDash \oslash\oslash\neg$ fence.

We can add a connective $\Box$ which ignores all the double arrows and all annotations. Thus we have

$$t \vDash \Box\Box\neg\text{Fence} \text{ iff } s \vDash \Box\neg\text{Fence} \text{ iff } (x \vDash \neg\text{Fence} \text{ and } y \vDash \neg\text{Fence})$$

Since in Figure 11.1 $x \vDash$ Fence we get that $t \nvDash \Box\Box\neg$Fence .

Our logic **SDLR1** is the logic with $\Box$ and $\oslash$. We introduce it formally in the next section. Then, in sections 11.3– 11.5 we shall translate the deontic paradoxes (such as clauses (1)–(4) of the Chisholm set and (HC1) and (HC2) of Figure 11.3 below) into it.

11.2 Formal properties of SDLR1

This section develops the formal technical machinery for our chapter. We give formal definitions of the reactive models we use. We also investigate an axiomatisation of our logic relative to the proposed semantics.

11.2.1 Introducing and discussing the semantics

Definition 11.1 (Reactive K model).

1. *Let S be a non-empty set. A reactive relation on S has the form*

$$R = R_1 \cup R_2$$

where $R_1 \subseteq S^2$ and $R_2 \subseteq S^2 \times S^2$ such that the following holds

$$\textit{If } ((x,y),(w,z)) \in R_2, \textit{ then } (x,y) \in R_1 \textit{ and } (w,z) \in R_1.$$

(This means $R_2 \subseteq R_1 \times R_1$).
We can represent the elements of R_1 as '$x \rightarrow y$' and the elements of R_2 as '$(x \rightarrow y) \twoheadrightarrow (w \rightarrow z)$'. The elements of R_1 are called arrows and those of R_2 are called double arrows.
2. *A function $\mathbf{f} : R_1 \rightarrow \{0,1\}$ is called an activity function. When $\mathbf{f}(x,y) = 1$ we say $x \rightarrow y$ is 'on' or is 'active' and we write $x \overset{+}{\rightarrow} y$. When $\mathbf{f}(x,y) = 0$, we say $x \rightarrow y$ is 'off' or 'not active' and we write $x \overset{-}{\rightarrow} y$.*
3. *An assignment h is a function giving each atomic q a subset $h(q) \subseteq S$.*
4. *A model has the form $\mathbf{m} = (S, R_1, R_2, \mathbf{f}, a, h)$ where $a \in S$ is the initial world.*

Definition 11.2 (Satisfaction). *Our language contains the classical connectives $\neg, \wedge, \vee, \rightarrow, \top, \bot$ and the two modalities $\Box$ and $\oslash$. We define satisfaction in a model as follows.*

1. *Let $(S, R_1, R_2, \mathbf{f})$ be given. Where $(x,y) \in R_1$ and $x \overset{+}{\rightarrow} y$, we define a new function, called $\mathbf{f}[x,y]$, which is derived using $\mathbf{f}$ and (x,y), and is defined by indicating what values this function gives to arbitrary pairs (u,v) in R_1 as follows:*

$$\mathbf{f}[x,y](u,v) = \begin{cases} \mathbf{f}(u,v) \textit{ if } ((x,y),(u,v)) \notin R_2 \\ 1 - \mathbf{f}(u,v)) \textit{ if } ((x,y),(u,v)) \in R_2 \end{cases}$$

2. *Given a model $\mathbf{m} = (S, R_1, R_2, \mathbf{f}, a, h)$ we define the notion of $\mathbf{m} \vDash A$ by induction as follows:*
   - $\mathbf{m} \vDash q$ *iff* $a \in h(q)$, *for q atomic*
   - $\mathbf{m} \vDash A \wedge B$ *iff* $\mathbf{m} \vDash A$ *and* $\mathbf{m} \vDash B$
   - $\mathbf{m} \vDash \neg A$ *iff* $\mathbf{m} \nvDash A$
   - *Similarly for the other classical connectives*
   - $\mathbf{m} \vDash \Box A$ *iff for all y such that $(a,y) \in R_1$ we have that $\mathbf{n} \vDash A$, for $\mathbf{n} = (S, R_1, R_2, \mathbf{f}, y, h)$*
   - $\mathbf{m} \vDash \oslash A$ *iff for all y such that $(a,y) \in R_1$ and $\mathbf{f}(a,y) = 1$ we have that $\mathbf{n}' \vDash A$ for $\mathbf{n}' = (S, R_1, R_2, \mathbf{f}[a,y], y, h)$.*
3. *The logic* **KR** *(reactive* **K***) is defined as the set of all wffs A such that for all models $\mathbf{m}$ we have $\mathbf{m} \vDash A$.*
4. *The logic* **SDLR1** *is defined analogously, except that R_1 is required to be serial.*

5. *Note that we have frames of the form* $(S, R_1, R_2, \mathbf{f})$. *The logic is normal, as a and h can be chosen arbitrarily.*

Remark 11.3. Note that the function **f** introduced in the previous Definition 11.1 actually defines another relation R_0 which is a subset of R_1: namely the active arrows in R_1. When **f** changes to $\mathbf{f}[x, y]$, as in item 1 of Definition 11.2, then equivalently R_0 changes to $R_0[x, y]$. We use the notation with the function **f** to stress the fact that all reactivity changes arise from arrow connections from R_1.

We now address the task of axiomatising the logic **KR**. The axiomatisation of **SDLR1** strengthens **KR** by **D**. To be able to do this effectively, let us familiarise ourselves better with **KR** and its unique features.

Consider Figure 11.2.

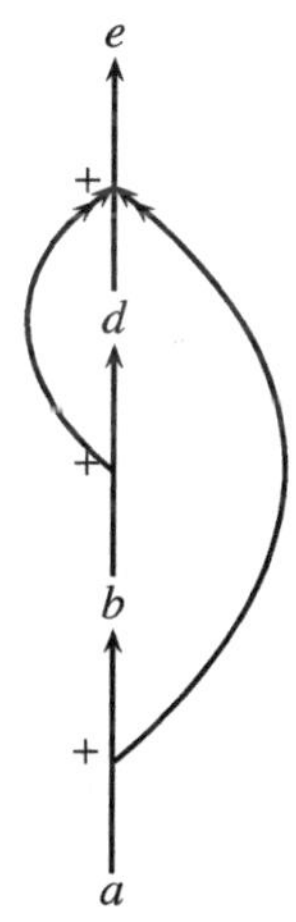

Fig. 11.2: A model for **KR**

Let our starting point be a. As we traverse the arc $a \to b$, there are two possibilities.

1. We ignore all double arrows, and what they do.
2. We take account of what the double arrows say and what they do.

Let possibility 1 be formalised by $\Diamond$ and let possibility 2 be formalised by $\mathbb{P}$. So we can write: $a \vDash \Diamond q$ to mean traverse from a to the accessible world b and ignore all double arrows and $a \vDash \mathbb{P}q$ to mean traverse from a to the accessible world b and take account of the double arrows.

Consider now

$$a \vDash \mathbb{P}\Diamond\Diamond\top$$

Here we move from a to b and activate the double arrow which disconnects the arc $d \to e$. We check:

$$b \vDash \Diamond\Diamond\top$$

and then

$$d \vDash \Diamond\top$$

Now the arc $d \to e$ is disconnected by a double arrow but as we are evaluating $\Diamond$ at node d and $\Diamond$ ignores all double arrows and their effects, as far as $\Diamond$ is concerned $d \to e$ is connected and hence $d \vDash \Diamond\top$ holds and therefore $a \vDash \mathbb{P}\Diamond\Diamond\top$ also holds.

The situation is different with

$$a \vDash \mathbb{P}\Diamond\mathbb{P}\top$$

Evaluating here takes us to $d \vDash \mathbb{P}\top$, but $\mathbb{P}$ is affected and takes account of double arrows and so $d \vDash \mathbb{P}\top$ does not hold and hence

$$a \vDash \mathbb{P}\Diamond\mathbb{P}\top$$

does not hold.

Consider now

$$a \vDash \mathbb{P}\mathbb{P}\mathbb{P}\top.$$

We can go from a to e taking account of double arrows. Moving from a to b switches the connection $d \to e$ off but continuing from b to d switches it on again and therefore $d \vDash \mathbb{P}\top$ holds.

Note that we cannot evaluate any $x \vDash \mathbb{P}q$ without knowing how we got to x and which double arrows we activated along the path.

So asking does $d \vDash ?\mathbb{P}\top$ hold cannot guarantee an answer. We need to ask something like

$$a \xrightarrow{\Diamond} b \xrightarrow{\mathbb{P}} d \vDash ?\mathbb{P}\top$$

or

$$a \xrightarrow{\mathbb{P}} b \xrightarrow{\Diamond} d \vDash ?\mathbb{P}\top$$

or in general

$$(x_0 \xrightarrow{M_1} x_1 \xrightarrow{M_2} x_2 \to \ldots \xrightarrow{M_n} x_n) \vDash ?A$$

where $M_i \in \{\Diamond, \mathbb{P}\}$.

Note that we need a notion of *legitimate sequences*. The sequence

$$a \xrightarrow{\mathbb{P}} b \xrightarrow{\Diamond} d \xrightarrow{\mathbb{P}} e$$

is not legitimate because by the time we get to d, $d \to e$ is disconnected. So to get to e from d we need to use $\Diamond$.

We thus get a bimodal logic with two modalities $\Diamond$ and $\mathbb{P}$ with evaluation of formulas depending on paths $(x_0, \ldots, x_n)$ but not ordinary paths. We also need to know how we traverse them. So our paths have the form $\mathbf{t} = (x_0, M_1x_1, M_2x_2, \ldots, M_nx_n)$, where $\mathbf{t}$ is a legitimate sequence.

Our task in this section is to find axioms and prove completeness for **KR**. We might have thought that it would be useful to consider **KR** as a bimodal logic with two accessibility relations $R_{\mathbb{P}}$ and $R_{\Diamond}$, and see what properties they have. The problem is that the accessible worlds are the same for both relations, the difference being the way we traverse them. So really our most reasonable option is to consider a transformed Kripke model where the points/worlds are legitimate paths of the form

$$\mathbf{t} = (a, M_1 x_1, \ldots, M_n x_n)$$

such that $aR_1x_1R_1x_2, \ldots, x_{n-1}R_1x_n$ holds.

Thus we start with

$$\mathbf{m} = (S, R_1, R_2, a, h)$$

for example (S, R_1, R_2, a) as in Figure 11.2, and then transform it into a model

$$\mathbf{m}^* = (S^*, R_{\Diamond}, R_{\mathbb{P}}, (a), h)$$

where S^* is the set of all legitimate sequences of the form $\mathbf{t} = (a, M_1 x_1, \ldots, M_n x_n)$ where $aR_1x_1 \wedge x_1R_1x_2\wedge, \ldots, \wedge x_{n-1}R_1x_n$.

We define $R_{\Diamond}$ and $R_{\mathbb{P}}$ by

$$\mathbf{t}R_{\Diamond}\mathbf{t} * (\Diamond y), \text{ when } x_nR_1y,$$
$$\mathbf{t}R_{\mathbb{P}}\mathbf{t} * (\mathbb{P}y), \text{ when } x_nR_1y$$

$*$ is concatenation of sequences.

Let $\mathbf{t} \vDash q$ iff $x_n \vDash q$, for q atomic.

Note that R_1 and R_2 are replaced by $R_{\Diamond}$ and $R_{\mathbb{P}}$. R_1 is used directly in the definition of the sequences $\mathbf{t}$ which go into S^* and R_2 is used in the notion of legitimate sequences.

So $\mathbf{m}$ is replaced by $\mathbf{m}^*$.

If we do this we can get an axiomatisation and a completeness theorem after some hard work, as we have seen in Chapter 5.

11.3 Contrary-to-duty in SDLR1

This section presents our reactive model. We do it in stages, presenting several intermediate models and modifying them.

We use a story (Example 11.4) to motivate our model. We offer the model of Figure 11.4 below as a first approximation, and then we move to the better model of Figure 11.7 and then, for technical reasons, we adopt the slightly better pre-final model of Figure 11.9. This is not the final model, but we stick with it for a while.

We then express the fence example in this model, and discuss it at length.

The final model, which is a slight modification of the pre-final model, is given formally in Section 5.

We want to alert the reader to some disadvantages of our model, and urge her or him to nevertheless continue to read.

The translation turns out to be syntactically complicated, even though the idea is semantically very simple. This can be a disappointment for the syntactically/Hilbert syste- minded reader, but there are two reasons for it:

1. The subtle complexity of the CTD paradox itself. The obligation not to have a fence is in one context while the CTD kicks in in another context. This subtle complexity is unavoidable.
2. The second reason for the complexity is the fact that semantically we are relying on the reactivity of the double arrows to solve the paradoxes while the syntax does not give us connectives which talk directly about the double arrows. This makes the model more complex, both in translating the CTD expressions and, as already mentioned, in axiomatising the logic.

The perceptive reader is sure to detect another problem. The translation of the Deontic modality "Obligatory" is not a pure connective but is done as a combination of the two modalities $\Box$ and $\oslash$ of the system. This is necessitated by the lack of a direct connective in the logic representing the double arrows, despite the fact that semantically the obligation is expressed by the double arrows!

We are thus forced to go in a roundabout way to express the obligations and thus make a mixed use of the modalities of the system. The weakness of the translation is compounded by the need to make it dependent on the structure of the model.

So why consider our system at all?

The answer is that it is semantically very intuitive and has the advantages of offering a systematic solution to the so-called paradoxes and can do multiple CTD sentences all at once. I put forward to the reader that the call for axiomatisation (more specifically, a Hilbert style axiom system) is partly a cultural remnant from the time when modal logic had no satisfactory semantics and when most logics were presented Hilbert style. See however paper [81], where a long chain of Chisholm type sets is considered and modelled in a Hilbert axiom system with a reactive semantical interpretation.

Furthermore, the above difficulties will disappear once we allow a direct syntactic counterpart to our double arrows in the axiomatic system. We can in fact give a really simple axiomatisation as a Labelled Deductive System, see [64] and [55].

We shall address these problems in the concluding section.

We now motivate, through examples, how we treat CTDs in **SDLR1**.

Example 11.4 (UK gas boiler). Many homes in the UK have heating with gas boilers. The technology of such boilers keeps on improving both in efficiency and safety. British Gas inspects each boiler once a year and makes recommendations. In many cases a slightly older model boiler no longer complies with new safety regulations, but of course one cannot ask for the boiler to be replaced just on the strength of that, so some additional safety measures are recommended. However, if one wants to install a new boiler then only the new models can be installed.

Consider now a family who have a very old boiler. This boiler breaks down. It is dead and finished. They want to get a new boiler. The law says only the new models are allowed to be installed. However, the slightly older models are still good, they cost a lot less and still have some good years of service in them. They are perfectly safe with the new additional safety measures added to them. The only problem is that the law does not allow certified engineers to install them. They can only modify older existing boilers but not install them!

Our family a wants to buy a boiler. They go to an agent e, who offers them options. They can get an old model or a ¬old model (= new model). The old model cannot be installed by a certified engineer but unofficially it can be installed and then the family can pretend that this is an existing old model and ask a certified engineer to add the safety modifications.

The clever agent e solves the problem of how to work around the regulations as follows: the sale of new models he assigns to a special department e_2 and the sale of old models he assigns to a separate department e_1. If e_1 is investigated and is caught trying to go around the rules, then that is a local problem. So in practice the agent will sell both boilers himself through different departments, the new boilers department and the old boilers department. Figure 11.3 shows the analogy between the fence example and the boiler example.

The only difference between the two examples is the incentive to the agent e to work through specialised departments e_1 and e_2. There is no such incentive in fences.

So let us go back to our family who want to buy a boiler.

Fence	Boiler
HC1. There should be no fence.	There should be no old boiler (installed).
HC2. If there is a fence it should be white.	If there is an existing old boiler it should be modified with additional safety measures.

Fig. 11.3: Fence and boiler problems compared

They approach agents $e_1, \ldots, e_m$ and each one of these agents offers a list of possible boilers for sale. Let $x^i_1, \ldots, x^i_{m(i)}, y^i_1, \ldots, y^i_{n(i)}$ be the list of boilers offered by e_i. Assume that x^i_j are old boilers , and y^i_j are not old boilers.

The family may observe the rule (HC1). Therefore they may consider only the new boilers. In fact, the agent will probably show them only the new boilers. This means that the story is represented as in Figure 11.4. This figure shows only two agents e_1, e_2, each showing one old boiler and one new boiler. They may be showing the same boilers in which case we have Figure 11.6a. The node a represents the family. The arrow from a to e_1 symbolises the family a going to e_1. (Note that according to Figure 11.4, the family goes directly to agents e_1, e_2. This scenario shall be modified in Figure 11.7, in which the family goes to agent e and agent e refers them to departments e_i.) The nodes x_i, y_i denote boilers for sale (x_i old boilers and y_i new boilers).

The '−' label on $e_1 \to x_1$ switches the link from e_1 to x_1 off. This blockage of access to the old boiler is in accordance with (HC1). Similarly, the arrow $e_2 \to x_2$ is labeled by '−' and hence switched off. Both dealers only present new boilers. Although they have the old boilers (x_1, x_2), they do not show them. Now if our family wants to be presented with old boilers they need to explicitly ask the dealer

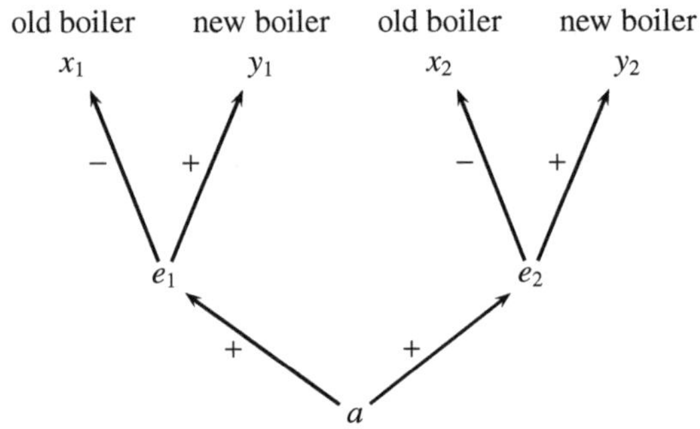

Fig. 11.4: Agents showing only new boilers

for old boilers. This is realized by means of double arrows and illustrated in Figure 11.5.

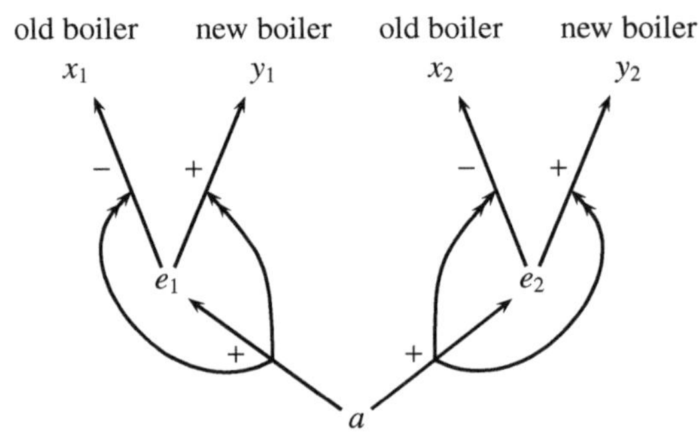

Fig. 11.5: The family asks the dealers to show only old boilers

Hence, if our family wishes to see the old boilers despite (HC1) they can force their way to x_1 and x_2 by changing the '−' label of the arcs via double arrows. They pass via $\mathbb{P}$ from a to the dealer e_i. This activates the double arrows $(a \to e_i) \twoheadrightarrow (e_i \to x_i)$ and $(a \to e_i) \twoheadrightarrow (e_i \to y_i)$. The double arrows invert the labels of the arcs $e_i \to x_i$ and $e_i \to y_i$ resulting in the '+'-labeling of $e_i \to x_i$ and the '−'-labelling of $e_i \to y_i$. Hence, when our family arrives at e_i they see Figure 11.6b.

We now consider Figure 11.5 as a reactive Kripke model.

By evaluating $a \vDash \Box \oslash \neg$ old boiler in Figure 11.5 we are saying that at all agents e_i, $\oslash\neg$ old boiler holds. Hence, (HC1) holds and the agents do not offer any old boiler. This concerns the case when the double arrows do not fire since we move from a to e_i via $\Diamond$. We take the wish of our family to see only the old boilers into account when we move from a to e_i via $\mathbb{P}$. In this case the double arrows fire and the dealers show only the old models. This is expressed by $a \vDash \oslash\oslash$ old boiler.

Thus we may translate

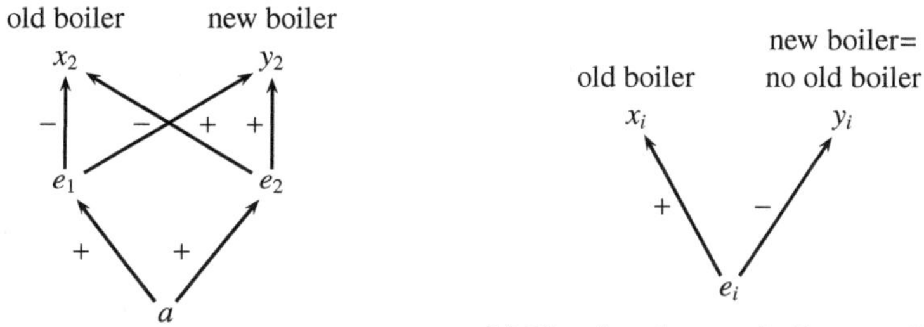

(a) Two agents showing new boilers only

(b) The view that new boiler = ¬ old boiler

Fig. 11.6

(HC1): There should be no old boiler.

as

(HC1)*: $\Box \oslash \neg$ old boiler.

The fact 'old boiler', which we interpret as what the family wants to buy, may be expressed by $\oslash\oslash$ old boiler.

To make the translation clear, let us evaluate $a \vDash \oslash\oslash$ old boiler.

We have $a \vDash \oslash\oslash$ old boiler iff for all e_i that we access via $\mathbb{P}$, $e_i \vDash \oslash$ old boiler.

Since we are arriving at e_i along the arrow $a \to e_i$ via $\mathbb{P}$ the double arrows fire (in Figure 11.5) and hence what we see at e_i is Figure 11.6b.

We now evaluate $e_i \vDash \oslash$ old boiler, and evaluating $\oslash$ does take account of the annotations and so we get:

$$e_i \vDash \oslash \text{ old boiler iff for all } x_i, x_i \vDash \text{ old boiler},$$

which holds.

How do we translate (HC2)? It is not convenient to use Figure 11.5 to model (HC2). The scenario of Figure 11.7 is better. The difference is that we use only one agent node e.

In Figure 11.4 we have two agents, each agent is selling both types of boilers. In Figure 11.7, there is one agent, agent e, who has two departments, one selling old boilers and one selling new boilers, denoted by e_1 and e_2. respectively.

The path from e to the department that sells the old boilers e_1 is labelled '−' and the path to the department that sells the new boilers e_2 is labeled '+'. This is in accordance with (HC1). That means, by default our family is referred to the department that sells only new boilers. (HC1) may be translated according to Figure 11.7 as $\Box \oslash \oslash \neg$ old boiler. However, we will fine-tune this further below.

As before, were our family to wish to see old boilers instead, they would have to request so. This is represented by means of double arrows that emanate from $a \to e$ and target both $e \to e_1$ and $e \to e_2$ (see Figure 11.8a). As the family passes through $a \to e$, a signal is sent to invert the labels of $e \to e_1$ and $e \to e_2$. Note that if the family moves through $a \to e$ via $\mathbb{P}$ then the switching takes place. However, if they move through $a \to e$ via $\Diamond$ then no switching takes place. The latter represents the situation in which we disregard the wishes of the family: then the labels that represent the obligations determine the available paths. The translation of the fact "old boiler" which we interpret as what the family wants to buy is $\oslash \oslash \oslash$ old boiler according to Figure 11.8a.

Now suppose our family is at department e_1 that sells the old boilers. (HC2) is realized by means of the labelling of the arcs $e_1 \to x'$ and $e_1 \to x$. The former leads to the node with the boilers that have added safety x' and hence it is labeled '+' in accordance with (HC2). The latter arc leading to the boilers without added safety x is labeled by '−'. Hence this path is blocked. So the translation of (HC2) will involve $\Box \oslash \oslash$ old boiler with safety.

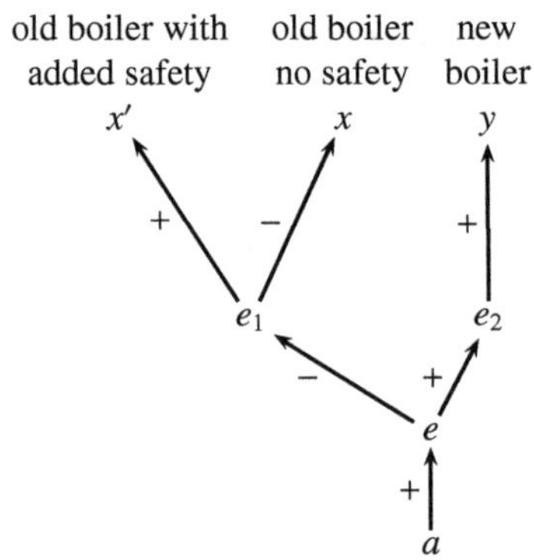

Fig. 11.7: You ought to have a new boiler but if not then you should have a safe old boiler

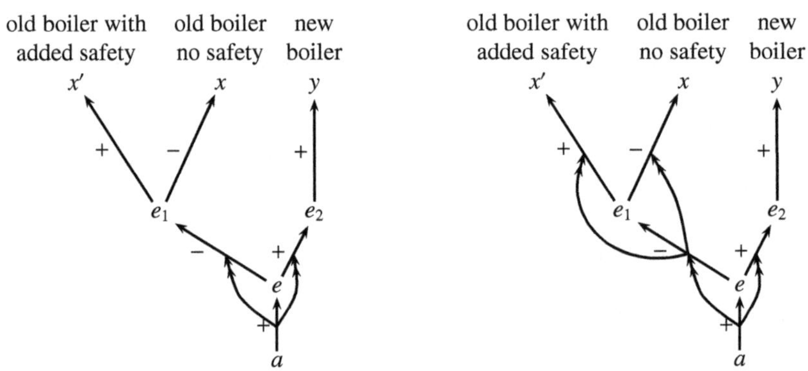

(a) The family wants to see old boilers with added safety.

(b) The family wants to see old boilers without added safety.

Fig. 11.8

Example 11.4 motivated a sub-tree of the reactive tree that can be used for any contrary-to-duty situation. We will enrich it stepwise by adding to the complexity of the considered sets of conditional obligations.

We needed an explanation for our choice of points e, e_1, e_2. If we were to say directly without explanation that we want semantics where every accessibility configuration of the form aRx_i and aRy_i is replaced by $aRe, eRe_1, eRe_2, e_1Rx_i, e_2Ry_j$, the reader would have wondered where these auxiliary e, e_1, e_2 pop up from and why! We have our explanation now.

We now summarise our graph model and analyse it. Once we do that we present the logical system formally and study its properties. We shall see that for technical reasons the pre-final model we want to use is Figure 11.9. This figure is the same as Figure 11.8a, except that we made the top nodes of the tree reflexive. The reasons for this are technical, having to do with substitution, as we shall discuss below.

Remark 11.5 (Summary of the proposed model). We have the CTD obligations of the form below (think of A=fence, B = white fence), where A and B are classical logic formulas:

(A1) Obligatory that $\neg A$
(A2) If it is the case that A then we should have $A \wedge B$ (in the context of A)
(A3) It is the case that A

Figure 11.9 is the graph corresponding to (A1)–(A3) which, when viewed as a reactive Kripke model, is supposed to encode the logical information expressed by the set (A1)–(A3) and thus model it. We first explain Figure 11.9 and then describe how it is done.

The story behind Figure 11.9 is that we have a family standing at node a intending to buy a boiler. We also have an agent e selling boilers, having two departments, department e_1 selling old boilers A, which can be safe B, or not safe $\neg B$, and department e_2 selling new boilers $\neg A$.

The arcs represent paths which the family can take moving from their initial position to the boiler they want to get.

We want to use this setup to represent the contrary-to-duty set (A1), (A2) and (A3). The representation must be intuitive and make (A1),(A2) and (A3) all completely independent of one another.

(A1) is what is obligatory. We represent it by labelling the arcs leading to the departments e_1 and e_2: $e \to e_1$ is labelled '–' and $e \to e_2$ is labeled '+'. This intuitively directs the family from node a to the node e_2 from which $\neg A$ –the new boiler– is accessible.

Similarly, the contrary-to-duty obligation (A2) is represented by labelling the arcs. (A2) concerns what boilers are presented by the department e_1. In accordance with (A2) the arc $e_1 \to x_i'$ is labeled '+' while the arc $e_1 \to x_j$ is labeled '–'.

(A3) are the facts which we interpret as what the family wants to buy, which means the path they want to follow. We give the family control of the double arrows so that they can force their way through the tree even when it means breaking obligations. In that case they need to invert the '–'-label of some arc by means of a double arrow. See Figures 11.8a and 11.8b. In both cases their wish is in conflict with (HC1). Hence, they use the double arrows to block the way to the dealer e_2 with the new boilers and instead free their way to the dealer e_1 who sells the old boilers. In Figure 11.8a their wish is to see old boilers with added safety and hence in coherence with (HC2). Thus, no additional double arrows are needed. The situation is different in Figure 11.8b where they want to see old boilers without added safety. Here they have to use the double arrows to invert the labels of $e_1 \to x$ and $e_1 \to x'$.

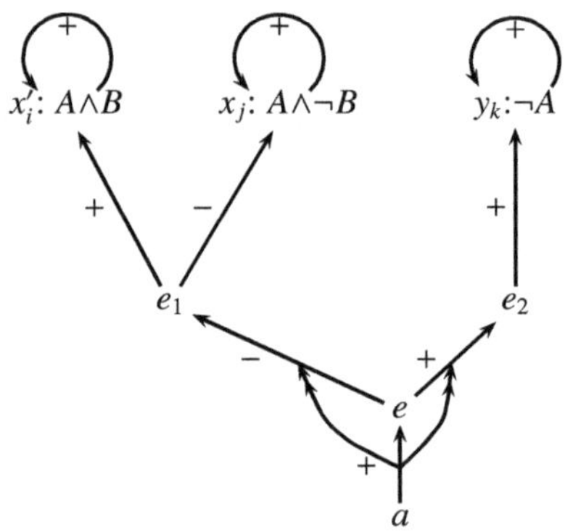

Fig. 11.9: A model for Remark 11.5

$x'_i, i = 1, 2, 3, \ldots$ are all points where $A \wedge B$ holds.
$x_j, j = 1, 2, 3, \ldots$ are all points where $A \wedge \neg B$ holds.
$y_k, k = 1, 2, 3, \ldots$ are all points where $\neg A$ holds.

Note that we added reflexive arrows to the top points, x'_i, x_j and y_k. The reason is that Standard Deontic Logic has the condition that every point has a world accessible to it. So we made the top points reflexive.

Let us now examine how we can express the ideas presented graphically by means of formulas. Let Q be a finite set of atoms, and $n = |Q|$. In the following we presuppose a semantic frame in the form of a tree analogous to the model in Figures 11.9 which has a maximal depth of n+1 from the root node a to the top node. The top nodes are reflexive, as remarked above. In Section 11.5 we will show how this frame can be characterised axiomatically.

We define:

$$\Diamond X =_{\mathrm{df}} \neg\Box\neg X$$
$$\mathbb{P}X =_{\mathrm{df}} \neg \oslash \neg X$$

The position of a node in the graph indicates what is to be considered as being settled when our family reaches the node. For instance, the position of e_1 indicates that the family settled for buying an old boiler, or, in the fence example, it indicates that there is a fence. We can express this as follows: where

$$\mathsf{context}(A) =_{\mathrm{df}} \Box^n A,$$

we have e.g., $e_1 \models \mathsf{context}(A)$, $e_2 \models \mathsf{context}(\neg A)$, $x'_i \models \mathsf{context}(B)$, and $x_j \models \mathsf{context}(\neg B)$. Note that the context of a node is established independent of the labels of the arcs. Note that due to the reflexivity of the top nodes the evaluation of $w \models \mathsf{context}(A)$ always checks whether A is valid at all accessible top models from w. There is no need to keep track of the level of the tree at which the evaluation takes place.

Another important property of nodes is that they provide us with specific choices. For instance, at node e our family has the choice between settling for buying an old (e_1) resp. new boilers (e_2). Similarly, at e_1 they have the choice between going for a boiler with or without added safety. This is expressed as follows: where

$$\mathsf{switch}(A) =_{\mathrm{df}} \Diamond\mathsf{context}(\neg A) \wedge \Diamond\neg\mathsf{context}(\neg A),$$

we have e.g., $e \models \mathsf{switch}(A)$, $e_1 \models \mathsf{switch}(B)$, and $e_1 \models \mathsf{switch}(\neg A \vee B)$.

A conditional obligation $\bigcirc_A B$ expresses that whenever our family stands at a node where A is considered to be settled and they have the choice between a node in a context that is consistent with B and a node where $\mathsf{context}(\neg B)$ holds, they should opt for the former node.

Translated into our language this can be expressed in the following way:

$$\pi^i =_{\mathrm{df}} \Box \ldots \Box \ (i \text{ times}) \text{ where } \pi \in \{\Box, \Diamond, \oslash, \mathbb{P}\}$$
$$\mathsf{opt}(A) =_{\mathrm{df}} \mathbb{P}\neg\mathsf{context}(\neg A) \wedge \neg\mathbb{P}\mathsf{context}(\neg A),$$
$$\mathsf{obliged}_A B =_{\mathrm{df}} \left[\bigwedge_{i=1}^{n} \Box^i(\mathsf{context}(A) \wedge \mathsf{switch}(B) \to \mathsf{opt}(B))\right] \wedge \left[\bigvee_{i=1}^{n} \Diamond^i(\mathsf{context}(A) \wedge \mathsf{switch}(B))\right], \text{ and}$$
$$\mathsf{obliged} B =_{\mathrm{df}} \mathsf{obliged}_{\top} B$$

The conditional obligation is expressed by $a \models \mathsf{obliged}_A B$. Note that in Figure 11.9 we have, for instance,

(*A1) $a \models \mathsf{obliged}_{\top}\neg A$, and
(*A2) $a \models \mathsf{obliged}_A A \wedge B$.

The first conjunct of $\mathsf{obliged}_A B$, namely $\bigwedge_{i=1}^{n} \Box^i(\mathsf{context}(A) \wedge \mathsf{switch}(B) \to \mathsf{opt}(B))$ expresses that, whenever our family is at a node where A is settled and where they have the choice between a path that is consistent with B and a path that leads to $\neg B$, they are supposed to opt for the former.

Note that a given conditional obligation $\bigcirc_A B$ imposes the following succession: if A is considered as settled, B is obliged. This is mirrored in the tree: first it splits between A (node e_1) and $\neg A$ (node e_2), then it splits between B (node x'_i) and $\neg B$ (node x_j). The second conjunction of $\mathsf{obliged}_A B$, namely $\bigvee_{i=1}^{n} \Diamond^i(\mathsf{context}(A) \wedge \mathsf{switch}(B))$, enforces the correct splitting in the tree. Were we to skip the second conjunct we would get pragmatic oddities. For instance, the model in Figure 11.9 would also verify $\mathsf{obliged}_B A$ and $\mathsf{obliged}_B \neg A$.

We have already mentioned that facts represent the wishes of our family. That means, whenever our family has the choice between A and $\neg A$ they "use" the double arrow to block the way to $\neg A$ and opt for A if they wish for A. For instance, when our family approaches the boiler shop e, they activate the double arrows $(a \to e) \twoheadrightarrow (e \to e_1)$ and $(a \to e) \twoheadrightarrow (e \to e_2)$ which inverts the labels on the arcs $e \to e_1$ and $e \to e_2$. After all, their wish is to buy an old boiler (despite (HC1)). This is expressed as follows:

(*A3) $\oslash^{n+1} A$

For instance we have $a \vDash \oslash^3 A$ and $a \vDash \oslash^3 B$ in Figure 11.9. In Section 11.5 it is demonstrated how we can axiomatically ensure that in each model our family takes a unique path to exactly one of the models at the top of the tree.

Remark 11.6. 1. It is very important to notice that if we substitute (A3*) for A inside (A1)* we get (A1)* back. This is because in Figure 11.9 the points x'_i and x_j and y_k are all reflexive, i.e. they are accessible to themselves and hence all modalities collapse.
Therefore, for example, $\Box\Box\oslash(\oslash\Box\oslash A) \equiv \Box\Box\oslash A$, and similarly $\oslash\oslash\oslash(\oslash\Box\oslash A \wedge \oslash\Box\oslash B)$ is equivalent to $\oslash\oslash\oslash(A \wedge B)$. Therefore we can say that first we translate A and B and then substitute the translation in (A2) and (A1) and get (A2)* and (A1)*.

2. Also note that the translations of $A \to \bigcirc B$ is taken essentially as the translation of $\bigcirc_A B$. Therefore we represent the linguistic CTD 'If A then obligatory B' in our logic as saying $\bigcirc_A B$. To see this consider Figure 11.9. The labels implementing $A \to \bigcirc B$ are blocking the arc $e_1 \to x_j$ to the point where $\neg B$ holds by '−' and give access to the point where B holds by means of the '+'-label on $e_1 \to x'_i$. Compare this with the original obligation $\bigcirc\neg A$. The labels that implement this obligation are the '−'-label on $e \to e_1$ blocking access to e_1 and hence to the points where A holds, and the '+'-label on $e \to e_2$ giving access to e_2 where $\neg A$ holds.
3. Note that the assumption is that A and B are classical wffs and are considered complete units for the purpose of translation. Thus for example if X is a wff of classical logic and is translated into say $\oslash\Box\oslash X$ for whatever reason, when $\neg X$ needs to be translated for the same reason, then $\neg X$ is translated as a unit into $\oslash\Box\oslash\neg X$ and not into $\neg\oslash\Box\oslash X$.
4. Note that the translation of the simple fact
(A3) It is the case that A
is rather complex. This happens because the node in the model where (A3) is "committed" and thus a violation of (A1) becomes inevitable is the node e_2. The node where the obligation (A1) is put forward is the node a. One-dimensional Kripke semantics allows us to evaluate at one point only. So we are forced to describe the entire set up from the context of a single point. It may be natural to take this point as e_2, where the violation is, but then we have no backward connective to take us back to point a, where the obligation is. Thus we describe all aspects of the model from point a, because our modalities allow us to go forward. So from point a we must isolate and talk about point e_2 where the violation occurs and make sure it is not "confused" with point e_1.
Expressing all of this becomes complicated.

Note that if we do not make these distinctions we get the paradoxes.
So the naive reader should NOT ask:

- "why is the translation of (A3) complex, after all what (A3) says is just a simple fact?!"

My answer would be:

"yes, it is because you keep it simple that you get paradoxes."

Note that in our paper [28] we keep it slightly more simple by using two-dimensional modal logic, and so we can keep track of both points, the point of obligation and the point of violation.

Remark 11.7 (Dog example, see [135]).
The CTD with the dog is slightly different from that of the fence. It is

(D1) There should be no dog.
(D2) If there is a dog, there should be a sign.
(D3) There is a dog.

Both examples have the form

(X1) $\bigcirc\neg X$
(X2) $X \to \bigcirc Y$
(X3) X

In the fence case we have $\vdash Y \to X$, which we do not have in the dog case.

We have *not* used the special feature $\vdash Y \to X$ in our solution in Remark 11.5. The translation is the same. Figure 11.9 becomes Figure 11.10.

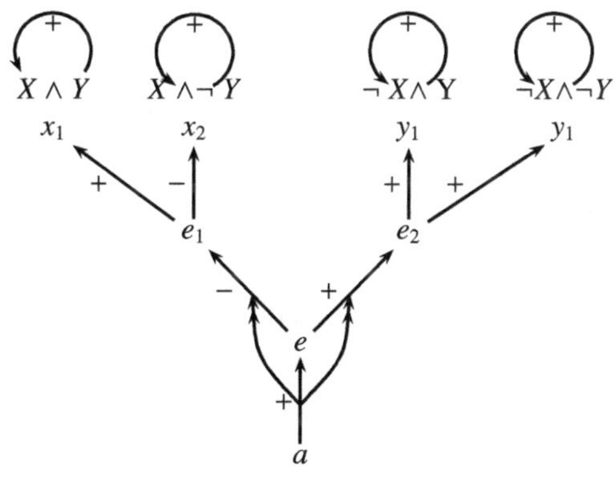

Fig. 11.10: Modelling the dog example

11.4 Checking for Paradoxes

Remark 11.8 (Rationality conditions). When we offer a translation τ into some deontic logic **L** and hope to show that we have no paradoxes, we must satisfy some rationality conditions for **L** and τ.

($\mathbb{R}1$) The semantics for **L** must be compatible in its spirit with our intuitive grasp of obligations and contrary-to-duties.

($\mathbb{R}2$) The translation of Δ into **L**, must be consistent in **L** with no pragmatic oddities (i.e. the translations of Obligatory A and Obligatory $\neg A$ are consistent in **L**). Furthermore, the translation must be modular and incremental. For example, if Δ_1 is a set of obligations (which might be a paradox) and Δ_2 is another set, then
(a) $\tau(\Delta_1 \cup \Delta_2) = \tau(\Delta_1) \cup \tau(\Delta_2)$
(b) If $\Delta\theta$ is obtained from Δ by substitution θ of classical formulas for the atoms in Δ then $\tau(\Delta\theta) = \tau(\Delta)\theta$.[3]

Example 11.9 (Chisholm paradox and rationality of translation). We illustrate our rationality conditions by applying them to the Chisholm set.

1. It ought to be that a certain man goes to the assistance of his neighbour. Written as $\bigcirc H$.
2. It ought to be that if he does go he tells him he is coming, written as $H \to \bigcirc T$.
3. If he does not go then he ought not to tell him he is coming, written $\neg H \to \bigcirc\neg T$.
4. He does not go, written $\neg H$.

Note that if we take in this Chisholm set the subset comprised of (1), (3) and (4) and substitute for

$$\neg H = F$$
$$\neg T = F \wedge W$$

i.e. we let θ be the substitution $H = \neg F$ and $T = \neg(F \wedge W)$, we get the fence example:

1. $\theta : \bigcirc\neg F$
2. $\theta : F \to \bigcirc(F \wedge W)$
3. $\theta : F$

Therefore any solution to the Chisholm paradox (1)–(4) via a logic **L** and a translation τ must also give a solution to the fence paradox by looking at $\tau(1)\theta, \tau(3)\theta$ and $\tau(4)\theta$.

We shall see that the Jones-Pörn solution in [38] does not satisfy the rationality condition.

---

[3] Carmo and Jones [35, p. 275] give the same rationality conditions except that they do not ask for condition $\mathbb{R}$2b. Indeed their system **DL** of [115] is not closed under substitution. See Example 11.9 below, and the discussion in Section 6 below, and see [35]. Our system does satisfy $\mathbb{R}2(b)$ as shown in Remark 11.6.

Example 11.10 (Translation of the Chisholm paradox). We now give the reader an intuitive idea for a rational solution to the Chisholm paradox. A proper solution will be given in Section 5, once we have a precise mathematical basis for our modelling.

Take the set

1. $\bigcirc H$
2. $H \to \bigcirc T$
3. $\neg H \to \bigcirc\neg T$
4. $\neg H$

We use Figure 11.11a for the translation. Note that it satisfies the rationality conditions.

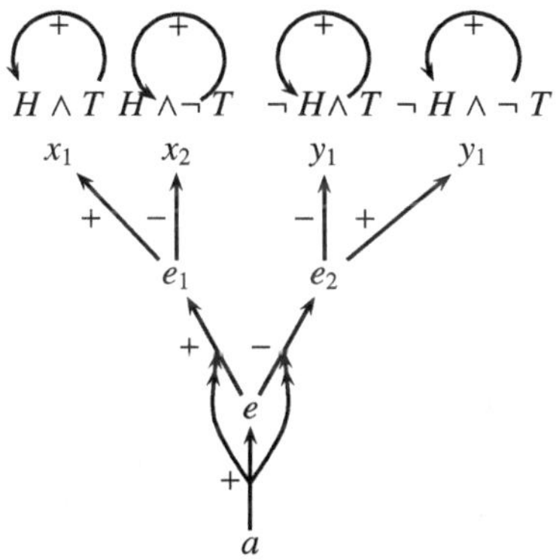

(a) Chisholm paradox

Figure 11.11a	Figure 11.9
H	$\neg A$
T	$\neg B$
x_1	y_k
y_1	x_j
y_2	x'_i
x_2	does not exist

(b)

Fig. 11.11

Let us now reduce Figure 11.11a to Figure 11.9 (the fence example) by looking at clauses (1), (3), (4) of the Chisholm set and by using the substitution θ, where $\theta(H) = \neg F = \neg$Fence ($A$ in Figure 11.9) and $\theta(T) = \neg(F \wedge W)(\neg(A \wedge B)$ in Figure 11.9). Note that this substitution complies with $\mathbb{R}2b$ of the Rationality Condition.

We get

$$x_1 = \neg F \wedge \neg(F \wedge W) = \neg F$$
$$x_2 = \neg F \wedge F \wedge W = \bot$$
$$y_1 = F \wedge \neg(F \wedge W) = F \wedge \neg W$$
$$y_2 = F \wedge (F \wedge W) = F \wedge W$$

We therefore get that after substitution Figure 11.11a becomes Figure 11.9 with the correspondence table as given in Figure 11.11b. Note that in Figure 11.11a we not only use the substitution θ, but we also drop out clause (2).

We have seen that our translation model makes the fence paradox and the Chisholm paradox disappear.

11.5 General theory of CTDs in SDLR1

This section describes a final general model for a general characterization of single CTD sets of arbitrary size (such as all previously considered examples, see the definition below). In the next section we give the most generic characterization that is able to deal with any sets of conditional obligations and facts.

Definition 11.11 (Generalised single CTD set). *Let $Q = \{p_1, \ldots, p_n\}$ be a finite set of atoms and $x_i \in \{p_i, \neg p_i\}$ for each $i \leq n$.*

A general single CTD set Δ of depth n contains

- $\Delta_0 = \{\bigcirc x_1\}$
- *a non-empty subset Δ^i_{CTD} of $\{x_i \rightarrow \bigcirc x_{i+1}, \mathsf{co}x_i \rightarrow \bigcirc \mathsf{co}x_{i+1}\}$, where $\mathsf{co}x_i = \neg p_i$ if $x_i = p_i$ and $\mathsf{co}x_i = p_i$ otherwise*
- *a set of facts $\Delta_F \subseteq \{\tau x_i \mid \tau \in \{\varepsilon, \mathsf{co}\}, i \leq n\}$ where ε is the empty string*

Hence, Δ is a triple $\langle \Delta_0, \Delta_{CTD}, \Delta_F \rangle$ where $\Delta_{CTD} = \bigcup_{i=1}^{n-1} \Delta^i_{CTD}$.

Example 11.12. The dog example and the Chisholm set fall under Definition 11.11.

Remark 11.13. For some CTD sets there are constraints. For instance in the fence example we have that w implies f (a white fence is a fence). We will first disregard such constraints but will later come back to them (see Remark 11.19).

When constructing a model for a CTD set Δ we have two degrees of freedom. On the one hand Δ_F may be incomplete in the sense that for some $p \in Q$ neither p nor $\neg p$ is in Δ_F. We want our models factually complete (so, for each $p \in \Delta_F$ either p is modelled or $\neg p$). Recall that the facts are taken to be the wishes of our family: they reflect which of the top models they want to reach. Hence, in each model M our family will reach a unique top model which determines the facts that are modelled by M. That means that for instance for the Chisholm set with fact g there are two corresponding models: one with fact t (where our family reaches the top model that verifies g and t), and one with $\neg t$ (where our family reaches the top model that verifies g and $\neg t$).

The second degree of freedom concerns cases in which for some $i < n$ e.g. $x_i \rightarrow \bigcirc x_{i+1} \notin \Delta_{CTD}$. There are different options to deal with this situation in a model of Δ: e.g., we could '+'-mark both arcs, the one leading from an x_i-node to an x_{i+1}-node and the one leading from an x_i-node to a $\mathsf{co}x_{i+1}$-node. Or we could validate $\mathsf{obliged}_{x_i} \mathsf{co}x_{i+1}$ in the model (in which case the former arc is labeled '−' and the latter is labeled '+').

For the sake of simplicity, in the following we will presuppose that Δ is complete in the sense that for each $i < n$, $x_i \rightarrow \bigcirc x_{i+1}, \mathsf{co}x_i \rightarrow \bigcirc \mathsf{co}x_{i+1} \in \Delta_{CTD}$.

Definition 11.14 (CTD model template for Δ). *Let Q be a finite set of atoms, and let Δ be a generalised single CTD set based on Q. Let $\Delta^*_F \supseteq \Delta_F$ such that for each p_i, $\{p_i, \neg p_i\} \cap \Delta^*_F = \emptyset$. We define a CTD reactive model template for Δ and Δ^*_F as follows:*

1. *The frame has the form $(S, R_1, R_2, \mathbf{f}, a)$*

2. *Let* $\mathsf{Q}_\Delta = \{\langle\tau_1 x_1, \ldots, \tau_k x_k\rangle \mid k \leq n, \tau_i \in \{\varepsilon, \mathsf{co}\}\}$. *For each* $s \in \mathsf{Q}_\Delta$, $\overline{s}$ *is the set of all members of s.*
3. $S = \{a, e, e_s \mid s \in \mathsf{Q}_\Delta\}$
4. R_1 *contains the following pairs:*
   - (a, e)
   - (e, e_{x_1}) *and* $(e, e_{\mathsf{co}x_1})$
   - $(e_s, e_{s \circ x_i})$ *where* $s, s \circ x_i \in \mathsf{Q}_\Delta$ *and* $\circ$ *is the concatenation function*
   - $(e_s, e_{s\circ\mathsf{co}x_i})$ *where* $s, s \circ \mathsf{co}x_i \in \mathsf{Q}_\Delta$
   - (e_s, e_s) *where s has length n*
5. *the graph has the following labels (where* $s \in \mathsf{Q}_\Delta$*):*
   - $\mathbf{f}(e \to e_{\mathsf{co}x_1}) = 0$
   - $\mathbf{f}(e_{s\circ x_i} \to e_{s\circ x_i \circ \mathsf{co}x_{i+1}}) = 0$
   - $\mathbf{f}(e_{s\circ\mathsf{co}x_i} \to e_{s\circ\mathsf{co}x_i\circ x_{i+1}}) = 0$
   - *all other labels are 1*
6. R_2 *contains the following double arrows as:*
   - $(a \to e) \twoheadrightarrow (e \to e_{x_1})$ *and* $(e \to e) \twoheadrightarrow (e \to e_{\mathsf{co}x_1})$ *iff* $\mathsf{co}x_1 \in \Delta_F^*$
   - $(e_s, e_{s'}) \twoheadrightarrow (e_{s'} \to e_{s'\circ y})$ *iff* $\overline{s'} \subseteq \Delta_F^*$ *and (i)* $\mathbf{f}(e_{s'} \to e_{s'\circ y}) = 1$ *and* $\mathsf{co}y \in \Delta_F^*$, *or (ii)* $\mathbf{f}(e_{s'} \to e_{s'\circ y}) = 0$ *and* $y \in \Delta_F^*$.
7. *for the assignment function* $v : S \to 2^Q$ *we have the following requirement for the models* e_s *on top of our tree* $v(e_s) = \overline{s} \cap Q$.

Remark 11.15. Item 4 in Definition 11.14 defines a modal frame in the form of a tree. Figure 11.12 illustrates this for $n = 3$. Figure 11.13 applies the labelling as in item 5 and the double arrows as in item 6 for the CTD set

$$\Delta_0 = \{\bigcirc x_1\},$$
$$\Delta_{CTD} = \{x_1 \to \bigcirc x_2, \mathsf{co}x_1 \to \bigcirc\mathsf{co}x_2, x_2 \to \bigcirc x_3, \mathsf{co}x_2 \to \bigcirc\mathsf{co}x_3\},$$
$$\Delta_F = \{\neg x_1, \neg x_2, x_3\}.$$

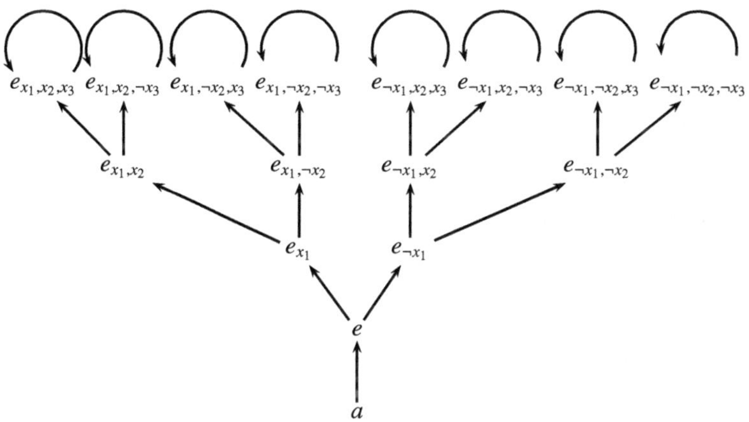

Fig. 11.12: The tree template with $n = 3$

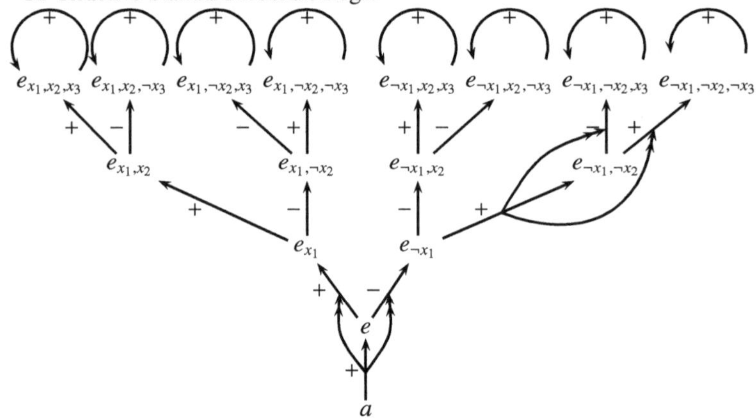

Fig. 11.13: The model template with $n = 3$ (incl. labels and double arrows)

Example 11.16 (Chisholm set). Compare with Example 11.10. We have $Q = \{g, t\}$.

$$\begin{aligned} \Delta_0 &= \{\bigcirc g\} \\ \Delta_{CTD} &= \{g \rightarrow \bigcirc t, \neg g \rightarrow \bigcirc \neg t\} \\ \Delta_F &= \{\neg g\}. \end{aligned}$$

Our models are illustrated in Figures 11.14a and 11.14b (the former for $\Delta_F^* = \{\neg g, \neg t\}$ and the latter for $\Delta_F^* = \{\neg g, t\}$).

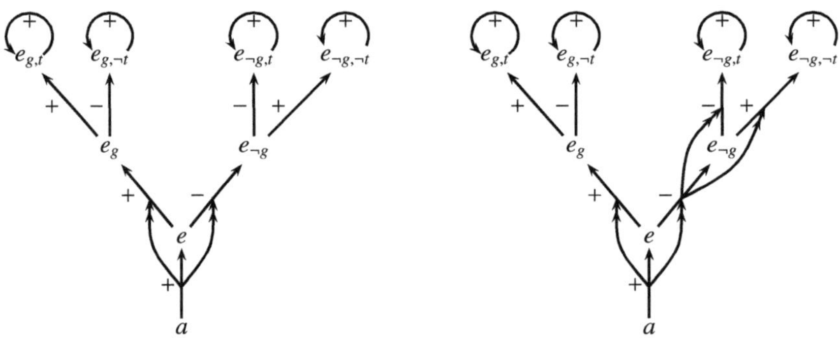

(a) Chisholm set with facts $\neg g, \neg t$

(b) Chisholm set with facts $\neg g, t$

Fig. 11.14

Some authors use $O(g \rightarrow t)$ in the Chisholm set. This is no problem for us, we can translate it, as we explained in item 5 of Remark 11.6.

Example 11.17 (Chisholm with $n = 3$). We now enhance our Chisholm set. g stands for "going", t stands for "telling", o stands for "dressing in your overalls". We enhance the usual Chisholm set $\{\bigcirc g, g \rightarrow \bigcirc t, \neg g \bigcirc \neg t\}$ by $\{t \rightarrow \bigcirc o, \neg t \rightarrow \bigcirc \neg o\}$: if you tell you should dress in your overalls, if you don't tell you should not dress in our overalls.

Obviously this set is in accordance with Definition 11.11. In Figure 11.15 we have a model for $\Delta^*_F = \{\neg g, t, \neg o\}$ constructed as in Definition 11.14.

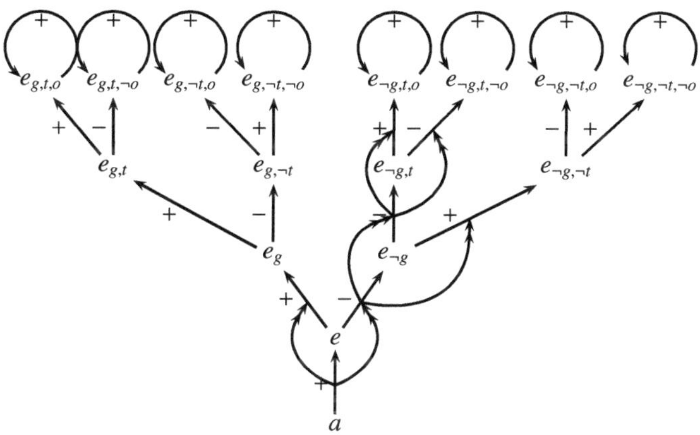

Fig. 11.15: Chisholm with $n = 3$

Example 11.18 (Expressive power of □ and ⊘). We try to characterise the graph axiomatically (as a Kripke frame) using □ by writing the modal axioms holding at a. The logic is non-normal, so the axioms hold at a only. If we want an axiom to hold at level two points such as e_g and $e_{\neg g}$ in Figure 11.15 we need to put a □ in front. The wffs X in the axioms do not contain ⊘ .

The following are the axiom schemas:

(*1) $\quad \Diamond X \to \Box X$

(*2) $\quad \Box X \to \Diamond X$

(*1) and (*2) say that a has exactly one accessible possible world e.

(*3) $\quad \bigwedge_{j=1}^{n} \Box^j (\bigwedge_{i=1}^{3} \Diamond X_i \to \bigvee_{i \neq j \leq 3} \Diamond (X_i \wedge X_j))$

(*3) says that each e_s has at most 2 accessible worlds.

(*4) $\quad \Box^{n+1}(X \leftrightarrow \Box X)$

(*4) expresses the reflexivity of our top nodes.

We now turn to our translation of Δ. While discussing the various parts of Δ, namely Δ_{CTD} and Δ_F, we will still add some more axioms in order to characterise our frame.

Now we make sure that the labels representing our obligations are placed correctly and the splitting of the branches of the tree corresponds to the violation order imposed by Δ. Each $x \to \bigcirc y \in \Delta_{CTD}$ is translated by:

$$\left[\bigwedge_{i=1}^{n} \Box^i(\mathsf{context}(x) \wedge \mathsf{switch}(y) \rightarrow \mathsf{opt}(y))\right] \wedge \left[\bigvee_{i=1}^{n} \Diamond^i(\mathsf{context}(x) \wedge \mathsf{switch}(y))\right] \qquad (\delta_{CTD})$$

(δ_{CTD}) expresses that whenever at some point e_s at which context(x) holds there is an arc towards a node where context($\neg y$) holds and another one towards a node where $\neg$context($\neg y$) holds, then the path towards context($\neg y$) is blocked and the path towards $\neg$context($\neg y$) is accessible. This realizes the obligation to y in the context x. The second conjunct of (δ_{CTD}) makes sure that the splitting of the branches of the tree are as expected (i.e., in accordance with the violation order).

The facts are modeled by means of the following formulas. Where x is a fact we add:

$$\oslash^{n+1} x \qquad (\delta_F)$$

Furthermore we add:

(*5) $\neg \mathbb{P}^{n+1} \bot$
(*6) $\bigwedge_{i=1}^{n+1} (\mathbb{P}^i X \wedge \mathbb{P}^i X' \rightarrow \mathbb{P}^i (X \wedge X'))$

These make sure that our family takes one unique path to a top node in each model.

(*7) $\bigwedge_{i=0}^{n-2} \Box^i \neg \oslash \bigvee_{j=1}^{n-1-i} \Diamond^j (\Diamond X \wedge \neg \mathbb{P} X)$

This formula makes sure that all double arrows are of the type $(e_s \rightarrow e_{s'}) \twoheadrightarrow (e_{s'} \rightarrow e_{s''})$ (i.e., double arrows always target arcs in the next level in the graph).

Hence, where (*1)–(*7) axiomatize our frame, (δ_{CTD}) and (δ_F) provide the translation of our CTD set Δ.

Remark 11.19 (Constraints). Suppose we have a set of constraints $\Delta_C = \{\neg \bigwedge \mathcal{E} \mid \mathcal{E} \subset Q^l\}$ such as $\neg(w \wedge \neg f)$ in the fence example (a white fence is a fence). In this case we have to change the construction of our model template slightly. For instance we adjust $\mathsf{Q}_\Delta = \{\langle \tau_1 x_1, \ldots, \tau_k x_k \rangle \mid k \leq n, \tau_i \in \{\varepsilon, \mathsf{co}\}$, there is no $\mathcal{E} \subseteq \{\tau_i x_i \mid i \leq k\}$ such that $\neg \bigwedge \mathcal{E} \in \Delta_C\}$.

Constraints can be translated into formulas by $\neg \Diamond^{n+1} \bigwedge \mathcal{E}$.

Remark 11.20 (Independence). Both (a) the translations of the members of CTD-sets and (b) the instructions for attaching the labels to the arcs and for drawing the arrows are fully independent in our treatment. There are no logical dependencies between the translations of the obligations and the facts of a given CTD set.[4]

[4] Note that the placement of the double arrows as presented in Definition 11.14 item 6 is a function of the labelling of the tree. This should not be taken as an indicator of dependence. The reason is that the instruction is generic and independent of the concretely given (obligations that are represented by) labels.

Our perspective is that the facts reflect the wishes of our family. The double arrows enable them to force their way to achieve what they want even if it violates obligations. If what they want violates an obligation the path will be labeled '−'. In this case a double arrow will free the way for them. This can be phrased in a generic way that is independent of the concrete obligations that are in place. E.g., if they want A such a generic instruction

Remark 11.21 (Solution to Chisholm). Consider the Chisholm set of Examples 11.10 and 11.16. We are seeking a logic **L** and a translation τ of the Chisholm sentences into **L** such that

1. All translations are logically independent in **L**.
2. The translation is consistent in **L**.
3. The translation satisfies the rationality conditions of Remark 11.8.

Let the logic **L** be characterised as in Example 11.18, then (δ_{CTD}) and (δ_F) give the translation of the Chisholm set.

11.6 Towards more generality: Permissions, separable conjunctive obligations, specificity cases and CTD cycles

While in the last section we gave a detailed account of how we model a generalized single CTD set in **SDLR1**, in this section we allow for less restrictive configurations. Namely, we will also allow for cycles in CTD sequences and specifically for specificity cases. Moreover, we show how different concepts of obligations as well as permissions can be expressed within our logic.

Example 11.22 (Extended Chisholm set with cycle). We now add $\bigcirc_t g$ and $\bigcirc_{\neg t} g$ to our Chisholm configuration. Hence, we have a cycle: $\bigcirc g, \bigcirc_g t, \bigcirc_t g$. The new conditional obligations $\bigcirc_t g$ and $\bigcirc_{\neg t} g$ introduce a new progression: namely the one where we first consider whether t resp. $\neg t$ is settled and then are concerned with the question whether we should do g. This is modelled semantically by introducing this sequence in the tree. See Figure 11.16. While the left-hand side of the tree is as before, we add the progressions on the right-hand side (i.e., the sub-trees $\langle a, e, e_t, \ldots\rangle$ and $\langle a, e, e_{\neg t}, \ldots\rangle$).[5]

Example 11.23 (Cycle with $n = 3$). We consider the following configuration

$$\{\bigcirc p_1, \bigcirc_{p_1} p_2, \bigcirc_{\neg p_1} \neg p_2, \bigcirc_{p_2} p_3, \bigcirc_{\neg p_2} \neg p_3, \bigcirc_{p_3} p_1, \bigcirc_{\neg p_3} p_1\}$$

See Figure 11.17 for the modeling which is analogous to the previous example.

which is independent of whether we have $\bigcirc\neg A$ or not is: if A is illegal and hence the path is barred by '–' then use a double arrow to force your way. For more complex facts that instruction is the generic case distinction offered in item 6 of our Definition.

Another more subtle worry is the following. By means of the double arrows we allow our family to react to the given obligations and force their way to reach what they want. One may argue that the norm giver may use a similar strategy: depending on what the individuals that are subjected to the norms do he may change the norms. E.g., if there is a certain loophole in the tax system and too many individuals use it in order to avoid taxation the norm giver may adjust the law accordingly in order to close the loophole. Our reply is that in this chapter we consider the norms as statically given and we do not study **SDLR1** as a system to model norm change.

[5] Hence, if we want our logic to model these more general cases we have to drop (*3) from our axiomatisation in Example 11.18.

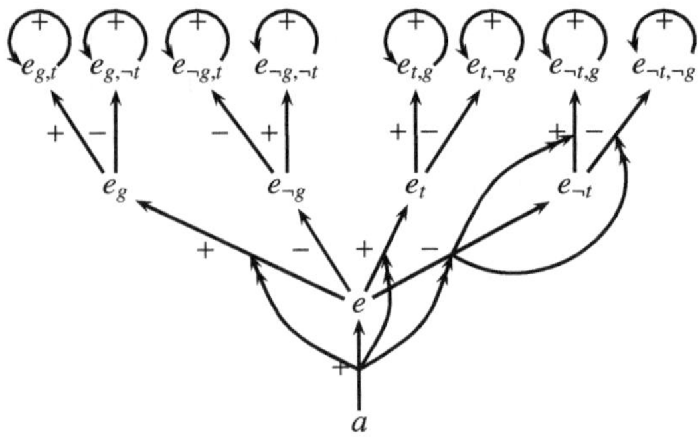

Fig. 11.16: Extended Chisholm with facts $\neg t, \neg g$ and obligations $\bigcirc_{\neg t} g, \bigcirc_t g$

Example 11.24 (Makinson's Moebius example). Makinson gave the following cyclic example:

$$\{\bigcirc_{p_1} p_2, \bigcirc_{p_2} p_3, \bigcirc_{p_3} \neg p_1\}$$

See Figure 11.18 for a model of this set.

Remark 11.25 (Parallel CTD sets, the problem with specificity). Another question is: how to deal with parallel CTD sets such as

$$\{x_i^j \to \bigcirc x_{i+1}^j, \mathsf{co}x_i^j \to \bigcirc \mathsf{co}x_{i+1}^j \mid i \leq n(j) - 1, j \leq m\}$$

where for each $j \leq m$ the set $\{x_i^j \to \bigcirc x_{i+1}^j, \mathsf{co}x_i^j \to \bigcirc \mathsf{co}x_{i+1}^j \mid i \leq n(j) - 1\}$ is a CTD set as in Definition 11.11.

If some of the CTD sets share a member, it may give rise to specificity cases. A case of specificity occurs whenever we have a conditional obligation $\bigcirc_A B$ and a more specific context $A \wedge A'$ (where $A \wedge A' \vdash A$ and $A \nvdash A \wedge A'$) such that $\neg \bigcirc_{A \wedge A'} B$, and A' is consistent with B ($\nvdash \neg(A' \wedge B)$). It is hence a specific failure of monotonicity in the condition of the obligation.

For instance, we can give another twist to the Chisholm set by introducing a case of specificity. While in Example 11.22 we had $\bigcirc_{\neg t} g$, we could instead add $\bigcirc_{\neg t} \neg g$ (after all, as a rule of politeness, you are supposed never to visit unannounced). Note that this set cannot be modelled by means of the translation we offered for conditional obligations in Sections 11.4 and 11.5. The reason is that: (a) $\mathsf{obliged}_{\neg t} \neg g$ enforces that at some node e_s in the tree we have $e_s \vDash \mathsf{context}(\neg t) \wedge \mathsf{switch}(\neg g)$, and (b) that $e_s \vDash \mathsf{opt}(\neg g)$. However, by $\mathsf{obliged} g$ we also have $e_s \vDash \mathsf{opt}(g)$ (in contradiction to (b)) since trivially $e_s \vDash \mathsf{context}(\top) \wedge \mathsf{switch}(\neg g)$.

Specificity can easily arise when we combine two CTD sets that share members. Take as a simple example the two CTD sets $\Delta_1 = \{\bigcirc a, a \to \bigcirc b, \neg a \to \bigcirc \neg b\}$ and $\Delta_2 = \{\bigcirc c, c \to \bigcirc b, \neg c \to \bigcirc \neg b\}$. Now suppose $a \wedge \neg c$: due to $a \to \bigcirc b$ we expect $(a \wedge \neg c) \to \bigcirc b$, however, due to $\neg c \to \bigcirc \neg b$ we expect $(a \wedge \neg c) \to \bigcirc \neg b$. Hence, we either get a violation of the monotonicity of $a \to \bigcirc b$ or a violation of the monotonicity of $\neg c \to \bigcirc \neg b$. Since so far we have not introduced means to deal with specificity cases, we cannot represent such combinations of CTD sets.

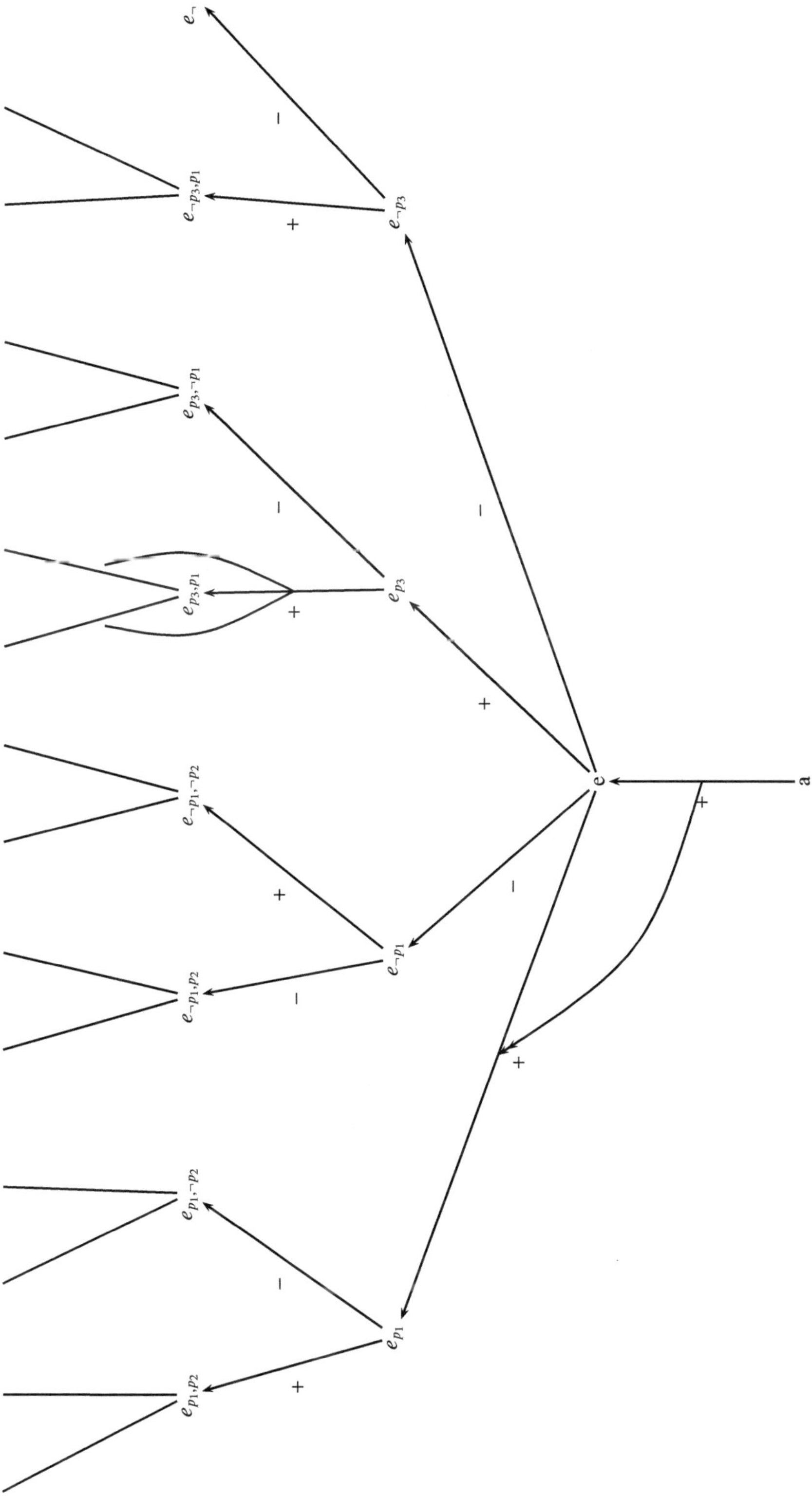

Fig. 11.17: A model with a cycle for $n = 3$ and with facts $p_3, p_1, \neg p_2$

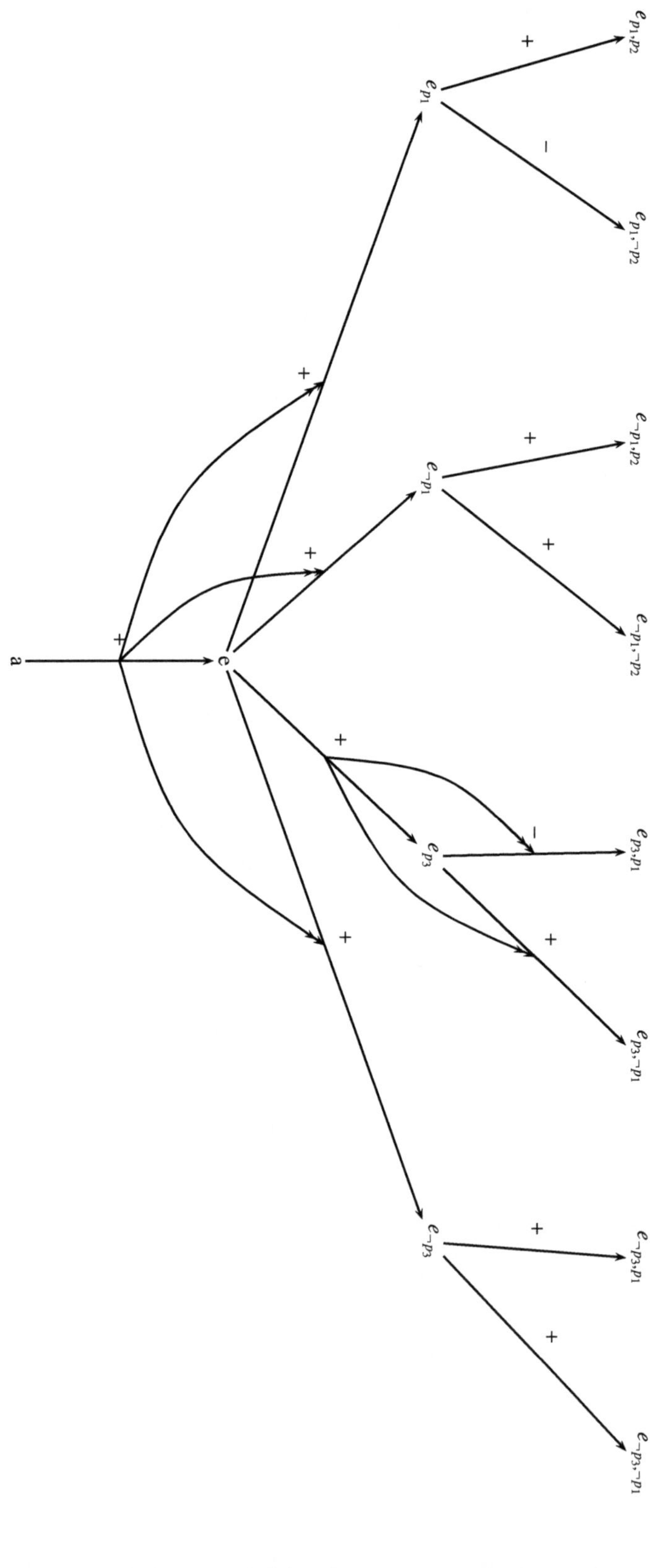

Fig. 11.18: A model for Makinson's Moebius example with facts p_3, p_1, p_2

However, if the various CTD sets do not share any literals (i.e., $\{x_i^j, \mathsf{co}x_i^j \mid i \leq n(j)\} \cap \{x_i^k, \mathsf{co}x_i^k \mid i \leq n(k)\} = \emptyset$ for all $j \neq k$ where $j, k \leq m$) we can easily construct models such as the one in Figure 11.19 for the combination of the two CTD sets $\Delta_1 = \{\bigcirc a, a \rightarrow \bigcirc b, \neg a \rightarrow \bigcirc\neg b\}$ and $\Delta_2 = \{\bigcirc c, c \rightarrow \bigcirc d, \neg c \rightarrow \bigcirc\neg d\}$.

Example 11.26 (Ross' Paradox). The Ross paradox concerns the inference from $\bigcirc l$ ("You ought to post the letter.") to $\bigcirc(l \vee b)$ ("You ought to post the letter or burn it."). By many authors this is considered as being not in coherence with our everyday usage of disjunctive obligations. In our semantics we can embed the Ross set paradox-free.

Figure 11.20 features a model for $\{\bigcirc l, \neg\bigcirc(l \vee b), \bigcirc\neg b, \neg l, \neg b\}$.

Remark 11.27 (Deliberative and non-deliberative obligations). Note that our translation of obligations has a deliberative flavour. Take for instance the fence example. Here we have the constraint $\neg(w \wedge \neg f)$ (a white fence is a fence, see Remark 11.19). Hence, in Figure 11.9 (where A is substituted by f and B is substituted by w), all the worlds above e_2 verify $\neg w \wedge \neg f$. Nevertheless, we do not have $\mathsf{obliged}_{\neg f}\neg w$. The reason is as follows. We have $e_2 \models \mathsf{context}(\neg f)$. However, we do not have $e_2 \models \mathsf{switch}(\neg w)$ since we do not have $e_2 \models \Diamond w$. Thus, from e_2 on $\neg w$ is necessary.

We can also express non deliberative obligations in our framework:

$$\begin{aligned}\mathsf{obliged}^*_A B =_{\mathrm{df}} & \Big[\Big(\bigwedge_{i=1}^n \Box^i(\mathsf{context}(A) \wedge \mathsf{switch}(B) \rightarrow \mathsf{opt}(B))\Big) \wedge \\ & \Big(\bigvee_{i=1}^n \Diamond^i(\mathsf{context}(A) \wedge \mathsf{switch}(B))\Big)\Big] \vee \\ & \Big[\bigwedge_{i=1}^n \Box^i(\mathsf{context}(A) \rightarrow \mathsf{context}(B)) \wedge \bigvee_{i=1}^n \Diamond^i\mathsf{context}(A)\Big]\end{aligned}$$

What is new in comparison to the definition of $\mathsf{obliged}_A B$ is the third line. It gives the non-deliberative flavor to obligations and expresses that $\mathsf{obliged}^*_A B$ holds also if any path leading to A leads also to B.

Note that the model in Figure 11.9 verifies $\mathsf{obliged}^*_{\neg f}\neg w$.

Remark 11.28. Note that $\mathsf{obliged}_C(A \rightarrow B)$ is not equivalent to $\mathsf{obliged}_{C \wedge A} B$ in our logic. This is demonstrated in Figure 11.21.

Remark 11.29 (Permissions). A permission $P_A B$ can be expressed by means of

$$\begin{aligned}\mathsf{permitted}^*_A B =_{\mathrm{df}} & \Big[\Big(\bigwedge_{i=1}^n \Box^i(\mathsf{context}(A) \wedge \mathsf{switch}(B) \rightarrow \mathbb{P}\neg\mathsf{context}(\neg B))\Big) \wedge \\ & \Big(\bigvee_{i=1}^n \Diamond^i(\mathsf{context}(A) \wedge \mathsf{switch}(B))\Big)\Big] \vee \\ & \Big[\bigwedge_{i=1}^n \Box^i(\mathsf{context}(A) \rightarrow \mathsf{context}(B)) \wedge \bigvee_{i=1}^n \Diamond^i\mathsf{context}(A)\Big]\end{aligned}$$

The first line expresses that whenever our family has the choice between $\neg B$ and a path that is compatible with B, they are free to go for the second since the corresponding arc is labeled by '+'.

The second line enforces the correct succession in the tree (see also the discussion of the definition of $\mathsf{obliged}$ in Section 11.3). Were we to skip this requirement we would face pragmatic oddities: e.g., the model in Figure 11.14a would verify $\mathsf{permitted}_t\neg g$.

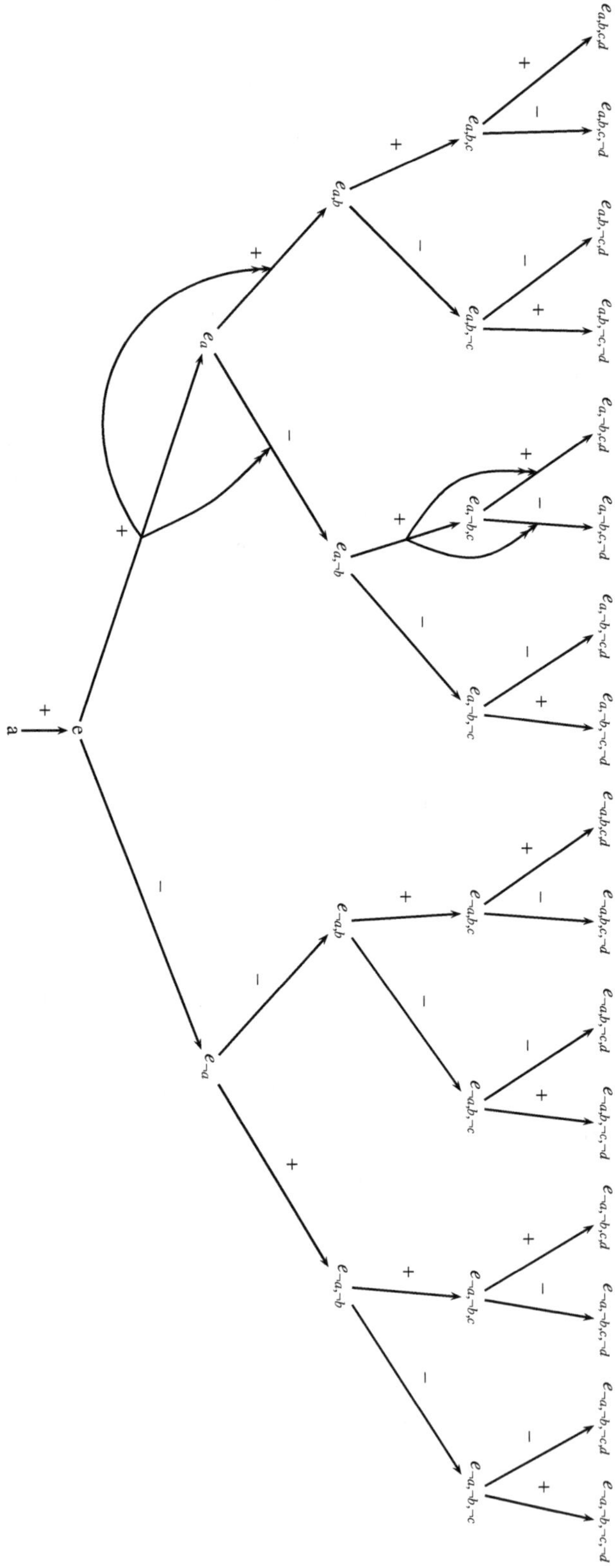

Fig. 11.19: Modeling multiple CTDs with facts $a, \neg b, c, \neg d$

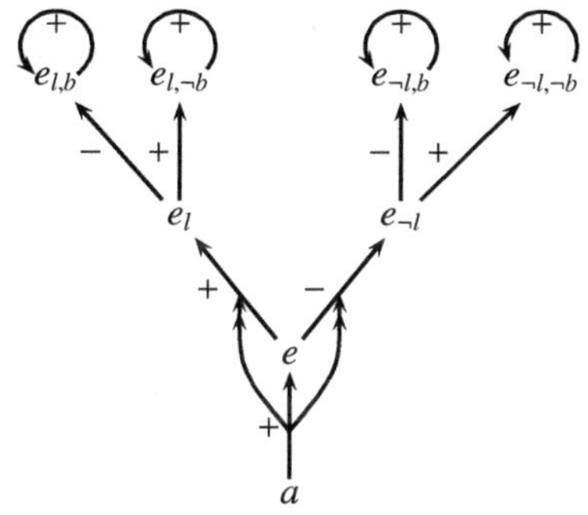

Fig. 11.20: Ross' Paradox with facts $\neg l, \neg b$

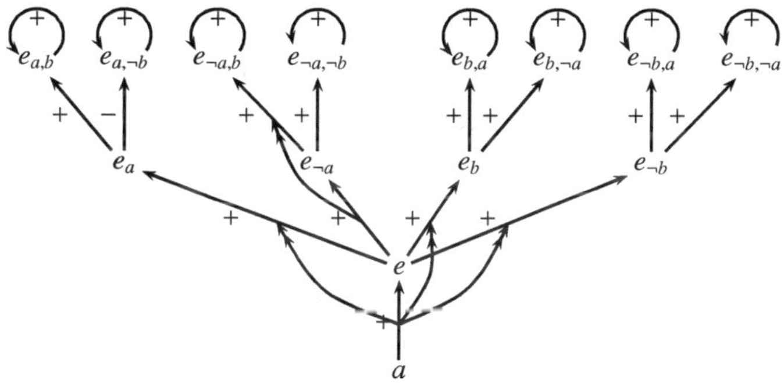

Fig. 11.21: A model that verifies $\neg\mathsf{obliged}(a \rightarrow b)$ and $\mathsf{obliged}_a b$ for the facts $\neg a, \neg b$

The third line expresses that in case every path leading to A is also leading to B, our family is allowed to bring about B in the context A (after all, in this context they have no other choice).

Note that weak permissions $PA =_{\mathrm{df}} \neg \bigcirc \neg A$ are problematic for logics that block the Ross-inference from $\bigcirc l$ to $\bigcirc(l \vee b)$ (such as **SDLR1**). The reason is that in such logics $\bigcirc l, \neg \bigcirc (l \vee b)$ is consistent. Now given that $\bigcirc(l \vee b)$ is equivalent to $\bigcirc\neg(\neg l \wedge \neg b)$, we get $P(\neg l \wedge \neg b)$, which is obviously counter-intuitive. Hence, our permission is not weak.

Note that the model in Figure 11.20 verifies: $\mathsf{permission}_\top(l \wedge \neg b)$, $\mathsf{permission}_{\neg l} \neg b$, $\neg\mathsf{permission}_\top(\neg l \wedge b)$, $\neg\mathsf{permission}_\top(\neg l \wedge \neg b)$, and $\neg\mathsf{permission}_\top \neg l$.

Remark 11.30 (Separable and inseparable conjunctive obligations). Note that in our translation of conditional obligations we do not get

$$\mathsf{obliged}_A(B \wedge C) \equiv \mathsf{obliged}_A B \wedge \mathsf{obliged}_A C. \qquad \text{(Sep)}$$

Conjunctive norms of this kind are often referred to as "inseparable conjunctive norms". Consider the following example:

Figure 11.22 is a model of

$$\{\mathsf{obliged}(a \wedge b), \mathsf{obliged}_{\neg a} \neg b, \mathsf{obliged} a, \neg\mathsf{obliged} b, \mathsf{obliged}_{\neg b} a, \neg\mathsf{obliged}_{\neg b} \neg a\}$$

Suppose we are supposed to buy ingredients for a strawberry cake which is to be baked tonight for the visit of a good friend. a stands for "buying flour" and b stands for "buying strawberries". Suppose the shop is out of flour and we have no time left to go to another shop. In this case we have to cancel our plan to bake the cake and are supposed not to buy strawberries since we have no use for them: hence $\bigcirc_{\neg a}\neg b$. Now suppose the shop ran out of strawberries but still has flour. Since the flour can still be used for baking fresh bread, we are still supposed to buy it: $\bigcirc_{\neg b} a$. This situation is modelled by the model in Figure 11.22. Note that this cannot be modelled if we have (Sep).

Now the question arises whether our logic is expressive enough also to represent separable conjunctive norms for which (Sep) holds. This can be achieved as follows: where $\mathcal{E} \subseteq Q^l$, we define $\mathsf{obliged}_A^{\mathsf{sep}} \bigvee \mathcal{E}$ just as $\mathsf{obliged}_A \bigvee \mathcal{E}$ by

$$\left[\bigwedge_{i=1}^{n} \Box^i(\mathsf{context}(A) \wedge \mathsf{switch}(B) \rightarrow \mathsf{opt}(B))\right] \wedge \left[\bigvee_{i=1}^{n} \Diamond^i(\mathsf{context}(A) \wedge \mathsf{switch}(B))\right]$$

Moreover, we define:

$$\begin{aligned}
\mathsf{obliged}_A^{\mathsf{sep}}(B \wedge C) &=_{\mathrm{df}} \mathsf{obliged}_A^{\mathsf{sep}} B \wedge \mathsf{obliged}_A^{\mathsf{sep}} C \\
\mathsf{obliged}_A^{\mathsf{sep}} \neg\neg B &=_{\mathrm{df}} \mathsf{obliged}_A^{\mathsf{sep}} B \\
\mathsf{obliged}_A^{\mathsf{sep}} \neg(B \vee C) &=_{\mathrm{df}} \mathsf{obliged}_A^{\mathsf{sep}}(\neg B \wedge \neg C) \\
\mathsf{obliged}_A^{\mathsf{sep}} \neg(B \wedge C) &=_{\mathrm{df}} \mathsf{obliged}_A^{\mathsf{sep}}(\neg B \vee \neg C)
\end{aligned}$$

Figure 11.23 is a model of $\mathsf{obliged}_\top^{\mathsf{sep}}(a \wedge b)$.

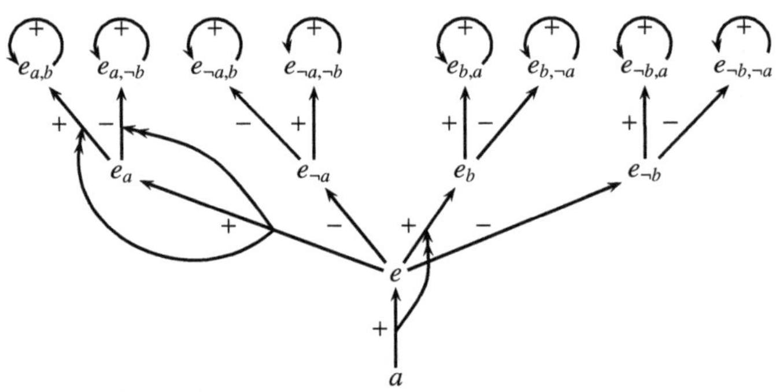

Fig. 11.22: Inseparable conjunctive norms: a model of $\mathsf{obliged}_\top(a \wedge b)$, $\neg\mathsf{obliged}_\top^{\mathsf{sep}}(a \wedge b)$, $\mathsf{obliged}_\top a$, $\neg\mathsf{obliged}_\top b$, and facts a, $\neg b$

Remark 11.31 (Temporal distinctions). The sequences encoded in the tree structure of our models offer means to express temporal successions. There are various interpretations. For instance we could interpret the fact that our family stands at a node with context A by A being the case. Another, less strict reading is that at this point our family considers the fact that they are going to realize A as being settled (without

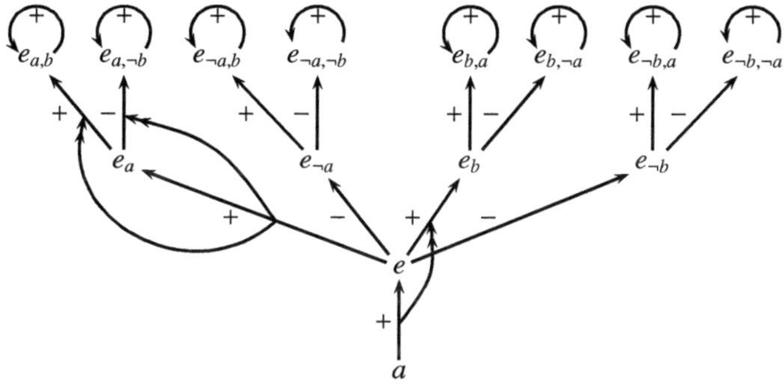

Fig. 11.23: Separable conjunctive norms: a model of $\mathsf{obliged}_{\top}(a \wedge b), \mathsf{obliged}_{\top}^{\mathsf{sep}}(a \wedge b)$, and facts $a, \neg b$

it necessarily already being realized). So in the Chisholm example they may stand at node e_g which means that they decided to go (for instance they all agreed that they will go). In this case they may next have to consider whether they tell or whether they do not tell that they are going.

The second reading is more apt for Forrester's gentle murderer example.

- $\bigcirc\neg k$ "Smith ought not to kill Jones."
- $k \to \bigcirc g$ "If Smith kills Jones he ought to do it gently."

See Figure 11.24. Of course, if the killing already took place it is also already a matter of fact whether the killing was gentle or not. However, if we go with the second reading we only consider it as being settled whether Smith decided to kill Jones. So if Smith reaches node e_k it still makes sense to make a decision between $e_{k,g}$ and $e_{k,\neg g}$ and hence to consider the CTD obligation to kill gently.

Altogether, this gives us additional expressivity which can be exploited. For instance, we can express conditional obligations that are sensitive to the exact sequence of decisions taken by the agent in question:

$$\mathsf{obliged}_{(A_1,\ldots,A_m)}B =_{\mathrm{df}} \left[\bigwedge_{i_1=1}^{n+1-m} \square^{i_1}\left(\mathsf{context}(A_1) \wedge \bigwedge_{1<j\leq m} \neg\mathsf{context}(A_j) \to \right.\right.$$
$$\left(\bigwedge_{i_2=1}^{n+1-m-(i_1-1)} \square^{i_2}\left(\mathsf{context}(A_2) \wedge \bigwedge_{2<j\leq m} \neg\mathsf{context}(A_j) \to \right.\right.$$
$$\left.\left.\left.\left(\ldots\left(\bigwedge_{i_m=1}^{n+1-m-(i_1-1)-\ldots-(i_{m-1}-1)} \square^{i_m}(\mathsf{context}(A_m) \wedge \mathsf{switch}(B) \to \mathsf{opt}(B))\right)\ldots\right)\right)\right)\right)\right] \wedge$$
$$\left[\bigwedge_{i_1=1}^{n+1-m} \diamond^{i_1}\left(\mathsf{context}(A_1) \wedge \bigwedge_{1<j\leq m} \neg\mathsf{context}(A_j) \to \right.\right.$$

$$\Bigg(\bigwedge_{i_2=1}^{n+1-m-(i_1-1)}\Bigg(\Diamond^{i_2}\mathsf{context}(A_2)\wedge\bigwedge_{2<j\leq m}\neg\mathsf{context}(A_j)\rightarrow$$

$$\Bigg(\ldots\Bigg(\bigwedge_{i_m=1}^{n+1-m-(i_1-1)-\ldots-(i_{m-1}-1)}\Diamond^{i_m}(\mathsf{context}(A_m)\wedge\mathsf{switch}(B))\Bigg)\ldots\Bigg)\Bigg)\Bigg)\Bigg)\Bigg)\Bigg]$$

The first "[...]"-conjunct expresses that whenever our family traverses the sequence $A_1,\ldots,A_m$ and faces the choice between an arc leading to $\neg B$ and an arc leading to a node that is compatible with B, the former is labeled '−' while the latter is labeled '+'. The second "[...]"-conjunct makes sure that the each model offers a path with the sequence $A_1,\ldots,A_m$ and a choice between $\neg B$ and a path that is consistent with B.

Of course, we can also give a temporal account of the facts. For instance our family may settle first A_1, then A_2, ..., then A_n. This can be expressed by

$$\oslash^2 A_1 \wedge \ldots \wedge \oslash^{n+1} A_n$$

Compare for instance Figures 11.16 and 11.25. Both models verify $\neg t$ and $\neg g$. However, in the former we have the sequence $(\neg t, \neg g)$ and hence $\oslash^2\neg t \wedge \oslash^3\neg g$ while in the latter we have the sequence $(\neg g, \neg t)$ and hence $\oslash^2\neg g \wedge \oslash^3\neg t$.

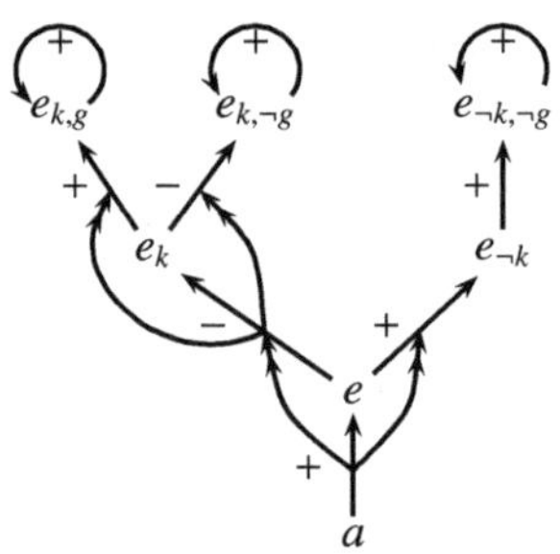

Fig. 11.24: The Forrester set with facts $k, \neg g$

Example 11.32 (Extended Chisholm with specificity). We have already discussed why our translation cannot model specificity cases (see Remark 11.25).

Hence, in order to model specificity cases we need to alter our translation of conditional obligations. As the examples in Remark 11.25 indicate, the translation needs to be more sensitive concerning the specificity of the given context. This can be achieved as follows. Where $\mathcal{E}\subseteq Q^l$ and

$$\mathsf{scontext}(\bigwedge\mathcal{E}) =_{\mathrm{df}} \mathsf{context}(\bigwedge\mathcal{E})\wedge\bigwedge_{p\in Q^l\setminus\{q,\mathsf{co}(q)\mid q\in\mathcal{E}\}}\neg\mathsf{context}(\bigwedge\mathcal{E}\wedge p)$$

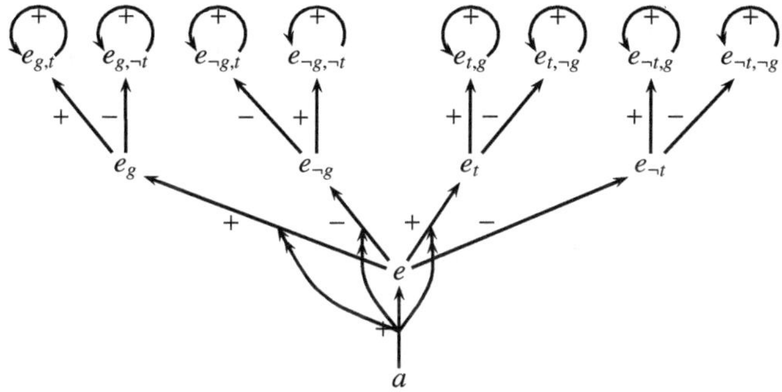

Fig. 11.25: Extended Chisholm with facts $\neg t, \neg g$ in the sequence $(\neg g, \neg t)$

we translate $\bigcirc_{\bigwedge \mathcal{E}} A$ by

$$\mathsf{obliged}^{\mathsf{spe}}_{\bigwedge \mathcal{E}} A =_{\mathrm{df}} \bigwedge_{i=1}^{n} \Box^{i}(\mathsf{scontext}(\bigwedge \mathcal{E}) \wedge \mathsf{switch}(A) \rightarrow \mathsf{opt}(A))$$

and define

$$\begin{aligned}
&\mathsf{obliged}^{\mathsf{spe}}_{A\vee B} C =_{\mathrm{df}} \mathsf{obliged}^{\mathsf{spe}}_{A} C \wedge \mathsf{obliged}^{\mathsf{spe}}_{B} C\\
&\mathsf{obliged}^{\mathsf{spe}}_{\neg\neg A} B =_{\mathrm{df}} \mathsf{obliged}^{\mathsf{spe}}_{A} C\\
&\mathsf{obliged}^{\mathsf{spe}}_{\neg(A\wedge B)} C =_{\mathrm{df}} \mathsf{obliged}^{\mathsf{spe}}_{\neg A\vee\neg B} C\\
&\mathsf{obliged}^{\mathsf{spe}}_{\neg(A\vee B)} C =_{\mathrm{df}} \mathsf{obliged}^{\mathsf{spe}}_{\neg A\wedge\neg B} C
\end{aligned}$$

Given this translation, the model in Figure 11.26 is for the set

$$\{\mathsf{obliged}^{\mathsf{spe}}_{\top} g, \mathsf{obliged}^{\mathsf{spe}}_{\top} t, \mathsf{obliged}^{\mathsf{spe}}_{\neg g} \neg t, \mathsf{obliged}^{\mathsf{spe}}_{g} t, \mathsf{obliged}^{\mathsf{spe}}_{t} g, \mathsf{obliged}^{\mathsf{spe}}_{\neg t} \neg g, \neg t, \neg g\}$$

Note that this model does not verify $\mathsf{obliged}_{\top} g$. The reason is that the arc $e_t \rightarrow e_{t,g}$ is '+'-labeled and the arc $e_t \rightarrow e_{t,\neg g}$ is '−'-labeled. Since we have $e_t \vDash \mathsf{context}\top$, if $\mathsf{obliged}_{\top} g$ were to hold we would have $e_t \vDash \mathsf{opt}(g)$. However, due to the labels of the two arcs that emanate from e_t we have $e_t \nvDash \mathsf{opt}(g)$. Note also that despite the labels of these arcs we have $\mathsf{obliged}^{\mathsf{spe}}_{\top} g$. The reason is that $e_t \nvDash \mathsf{scontext}\top$. Indeed, the only node where $\mathsf{scontext}\top$ holds is e. Since we have $e \vDash \mathsf{opt}(g)$, we also have $\mathsf{obliged}^{\mathsf{spe}}_{\top} g$.

Another possibility is to alter the Chisholm set by adding $\neg \bigcirc_{\neg t} \neg g$ and $\neg \bigcirc_{\neg t} g$. This expresses that in case you don't tell, you are neither obliged to come nor are you obliged not to come. Neither option is preferable to the other. The model in Figure 11.27 is for the set:

$$\{\mathsf{obliged}^{\mathsf{spe}}_{\top} g, \mathsf{obliged}^{\mathsf{spe}}_{\top} t, \mathsf{obliged}^{\mathsf{spe}}_{\neg g} \neg t, \mathsf{obliged}^{\mathsf{spe}}_{g} t, \mathsf{obliged}^{\mathsf{spe}}_{t} g,$$
$$\neg\mathsf{obliged}^{\mathsf{spe}}_{\neg t} \neg g, \neg\mathsf{obliged}^{\mathsf{spe}}_{\neg t} g, \neg t, \neg g\}$$

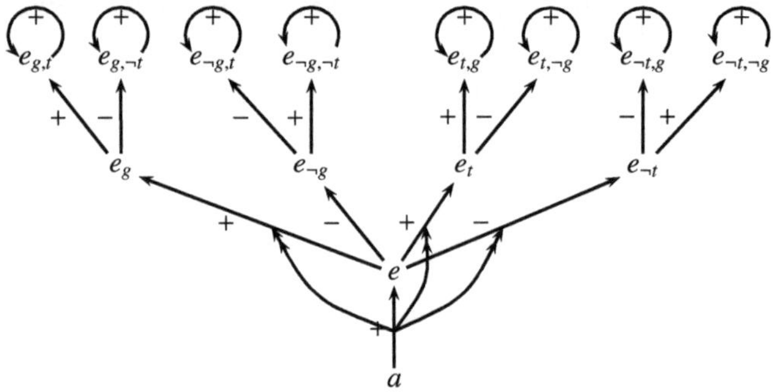

Fig. 11.26: Extended Chisholm with facts $\neg t$, $\neg g$ and $\bigcirc_{\neg t}\neg g$

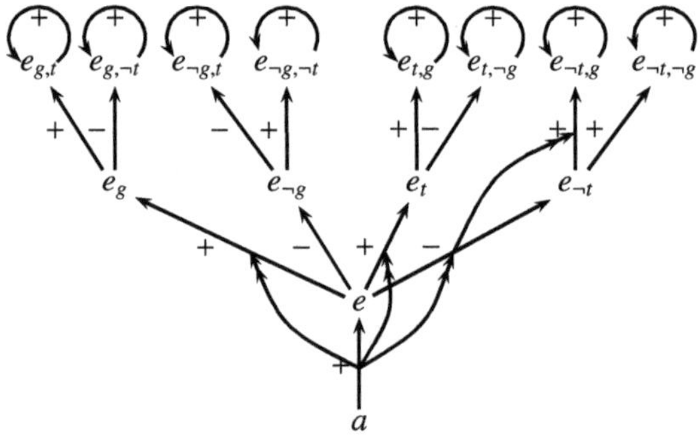

Fig. 11.27: Extended Chisholm with facts $\neg t$, $\neg g$ and $\neg \bigcirc_{\neg t} \neg g$, $\neg \bigcirc_{\neg t}\, g$

Example 11.33 (Horty's Asparagus, Specificity). Another well-known case of specificity is Horty's example $\bigcirc\neg f$ and $\bigcirc_a f$ where f is "eating with fingers" and a is "eating asparagus". The model in Figure 11.28 is for the set

$$\{\mathsf{obliged}^{\mathsf{spe}}\neg f, \mathsf{obliged}_a^{\mathsf{spe}} f, \neg f, a\}$$

Example 11.34 (Parallel CTD sets). By means of obliged we can only model combinations of CTD sets that do not include any cases of specificity (see Remark 11.25). However, since obliged^{spe} is able to tolerate specificity cases, we are now in a position to model combinations of any number of CTD sets. E.g., given

$$\Delta = \left\{\left(\bigwedge_{k\le i} x_k^j\right) \to \bigcirc x_{i+1}^j, \left(\bigwedge_{k\le i} \mathsf{co}x_k^j\right) \to \bigcirc \mathsf{co}x_{i+1}^j \mid i \le n(j)-1, j \le m\right\}$$

we only need to merge the semantic trees for each $\{(\bigwedge_{k\le i} x_k^j) \to \bigcirc x_{i+1}^j, (\bigwedge_{k\le i} \mathsf{co}x_k^j) \to \bigcirc \mathsf{co}x_{i+1}^j \mid i \le n(j)\}$ so that they share the nodes a and e in order to get a model of Δ.

Figure 11.29 is a model of the two CTD sets $\Delta_1 = \{\bigcirc a, a \to \bigcirc b, \neg a \to \bigcirc\neg b\}$ and $\Delta_2 = \{\bigcirc c, c \to \bigcirc b, \neg c \to \bigcirc\neg b\}$. As noted in Remark 11.25, this example

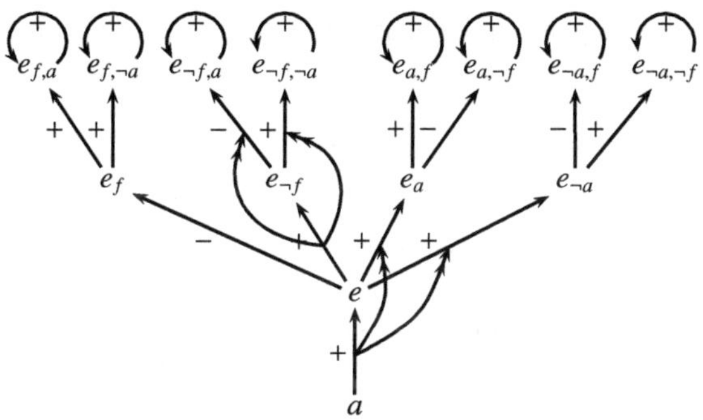

Fig. 11.28: The asparagus example with facts $\neg f, a$

cannot be modelled by means of obliged since it includes a specificity case. The model in Figure 11.29 verifies

$$\{\mathsf{obliged}^{\mathsf{spe}} a, \mathsf{obliged}_a^{\mathsf{spe}} b, \mathsf{obliged}_{\neg a}^{\mathsf{spe}} \neg b\} \cup \{\mathsf{obliged}^{\mathsf{spe}} c, \mathsf{obliged}_c^{\mathsf{spe}} b, \mathsf{obliged}_{\neg c}^{\mathsf{spe}} \neg b\}$$

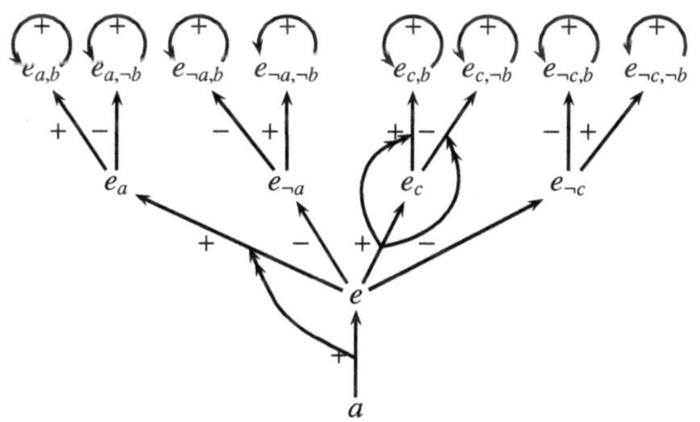

Fig. 11.29: Parallel CTDs with facts $c, \neg b$

11.7 Comparison with the Jones–Pörn system DL[6]

It is illuminating to compare our system with the system **DL** presented in the classic paper [115] of Jones and Pörn in 1985.

[6] The perceptive reader might ask why are we comparing with the Jones and Pörn system, when there are many other systems to discuss, especially those which use preferential ordering, a device which can also be simulated by reactivity. The answer is simply pragmatic: Proper systematic comparison with other systems requires technical development and can be the subject of a subsequent paper. The comparison with Jones and Pörn is simple and quick and we already studied this system in [28].

They focus in their paper on the Chisholm paradox and offer on page 278 semantics of the form $(S, R_0, R_{0'}, h)$, where $R_0, R_{0'} \subseteq S^2$ are serial and

$$C1 : R_0 \cap R_{0'} = \varnothing$$
$$C2 : \{(x, x) | x \in S\} \subseteq R_0 \cup R_{0'}$$

The two relations correspond to modalities O and O'.

Let $\square A =_{\text{df}} \bigcirc A \wedge \bigcirc' A$. After some serious and illuminating discussion Jones and Pörn translate the Chisholm set into their system as follows (see page 284):

Let Ought $p =_{\text{df}} \bigcirc p \wedge \neg \bigcirc' p$. Then the Chisholm set becomes (note that Ought p is also equivalent to $\bigcirc p \wedge \neg \square p$, and thus the translation below does not use $\bigcirc'$).

1. Ought p
2. $\square(p \rightarrow$ Ought $q)$
3. $\square(\neg p \rightarrow$ Ought $\neg q)$
4. $\neg p$

Their proposal was criticised by Hanson [103] and replied to immediately in [110].

We will continue the criticism in the spirit of Hanson and show that the solution proposed by Jones and Pörn to the Chisholm paradox is not closed under substitution.

Remark 11.35 (Substitution in the Jones-Pörn system). We show that the solution of the Chisholm paradox proposed in the Jones-Pörn system **DL** in [115] is not closed under substitution.

We observed that if we take the Chisholm set (see example 11.10) and substitute $H = \neg F$ and $T = \neg(F \wedge W)$, we get the fence paradox if we ignore the CTD linguistic formula $\neg H \rightarrow O(\neg T)$. Thus if the Jones-Pörn solution is consistent and independent it must solve the fence paradox under the above substitution. However, it does not. We get an inconsistent set under the above substitution.

The Jones-Pörn translation is into the modal logic with $R_\square$ and R_0 as we have shown at the beginning of this section. By substituting the fence variables we get

1. Ought$\neg$Fence
2. $\square(\neg$Fence $\rightarrow$ Ought$(\neg($Fence $\wedge$ white$)))$
3. $\square($Fence $\rightarrow$ Ought$($Fence $\wedge$ white$))$
4. Fence

Recall that Ought $X \equiv OX \wedge \neg\square X$.

Let F = Fence , W = white we get

1. $O\neg F \wedge \Diamond F$
2. $\square(\neg F \rightarrow O(\neg(F \wedge W)) \wedge \Diamond(F \wedge W))$
3. $\square(F \rightarrow O(F \wedge W) \wedge \Diamond\neg(F \wedge W))$
4. F

Let us check whether the above is consistent.

Let $(S, R_\square, R_0, a, h)$ be a model of (1)–(4) above, with $R_0 \subseteq R_\square$ and $R_\square$ reflexive and R_0 a **KD** modality.

So under h assume (1)–(4) hold at a. We claim (1), (3) and (4) (the fence example) is already inconsistent.

Use Figure 11.30 for help.

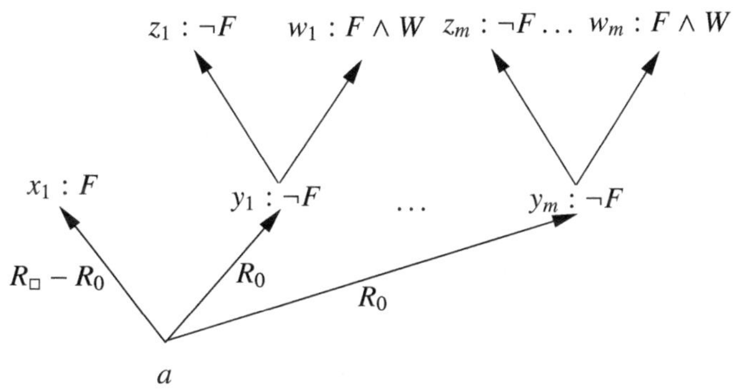

Fig. 11.30

From (1) we get that $a \vDash \Diamond F \wedge O\neg F$, hence for some x_1 such that $aR_{\Box}x_1 \wedge \neg aR_0x_1$ we have $x_1 \vDash F$. We say $\neg aR_0x_1$ because $a \vDash O\neg F$. Let $y_1, \ldots, y_m$ in the figure be all R_0-accessible points to a. $m \geq 1$ because O is a **D** modality. We have $y_i \vDash \neg F$.

From (4) we get $a \vDash F$. From (3) and reflexivity we get that $O(F \wedge W)$ must hold at a but this is not possible since $m \geq 1$ and $\neg F$ holds at y_1. Note that (2) can hold in the model if we add points z_i, w_i as indicated!

Note that Remark 11.35 could have been written in 1985. It has nothing to do with reactivity. It is a straightforward calculation in the same spirit as Hansson 1989, see [103]. For a serious analysis of the Jones and Pörn 1985 system using two-dimensional modal logic and other pre-1985 methods, see [28].

11.8 Conclusion and comparison

We would like to compare this chapter with several papers we wrote which make use of the reactive idea. We use the Chisholm paradox as a case study for comparison.

In Section 11.6 (see Remark 11.25) we have offered the following generic characterisation of combinations of various Chisholm sets of arbitrary depth:

$$\{x_i^j \to \bigcirc x_{i+1}^j, \mathsf{co}x_i^j \to \bigcirc \mathsf{co}x_{i+1}^j \mid i \leq n(j) - 1, j \leq m\}$$

The following are problems to be solved:

GP Give a logic and reactive semantics for the general problem, for m and n arbitrary.

We have two options for the logic we use:

1. a traditional Hilbert System with Reactive Kripke Semantics, or
2. a most general Labelled Deductive System for General Reactive Kripke Semantics.

We can break the general problem **GP** into several particular ones **PP**, which we may solve for the time being, using simpler models.

PP1 Solve **GP** for the case $m = 1$, and n arbitrary, i.e., a single arbitrarily long chain of Chisholm-like contrary-to-duties, only paying attention to the violation ordering while neglecting temporal considerations.
PP2 Solve **GP** for the case n and m arbitrary, i.e. for several Chisholm-like sets, taking into account both the violation order (indicated by the sequence $1, \ldots, n(j)$ for each $j \leq m$) and temporal considerations.
PP3 Solve **GP** by giving a general Labelled Deductive system semantically, without giving any proof theory for it.
PP4 Add a labelled proof system to **PP3**.
PP5 Offer a solution with a new type of modelling, like argumentation or automata theory.

We now make our comments:

1. In the present chapter we first offered a solution to **PP1** (see Section 11.5) by using a semantic approach. We offered a reactive model where the annotations to arcs mirror the obligations and where there are double arrows which can change the annotations. The latter represent the given facts. Using such semantics we can translate the traditional Chisholm set into a language with the two modalities $\square$ and $\oslash$.
   In Section 11.6 we demonstrated how the present approach can be used to tackle **PP2**. We discussed combinations of CTD sets. We pointed out how interactions between various CTD sets can lead to complications such as specificity cases and how to deal with them in our semantics. We also demonstrated how **SDLR1** can be used to model temporal considerations.

2. We now compare with other chapters and other published papers using the reactive idea.
   a) Chapter 2 of this book is a general exposition, introducing for the first time the idea of reactivity.
   b) Papers [102, 64] use a similar semantic approach to deal with the single Chisholm paradox with two letters (go, tell) where the temporal aspect is also taken into account. Part A ([64]) deals with the semantics, i.e. it solves **PP3**. Part B ([64]) deals with the proof theory, i.e. papers [64] address **PP4** by means of the Labelled Deductive System approach. In fact, we can deal with several letters. Once we have semantics, we can deal with a family of general Chisholm sets, because we can evaluate them individually. The problem is that they may interact and so we need to process them first and this requires proof theory.

c) Paper [81] uses an axiomatic Hilbert type approach, to axiomatise the reactive semantics and is capable of translating any single Chisholm set with an arbitrary number of letters ($n > 2$), including an arbitrary temporal order (different from the obligation/violation order).
We believe the axiomatic approach can handle several Chisholm sets (m, n arbitrary), but a complete temporal order for all sets should be given. So for example the case of $m = 1, n = 2$ would require a complete temporal order, say $x_1 < x_2 < z_1 < z_2$, and the resulting translation and modelling will be different from what we proposed in (1) above.
d) Paper [28] is based on Jones and Pörn [115] and addresses the problem discussed in Remark 11.35 of the present chapter, the solution of which requires a modified model. We use axiomatic two-dimensional temporal logic to model the Chisholm case ($m = 1, n = 2$). Strangely enough we do not model the temporal order. We do not think it is possible to do so in this approach. We use the obligation/violation progression (go < tell) as our "virtual time", and it is not clear how to account for real time as well.
We believe that combining the approach of the present chapter with that of [81] will allow us to model the most general family of Chisholm sets (i.e. m, n arbitrary with arbitrary temporal order) without using the theory of Labelled Deductive Systems . This is probably possible by using multimodal logic. IIowever, this way we may not achieve an intuitive model but rather a purely technical one.

3. Completely new ideas tackling **PP5** can be found in papers [32, 80]. The direction in which they go is clear from their titles.
4. The reader is invided to look forward to our book [96].

12

Reactive Preferential Structures and Non-monotonic Consequence

12.1 Introduction

12.1.1 Overview

Our aim is twofold here. We want to:

1. introduce IBRS (Information Bearing System, defined in Definition 12.1), an abstraction of many semantical structures in nonclassical logics,
2. use IBRS as a generalization of preferential structures to give semantics for logics weaker than preferential logics.

Section 12.1 and Section 12.2 discuss general IBRS, Section 12.3 discusses generalized preferential structures. The main technical developments are in Section 3, the main conceptual discussions are in Sections 1 and 2.

At the beginning of Section 12.3, we briefly argue why we think that the results of this Section are important (they show independence of certain logical rules), and, consequently, why the techniques used to obtain them are important. Section 12.3 is thus a technical justification of the concept of IBRS. (The results are briefly put into perspective at the end of this overview.)

Section 12.1 gives a general framework within which we place ourselves, and then gives the formal definition of IBRS. We then show (not exhaustively!) how our definition covers many different formal logics and their semantics. We also stress the main idea of IBRS, reactivity, whose base is to have not only arrows between points, but also "higher" arrows between points and arrows. This is also basic for the concept of generalized preferential structures, discussed in Section 12.3.

In Section 12.2, we present the outlines of a semantics for general IBRS, together with a circuit semantics for a special case. The main result of Section 12.2 is perhaps a warning: one has to be careful about transmission times in real-life scenarios. Depending on those times, a network may or may not oscillate.

For the reader's convenience, we now describe the main technical results from Section 12.3.

D.M. Gabbay, *Reactive Kripke Semantics*, Cognitive Technologies,
DOI 10.1007/978-3-642-41389-6_12, © Springer-Verlag Berlin Heidelberg 2013

The model choice functions μ of usual preferential structures satisfy two basic laws, and if they are smooth, they also satisfy a third law:

1. $(\mu \subseteq)$ $\mu(X) \subseteq X$ - this is trivial,
2. (μPR) $X \subseteq Y \Rightarrow \mu(Y) \cap X \subseteq \mu(X)$ - anything which is not minimal in X will a fortiori not be minimal in Y, if $X \subseteq Y$,
3. (μCUM) $\mu(X) \subseteq Y \subseteq X \Rightarrow \mu(X) = \mu(Y)$ - which holds in smooth structures, but not necessarily in non-smooth structures.

(μPR) is closely linked to the logical rule
(OR) $\alpha \mid\!\sim \beta, \alpha' \mid\!\sim \beta \Rightarrow \alpha \vee \alpha' \mid\!\sim \beta$
(μCUM) to the logical rule
(CUM) $\alpha \mid\!\sim \beta, \alpha \mid\!\sim \beta' \Rightarrow \alpha \wedge \beta \mid\!\sim \beta'$.

Thus, in usual preferential structures (OR) will hold, and if they are smooth, so will (CUM), but we cannot have (CUM) without (OR). The main result of Section 12.3 is now to fully separate them. We can have (OR) without (CUM), but also (CUM) without (OR). The "trick" is that reactive structures allow us to break the coherence condition of (μPR), as a bigger set allows us to "destroy destruction". For all details, the reader is referred to Section 12.3.

We now conclude this overview with an informal motivation, showing that IBRS are a very natural generalisation of existing structures.

Motivation for IBRS and generalized preferential structures

Many semantical structures for nonclassical logics use sets of models, called possible worlds, together with a relation, e.g. accessibility in the case of Kripke models for modal logic, distance in the case of Stalnaker-Lewis semantics for counterfactual conditionals, or comparison of "normality" in the case of preferential structures. IBRS allow not only relations between objects, i.e. possible worlds, but also "higher" relations, where e.g. an object can be in relation with a pair of objects (which are in a "lower" relation), etc.

It is sometimes natural to see the (basic) relations as attacks: if $a \prec b$, then a is considered more "normal" (or less "abnormal") than b, or, we may say that a attacks b's normality, and we may write this as an arrow from a to b. Now when we have an attack from c against this arrow, i.e. the relation $a \prec b$, it is natural to see now the original attack as destroyed, i.e. a does not attack b any more. So $a \prec b$ is still there, but without any effect. This is possible in an IBRS, see Figure 12.1.

The decisive property of preferential structures is the trivial fact that if $a, b \in X \subseteq Y$, and $a \prec b$ "in X", then this will also hold "in Y". This creates strong coherence properties. IBRS can break them: if $c \in Y - X$, and c attacks $a \prec b$, then $a \prec b$ is still there in Y, but not effective any more. Thus, IBRS, or generalized preferential structures, as we will call this special case of IBRS, can give semantics for logics weaker than preferential logics.

Preferential structures themselves were introduced as abstractions of Circumscription independently in [148] and [33]. A precise definition of these structures is given below in Definition 12.12.

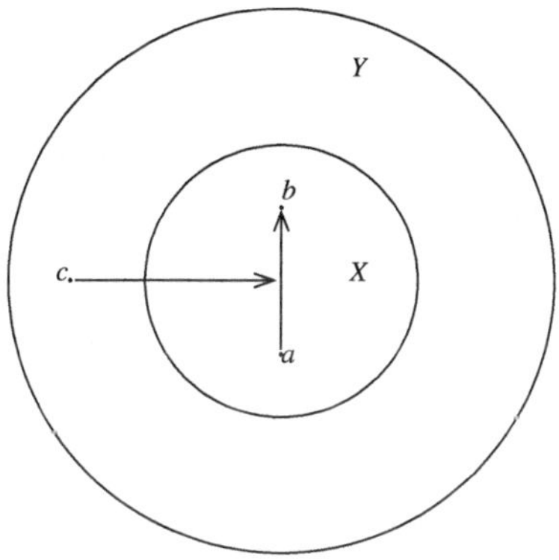

Fig. 12.1: Attacking an arrow

In an abstract consideration of desirable properties a logic might have, [51] examined rules a nonmonotonic consequence relation $\mid\sim$ should satisfy:

1. $\Delta, \alpha \mid\sim \alpha$,
2. $\Delta \mid\sim \alpha \Rightarrow (\Delta \mid\sim \beta \Leftrightarrow \Delta, \alpha \mid\sim \beta)$.

The semantic and the syntactic approaches were connected in [119], where a representation theorem was proved, showing that the system P (which is stronger than Gabbay's system) corresponds to "smooth" preferential structures. System P consists of

1. $\phi \mid\sim \psi, \phi \mid\sim \psi' \Rightarrow \phi \mid\sim \psi \wedge \psi'$,
2. $\phi \mid\sim \psi, \phi' \mid\sim \psi \Rightarrow \phi \vee \phi' \mid\sim \psi$,
3. $\vdash \phi \leftrightarrow \phi' \Rightarrow (\phi \mid\sim \psi \Leftrightarrow \phi' \mid\sim \psi)$,
4. $\phi \mid\sim \psi, \vdash \psi \rightarrow \psi' \Rightarrow \phi \mid\sim \psi'$,
5. $\vdash \phi \rightarrow \phi' \Rightarrow \phi \mid\sim \phi'$,
6. $\phi \mid\sim \psi \Rightarrow (\phi \mid\sim \psi' \Leftrightarrow \phi \wedge \psi \mid\sim \psi')$.

where $\vdash$ is classical provability.

Details can be found in Definition 12.9.

To the authors' knowledge, a precise semantics for Gabbay's system was still lacking. We will give it here, applying the idea of IBRS, which allow us to "break" the relations of preferential structures, and thus their strong coherence conditions, as was already illustrated above.

We will also see that the usual definition of "smoothness" - for every arrow from a to b, there has to be an arrow from a' to b, where a' is a minimal element - is stronger than needed, and a weaker variant, "essential smoothness" (see Definition 12.30 is sufficient to generate the desired property of cumulativity on the logical side.

12.1.2 Introduction to IBRS

The human agent in his daily activity has to deal with many situations involving change. Chief among them are the following:

1. Common sense reasoning from available data. This involves prediction of what unavailable data is supposed to be (non-monotonic deduction) but it is a defeasible prediction, geared towards immediate change. This is formally known as non-monotonic reasoning and is studied by the non-monotonic community.
2. Belief revision, studied by a very large community. The agent is unhappy with the totality of his beliefs which he finds internally unacceptable (usually logically inconsistent but not necessarily so) and needs to change/revise it.
3. Receiving and updating his data, studied by the update community.
4. Making morally correct decisions, studied by the deontic logic community.
5. Dealing with hypothetical and counterfactual situations. This is studied by a large community of philosophers and AI researchers.
6. Considering temporal future possibilities; this is covered by modal and temporal logic.
7. Dealing with properties that persist through time in the near future and with reasoning that is constructive. This is covered by intuitionistic logic.

All the above types of reasoning exist in the human mind and are used continuously and coherently every hour of the day. The formal modelling of these types is done by diverse communities which are largely distinct with no significant communication or cooperation. The formal models they use are very similar and arise from a more general theory, what we might call: "Reasoning with information-bearing binary relations".

Definition 12.1.

1. *An information-bearing binary relation frame IBR, has the form* $(S, \mathfrak{R})$, *where* S *is a nonempty set and* $\mathfrak{R}$ *is a subset of* S_ω, *where* S_ω *is defined by induction as follows:*
   a) $S_0 := S$
   b) $S_{n+1} := S_n \cup (S_n \times S_n)$

c) $S_\omega = \bigcup\{S_n : n \in \omega\}$.

We call elements from S *points or nodes, and elements from* $\mathfrak{R}$ *arrows. Given* $(S, \mathfrak{R})$, *we also set* $\mathbf{P}((S, \mathfrak{R})) := S$ *and* $\mathbf{A}((S, \mathfrak{R})) := \mathfrak{R}$.

If α *is an arrow, the origin and destination of* α *are defined as usual, and we write* $\alpha : x \to y$ *when* x *is the origin and* y *the destination of the arrow* α. *We also write* $o(\alpha)$ *and* $d(\alpha)$ *for the origin and destination of* α.

2. *Let* Q *be a set of atoms, and* $\mathbf{L}$ *be a set of labels (usually* $\{0, 1\}$ *or* $[0, 1]$*). An information assignment* h *on* $(S, \mathfrak{R})$ *is a function* $h : Q \times \mathfrak{R} \to \mathbf{L}$.
3. *An information-bearing system IBRS has the form* $(S, \mathfrak{R}, h, Q, \mathbf{L})$, *where* S, $\mathfrak{R}$, h, Q, $\mathbf{L}$ *are as above.*

See Figure 12.2 for an illustration.

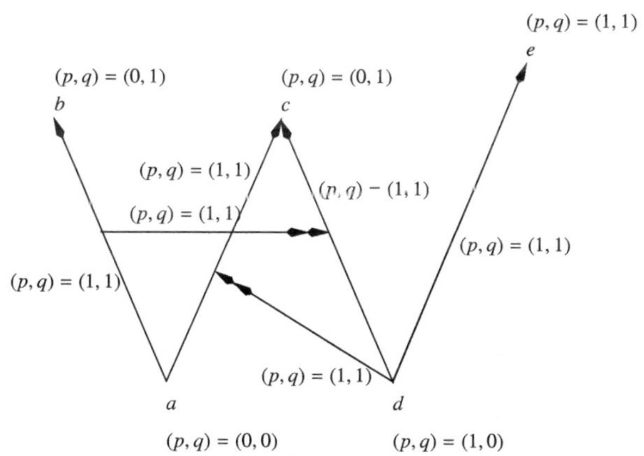

Fig. 12.2: A simple example of an information-bearing system

We have here:

$$S = \{a, b, c, d, e\},$$
$$\mathfrak{R} = S \cup \{(a, b), (a, c), (d, c), (d, e)\} \cup \{((a, b), (d, c)), (d, (a, c))\},$$
$$Q = \{p, q\}.$$

The values of h for p and q are as indicated in the figure. For example $h(p, (d, (a, c))) = 1$.

Thus, IBRS can be used as a source of information for various logics based on the atoms in Q. We now illustrate by listing several such logics.

Modal logic

One can consider the figure as giving rise to two modal logic models, one with actual world a and one with d, these being the two minimal points of the relation. Consider a language with $\Box q$. How do we evaluate $a \models \Box q$?

The modal logic will have to give an algorithm for calculating the values.

Say we choose algorithm $\mathcal{A}_1$ for $a \models \Box q$, namely:

[$\mathcal{A}_1(a, \Box q) = 1$] iff for all $x \in S$ such that $a = x$ or $(a, x) \in \mathfrak{R}$ we have $h(q, x) = 1$.

According to $\mathcal{A}_1$ we get that $\Box q$ is false at a. $\mathcal{A}_1$ gives rise to a T-modal logic. Note that the reflexivity is not anchored in the relation $\mathfrak{R}$ of the network but in the algorithm $\mathcal{A}_1$ in the way we evaluate. We say $(S, \mathfrak{R}, \ldots.) \models \Box\, q$ iff $\Box q$ holds at all minimal points of $(S, \mathfrak{R})$.

For orderings without minimal points we may choose a subset of distinguished points.

Non-monotonic deduction

We can ask whether $p \mid\!\sim q$ according to algorithm $\mathcal{A}_2$ defined below. $\mathcal{A}_2$ says that $p \mid\!\sim q$ holds iff q holds in all minimal models of p. Let us check the value of $\mathcal{A}_2$ in this case:

Let $S_p = \{s \in S \mid h(p, s) = 1\}$. Thus $S_p = \{d, e\}$.

The minimal points of S_p are $\{d\}$. Since $h(q, d) = 0$, we have that $p \mid\!\not\sim q$.

Note that in the cases of modal logic and non-monotonic logic we ignored the arrows $(d, (a, c))$ (i.e. the double arrow from d to the arrow (a, c)) and the h values to arrows. These values do not play a part in the traditional modal or non-monotonic logic. They do play a part in other logics. The attentive reader may already suspect that we have here an opportunity for generalisation of, say, non-monotonic logic, by giving a role to arrow annotations.

Argumentation nets

Here the nodes of S are interpreted as arguments. The atoms $\{p, q\}$ can be interpreted as types of arguments and the arrows, e.g. $(a, b) \in \mathfrak{R}$, as indicating that the argument a is attacking the argument b.

So, for example, let

a = We must win votes.
b = Death sentence for murderers.
c = We must allow abortion for teenagers.
d = Bible forbids taking of life.
q = The argument is a social argument.
p = The argument is a religious argument.
$(d, (a, c))$ = There should be no connection between winning votes and abortion.

$((a, b), (d, c))$ = If we attack the death sentence in order to win votes then we must stress (attack) that there should be no connection between religion (Bible) and social issues.

Thus we have according to this model that supporting abortion can lose votes. The argument for abortion is a social one and the argument from the Bible against it is a religious one.

We can extract information from this IBRS using two algorithms. The modal logic one can check whether for example every social argument is attacked by a religious argument. The answer is no, since the social argument b is attacked only by a which is not a religious argument.

We can also use algorithm $\mathcal{A}_3$ (following Dung) to extract the winning arguments of this system. The arguments a and d are winning since they are not attacked. d attacks the connection between a and c (i.e. stops a attacking c).

The attack of a on b is successful and so b is out. However the arrow (a, b) attacks the arrow (d, c). So c is not attacked at all as both arrrows leading into it are successfully eliminated. So c is in. e is out because it is attacked by d.

So the winning arguments are $\{a, c, d\}$.

In this model we ignore the annotations on arrows. To be consistent in our mathematics we need to say that h is a partial function on $\mathfrak{R}$. The best way is to give a more specific definition of IBRS to make it suitable for each logic.

See also [66] and [14].

Counterfactuals

The traditional semantics for counterfactuals involves closeness of worlds. The clause $y \models p \hookrightarrow q$, where $\hookrightarrow$ is a counterfactual implication is that q holds in all worlds y' "near enough" to y in which p holds. So if we interpret the annotation on arrows as distances then we can define "near" as distance ≤ 2, and we get: $a \models p \hookrightarrow q$ iff in all worlds of p-distance ≤ 2 if p holds so does q. Note that the distance depends on p.

In this case we get that $a \models p \hookrightarrow q$ holds. The distance function can also use the arrows from arrows to arrows, etc. There are many opportunities for generalisation in our IBRS setup.

Intuitionistic persistence

We can get an intuitionistic Kripke model out of this IBRS by letting, for $t, s \in S$, $t\rho_0 s$ iff $t = s$ or $[tRs \wedge \forall q \in Q(h(q, t) \leq h(q, s))]$. We get that

$[\ r_0 = \{(y, y) \mid y \in S\} \cup \{(a, b), (a, c), (d, e)\}.\ ]$

Let ρ be the transitive closure of ρ_0. Algorithm $\mathcal{A}_4$ evaluates $p \Rightarrow q$ in this model, where $\Rightarrow$ is intuitionistic implication.

$\mathcal{A}_4$: $p \Rightarrow q$ holds at the IBRS iff $p \Rightarrow q$ holds intuitionistically at every ρ-minimal point of (S, ρ).

12.1.3 Purpose of this chapter

In this chapter, we will not cover all these applications of the above abstract definition of IBRS, but will

(1) give an abstract and also a very concrete semantics to IBRS

(2) show that a special case of IBRS generalizes preferential semantics in a very natural way and solves open representation problems for weak logical systems. This is possible, as we can "break" the strong coherence properties of preferential structures by higher arrows, i.e. arrows, which do not go to points, but to arrows themselves.

The notion of IBRS is within the Reactive approach in applied logic. The reader will find more on the Reactive approach in Section 12.1.4.

12.1.4 Concluding discussion - the reactive idea

This section introduces the reactive idea for algorithmic systems. This idea is the intuition behind the general notion of IBRS. We will explain it by looking at a fairly general example involving possible world semantics.

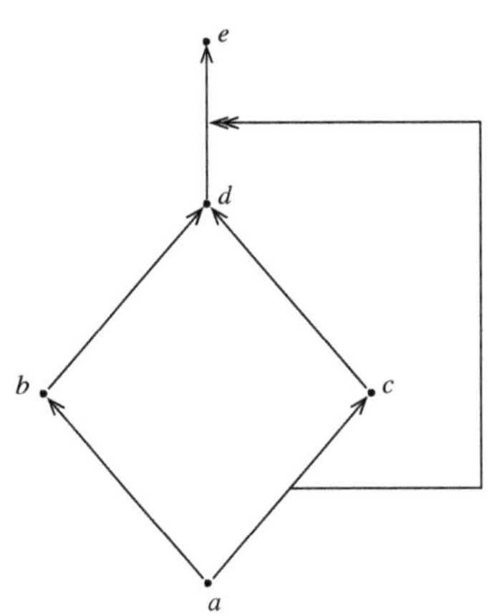

Fig. 12.3

Consider the situation in Figure 12.3. Consider this figure with a starting point at state a. The arrows indicate non-deterministic options for moving along the graph. From a we can reach d in two ways, one through b and one through c. If we move to d through c, passing from a to c (through the arc $a \to c$) sends a signal through the double arrow that disconnects the arc $a \to e$. The idea of these double arrows is the reactive idea.

We use an algorithmic system, and depending on the path the algorithm takes, the system changes.

Many systems can be represented in a graph form and therefore may become reactive.

Here are some examples.

1. Kripke semantics, see Chapter 2. The model can change as we traverse the nodes in the evaluation process.
2. Context free grammars, see Chapter 9. The set of rules available to us can change depending which rules we use (i.e. rules can be active or non-active depending on the use of the other rules).
3. Finite automata, see Chapter 8. The automaton can change its transition table as it changes state.
4. Reactive tableaux or proof theory. The rules can change as some rules are applied.
5. The system of this chapter. With preferential models, a set A has minimal points. When A is enlarged to A', more points and more preferential connections are added but no existing connections are cancelled. Thus minimal points can only be cancelled by addition. The reactive double arrow idea allows us to cancel connections when we enlarge. Figure 12.2 illustrates the situation.

Intuitively, it is easy to see that any clear-cut algorithmic system can be made reactive. The system has to tell us what it means to make a move and so we can change the system as we make such a move. The following question may be asked: can we gain anything by introducing reactivity? Is there any gain in expressive power in addition to a better intuitive representation? The answer is yes.

1. There are logical systems with one modal operator $\Box$ which can be characterized by a class of reactive Kripke frames but cannot be characterized by any class of ordinary Kripke frames.
2. There is no context-free grammar which can recognize the set of words of the form $a^n b^n c^n$, $n \in \omega$. However, there is a reactive context-free grammar which can do the job.
3. Given an ordinary finite automaton with k^n (respectively $k * n$) states, there exists a reactive automaton with $k - n$ (respectively $k + n$) states which recognizes the same inputs.

Let us now gain a bit more insight into what is going on in the Kripke model case, especially with regards to tableaux and axiomatisations. This should be of interest to our readers, since this chapter deals with (preferential) possible worlds.

Consider again Figure 12.3. What we see at world d depends on how we got there. So Figure 12.3 is equivalent to Figure 12.4. The points here are paths. The rectangle around abd and acd indicates these points should have the same assignment (labels) for atoms. We can indicate this by an equivalence relation $\equiv$, $abd \equiv acd$.

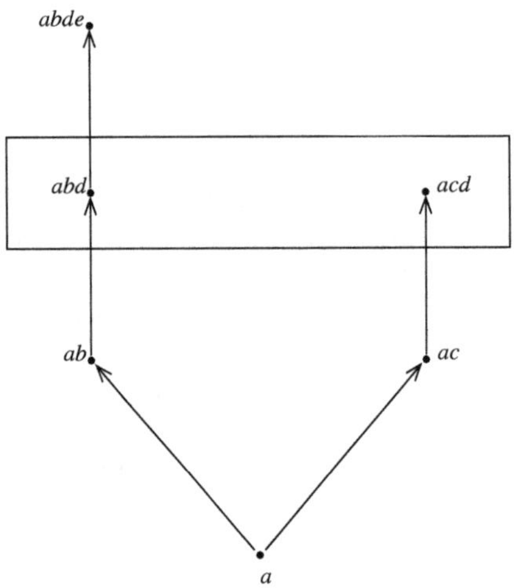

Fig. 12.4

Thus essentially a reactive Kripke model with one-level double arrows for one modality $\Box\phi$ is like an ordinary Kripke model with two modalities $\Box\phi$ and $K\phi$ (necessity and knowledge). This is so because of the unfolding of Figure 12.3 to Figure 12.4. Knowledge is represented by an equivalence relation $\equiv$, $t \models K\phi$ iff for all $s \equiv t$, we have $s \models \phi$. Thus in a reactive Kripke model for $\Box$ with a starting point t_0, we can model knowledge by $K\phi$ holds at t iff ϕ holds at t no matter how we get to t from t_0.

In fact for every t_0 we get a different knowledge operator $K(t_0)$.

How about proof theory? In Figure 12.3 let us assume that q_x holds at x. Thus at a we have

1. $\Diamond\Diamond\Box\bot$ (path through c)
2. $\Diamond\Diamond\Diamond q_e$ (path through b)
3. $\Diamond\Diamond q \to \Box\Box q$ for all atomic q (because there is only one point d).

These hold in Figure 12.3 as a frame. They don't hold in Figure 12.4 as a frame, only if we restrict the assignment for atoms.

12.2 A semantics for IBRS

12.2.1 Introduction

We give here a rough outline of a formal semantics for IBRS. It consists more of some hints where difficulties are, than a finished construction. Still, the authors feel that this is sufficient to complete the work, and, on the other hand, our remarks might be useful to those who intend to finalize the construction.

(1) Nodes and arrows

As we may have counterarguments not only against nodes, but also against arrows, they must be treated basically the same way, i.e. in some way there has to be a positive, but also a negative influence on both. So arrows cannot just be concatenation between the contents of nodes.

We will differentiate between nodes and arrows by labelling arrows in addition with a time delay. We see nodes as situations, where the output is computed instantanously from the input, whereas arrows describe some "force" or "mechanism" which may need some time to "compute" the result from the input.

Consequently, if α is an arrow, and β an arrow pointing to α, then it should point to the input of α, i.e. before the time lapse. Conversely, any arrow originating in α should originate after the time lapse.

Apart from this distinction, we will treat nodes and arrows the same way, so the following discussion will apply to both - which we call just "objects".

(2) Defeasibility

The general idea is to code each object, say X, by $I(X) : U(X) \to C(X)$: If $I(X)$ holds then, unless $U(X)$ holds, consequence $C(X)$ will hold. (We adopted Reiter's notation for defaults, as IBRS have common points with the former.)

The situation is slightly more complicated, as there can be several counterarguments, so $U(X)$ really is an "or". Likewise, there can be several supporting arguments, so $I(X)$ also is an "or".

A counterargument must not always be an argument against a specific supporting argument, but it can be. Thus, we should admit both possibilities. As we can use arrows to arrows, the second case is easy to treat (as is the dual, a supporting argument can be against a specific counterargument). How do we treat the case of unspecific pro- and counterarguments? Probably the easiest way is to adopt Dung's idea: an object is in, if it has at least one support, and no counterargument — see [45]. Of course, other possibilities may be adopted, counting, use of labels, etc., but we just consider the simple case here.

(3) Labels

In the general case, objects stand for some kind of defeasible transmission. We may in some cases see labels as restricting this transmission to certain values. For instance, if the label is $p = 1$ and $q = 0$, then the p-part may be transmitted and the q-part not.

Thus, a transmission with a label can sometimes be considered as a family of transmissions; which ones are active is indicated by the label.

Example 12.2.

In fuzzy Kripke models, labels are elements of $[0, 1]$. $p = 0.5$ as label for a node m' which stands for a fuzzy model means that the value of p is 0.5. $p = 0.5$ as label for an arrow from m to m' means that p is transmitted with value 0.5. Thus, when we look from m to m', we see p with value $0.5 * 0.5 = 0.25$. So, we have $\Diamond p$ with value 0.25 at m - if, e.g. m, m' are the only models.

(4) Putting things together

If an arrow leaves an object, the object's output will be connected to the (only) positive input of the arrow. (An arrow has no negative inputs from objects it leaves.) If a positive arrow enters an object, it is connected to one of the positive inputs of the object, analogously for negative arrows and inputs.

When labels are present, they are transmitted through some operation.

In slightly more formal terms, we have:

Definition 12.3. *In the most general case, objects of IBRS have the form:* $(\langle I_1, L_1\rangle, \ldots, \langle I_n, L_n\rangle) : (\langle U_1, L'_1\rangle, \ldots, \langle U_n, L'_n\rangle)$, *where the* L_i, L'_i *are labels and the* I_i, U_i *might be just truth values, but can also be more complicated, a (possibly infinite) sequence of some values. Connected objects have, of course, to have corresponding sequences. In addition, the object X has a criterion for each input, whether it is valid or not (in the simple case, this will just be the truth value "TRUE"). If there is at least one positive valid input* I_i, *and no valid negative input* U_i, *then the output* $C(X)$ *and its label are calculated on the basis of the valid inputs and their labels. If the object is an arrow, this will take some time, t, otherwise, this is instantaneous.*

Evaluating a diagram

An evaluation is relative to a fixed input, i.e. some objects will be given certain values, and the diagram is left to calculate the others. It may well be that it oscillates, i.e. shows a cyclic behaviour. This may be true for a subset of the diagram, or the whole diagram. If it is restricted to an unimportant part, we might neglect this. Whether it oscillates or not can also depend on the time delays of the arrows (see Example 12.4.

We therefore define for a diagram Δ

$\alpha \mid\!\sim_{\Delta} \beta$ iff

(a) α is a (perhaps partial) input — where the other values are set "not valid"

(b) β is a (perhaps partial) output

(c) after some time, β is stable, i.e. all still possible oscillations do not affect β

(d) the other possible input values do not matter, i.e. whatever the input, the result is the same.

In the cases examined here more closely, all input values will be defined.

12.2.2 A circuit semantics for simple IBRS without labels

It is natural to implement IBRS by (modified) logical circuits. The nodes will be implemented by sub-circuits which can store information, and the arcs by connections between them. As connections can connect connections, connections will not just be simple wires. The objective of the present discussion is to warn the reader that care has to be taken about the temporal behaviour of such circuits, especially when feedback is allowed.

All details are left to those interested in a realization.

Background: It is standard to implement the usual logical connectives by electronic circuits. These components are called gates. Circuits with feedback sometimes show undesirable behaviour when the initial conditions are not specified. (When we switch a circuit on, the outputs of the individual gates can have arbitrary values.) The technical realization of these initial values shows the way to treat defaults. The initial values are set via resistors (on the order of 1 $k\Omega$) between the point in the circuit we want to initialize and the desired tension (say 0 Volt for FALSE, 5 Volt for TRUE). They are called pull-down or pull-up resistors (for default 0 or 5 Volt). When a "real" result comes in, it will override the tension applied via the resistor.

Closer inspection reveals that we have here a three-level default situation: The initial value will be the weakest, which can be overridden by any "real" signal, but a positive argument can be overridden by a negative one. Thus, the biggest resistor will be for the initialization, the smaller ones for the supporting arguments, and the negative arguments have full power. Technical details will be left to the experts.

We give now an example which shows that the delays of the arrows can matter. In one situation, a stable state is reached, in another, the circuit begins to oscillate.

Example 12.4. (In engineering terms, this is a variant of a JK flip-flop with $R*S = 0$, a circuit with feedback.)

We have eight measuring points.

$In1, In2$ are the overall input, $Out1, Out2$ the overall output, $A1, A2, A3, A4$ are auxiliary internal points. All points can be TRUE or FALSE.

The logical structure is as follows:

$A1 = In1 \wedge Out1$, $A2 = In2 \wedge Out2$,

$A3 = A1 \vee Out2$, $A4 = A2 \vee Out1$,

$Out1 = \neg A3$, $Out2 = \neg A4$.

Thus, the circuit is symmetrical, with $In1$ corresponding to $In2$, $A1$ to $A2$, $A3$ to $A4$, $Out1$ to $Out2$.

The input is held constant. See Figure 12.5.

We suppose that the output of the individual gates is present n time slices after the input was present. In the first circuit n will be equal to 1 for all gates, in the second circuit equal to 1 for all but the AND gates, which will take 2 time slices. Thus, in both cases, e.g. $Out1$ at time t will be the negation of $A3$ at time $t-1$. In the first case, $A1$ at time t will be the conjunction of $In1$ and $Out1$ at time $t-1$, and in the second case the conjunction of $In1$ and $Out1$ at time $t-2$.

We initialize $In1$ as TRUE, all others as FALSE. (The initial value of $A3$ and $A4$ does not matter, the behaviour is essentially the same for all such values.)

The first circuit will oscillate with a period of 4, the second circuit will go to a stable state.

We have the following transition tables (time slice shown at left):

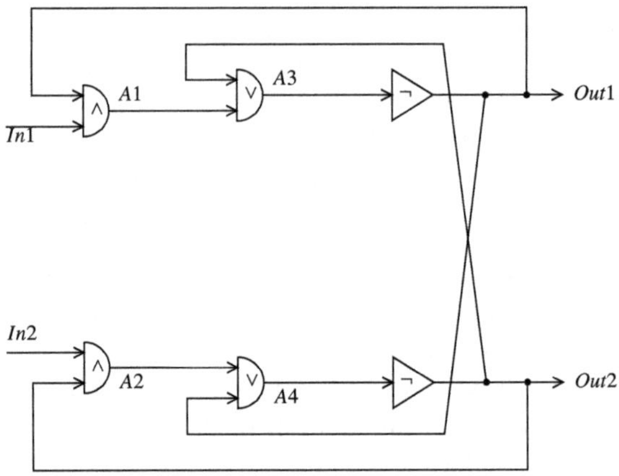

Fig. 12.5: Gate semantics

Table 12.1: Oscillating variant

	*In*1	*In*2	*A*1	*A*2	*A*3	*A*4	*Out*1	*Out*2	
1:	T	F	F	F	F	F	F	F	
2:	T	F	F	F	F	F	T	T	
3:	T	F	T	F	T	T	T	T	
4:	T	F	T	F	T	T	F	F	
5:	T	F	F	F	T	F	F	F	oscillation starts
6:	T	F	F	F	F	F	F	T	
7:	T	F	F	F	T	F	T	T	
8:	T	F	T	F	T	T	F	T	
9:	T	F	F	F	T	F	F	F	back to start of oscillation

Circuit 1, *delay* = 1 everywhere:

Circuit 2, *delay* = 1 everywhere, except for AND with *delay* = 2.

(Thus, $A1$ and $A2$ are held at their intial value up to time 2, then they are calculated using the values of time $t-2$.)

Table 12.2: No oscillation

	*In*1	*In*2	*A*1	*A*2	*A*3	*A*4	*Out*1	*Out*2	
1:	T	F	F	F	F	F	F	F	
2:	T	F	F	F	F	F	T	T	
3:	T	F	F	F	T	T	T	T	
4:	T	F	T	F	T	T	F	F	
5:	T	F	T	F	T	F	F	F	
6:	T	F	F	F	T	F	F	T	stable state reached
7:	T	F	F	F	T	F	F	T	
8:	T	F	F	F	T	F	F	T	

Note that state 6 of circuit 2 is also stable in circuit 1, but it is never reached in that circuit.

12.3 IBRS as generalized preferential structures

12.3.1 Introduction

Overview

The main objective of this Section is to show independence results for certain logical and algebraic rules, which are usually treated together in the system P (see below for P and other concepts mentioned here), and which are all valid in usual preferential structures.

The basic idea is to use generalized preferential structures, i.e. special IBRS, to break the strong coherence conditions of usual preferential structures. This allows us to work without the basic condition (μPR), and we are left with only $(\mu \subseteq)$, or can add independently (μCUM), respectively Cumulativity or Cautious Monotony. Thus we can give, for the first time to our knowledge, a structural semantics to certain weak systems of nonmonotonic logics.

We will also modify the usual smoothness condition to "essential smoothness", sufficient for obtaining Cautious Monotony, and thus contribute to a clarification of the concept of smoothness.

In a certain way, our present work goes the opposite way to our [74], where we showed strong conceptual connections between rules like (AND), (OR), Cautious Monotony, etc. There, we worked with an abstract algebraic semantics based on manipulation of the concept of size, here we work with the structural semantics of (generalized) preferential structures. Thus, the present chapter and [74] can be seen as complementary.

To summarize, the main result of this chapter is to show how to work semantically without (μPR), and in particular, how to have (μCUM) and related properties without (μPR). The corresponding logical properties are then easily shown.

Basic definitions, fundamental results

Definition 12.5. *1. We use* $:=$ *and* $:\Leftrightarrow$ *to define the left-hand side by the right-hand side, as in the following two examples:*
$X := \{x\}$ *defines* X *as the singleton with element* x.
$X < Y :\Leftrightarrow \forall x \in X \forall y \in Y (x < y)$ *extends the relation* $<$ *from elements to sets.*

2. *We use* $\mathcal{P}$ *to denote the power set operator.*
$\Pi\{X_i : i \in I\} := \{g : g : I \rightarrow \bigcup\{X_i : i \in I\}, \forall i \in I. g(i) \in X_i\}$ *is the general Cartesian product,* $X \times X'$ *is the binary Cartesian product.*
card(X) *shall denote the cardinality of* X, *and* V *the set-theoretic universe we work in — the class of all sets.*
Given a set of pairs $\mathcal{X}$, *and a set* X, *we let* $\mathcal{X} \upharpoonright X := \{\langle x, i\rangle \in \mathcal{X} : x \in X\}$. *(When the context is clear, we will sometimes simply write* X *for* $\mathcal{X} \upharpoonright X$.*)*
We will use the same notation $\upharpoonright$ *to denote the restriction of functions and in particular of sequences to a subset of the domain.*
If Σ *is a set of sequences over an index set* X, *and* $X' \subseteq X$, *we will abuse notation and write* $\Sigma \upharpoonright X'$ *for* $\{\sigma \upharpoonright X' : \sigma \in \Sigma\}$.
Concatenation of sequences, e.g. of σ *and* σ', *will be denoted by juxtaposition:* $\sigma\sigma'$.

3. $A \subseteq B$ *will denote that* A *is a subset of* B *or equal to* B, *and* $A \subset B$ *that* A *is a proper subset of* B; *likewise for* $A \supseteq B$ *and* $A \supset B$.
Given some fixed set U *we work in, and* $X \subseteq U$, $\mathbf{C}(X) := U - X$.

4. *If* $\mathcal{Y} \subseteq \mathcal{P}(X)$ *for some* X, *we say that* $\mathcal{Y}$ *satisfies*
$(\cap)$ *iff it is closed under finite intersections,*
$(\bigcap)$ *iff it is closed under arbitrary intersections,*
$(\cup)$ *iff it is closed under finite unions,*
$(\bigcup)$ *iff it is closed under arbitrary unions,*
$(\mathbf{C})$ *iff it is closed under complementation,*
$(-)$ *iff it is closed under set difference.*

5. *We will sometimes write* $A = B \parallel C$ *for:* $A = B$, *or* $A = C$, *or* $A = B \cup C$.
We make ample and tacit use of the Axiom of Choice.

Definition 12.6. $<^*$ *will denote the transitive closure of the relation* $<$. *If* $<$, $\prec$, *or a similar relation is given,* $a \perp b$ *will express that* a *and* b *are* $<$*-incomparable (or* $\prec$*-incomparable); context will tell. Given any relation* $<$, $\leq$ *will stand for* $<$ *or* $=$; *conversely, given* $\leq$, $<$ *will stand for* $\leq$, *but not* $=$; *similarly for* $\prec$, *etc.*

Definition 12.7.

1. *V will be the set of truth values when there are more than the classical ones, TRUE and FALSE.*

We work here in a classical propositional language $\mathcal{L}$; a theory T will be an arbitrary set of formulas. Formulas will often be named ϕ, ψ, etc., theories T, S, etc.
$v(\mathcal{L})$ or simply L will be the set of propositional variables of $\mathcal{L}$.
$F(\mathcal{L})$ will be the set of formulas of $\mathcal{L}$.
A propositional model m will be a function from the set of propositional variables to the set of truth values V when we have more than two truth values.
$M_{\mathcal{L}}$, or simply M when the context is clear, will be the set of (classical) models for $\mathcal{L}$; $M(T)$ or M_T is the set of models of T; likewise for $M(\phi)$ for a formula ϕ.

2. *$\mathbf{D}_{\mathcal{L}} := \{M(T) : T$ a theory in $\mathcal{L}\}$, the set of definable model sets.*
   Note that, in classical propositional logic, $\emptyset, M_{\mathcal{L}} \in \mathbf{D}_{\mathcal{L}}$, $\mathbf{D}_{\mathcal{L}}$ contains singletons, and is closed under arbitrary intersections and finite unions.
   An operation $f : \mathcal{Y} \to \mathcal{P}(M_{\mathcal{L}})$ for $\mathcal{Y} \subseteq \mathcal{P}(M_{\mathcal{L}})$ is called definability preserving, (dp) or (μdp) for short, iff for all $X \in \mathbf{D}_{\mathcal{L}} \cap \mathcal{Y}$, $f(X) \in \mathbf{D}_{\mathcal{L}}$.
   We will also use (μdp) for binary functions $f : \mathcal{Y} \times \mathcal{Y} \to \mathcal{P}(M_{\mathcal{L}})$ — as needed for theory revision — with the obvious meaning.
3. *$\vdash$ will be classical derivability, and*
   $\overline{T} := \{\phi : T \vdash \phi\}$, the closure of T under $\vdash$.
4. *$Con(.)$ will stand for classical consistency, so $Con(\phi)$ will mean that ϕ is classically consistent; likewise for $Con(T)$. $Con(T, T')$ will stand for $Con(T \cup T')$, etc.*
5. *Given a consequence relation $\mid\sim$, we define*
   $\overline{\overline{T}} := \{\phi : T \mid\sim \phi\}$.
   (There is no fear of confusion with $\overline{T}$, as it just is not useful to close twice under classical logic.)
6. *$T \vee T' := \{\phi \vee \phi' : \phi \in T, \phi' \in T'\}$.*
7. *If $X \subseteq M_{\mathcal{L}}$, then $Th(X) := \{\phi : X \models \phi\}$; likewise for $Th(m)$, $m \in M_{\mathcal{L}}$. ($\models$ will usually be classical validity.)*

We recollect and note:

Fact 12.8 *Let $\mathcal{L}$ be a fixed propositional language, $\mathbf{D}_{\mathcal{L}} \subseteq X$, $\mu : X \to \mathcal{P}(M_{\mathcal{L}})$, for a $\mathcal{L}$-theory T. Suppose $\overline{\overline{T}} = Th(\mu(M_T))$, let T, T' be arbitrary theories, then:*
(1) $\mu(M_T) \subseteq M_{\overline{\overline{T}}}$,
(2) $M_T \cup M_{T'} = M_{T \vee T'}$ and $M_{T \cup T'} = M_T \cap M_{T'}$, moreover, if T and T' are deductively closed, then also $M_T \cup M_{T'} = M_{T \cap T'}$, (3) $\mu(M_T) = \emptyset \Leftrightarrow \bot \in \overline{\overline{T}}$.
If μ is definability preserving or $\mu(M_T)$ is finite, then the following also hold:
(4) $\mu(M_T) = M_{\overline{\overline{T}}}$,
(5) $T' \vdash \overline{\overline{T}} \Leftrightarrow M_{T'} \subseteq \mu(M_T)$,
(6) $\mu(M_T) = M_{T'} \Leftrightarrow \overline{T'} = \overline{\overline{T}}$. □

Definition 12.9. *The definitions are given in Tables 12.3, "Logical rules, definitions and connections Part I" and 12.4, "Logical rules, definitions and connections Part*

II", which also show connections between different versions of rules and the semantics. (The tables are split in two, as they would not fit onto one page otherwise.)

Explanation of the tables:

1. *The first table gives the basic properties, the second table those for Cumulativity and Rational Monotony.*
2. *The difference between the first two columns is that the first column treats the formula version of the rule, the second the more general theory (i.e. set of formulas) version.*
3. *"Corr." stands for "Correspondence".*
4. *The third column, "Corr.", is to be understood as follows:*
   Let a logic $\mid\sim$ satisfy (LLE) and (CCL), and define a function $f : \mathbf{D}_{\mathcal{L}} \to \mathbf{D}_{\mathcal{L}}$ by $f(M(T)) := M(\overline{\overline{T}})$. Then f is well defined and satisfies (μdp), and $\overline{\overline{T}} = Th(f(M(T)))$.
   If $\mid\sim$ satisfies a rule on the left-hand side, then — provided the additional properties noted in the middle for $\Rightarrow$ hold, too — f will satisfy the property on the right-hand side.
   Conversely:
   If $f : \mathcal{Y} \to \mathcal{P}(M_{\mathcal{L}})$ is a function, with $\mathbf{D}_{\mathcal{L}} \subseteq \mathcal{Y}$, and we define a logic $\mid\sim$ by $\overline{\overline{T}} := Th(f(M(T)))$, then $\mid\sim$ satisfies (LLE) and (CCL). If f satisfies (μdp), then $f(M(T)) = M(\overline{\overline{T}})$.
   If f satisfies a property on the right-hand side, then — provided the additional properties noted in the middle for $\Leftarrow$ hold, too — $\mid\sim$ will satisfy the property on the left-hand side.
5. *We use the following abbreviations for those supplementary conditions in the "Correspondence" columns:*
   "$T = \phi$" means that if one of the theories (the one named the same way in Definition 12.9 is equivalent to a formula, we do not need (μdp).
   $-(\mu dp)$ stands for "without (μdp)".
6. *$A = B \parallel C$ will abbreviate $A = B$, or $A = C$, or $A = B \cup C$.*

Further comments:

1. *(PR) is also called infinite conditionalization We choose this name for its central role for preferential structures (PR) or (μPR).*
2. *The system of rules (AND) (OR) (LLE) (RW) (SC) (CP) (CM) (CUM) is also called system P (for preferential). Adding (RatM) gives the system R (for rationality or rankedness).*
   Roughly, smooth preferential structures generate logics satisfying system P, while ranked structures generate logics satisfying system R.
3. *A logic satisfying (REF), (ResM), and (CUT) is called a consequence relation.*
4. *(LLE) and (CCL) will hold automatically, whenever we work with model sets.*
5. *(AND) is obviously closely related to filters, and corresponds to closure under finite intersections. (RW) corresponds to upward closure of filters.*

More precisely, validity of both depend on the definition, and the direction we consider.
Given f *and* $(\mu \subseteq)$, $f(X) \subseteq X$ *generates a principal filter,* $\{X' \subseteq X : f(X) \subseteq X'\}$, *with the definition, if* $X = M(T)$, *then* $T{\mid\!\sim}\phi$ *iff* $f(X) \subseteq M(\phi)$. *Validity of* (AND) *and* (RW) *are then trivial.*
Conversely, we can define for $X = M(T)$
$\mathcal{X} := \{X' \subseteq X : \exists\phi(X' = X \cap M(\phi)$ *and* $T{\mid\!\sim}\phi)\}$.
(AND) *then makes* $\mathcal{X}$ *closed under finite intersections, and* (RW) *makes* $\mathcal{X}$ *upward closed. In the infinite case this is usually not yet a filter, as not all subsets of* X *need to be definable this way. In this case, we complete* $\mathcal{X}$ *by adding all* X'' *such that there is* $X' \subseteq X'' \subseteq X$, $X' \in \mathcal{X}$.
Alternatively, we can define
$\mathcal{X} := \{X' \subseteq X : \bigcap\{X \cap M(\phi) : T{\mid\!\sim}\phi\} \subseteq X'\}$.

6. (SC) *corresponds to the choice of a subset.*
7. (CP) *is somewhat delicate, as it presupposes that the chosen model set is nonempty. This might fail in the presence of ever better choices, without ideal ones; the problem is addressed by the limit versions.*
8. (PR) *is an infinitary version of one half of the deduction theorem: Let* T *stand for* ϕ, T' *for* ψ, *and* $\phi \wedge \psi{\mid\!\sim}\sigma$, *so* $\phi{\mid\!\sim}\psi \to \sigma$, *but* $(\psi \to \sigma) \wedge \psi \vdash \sigma$.
9. (CUM) *(whose more interesting half in our context is* (CM)*) may best be seen as a normal use of lemmas: We have worked hard and found some lemmas. Now we can take a rest, and come back again with our new lemmas. Adding them to the axioms will neither add new theorems, nor prevent old ones from holding. (This is, of course, a meta-level argument concerning an object-level rule. But also object-level rules should — at least generally — have an intuitive justification, which will then come from a meta-level argument.)*

The following results are mentioned here to put our work into perspective. The (somewhat lengthy) proofs can be found in [85].

Fact 12.10 *Table 12.5, "Interdependencies of algebraic rules", is to be read as follows: If the left-hand side holds for some function* $f : \mathcal{Y} \to \mathcal{P}(U)$, *and the auxiliary properties noted in the middle also hold for* f *or* $\mathcal{Y}$, *then the right-hand side will hold, too - and conversely.*

"sing." will stand for: "$\mathcal{Y}$ *contains singletons"*

Proposition 12.11. *Table 12.6 "Logical and algebraic rules" is to be read as follows:*

The definitions are given in Definition 12.9.

Let a logic ${\mid\!\sim}$ *satisfy (LLE) and (CCL), and define a function* $f : \mathbf{D}_{\mathcal{L}} \to \mathbf{D}_{\mathcal{L}}$ *by* $f(M(T)) := M(\overline{\overline{T}})$. *Then* f *is well defined, satisfies* (μdp), *and* $\overline{\overline{T}} = Th(f(M(T)))$.

If ${\mid\!\sim}$ *satisfies a rule in the left-hand side, then — provided the additional properties noted in the middle for* $\Rightarrow$ *hold, too —* f *will satisfy the property in the right-hand side.*

Table 12.3: Logical rules, definitions and connections Part I

Logical rules, definitions and connections Part I			
Logical rule		Corr.	Model set
Basics			
(SC) Supraclassicality $\alpha \vdash \beta \Rightarrow \alpha \mid\!\sim \beta$	(SC) $\overline{T} \subseteq \overline{\overline{T}}$	$\Rightarrow$ $\Leftarrow$	$(\mu \subseteq)$ $f(X) \subseteq X$
(REF) Reflexivity $T \cup \{\alpha\} \mid\!\sim \alpha$			
(LLE) Left Logical Equivalence $\vdash \alpha \leftrightarrow \alpha', \alpha \mid\!\sim \beta \Rightarrow \alpha' \mid\!\sim \beta$	(LLE) $\overline{T} = \overline{T'} \Rightarrow \overline{\overline{T}} = \overline{\overline{T'}}$		(trivially true)
(RW) Right Weakening $\alpha \mid\!\sim \beta, \vdash \beta \rightarrow \beta' \Rightarrow \alpha \mid\!\sim \beta'$	(RW) $T \mid\!\sim \beta, \vdash \beta \rightarrow \beta' \Rightarrow T \mid\!\sim \beta'$		(upward closure)
(wOR) $\alpha \mid\!\sim \beta, \alpha' \vdash \beta \Rightarrow \alpha \vee \alpha' \mid\!\sim \beta$	(wOR) $\overline{\overline{T}} \cap \overline{T'} \subseteq \overline{\overline{T \vee T'}}$	$\Rightarrow$ $\Leftarrow$	(μwOR) $f(X \cup Y) \subseteq f(X) \cup Y$
$(disjOR)$ $\alpha \vdash \neg \alpha', \alpha \mid\!\sim \beta, \alpha' \mid\!\sim \beta \Rightarrow \alpha \vee \alpha' \mid\!\sim \beta$	$(disjOR)$ $\neg Con(T \cup T') \Rightarrow \overline{\overline{T}} \cap \overline{\overline{T'}} \subseteq \overline{\overline{T \vee T'}}$	$\Rightarrow$ $\Leftarrow$	$(\mu disjOR)$ $X \cap Y = \emptyset \Rightarrow f(X \cup Y) \subseteq f(X) \cup f(Y)$
(CP) Consistency Preservation $\alpha \mid\!\sim \bot \Rightarrow \alpha \vdash \bot$	(CP) $T \mid\!\sim \bot \Rightarrow T \vdash \bot$	$\Rightarrow$ $\Leftarrow$	$(\mu\emptyset)$ $f(X) = \emptyset \Rightarrow X = \emptyset$
			$(\mu\emptyset fin)$ $X \neq \emptyset \Rightarrow f(X) \neq \emptyset$ for finite X
	(AND_1) $\alpha \mid\!\sim \beta \Rightarrow \alpha \not\mid\!\sim \neg\beta$		
	(AND_n) $\alpha \mid\!\sim \beta_1, \ldots, \alpha \mid\!\sim \beta_{n-1} \Rightarrow \alpha \not\mid\!\sim (\neg\beta_1 \vee \ldots \vee \neg\beta_{n-1})$		
(AND) $\alpha \mid\!\sim \beta, \alpha \mid\!\sim \beta' \Rightarrow \alpha \mid\!\sim \beta \wedge \beta'$	(AND) $T \mid\!\sim \beta, T \mid\!\sim \beta' \Rightarrow T \mid\!\sim \beta \wedge \beta'$		(closure under finite intersection)
(CCL) Classical Closure	(CCL) $\overline{\overline{T}}$ classically closed		(trivially true)
(OR) $\alpha \mid\!\sim \beta, \alpha' \mid\!\sim \beta \Rightarrow \alpha \vee \alpha' \mid\!\sim \beta$	(OR) $\overline{\overline{T}} \cap \overline{\overline{T'}} \subseteq \overline{\overline{T \vee T'}}$	$\Rightarrow$ $\Leftarrow$	(μOR) $f(X \cup Y) \subseteq f(X) \cup f(Y)$
$\overline{\overline{\alpha \wedge \alpha'}} \subseteq \overline{\overline{\overline{\alpha}} \cup \{\alpha'\}}$	(PR) $\overline{\overline{T \cup T'}} \subseteq \overline{\overline{\overline{T}} \cup T'}$	$\Rightarrow$ $\Leftarrow (\mu dp) + (\mu \subseteq)$ $\not\Leftarrow \neg(\mu dp)$ $\Leftarrow (\mu \subseteq)$ $T' = \phi$	(μPR) $X \subseteq Y \Rightarrow f(Y) \cap X \subseteq f(X)$
		$\Leftarrow$ $T' = \phi$	$(\mu PR')$ $f(X) \cap Y \subseteq f(X \cap Y)$
(CUT) $T \mid\!\sim \alpha; T \cup \{\alpha\} \mid\!\sim \beta \Rightarrow T \mid\!\sim \beta$	(CUT) $T \subseteq \overline{T'} \subseteq \overline{\overline{T}} \Rightarrow \overline{\overline{T'}} \subseteq \overline{\overline{T}}$	$\Rightarrow$ $\Leftarrow$	(μCUT) $f(X) \subseteq Y \subseteq X \Rightarrow f(X) \subseteq f(Y)$

Conversely, if $f : \mathcal{Y} \rightarrow \mathcal{P}(M_{\mathcal{L}})$ is a function, with $\mathbf{D}_{\mathcal{L}} \subseteq \mathcal{Y}$, and we define a logic $\mid\!\sim$ by $\overline{\overline{T}} := Th(f(M(T)))$, then $\mid\!\sim$ satisfies (LLE) and (CCL). If f satisfies (μdp), then $f(M(T)) = M(\overline{\overline{T}})$.

If f satisfies a property in the right-hand side, then — provided the additional properties noted in the middle for $\Leftarrow$ hold, too — $\mid\!\sim$ will satisfy the property in the left-hand side.

If "$T = \phi$" is noted in the table, this means that, if one of the theories (the one named the same way in Definition 12.9 is equivalent to a formula, we do not need (μdp).

Definition 12.12. *Fix $U \neq \emptyset$, and consider arbitrary X. Note that this X has not necessarily anything to do with U, or the $\mathcal{U}$ below. Thus, the functions $\mu_{\mathcal{M}}$ below are in principle functions from V to V, where V is the set-theoretical universe we work in.*

Note that we work here often with copies of elements (or models). In other areas of logic, most authors work with valuation functions. The two definitions — copies and valuation functions — are equivalent; a copy $\langle x, i \rangle$ can be seen as a state $\langle x, i \rangle$ with valuation x. In the beginning of research on preferential structures, the notion of copies was widely used, whereas, e.g. [121] used that of valuation functions. There is perhaps a weak justification for the former terminology. In modal logic, even if two states have the same valid classical formulas, they might still be distinguishable by their valid modal formulas. But this depends on the modality being in the object

Table 12.4: Logical rules, definitions and connections Part II

Logical rules, definitions and connections Part II			
Logical rule		Corr.	Model set
Cumulativity			
(wCM) $\alpha \mid\!\sim \beta, \alpha' \vdash \alpha, \alpha \wedge \beta \vdash \alpha' \Rightarrow$ $\alpha' \mid\!\sim \beta$			
(CM_2) $\alpha \mid\!\sim \beta, \alpha \mid\!\sim \beta' \Rightarrow \alpha \wedge \beta \not\vdash \neg \beta'$			
(CM_n) $\alpha \mid\!\sim \beta_1, \ldots, \alpha \mid\!\sim \beta_n \Rightarrow$ $\alpha \wedge \beta_1 \wedge \ldots \wedge \beta_{n-1} \not\vdash \neg \beta_n$			
(CM) Cautious Monotony $\alpha \mid\!\sim \beta, \alpha \mid\!\sim \beta' \Rightarrow$ $\alpha \wedge \beta \mid\!\sim \beta'$	(CM) $T \subseteq \overline{T'} \subseteq \overline{\overline{T}} \Rightarrow$ $\overline{\overline{T}} \subseteq \overline{\overline{T'}}$	$\Rightarrow$ $\Leftarrow$	(μCM) $f(X) \subseteq Y \subseteq X \Rightarrow$ $f(Y) \subseteq f(X)$
or $(ResM)$ Restricted Monotony $T \mid\!\sim \alpha, \beta \Rightarrow T \cup \{\alpha\} \mid\!\sim \beta$		$\Rightarrow$ $\Leftarrow$	$(\mu ResM)$ $f(X) \subseteq A \cap B \Rightarrow$ $f(X \cap A) \subseteq B$
(CUM) Cumulativity $\alpha \mid\!\sim \beta \Rightarrow$ $(\alpha \mid\!\sim \beta' \Leftrightarrow \alpha \wedge \beta \mid\!\sim \beta')$	(CUM) $T \subseteq \overline{T'} \subseteq \overline{\overline{T}} \Rightarrow$ $\overline{\overline{T}} = \overline{\overline{T'}}$	$\Rightarrow$ $\Leftarrow$	(μCUM) $f(X) \subseteq Y \subseteq X \Rightarrow$ $f(Y) = f(X)$
	$(\subseteq\supseteq)$ $T \subseteq \overline{\overline{T'}}, T' \subseteq \overline{\overline{T}} \Rightarrow$ $\overline{\overline{T'}} = \overline{\overline{T}}$	$\Rightarrow$ $\Leftarrow$	$(\mu \subseteq\supseteq)$ $f(X) \subseteq Y, f(Y) \subseteq X \Rightarrow$ $f(X) = f(Y)$
Rationality			
$(RatM)$ Rational Monotony $\alpha \mid\!\sim \beta, \alpha \not\mid\!\sim \neg \beta' \Rightarrow$ $\alpha \wedge \beta' \mid\!\sim \beta$	$(RatM)$ $Con(T \cup \overline{\overline{T'}}), T \vdash T' \Rightarrow$ $\overline{\overline{T}} \supseteq \overline{\overline{\overline{T'}} \cup T}$	$\Rightarrow$ $\Leftarrow (\mu dp)$ $\not\Leftarrow -(\mu dp)$ $\Leftarrow T = \phi$	$(\mu RatM)$ $X \subseteq Y, X \cap f(Y) \neq \emptyset \Rightarrow$ $f(X) \subseteq f(Y) \cap X$
	$(RatM =)$ $Con(T \cup \overline{\overline{T'}}), T \vdash T' \Rightarrow$ $\overline{\overline{T}} = \overline{\overline{\overline{T'}} \cup T}$	$\Rightarrow$ $\Leftarrow (\mu dp)$ $\not\Leftarrow -(\mu dp)$ $\Leftarrow T = \phi$	$(\mu =)$ $X \subseteq Y, X \cap f(Y) \neq \emptyset \Rightarrow$ $f(X) = f(Y) \cap X$
	$(Log =')$ $Con(\overline{\overline{T'}} \cup T) \Rightarrow$ $\overline{\overline{T \cup T'}} = \overline{\overline{\overline{T'}} \cup T}$	$\Rightarrow$ $\Leftarrow (\mu dp)$ $\not\Leftarrow -(\mu dp)$ $\Leftarrow T = \phi$	$(\mu =')$ $f(Y) \cap X \neq \emptyset \Rightarrow$ $f(Y \cap X) = f(Y) \cap X$
(DR) $\alpha \vee \beta \mid\!\sim \gamma \Rightarrow$ $\alpha \mid\!\sim \gamma$ or $\beta \mid\!\sim \gamma$	$(Log \parallel)$ $\overline{\overline{T \vee T'}}$ is one of $\overline{\overline{T}}$, or $\overline{\overline{T'}}$, or $\overline{\overline{T}} \cap \overline{\overline{T'}}$ (by (CCL))	$\Rightarrow$ $\Leftarrow$	$(\mu \parallel)$ $f(X \cup Y)$ is one of $f(X)$, $f(Y)$ or $f(X) \cup f(Y)$
	$(Log\cup)$ $Con(\overline{\overline{T'}} \cup T), \neg Con(\overline{\overline{T'}} \cup \overline{\overline{T}}) \Rightarrow$ $\neg Con(\overline{\overline{T \vee T'}} \cup T')$	$\Rightarrow (\mu \subseteq) + (\mu =)$ $\Leftarrow (\mu dp)$ $\not\Leftarrow -(\mu dp)$	$(\mu\cup)$ $f(Y) \cap (X - f(X)) \neq \emptyset \Rightarrow$ $f(X \cup Y) \cap Y = \emptyset$
	$(Log\cup')$ $Con(\overline{\overline{T'}} \cup T), \neg Con(\overline{\overline{T'}} \cup \overline{\overline{T}}) \Rightarrow$ $\overline{\overline{T \vee T'}} = \overline{\overline{T}}$	$\Rightarrow (\mu \subseteq) + (\mu =)$ $\Leftarrow (\mu dp)$ $\not\Leftarrow -(\mu dp)$	$(\mu\cup')$ $f(Y) \cap (X - f(X)) \neq \emptyset \Rightarrow$ $f(X \cup Y) = f(X)$
			$(\mu \in)$ $a \in X - f(X) \Rightarrow$ $\exists b \in X. a \notin f(\{a, b\})$

language. In most work on preferential structures, the consequence relation is outside the object language, so different states with the same valuation are in a stronger sense copies of each other.

1. *Preferential models* or *structures*.
   a) The version without copies:
   A pair $\mathcal{M} := \langle U, \prec \rangle$ with U an arbitrary set and $\prec$ an arbitrary binary relation on U is called a *preferential model* or *structure*.
   b) The version with copies:
   A pair $\mathcal{M} := \langle \mathcal{U}, \prec \rangle$ with $\mathcal{U}$ an arbitrary set of pairs and $\prec$ an arbitrary binary relation on $\mathcal{U}$ is called a *preferential model* or *structure*.
   If $\langle x, i \rangle \in \mathcal{U}$, then x is intended to be an element of U, and i the index of the copy.
   We sometimes also need copies of the relation $\prec$. We will then replace $\prec$ by one or several arrows α attacking non-minimal elements, e.g. $x \prec y$ will be written $\alpha : x \to y$, $\langle x, i \rangle \prec \langle y, i \rangle$ will be written $\alpha : \langle x, i \rangle \to \langle y, i \rangle$, and finally we might have $\langle \alpha, k \rangle : x \to y$ and $\langle \alpha, k \rangle : \langle x, i \rangle \to \langle y, i \rangle$.
2. *Minimal elements*, the functions $\mu_{\mathcal{M}}$.
   a) The version without copies:
   Let $\mathcal{M} := \langle U, \prec \rangle$, and define

Table 12.5: Interdependencies of algebraic rules

	Interdependencies of algebraic rules		
	Basics		
(1.1)	(μPR)	$\Rightarrow (\cap) + (\mu \subseteq)$	$(\mu PR')$
(1.2)		$\Leftarrow$	
(2.1)	(μPR)	$\Rightarrow (\mu \subseteq)$	(μOR)
(2.2)		$\Leftarrow (\mu \subseteq) + (-)$	
(2.3)		$\Rightarrow (\mu \subseteq)$	(μwOR)
(2.4)		$\Leftarrow (\mu \subseteq) + (-)$	
(3)	(μPR)	$\Rightarrow$	(μCUT)
(4)	$(\mu \subseteq) + (\mu \subseteq\supseteq) + (\mu CUM)$ $+(\mu RatM) + (\cap)$	$\not\Rightarrow$	(μPR)
	Cumulativity		
(5.1)	(μCM)	$\Rightarrow (\cap) + (\mu \subseteq)$	$(\mu ResM)$
(5.2)		$\Leftarrow$ (infin.)	
(6)	$(\mu CM) + (\mu CUT)$	$\Leftrightarrow$	(μCUM)
(7)	$(\mu \subseteq) + (\mu \subseteq\supseteq)$	$\Rightarrow$	(μCUM)
(8)	$(\mu \subseteq) + (\mu CUM) + (\cap)$	$\Rightarrow$	$(\mu \subseteq\supseteq)$
(9)	$(\mu \subseteq) + (\mu CUM)$	$\not\Rightarrow$	$(\mu \subseteq\supseteq)$
	Rationality		
(10)	$(\mu RatM) + (\mu PR)$	$\Rightarrow$	$(\mu =)$
(11)	$(\mu =)$	$\Rightarrow$	$(\mu PR) + (\mu RatM)$
(12.1)	$(\mu =)$	$\Rightarrow (\cap) + (\mu \subseteq)$	$(\mu =')$
(12.2)		$\Leftarrow$	
(13)	$(\mu \subseteq) + (\mu =)$	$\Rightarrow (\cup)$	$(\mu \cup)$
(14)	$(\mu \subseteq) + (\mu \emptyset) + (\mu =)$	$\Rightarrow (\cup)$	$(\mu \Vert)$, $(\mu \cup')$, (μCUM)
(15)	$(\mu \subseteq) + (\mu \Vert)$	$\Rightarrow (-)$ of $\mathcal{Y}$	$(\mu =)$
(16)	$(\mu \Vert) + (\mu \in) + (\mu PR)+$ $(\mu \subseteq)$	$\Rightarrow (\cup) +$ sing.	$(\mu =)$
(17)	$(\mu CUM) + (\mu =)$	$\Rightarrow (\cup) +$ sing.	$(\mu \in)$
(18)	$(\mu CUM) + (\mu =) + (\mu \subseteq)$	$\Rightarrow (\cup)$	$(\mu \Vert)$
(19)	$(\mu PR) + (\mu CUM) + (\mu \Vert)$	$\Rightarrow$ sufficient, e.g. true in $\boldsymbol{D}_{\mathcal{L}}$	$(\mu =)$.
(20)	$(\mu \subseteq) + (\mu PR) + (\mu =)$	$\not\Rightarrow$	$(\mu \Vert)$
(21)	$(\mu \subseteq) + (\mu PR) + (\mu \Vert)$	$\not\Rightarrow$ (without $(-)$)	$(\mu =)$
(22)	$(\mu \subseteq) + (\mu PR) + (\mu \Vert)+$ $(\mu =) + (\mu \cup)$	$\not\Rightarrow$	$(\mu \in)$ (thus not representable by ranked structures)

$\mu_{\mathcal{M}}(X) := \{x \in X : x \in U \wedge \neg\exists x' \in X \cap U.x' \prec x\}$.
$\mu_{\mathcal{M}}(X)$ is called the set of *minimal elements* of X (in $\mathcal{M}$).
Thus, $\mu_{\mathcal{M}}(X)$ is the set of elements such that there is no smaller one in X.

b) The version with copies:
Let $\mathcal{M} := \langle \mathcal{U}, \prec \rangle$ be as above. Define
$\mu_{\mathcal{M}}(X) := \{x \in X : \exists \langle x, i \rangle \in \mathcal{U}.\neg\exists \langle x', i' \rangle \in \mathcal{U}(x' \in X \wedge \langle x', i' \rangle' \prec \langle x, i \rangle)\}$.
Thus, $\mu_{\mathcal{M}}(X)$ is the projection on the first coordinate of the set of elements such that there is no smaller one in X.
Again, by abuse of language, we say that $\mu_{\mathcal{M}}(X)$ is the set of *minimal elements* of X in the structure. If the context is clear, we will write just μ.
We sometimes say that $\langle x, i \rangle$ "*kills*" or "*minimizes*" $\langle y, j \rangle$ if $\langle x, i \rangle \prec \langle y, j \rangle$. By abuse of language we also say a set X *kills* or *minimizes* a set Y if for all $\langle y, j \rangle \in \mathcal{U}$, $y \in Y$ there exists $\langle x, i \rangle \in \mathcal{U}$, $x \in X$ s.t. $\langle x, i \rangle \prec \langle y, j \rangle$.

Table 12.6: Logical and algebraic rules

Logical and algebraic rules			
		Basics	
(1.1)	(OR)	$\Rightarrow$	(μOR)
(1.2)		$\Leftarrow$	
(2.1)	$(disjOR)$	$\Rightarrow$	$(\mu disjOR)$
(2.2)		$\Leftarrow$	
(3.1)	(wOR)	$\Rightarrow$	(μwOR)
(3.2)		$\Leftarrow$	
(4.1)	(SC)	$\Rightarrow$	$(\mu\subseteq)$
(4.2)		$\Leftarrow$	
(5.1)	(CP)	$\Rightarrow$	$(\mu\emptyset)$
(5.2)		$\Leftarrow$	
(6.1)	(PR)	$\Rightarrow$	(μPR)
(6.2)		$\Leftarrow (\mu dp) + (\mu\subseteq)$	
(6.3)		$\not\Leftarrow -(\mu dp)$	
(6.4)		$\Leftarrow (\mu\subseteq)$ $T' = \phi$	
(6.5)	(PR)	$\Leftarrow$ $T' = \phi$	$(\mu PR')$
(7.1)	(CUT)	$\Rightarrow$	(μCUT)
(7.2)		$\Leftarrow$	
		Cumulativity	
(8.1)	(CM)	$\Rightarrow$	(μCM)
(8.2)		$\Leftarrow$	
(9.1)	$(ResM)$	$\Rightarrow$	$(\mu ResM)$
(9.2)		$\Leftarrow$	
(10.1)	$(\subseteq\supseteq)$	$\Rightarrow$	$(\mu\subseteq\supseteq)$
(10.2)		$\Leftarrow$	
(11.1)	(CUM)	$\Rightarrow$	(μCUM)
(11.2)		$\Leftarrow$	
		Rationality	
(12.1)	$(RatM)$	$\Rightarrow$	$(\mu RatM)$
(12.2)		$\Leftarrow (\mu dp)$	
(12.3)		$\not\Leftarrow -(\mu dp)$	
(12.4)		$\Leftarrow$ $T = \phi$	
(13.1)	$(RatM =)$	$\Rightarrow$	$(\mu =)$
(13.2)		$\Leftarrow (\mu dp)$	
(13.3)		$\not\Leftarrow -(\mu dp)$	
(13.4)		$\Leftarrow$ $T = \phi$	
(14.1)	$(Log =')$	$\Rightarrow$	$(\mu =')$
(14.2)		$\Leftarrow (\mu dp)$	
(14.3)		$\not\Leftarrow -(\mu dp)$	
(14.4)		$\Leftarrow T = \phi$	
(15.1)	$(Log \parallel)$	$\Rightarrow$	$(\mu \parallel)$
(15.2)		$\Leftarrow$	
(16.1)	$(Log\cup)$	$\Rightarrow (\mu\subseteq) + (\mu =)$	$(\mu\cup)$
(16.2)		$\Leftarrow (\mu dp)$	
(16.3)		$\not\Leftarrow -(\mu dp)$	
(17.1)	$(Log\cup')$	$\Rightarrow (\mu\subseteq) + (\mu =)$	$(\mu\cup')$
(17.2)		$\Leftarrow (\mu dp)$	
(17.3)		$\not\Leftarrow -(\mu dp)$	

$\mathcal{M}$ is also called *injective* or 1-copy iff there is always at most one copy $\langle x, i\rangle$ for each x. Note that the existence of copies corresponds to a non-injective labelling function — as is often used in nonclassical logic, e.g. modal logic.

We say that $\mathcal{M}$ is *transitive, irreflexive*, etc., iff $\prec$ is.
Note that $\mu(X)$ might well be empty, even if X is not.

Definition 12.13. *We define the consequence relation of a preferential structure for a given propositional language $\mathcal{L}$.*

1. *a) If m is a classical model of a language $\mathcal{L}$, we say by abuse of language*
   $\langle m, i\rangle \models \phi$ *iff* $m \models \phi$,
   and if X is any set of such pairs, that
   $X \models \phi$ *iff for all* $\langle m, i\rangle \in X$ $m \models \phi$.
   b) If $\mathcal{M}$ is a preferential structure, and X is a set of $\mathcal{L}$-models for a classical propositional language $\mathcal{L}$, or is a set of pairs $\langle m, i\rangle$ where the m are such models, we call $\mathcal{M}$ a classical preferential structure or model.
2. *Validity in a preferential structure, or the semantical consequence relation defined by such a structure:*
   Let $\mathcal{M}$ be as above.
   We define
   $T \models_{\mathcal{M}} \phi$ *iff* $\mu_{\mathcal{M}}(M(T)) \models \phi$, *i.e.* $\mu_{\mathcal{M}}(M(T)) \subseteq M(\phi)$.
3. *$\mathcal{M}$ will be called definability preserving iff for all $X \in \boldsymbol{D}_{\mathcal{L}}$ we have $\mu_{\mathcal{M}}(X) \in \boldsymbol{D}_{\mathcal{L}}$.*

As $\mu_{\mathcal{M}}$ is defined on $\boldsymbol{D}_{\mathcal{L}}$, but need by no means always result in some new definable set, this is (and reveals itself as a quite strong) additional property.

Definition 12.14. *Let $\mathcal{Y} \subseteq \mathcal{P}(U)$. (In applications to logic, $\mathcal{Y}$ will be $\boldsymbol{D}_{\mathcal{L}}$.)*

A preferential structure $\mathcal{M}$ is called $\mathcal{Y}$-smooth iff for every $X \in \mathcal{Y}$ every element $x \in X$ is either minimal in X or above an element which is minimal in X. More precisely:

1. *The version without copies:*
   If $x \in X \in \mathcal{Y}$, then either $x \in \mu(X)$ or there is $x' \in \mu(X).x' \prec x$.
2. *The version with copies:*
   If $x \in X \in \mathcal{Y}$, and $\langle x, i\rangle \in \mathcal{U}$, then either there is no $\langle x', i'\rangle \in \mathcal{U}$, $x' \in X$, $\langle x', i'\rangle \prec \langle x, i\rangle$ or there is a $\langle x', i'\rangle \in \mathcal{U}$, $\langle x', i'\rangle \prec \langle x, i\rangle$, $x' \in X$, s.t. there is no $\langle x'', i''\rangle \in \mathcal{U}$, $x'' \in X$, with $\langle x'', i''\rangle \prec \langle x', i'\rangle$.
   (Writing down all details here again might make it easier to read applications of the definition later on.)

When considering the models of a language $\mathcal{L}$, $\mathcal{M}$ will be called smooth iff it is $\boldsymbol{D}_{\mathcal{L}}$-smooth; $\boldsymbol{D}_{\mathcal{L}}$ is the default.

Obviously, the richer the set $\mathcal{Y}$ is, the stronger the condition $\mathcal{Y}$-smoothness will be.

A remark for the intuition: Smoothness is perhaps best motivated through Gabbay's concept of reactive diagrams; see, e.g. [66] and [64], and also [85], [73]. In this concept, smaller, or "better", elements attack bigger, or "less good", elements. But when a attacks b, and b attacks c, then one might consider the attack of b against c weakened by the attack of a against b. In a smooth structure, for every attack against some element x, there is also an uncontested attack against x, as it originates in an element y, which is not attacked itself.

The proofs of the following results can best be found in [147].

Table 12.7, "Preferential representation", summarizes the more difficult half of a full representation result for preferential structures. It shows equivalence between certain abstract conditions for model choice functions and certain preferential structures. They are shown in the respective representation theorems.

"Singletons" means that the domain must contain all singletons, "1 copy" or "≥ 1 copy" means that the structure may contain only one copy for each point, or several, and "$(\mu\emptyset)$" for the preferential structure means that the μ-function of the structure has to satisfy this property.

We call a characterization "normal" iff it is a universally quantified Boolean combination (of any fixed, but perhaps infinite, length) of rules of the usual form. We do not go into details here.

In the second column from the left, "$\Rightarrow$" means, for instance for the smooth case, that for any $\mathcal{Y}$ closed under finite unions, and any choice function f which satisfies the conditions in the left-hand column, there is a (here $\mathcal{Y}$-smooth) preferential structure $\mathcal{X}$ which represents it, i.e. for all $Y \in \mathcal{Y}$, $f(Y) = \mu_{\mathcal{X}}(Y)$, where $\mu_{\mathcal{X}}$ is the model choice function of the structure $\mathcal{X}$. The inverse arrow $\Leftarrow$ means that the model choice function for any smooth $\mathcal{X}$ defined on such a $\mathcal{Y}$ will satisfy the conditions on the left.

12.3.2 Generalized preferential structures

Introduction

Definition 12.15. *An IBRS is called a generalized $preferential$ structure iff the origins of all arrows are points. We will usually write x, y, etc. for points, α, β, etc. for arrows.*

Definition 12.16. *Consider a generalized preferential structure $\mathcal{X}$.*

(1) Level n arrow :

Definition by upward induction.

If $\alpha : x \to y$, x, y are points, then α is a level 1 arrow.

If $\alpha : x \to \beta$, x is a point, β a level n arrow, then α is a level $n+1$ arrow. ($o(\alpha)$ is the origin, $d(\alpha)$ is the destination of α.)

$\lambda(\alpha)$ will denote the level of α.

(2) Level n structure :

$\mathcal{X}$ is a level n structure iff all arrows in $\mathcal{X}$ are at most level n arrows.

We consider here only structures of some arbitrary but finite level n.

Table 12.7: Preferential representation

Preferential representation				
μ-function		Pref. Structure		Logic
$(\mu\subseteq)$	$\Leftrightarrow$	reactive	$\Leftrightarrow$	$(LLE)+(CCL)+$ (SC)
$(\mu\subseteq)+(\mu CUM)$	$\Rightarrow (\cap)$	reactive + essentially smooth		
$(\mu\subseteq)+(\mu\subseteq\supseteq)$	$\Rightarrow$	reactive + essentially smooth	$\Leftrightarrow$	$(LLE)+(CCL)+$ $(SC)+(\subseteq\supseteq)$
$(\mu\subseteq)+(\mu CUM)+(\mu\subseteq\supseteq)$	$\Leftarrow$	reactive + essentially smooth		
$(\mu\subseteq)+(\mu PR)$	$\Leftarrow$	general	$\Rightarrow (\mu dp)$	$(LLE)+(RW)+$ $(SC)+(PR)$
	$\Rightarrow$		$\Leftarrow$	
			$\not\Rightarrow$ without (μdp)	
			$\not\Leftrightarrow$ without (μdp)	any "normal" characterization of any size
$(\mu\subseteq)+(\mu PR)$	$\Leftarrow$	transitive	$\Rightarrow (\mu dp)$	$(LLE)+(RW)+$ $(SC)+(PR)$
	$\Rightarrow$		$\Leftarrow$	
			$\not\Rightarrow$ without (μdp)	
			$\Leftrightarrow$ without (μdp)	using "small" exception sets
$(\mu\subseteq)+(\mu PR)+(\mu CUM)$	$\Leftarrow$	smooth	$\Rightarrow (\mu dp)$	$(LLE)+(RW)+$ $(SC)+(PR)+$ (CUM)
	$\Rightarrow (\cup)$		$\Leftarrow (\cup)$	
	$\not\Rightarrow$ without $(\cup)$		$\not\Rightarrow$ without (μdp)	
$(\mu\subseteq)+(\mu PR)+(\mu CUM)$	$\Leftarrow$	smooth+transitive	$\Rightarrow (\mu dp)$	$(LLE)+(RW)+$ $(SC)+(PR)+$ (CUM)
	$\Rightarrow (\cup)$		$\Leftarrow (\cup)$	
			$\not\Rightarrow$ without (μdp)	
			$\Leftrightarrow$ without (μdp)	using "small" exception sets
$(\mu\subseteq)+(\mu =)+(\mu PR)+$ $(\mu =')+(\mu \parallel)+(\mu\cup)+$ $(\mu\cup')+(\mu\in)+(\mu RatM)$	$\Leftarrow$	ranked, ≥ 1 copy		
$(\mu\subseteq)+(\mu =)+(\mu PR)+$ $(\mu\cup)+(\mu\in)$	$\not\Rightarrow$	ranked,		
$(\mu\subseteq)+(\mu =)+(\mu\emptyset)$	$\Leftrightarrow$, $(\cup)$	ranked, 1 copy + $(\mu\emptyset)$		
$(\mu\subseteq)+(\mu =)+(\mu\emptyset)$	$\Leftrightarrow$, $(\cup)$	ranked, smooth, 1 copy + $(\mu\emptyset)$		
$(\mu\subseteq)+(\mu =)+(\mu\emptyset fin)+$ $(\mu\in)$	$\Leftrightarrow$, $(\cup)$, singletons	ranked, smooth, ≥ 1 copy + $(\mu\emptyset fin)$		
$(\mu\subseteq)+(\mu PR)+(\mu \parallel)+$ $(\mu\cup)+(\mu\in)$	$\Leftrightarrow$, $(\cup)$, singletons	ranked, ≥ 1 copy	$\not\Rightarrow$ without (μdp)	$(RatM)$, $(RatM =)$, $(Log\cup)$, $(Log\cup')$
			$\not\Leftrightarrow$ without (μdp)	any "normal" characterization of any size

(3) We define for an arrow α by induction $O(\alpha)$ and $D(\alpha)$.
If $\lambda(\alpha) = 1$, then $O(\alpha) := \{o(\alpha)\}$, $D(\alpha) := \{d(\alpha)\}$.
If $\alpha : x \to \beta$, then $D(\alpha) := D(\beta)$, and $O(\alpha) := \{x\} \cup O(\beta)$.

Thus, for example, if $\alpha : x \to y$, $\beta : z \to \alpha$, then $O(\beta) := \{x, z\}$, $D(\beta) = \{y\}$. Consider also the arrow $\beta := \langle \beta', l' \rangle$ in Figure 12.7. There, $D(\beta) = \{\langle x, i \rangle\}$, $O(\beta) = \{\langle z', m' \rangle, \langle y, j \rangle\}$.

We will not consider here diagrams with arbitrarily high levels. One reason is that diagrams like the following will have an unclear meaning:

Example 12.17. $\langle \alpha, 1 \rangle : x \to y$,
$\langle \alpha, n+1 \rangle : x \to \langle \alpha, n \rangle$ $(n \in \omega)$.
Is $y \in \mu(X)$?

Definition 12.18. *Let $\mathcal{X}$ be a generalized preferential structure of (finite) level n.*

We define (by downward induction):

(1) Valid $X \mapsto Y$ arrow :

Let $X, Y \subseteq \mathbf{P}(\mathcal{X})$.

$\alpha \in \mathbf{A}(\mathcal{X})$ is a valid $X \mapsto Y$ arrow iff

(1.1) $O(\alpha) \subseteq X$, $D(\alpha) \subseteq Y$,

(1.2) $\forall \beta : x' \to \alpha.(x' \in X \Rightarrow \exists \gamma : x'' \to \beta.(\gamma$ is a valid $X \mapsto Y$ arrow)).

We will also say that α is a valid arrow in X, or just valid in X, iff α is a valid $X \mapsto X$ arrow.

(2) Valid $X \Rightarrow Y$ arrow :

Let $X \subseteq Y \subseteq \mathbf{P}(\mathcal{X})$.

$\alpha \in \mathbf{A}(\mathcal{X})$ is a valid $X \Rightarrow Y$ arrow iff

(2.1) $o(\alpha) \in X$, $O(\alpha) \subseteq Y$, $D(\alpha) \subseteq Y$,

(2.2) $\forall \beta : x' \to \alpha.(x' \in Y \Rightarrow \exists \gamma : x'' \to \beta.(\gamma$ is a valid $X \Rightarrow Y$ arrow)).

Thus, any attack β from Y against α has to be countered by a valid attack on β.

(Note that in particular $o(\gamma) \in X$, and that $o(\beta)$ need not be in X, but can be in the bigger Y.)

Remark 12.19. Note that, in the definition of valid $X \mapsto Y$ arrow, X and Y need not be related, but in the definition of valid $X \Rightarrow Y$ arrow, $X \subseteq Y$.

Let us assume now that $X \subseteq Y$, and look at the remaining differences.

In both cases, $D(\alpha) \subseteq Y$.

In the $X \mapsto Y$ case, $O(\alpha) \subseteq X$, and attacks from X are countered.

In the $X \Rightarrow Y$ case, $o(\alpha) \in X$, $O(\alpha) \subseteq Y$, and attacks from Y are countered.

So the first condition is stronger in the $X \mapsto Y$ case, the second in the $X \Rightarrow Y$ case.

Example 12.20.

(1) Consider the arrow $\beta := \langle \beta', l' \rangle$ in Figure 12.7. $D(\beta) = \{\langle x, i \rangle\}$, $O(\beta) = \{\langle z', m' \rangle, \langle y, j \rangle\}$, and the only arrow attacking β originates outside X, so β is a valid $X \mapsto \mu(X)$ arrow.

(2) Consider the arrows $\langle \alpha', k' \rangle$ and $\langle \gamma', n' \rangle$ in Figure 12.8. Both are valid $\mu(X) \Rightarrow X$ arrows.

Example 12.21. See Figure 12.6.

Let $X \subseteq Y$.

Consider the left-hand side of the diagram.

The fact that δ originates in Y, but not in X, establishes that α is not a valid $X \mapsto Y$ arrow, as the condition $O(\alpha) \subseteq X$ is violated. To be a valid $X \Rightarrow Y$ arrow, we have to show that all attacks on α originating from Y (not only from X) are countered by valid $X \Rightarrow Y$ arrows. This holds, as β_1 is countered by γ_1, β_2 by γ_2. All possible attacks on α, γ_1, γ_2 from outside Y, like ρ, need not be considered.

Consider the right-hand side of the diagram.

The fact that there is no valid counterargument to β_2' establishes that α' is not a valid $X \Rightarrow Y$ arrow. It is a valid $X \mapsto Y$ arrow, as counterarguments to α' like β_2', which do not originate in X, are not considered. The counterargument β_1' is considered, but it is destroyed by the valid $X \mapsto Y$ arrow γ_1'.

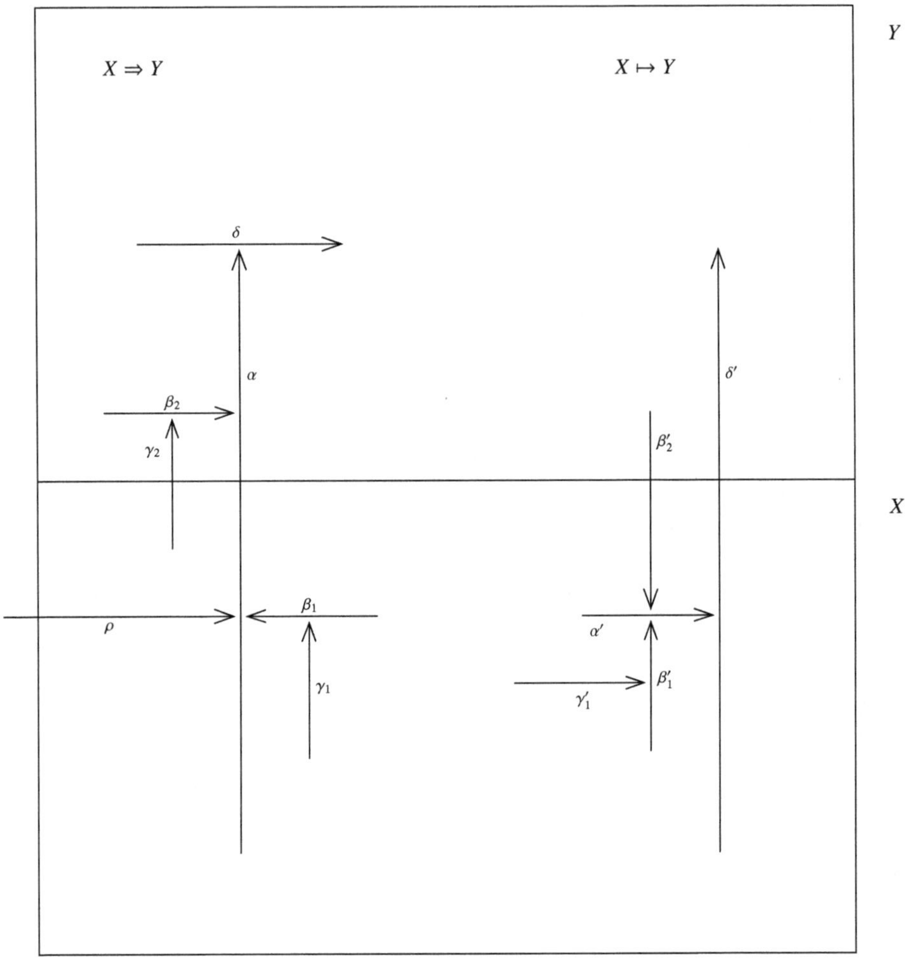

Fig. 12.6

Fact 12.22 *(1) If α is a valid $X \Rightarrow Y$ arrow, then α is a valid $Y \mapsto Y$ arrow.*

(2) If $X \subseteq X' \subseteq Y' \subseteq Y \subseteq \mathbf{P}(\mathcal{X})$ and $\alpha \in \mathbf{A}(\mathcal{X})$ is a valid $X \Rightarrow Y$ arrow, and $O(\alpha) \subseteq Y'$, $D(\alpha) \subseteq Y'$, then α is a valid $X' \Rightarrow Y'$ arrow.

Proof. Let α be a valid $X \Rightarrow Y$ arrow. We show (1) and (2) together by downward induction (both are trivial).

By the prerequisite, $o(\alpha) \in X \subseteq X'$, $O(\alpha) \subseteq Y' \subseteq Y$, $D(\alpha) \subseteq Y' \subseteq Y$.

Case 1: $\lambda(\alpha) = n$. So α is a valid $X' \Rightarrow Y'$ arrow and a valid $Y \mapsto Y$ arrow.

Case 2: $\lambda(\alpha) = n - 1$. So there is no $\beta : x' \to \alpha$, $y \in Y$, so α is a valid $Y \mapsto Y$ arrow. By $Y' \subseteq Y$, α is a valid $X' \Rightarrow Y'$ arrow.

Case 3: Let the result be shown down to m, $n > m > 1$; let $\lambda(\alpha) = m - 1$. So $\forall \beta : x' \to \alpha(x' \in Y \Rightarrow \exists \gamma : x'' \to \beta\ (x'' \in X$ and γ is a valid $X \Rightarrow Y$ arrow)). By induction hypothesis γ is a valid $Y \mapsto Y$ arrow, and a valid $X' \Rightarrow Y'$ arrow. So α is a valid $Y \mapsto Y$ arrow, and by $Y' \subseteq Y$, α is a valid $X' \Rightarrow Y'$ arrow.

Definition 12.23. *Let $\mathcal{X}$ be a generalized preferential structure of level n, $X \subseteq \mathbf{P}(\mathcal{X})$.*

$\mu(X) := \{x \in X : \exists\langle x, i\rangle.\neg\exists$ *valid* $X \mapsto X$ *arrow* $\alpha : x' \to \langle x, i\rangle\}$.

Definition 12.24. *Let $\mathcal{X}$ be a generalized preferential structure.*

$X \sqsubseteq X'$ iff

(1) $X \subseteq X' \subseteq \mathbf{P}(\mathcal{X})$,

(2) $\forall x \in X' - X\ \forall\langle x, i\rangle\ \exists \alpha : x' \to \langle x, i\rangle$ (α is a valid $X \Rightarrow X'$ arrow),

(3) $\forall x \in X\ \exists\langle x, i\rangle$

($\forall \alpha : x' \to \langle x, i\rangle\ (x' \in X' \Rightarrow \exists \beta : x'' \to \alpha.(\beta$ is a valid $X \Rightarrow X'$ arrow))).

Note that (3) is not simply the negation of (2):

Consider a level 1 structure. Thus all level 1 arrows are valid, but the source of the arrows must not be neglected.

(2) reads now: $\forall x \in X' - X\ \forall\langle x, i\rangle\ \exists \alpha : x' \to \langle x, i\rangle.x' \in X$

(3) reads: $\forall x \in X\ \exists\langle x, i\rangle\ \neg\exists \alpha : x' \to \langle x, i\rangle.x' \in X'$

This is intended: intuitively, read $X = \mu(X')$, and minimal elements must not be attacked at all, but non-minimals must be attacked from X — which is a modified version of smoothness. More precisely: non-minimal elements (i.e. from $X' - X$) have to be validly attacked from X, minimal elements must not be validly attacked at all from X' (only perhaps from the outside).

Remark 12.25. We note the special case of Definition 12.24 for level 3 structures. We also write it immediately for the intended case $\mu(X) \sqsubseteq X$, and explicitly with copies.

$x \in \mu(X)$ iff

(1) $\exists\langle x, i\rangle\forall\langle\alpha, k\rangle : \langle y, j\rangle \to \langle x, i\rangle$

$(y \in X \Rightarrow \exists\langle\beta', l'\rangle : \langle z', m'\rangle \to \langle\alpha, k\rangle.$

$(z' \in \mu(X) \wedge \neg\exists\langle\gamma', n'\rangle : \langle u', p'\rangle \to \langle\beta', l'\rangle.u' \in X))$.

See Figure 12.7.

$x \in X - \mu(X)$ iff

(2) $\forall\langle x, i\rangle\exists\langle\alpha', k'\rangle : \langle y', j'\rangle \to \langle x, i\rangle$

$(y' \in \mu(X) \wedge$

(a) $\neg\exists\langle\beta', l'\rangle : \langle z', m'\rangle \to \langle\alpha', k'\rangle.z' \in X$

or

(*b*) $\forall\langle\beta', l'\rangle : \langle z', m'\rangle \to \langle\alpha', k'\rangle$

$(z' \in X \Rightarrow \exists\langle\gamma', n'\rangle : \langle u', p'\rangle \to \langle\beta', l'\rangle.u' \in \mu(X)))$

See Figure 12.8.

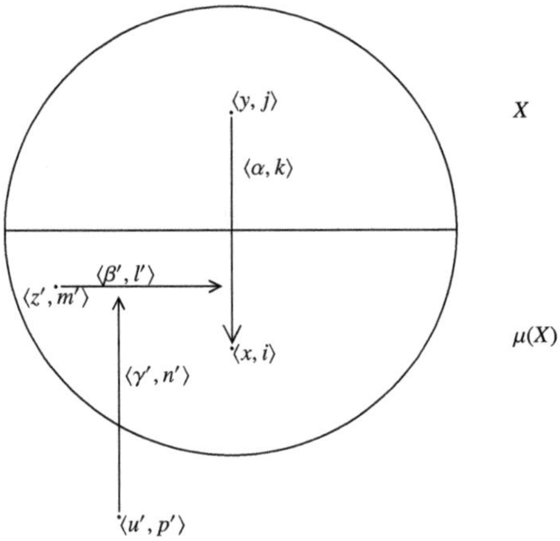

Fig. 12.7: Case 3-1-2

Fact 12.26 *(1) If* $X \sqsubseteq X'$*, then* $X = \mu(X')$,

(2) $X \sqsubseteq X', X \subseteq X'' \subseteq X' \Rightarrow X \sqsubseteq X''$. *(This corresponds to* (μCUM)*.)*

(3) $X \sqsubseteq X', X \subseteq Y', Y \sqsubseteq Y', Y \subseteq X' \Rightarrow X = Y$. *(This corresponds to* $(\mu \subseteq\supseteq)$*.)*

Proof. (1) Trivial by Fact 12.22 (1).

(2)

We have to show

(a) $\forall x \in X'' - X\ \forall\langle x, i\rangle\ \exists\alpha : x' \to \langle x, i\rangle$ (α is a valid $X \Rightarrow X''$ arrow), and

(b) $\forall x \in X\ \exists\langle x, i\rangle\ (\forall\alpha : x' \to \langle x, i\rangle\ (x' \in X'' \Rightarrow \exists\beta : x'' \to \alpha.$($\beta$ is a valid $X \Rightarrow X''$ arrow))).

Both follow from the corresponding condition for $X \Rightarrow X'$, the restriction of the universal quantifier, and Fact 12.22 (2).

(3)

Let $x \in X - Y$.

(a) By $x \in X \sqsubseteq X'$, $\exists\langle x, i\rangle$ s.t. $(\forall\alpha : x' \to \langle x, i\rangle\ (x' \in X' \Rightarrow \exists\beta : x'' \to \alpha.$($\beta$ is a valid $X \Rightarrow X'$ arrow))).

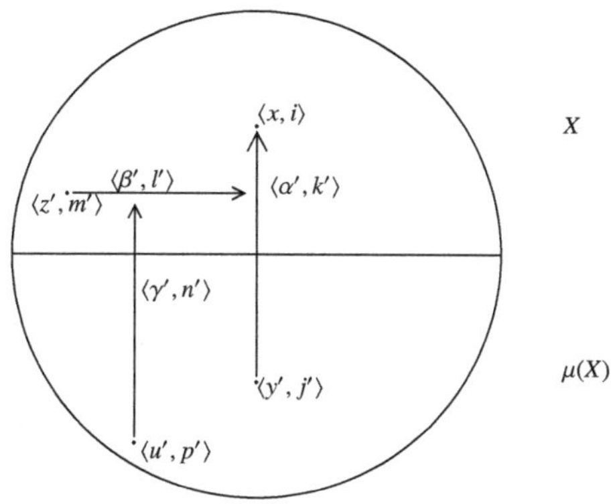

Fig. 12.8: Case 3-2

(b) By $x \notin Y \subseteq \exists \alpha_1 : x' \to \langle x, i\rangle$ α_1 is a valid $Y \Rightarrow Y'$ arrow; in particular $x' \in Y \subseteq X'$. Moreover, $\lambda(\alpha_1) = 1$.

So by (a) $\exists \beta_2 : x'' \to \alpha_1.(\beta_2$ is a valid $X \Rightarrow X'$ arrow), in particular $x'' \in X \subseteq Y'$; moreover $\lambda(\beta_2) = 2$.

It follows by induction from the definition of valid $A \Rightarrow B$ arrows that
$\forall n \exists \alpha_{2m+1}, \lambda(\alpha_{2m+1}) = 2m + 1, \alpha_{2m+1}$ a valid $Y \Rightarrow Y'$ arrow and
$\forall n \exists \beta_{2m+2}, \lambda(\beta_{2m+2}) = 2m + 2, \beta_{2m+2}$ a valid $X \Rightarrow X'$ arrow,
which is impossible, as $\mathcal{X}$ is a structure of finite level.

Definition 12.27. *Let $\mathcal{X}$ be a generalized preferential structure, $X \subseteq \mathbf{P}(\mathcal{X})$.*
$\mathcal{X}$ is called totally smooth for X iff
(1) $\forall \alpha : x \to y \in \mathbf{A}(\mathcal{X})(O(\alpha) \cup D(\alpha) \subseteq X \Rightarrow \exists \alpha' : x' \to y.x' \in \mu(X))$
(2) if α is valid, then there must also exist such α' which is valid
(y a point or an arrow).
If $\mathcal{Y} \subseteq \mathbf{P}(\mathcal{X})$, then $\mathcal{X}$ is called $\mathcal{Y}$ – totally smooth
iff for all $X \in \mathcal{Y}$ $\mathcal{X}$ is totally smooth for X.

Example 12.28. $X := \{\alpha : a \to b, \alpha' : b \to c, \alpha'' : a \to c, \beta : b \to \alpha'\}$ is not totally smooth,

$X := \{\alpha : a \to b, \alpha' : b \to c, \alpha'' : a \to c, \beta : b \to \alpha', \beta' : a \to \alpha'\}$ is totally smooth.

Example 12.29. Consider $\alpha' : a \to b, \alpha'' : b \to c, \alpha : a \to c, \beta : a \to \alpha$.

Then $\mu(\{a, b, c\}) = \{a\}$, $\mu(\{a, c\}) = \{a, c\}$. Thus, (μCUM) does not hold in this structure. Note that there is no valid arrow from $\mu(\{a, b, c\})$ to c.

Definition 12.30. *Let $\mathcal{X}$ be a generalized preferential structure, $X \subseteq \mathbf{P}(\mathcal{X})$.*

$\mathcal{X}$ is called essentially smooth for X iff $\mu(X) \sqsubseteq X$. If $\mathcal{Y} \subseteq \mathbf{P}(\mathcal{X})$, then $\mathcal{X}$ is called $\mathcal{Y}$ – essentially smooth

iff for all $X \in \mathcal{Y}$ $\mu(X) \sqsubseteq X$.

Example 12.31. It is easy to see that we can distinguish total and essential smoothness in richer structures, as the following Example shows:

We add an accessibility relation R, and consider only those models which are accessible.

Let e.g. $a \to b \to \langle c, 0\rangle, \langle c, 1\rangle$, without transitivity. Thus, only c has two copies. This structure is essentially smooth, but of course not totally so.

Let now mRa, mRb, $mR\langle c, 0\rangle$, $mR\langle c, 1\rangle$, $m'Ra$, $m'Rb$, $m'R\langle c, 0\rangle$.

Thus, seen from m, $\mu(\{a, b, c\}) = \{a, c\}$, but seen from m', $\mu(\{a, b, c\}) = \{a\}$, but $\mu(\{a, c\}) = \{a, c\}$, contradicting (CUM).

Results on not necessarily smooth structures

Example 12.32. We show here $(\mu \subseteq) + (\mu \subseteq\supseteq) + (\mu CUM) + (\mu RatM) + (\cap) \not\Rightarrow (\mu PR)$.

Let $U := \{a, b, c\}$. Let $\mathcal{Y} = \mathcal{P}(U)$. So $(\cap)$ is trivially satisfied. Set $f(X) := X$ for all $X \subseteq U$ except for $f(\{a, b\}) = \{b\}$. Obviously, this cannot be represented by a preferential structure and (μPR) is false for U and $\{a, b\}$. But it satisfies $(\mu \subseteq)$, (μCUM), $(\mu RatM)$. $(\mu \subseteq)$ is trivial. (μCUM) : Let $f(X) \subseteq Y \subseteq X$. If $f(X) = X$, we are done. Consider $f(\{a, b\}) = \{b\}$. If $\{b\} \subseteq Y \subseteq \{a, b\}$, then $f(Y) = \{b\}$, so we are done again. It is shown in Fact 12.10, (8) that $(\mu \subseteq\supseteq)$ follows. $(\mu RatM)$: Suppose $X \subseteq Y$, $X \cap f(Y) \neq \emptyset$, we have to show $f(X) \subseteq f(Y) \cap X$. If $f(Y) = Y$, the result holds by $X \subseteq Y$, so it does if $X = Y$. The only remaining case is $Y = \{a, b\}$, $X = \{b\}$, and the result holds again.

The idea to solve the representation problem illustrated by Example 12.32 is to use the points c and d as bases for counterarguments against $\alpha : b \to a$ - as is possible in IBRS. We do this now. We will obtain a representation for logics weaker than P by generalized preferential structures.

We will now prove a representation theorem, but will make it more general than for preferential structures only. For this purpose, we will introduce some definitions first.

Definition 12.33. *Let $\eta, \rho : \mathcal{Y} \to \mathcal{P}(U)$.*

(1) If $\mathcal{X}$ is a simple structure

$\mathcal{X}$ is called an attacking structure relative to η representing ρ iff

$\rho(X) = \{x \in \eta(X)$: there is no valid $X - to - \eta(X)$ arrow $\alpha : x' \to x\}$

for all $X \in \mathcal{Y}$.

(2) If $\mathcal{X}$ is a structure with copies

$\mathcal{X}$ is called an attacking structure relative to η representing ρ iff

$\rho(X) = \{x \in \eta(X)$: there is $\langle x, i\rangle$ and no valid $X - to - \eta(X)$ arrow $\alpha : \langle x', i'\rangle \to \langle x, i\rangle\}$

for all $X \in \mathcal{Y}$.

Obviously, in those cases $\rho(X) \subseteq \eta(X)$ *for all* $X \in \mathcal{Y}$.
Thus, $\mathcal{X}$ *is a preferential structure iff* η *is the identity.*
See Figure 12.9 .

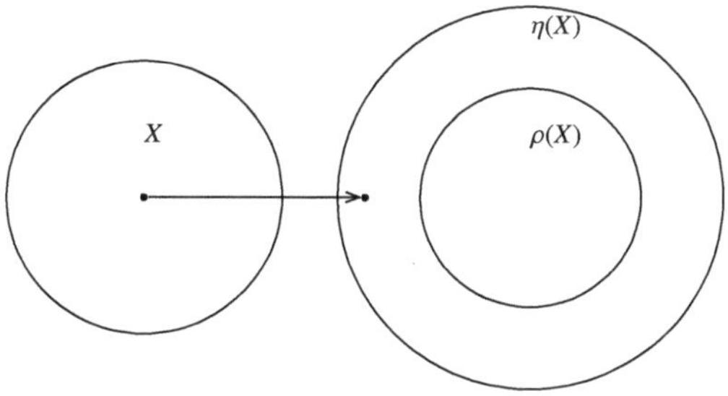

Fig. 12.9: Attacking structure

(Note that it does not seem very useful to generalize the notion of smoothness from preferential structures to general attacking structures, as, in the general case, the minimizing set X and the result $\rho(X)$ may be disjoint.)

The following result is the first positive representation result here, and shows that we can obtain (almost) anything with level 2 structures.

Proposition 12.34. *Let* $\eta, \rho : \mathcal{Y} \to \mathcal{P}(U)$. *Then there is an attacking level 2 structure relative to* η *representing* ρ *iff*
(1) $\rho(X) \subseteq \eta(X)$ *for all* $X \in \mathcal{Y}$,
(2) $\rho(\emptyset) = \eta(\emptyset)$ *if* $\emptyset \in \mathcal{Y}$.

(2) is, of course, void for preferential structures.

Proof. (A) The construction
We perform a two stage construction.
(A.1) Stage 1.
In stage 1, consider (almost as usual)
$\mathcal{U} := \langle \mathcal{X}, \{\alpha_i : i \in I\} \rangle$ where
$\mathcal{X} := \{\langle x, f \rangle : x \in U, f \in \Pi\{X \in \mathcal{Y} : x \in \eta(X) - \rho(X)\}\}$,
$\alpha : x' \to \langle x, f \rangle :\Leftrightarrow x' \in ran(f)$. Attention: $x' \in X$, not $x' \in \rho(X)$!
(A.2) Stage 2.
Let $\mathcal{X}'$ be the set of all $\langle x, f, X \rangle$ s.t. $\langle x, f \rangle \in \mathcal{X}$ and
(a) either X is some dummy value, say $*$
or
(b) all of the following (1) – (4) hold:

(1) $X \in \mathcal{Y}$,
(2) $x \in \rho(X)$,
(3) there is $X' \subseteq X$, $x \in \eta(X') - \rho(X')$, $X' \in \mathcal{Y}$, (thus $ran(f) \cap X \neq \emptyset$ by definition),
(4) $\forall X'' \in \mathcal{Y}.(X \subseteq X'', x \in \eta(X'') - \rho(X'') \Rightarrow (ran(f) \cap X'') - X \neq \emptyset)$.

(Thus, f chooses in (4) X'' also outside X. If there is no such X'', (4) is void, and only (1) – (3) need to hold, i.e. we may take any f with $\langle x, f \rangle \in \mathcal{X}$.)

See Figure 12.10.

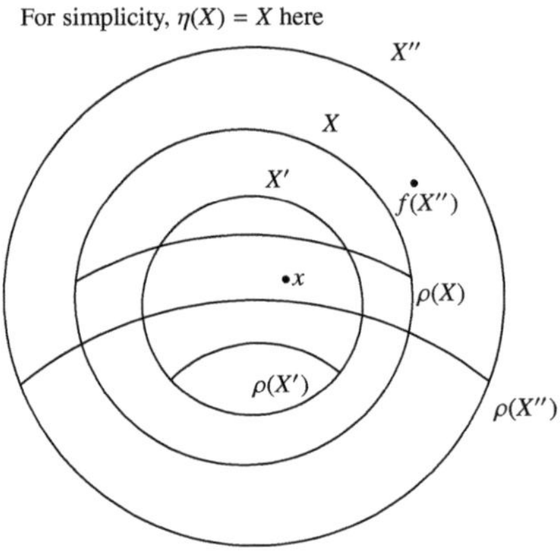

Fig. 12.10: The complicated case

Note: If (1) – (3) are satisfied for x and X, then we will find f s.t. $\langle x, f \rangle \in \mathcal{X}$, and $\langle x, f, X \rangle$ satisfies (1) – (4) : As $X \subset X''$ for X'' as in (4), we find f which chooses such X'' outside of X.

So for any $\langle x, f \rangle \in \mathcal{X}$, there is $\langle x, f, * \rangle$, and maybe also some $\langle x, f, X \rangle$ in $\mathcal{X}'$.

Again for any x', let $\langle x, f, X \rangle \in \mathcal{X}'$

$\alpha : x' \rightarrow \langle x, f, X \rangle :\Leftrightarrow x' \in ran(f)$

(A.3) Adding arrows.

Consider x' and $\langle x, f, X \rangle$.

If $X = *$, or $x' \notin X$, we do nothing, i.e. leave a simple arrow $\alpha : x' \rightarrow \langle x, f, X \rangle \Leftrightarrow x' \in ran(f)$.

If $X \in \mathcal{Y}$, and $x' \in X$, and $x' \in ran(f)$, we make card(X) many copies of the attacking arrow and have then: $\langle \alpha, x'' \rangle : x' \rightarrow \langle x, f, X \rangle$ for all $x'' \in X$.

In addition, we add attacks on the $\langle \alpha, x'' \rangle : \langle \beta, x'' \rangle : x'' \rightarrow \langle \alpha, x'' \rangle$ for all $x'' \in X$.

The full structure $\mathcal{Z}$ is thus:

$\mathcal{X}'$ is the set of elements.

If $x' \in ran(f)$, and $X = *$ or $x' \notin X$ then $\alpha : x' \rightarrow \langle x, f, X \rangle$.

If $x' \in ran(f)$, and $X \neq *$ and $x' \in X$ then

(a) $\langle \alpha, x'' \rangle : x' \rightarrow \langle x, f, X \rangle$ for all $x'' \in X$,

(b) $\langle\beta, x''\rangle : x'' \to \langle\alpha, x''\rangle$ for all $x'' \in X$.
See Figure 12.11.

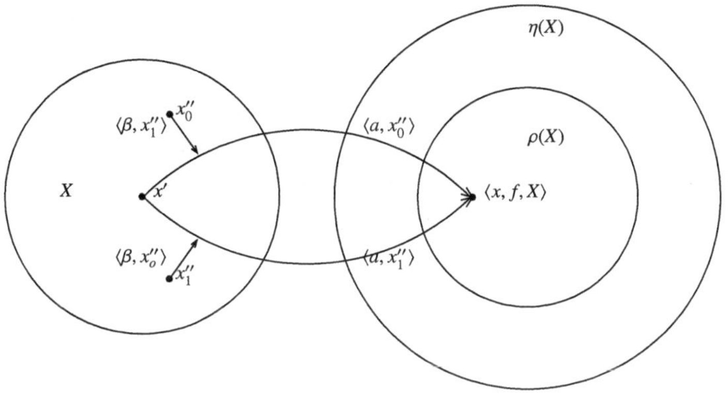

Fig. 12.11: Attacking structure

(B) Representation

We have to show that this structure represents ρ relative to η.

Let $y \in \eta(Y)$, $Y \in \mathcal{Y}$.

Case 1. $y \in \rho(Y)$.

We have to show that there is $\langle y, g, Y''\rangle$ s.t. there is no valid $\alpha : y' \to \langle y, g, Y''\rangle$, $y' \in Y$. In Case 1.1 below, Y'' will be $*$, in Case 1.2, Y'' will be Y, g will be chosen suitably.

Case 1.1. There is no $Y' \subseteq Y$, $y \in \eta(Y') - \rho(Y')$, $Y' \in \mathcal{Y}$.

So for all Y' with $y \in \eta(Y') - \rho(Y')$ $Y' - Y \neq \emptyset$. Let $g \in \Pi\{Y' - Y : y \in \eta(Y') - \rho(Y')\}$. Then $ran(g) \cap Y = \emptyset$, and $\langle y, g, *\rangle$ is not attacked from Y. ($\langle y, g\rangle$ was already not attacked in $\mathcal{X}$.)

Case 1.2. There is $Y' \subseteq Y$, $y \in \eta(Y') - \rho(Y')$, $Y' \in \mathcal{Y}$.

Now let $\langle y, g, Y\rangle \in \mathcal{X}'$, s.t. $g(Y'') \notin Y$ if $Y \subseteq Y''$, $y \in \eta(Y'') - \rho(Y'')$, $Y'' \in \mathcal{Y}$. As noted above, such g and thus $\langle y, g, Y\rangle$ exist. Fix $\langle y, g, Y\rangle$.

Consider any $y' \in ran(g)$. If $y' \notin Y$, y' does not attack $\langle y, g, Y\rangle$ in Y. Suppose $y' \in Y$. We had made card(Y) many copies $\langle\alpha, y''\rangle$, $y'' \in Y$ with $\langle\alpha, y''\rangle : y' \to \langle y, g, Y\rangle$ and had added the level 2 arrows $\langle\beta, y''\rangle : y'' \to \langle\alpha, y''\rangle$ for $y'' \in Y$. So all copies $\langle\alpha, y''\rangle$ are destroyed in Y. This was done for all $y' \in Y$, $y' \in ran(g)$, so $\langle y, g, Y\rangle$ is now not (validly) attacked in Y any more.

Case 2. $y \in \eta(Y) - \rho(Y)$.

Let $\langle y, g, Y' \rangle$ (where Y' can be $*$) be any copy of y, we have to show that there is $z \in Y$, $\alpha : z \to \langle y, g, Y' \rangle$, or some $\langle \alpha, z' \rangle : z \to \langle y, g, Y' \rangle$, $z' \in Y'$, which is not destroyed by some level 2 arrow $\langle \beta, z' \rangle : z' \to \langle \alpha, z' \rangle$, $z' \in Y$.

As $y \in \eta(Y) - \rho(Y)$, $ran(g) \cap Y \neq \emptyset$, so there is $z \in ran(g) \cap Y$. Fix such z. (We will modify the choice of z only in Case 2.2.2 below.)

Case 2.1. $Y' = *$.

As $z \in ran(g)$, $\alpha : z \to \langle y, g, * \rangle$. (There were no level 2 arrows introduced for this copy.)

Case 2.2. $Y' \neq *$.

So $\langle y, g, Y' \rangle$ satisfies the conditions (1) – (4) of (b) at the beginning of the proof.

If $z \notin Y'$, we are done, as $\alpha : z \to \langle y, g, Y' \rangle$, and there were no level 2 arrows introduced in this case. If $z \in Y'$, we had made card(Y') many copies $\langle \alpha, z' \rangle$, $\langle \alpha, z' \rangle : z \to \langle y, g, Y' \rangle$, one for each $z' \in Y'$. Each $\langle \alpha, z' \rangle$ was destroyed by $\langle \beta, z' \rangle : z' \to \langle \alpha, z' \rangle$, $z' \in Y'$.

Case 2.2.1. $Y' \not\subseteq Y$.

Let $z'' \in Y'$-Y, then $\langle \alpha, z'' \rangle : z \to \langle y, g, Y' \rangle$ is destroyed only by $\langle \beta, z'' \rangle : z'' \to \langle \alpha, z'' \rangle$ in Y', but not in Y, as $z'' \notin Y$, so $\langle y, g, Y' \rangle$ is attacked by $\langle \alpha, z'' \rangle : z \to \langle y, g, Y' \rangle$, valid in Y.

Case 2.2.2. $Y' \subset Y$ ($Y = Y'$ is impossible, as $y \in \rho(Y')$, $y \notin \rho(Y)$).

Then there was by definition (condition (b) (4)) some $z' \in (ran(g) \cap Y) - Y'$ and $\alpha : z' \to \langle y, g, Y' \rangle$ is valid, as $z' \notin Y'$. (In this case, there are no copies of α and no level 2 arrows.)

Corollary 12.35. *(1) We cannot distinguish general structures of level 2 from those of higher levels by their ρ-functions relative to η.*

(2) Let U be the universe, $\mathcal{Y} \subseteq \mathcal{P}(U)$, $\mu : \mathcal{Y} \to \mathcal{P}(U)$. Then any μ satisfying $(\mu \subseteq)$ can be represented by a level 2 preferential structure. (Choose η = identity.)

Again, we cannot distinguish general structures of level 2 from those of higher levels by their μ-functions.

A remark on the function η :

We can also obtain the function η via arrows. Of course, then we need positive arrows (not only negative arrows against negative arrows, as we first need to have something positive).

If η is the identity, we can make a positive arrow from each point to itself. Otherwise, we can connect every point to every point by a positive arrow, and then choose those we really want in η by a choice function obtained from arrows just as we obtained ρ from arrows.

Results on total smoothness

Fact 12.36 *Let $X, Y \in \mathcal{Y}$, $\mathcal{X}$ a level n structure. Let $\langle \alpha, k \rangle : \langle x, i \rangle \to \langle y, j \rangle$, where $\langle y, j \rangle$ may itself be (a copy of) an arrow.*

(1) Let $n > 1$, $X \subseteq Y$, $\langle \alpha, k \rangle \in X$ a level $n-1$ arrow in $\mathcal{X} \upharpoonright X$. If $\langle \alpha, k \rangle$ is valid in $\mathcal{X} \upharpoonright Y$, then it is valid in $\mathcal{X} \upharpoonright X$.

(2) Let $\mathcal{X}$ be totally smooth, $\mu(X) \subseteq Y$, $\mu(Y) \subseteq X$, $\langle \alpha, k \rangle \in \mathcal{X} \upharpoonright X \cap Y$, then $\langle \alpha, k \rangle$ is valid in $\mathcal{X} \upharpoonright X$ iff it is valid in $\mathcal{X} \upharpoonright Y$.

Note that we will also sometimes write X for $\mathcal{X} \upharpoonright X$, when the context is clear.

Proof. (1) If $\langle \alpha, k \rangle$ is not valid in $\mathcal{X} \upharpoonright X$, then there must be a level n arrow $\langle \beta, r \rangle : \langle z, s \rangle \to \langle \alpha, k \rangle$ in $\mathcal{X} \upharpoonright X \subseteq \mathcal{X} \upharpoonright Y$. $\langle \beta, r \rangle$ must be valid in $\mathcal{X} \upharpoonright X$ and $\mathcal{X} \upharpoonright Y$, as there are no level $n + 1$ arrows. So $\langle \alpha, k \rangle$ is not valid in $\mathcal{X} \upharpoonright Y$, *contradiction*.

(2) By downward induction. Case $n : \langle \alpha, k \rangle \in \mathcal{X} \upharpoonright X \cap Y$, so it is valid in both as there are no level $n + 1$ arrows. Case $m \to m - 1$: Let $\langle \alpha, k \rangle \in \mathcal{X} \upharpoonright X \cap Y$ be a level $m - 1$ arrow valid in $\mathcal{X} \upharpoonright X$, but not in $\mathcal{X} \upharpoonright Y$. So there must be a level m arrow $\langle \beta, r \rangle : \langle z, s \rangle \to \langle \alpha, k \rangle$ valid in $\mathcal{X} \upharpoonright Y$. By total smoothness, we may assume $z \in \mu(Y) \subseteq X$, so $\langle \beta, r \rangle \in \mathcal{X} \upharpoonright X$ is valid by induction hypothesis. So $\langle \alpha, k \rangle$ is not valid in $\mathcal{X} \upharpoonright X$, *contradiction*.

Corollary 12.37. *Let $X, Y \in \mathcal{Y}$, $\mathcal{X}$ a totally smooth level n structure, $\mu(X) \subseteq Y$, $\mu(Y) \subseteq X$. Then $\mu(X) = \mu(Y)$.*

Proof. Let $x \in \mu(X) - \mu(Y)$. Then by $\mu(X) \subseteq Y$, $x \in Y$, so there must be for all $\langle x, i \rangle \in \mathcal{X}$ an arrow $\langle \alpha, k \rangle : \langle y, j \rangle \to \langle x, i \rangle$ valid in $\mathcal{X} \upharpoonright Y$, without loss of generality $y \in \mu(Y) \subseteq X$ by total smoothness. So by Fact 12.36, (2), $\langle \alpha, k \rangle$ is valid in $\mathcal{X} \upharpoonright X$. This holds for all $\langle x, i \rangle$, so $x \notin \mu(X)$, *contradiction*.

Fact 12.38 *There are situations satisfying $(\mu \subseteq) + (\mu CUM) + (\cap)$ which cannot be represented by level 2 totally smooth preferential structures.*

The proof is given in the following example.

Example 12.39. Let $Y := \{x, y, y'\}$, $X := \{x, y\}$, $X' := \{x, y'\}$. Let $\mathcal{Y} := \mathcal{P}(Y)$. Let $\mu(Y) := \{y, y'\}$, $\mu(X) := \mu(X') := \{x\}$, and $\mu(Z) := Z$ for all other sets.

Obviously, this satisfies $(\cap)$, $(\mu \subseteq)$, and (μCUM).

Suppose $\mathcal{X}$ is a totally smooth level 2 structure representing μ.

So $\mu(X) = \mu(X') \subseteq Y - \mu(Y)$, $\mu(Y) \subseteq X \cup X'$. Let $\langle x, i \rangle$ be minimal in $\mathcal{X} \upharpoonright X$. As $\langle x, i \rangle$ cannot be minimal in $\mathcal{X} \upharpoonright Y$, there must be $\alpha : \langle z, j \rangle \to \langle x, i \rangle$, valid in $\mathcal{X} \upharpoonright Y$.

Case 1: $z \in X'$.

So $\alpha \in \mathcal{X} \upharpoonright X'$. If α is valid in $\mathcal{X} \upharpoonright X'$, there must be $\alpha' : \langle x', i' \rangle \to \langle x, i \rangle$, $x' \in \mu(X')$, valid in $\mathcal{X} \upharpoonright X'$, and thus in $\mathcal{X} \upharpoonright X$, by $\mu(X) = \mu(X')$ and Fact 12.36 (2). This is impossible, so there must be $\beta : \langle x', i' \rangle \to \alpha$, $x' \in \mu(X')$, valid in $\mathcal{X} \upharpoonright X'$. As β is in $\mathcal{X} \upharpoonright Y$ and $\mathcal{X}$ is a level ≤ 2 structure, β is valid in $\mathcal{X} \upharpoonright Y$, so α is not valid in $\mathcal{X} \upharpoonright Y$, *contradiction*.

Case 2: $z \in X$.

α cannot be valid in $\mathcal{X} \upharpoonright X$, so there must be $\beta : \langle x', i' \rangle \to \alpha$, $x' \in \mu(X)$, valid in $\mathcal{X} \upharpoonright X$. Again, as β is in $\mathcal{X} \upharpoonright Y$ and $\mathcal{X}$ is a level ≤ 2 structure, β is valid in $\mathcal{X} \upharpoonright Y$, so α is not valid in $\mathcal{X} \upharpoonright Y$, *contradiction*.

It is unknown to the authors whether an analogue is true for essential smoothness, i.e. whether there are examples of such a function μ which need at least level 3 essentially smooth structures for representation. Proposition 12.45 below shows that such structures suffice, but we do not know whether level 3 is necessary.

Fact 12.40 *The above Example 12.39 can be solved by a totally smooth level 3 structure:*

Let $\alpha_1 : x \to y, \alpha_2 : x \to y', \alpha_3 : y \to x, \beta_1 : y \to \alpha_2, \beta_2 : y' \to \alpha_1, \beta_3 : y \to \alpha_3, \beta_4 : x \to \alpha_3, \gamma_1 : y' \to \beta_3, \gamma_2 : y' \to \beta_4.$

See Figure 12.12.

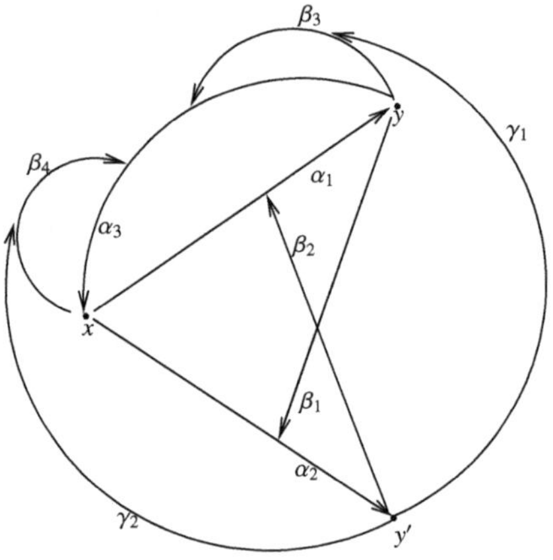

Fig. 12.12: Solution by smooth level 3 structure

The subdiagram generated by X contains $\alpha_1, \alpha_3, \beta_3$, and β_4. $\alpha_1, \beta_3, \beta_4$ are valid, so $\mu(X) = \{x\}$.

The subdiagram generated by X' contains α_2. α_2 is valid, so $\mu(X') = \{x\}$.

In the full diagram, $\alpha_3, \beta_1, \beta_2, \gamma_1$, and γ_2 are valid, so $\mu(Y) = \{y, y'\}$.

Remark 12.41. Example 12.42 together with Corollary 12.37 show that $(\mu \subseteq)$ and (μCUM) without $(\cap)$ do not guarantee representability by a level n totally smooth structure.

Example 12.42. We show here $(\mu \subseteq) + (\mu CUM) \not\Rightarrow (\mu \subseteq\supseteq)$.

Consider $X := \{a, b, c\}$, $Y := \{a, b, d\}$, $f(X) := \{a\}$, $f(Y) := \{a, b\}$, $\mathcal{Y} := \{X, Y\}$. (If $f(\{a, b\})$ were defined, we would have $f(X) = f(\{a, b\}) = f(Y)$, *contradiction*.)

Obviously, $(\mu \subseteq)$ and (μCUM) hold, but not $(\mu \subseteq\supseteq)$.

Results on essential smoothness

Definition 12.43. *Let* $\mu : \mathcal{Y} \to \mathcal{P}(U)$ *and* $\mathcal{X}$ *be given, let* $\alpha : \langle y, j\rangle \to \langle x, i\rangle \in \mathcal{X}$.

Define

$\boldsymbol{O}(\alpha) := \{Y \in \mathcal{Y} : x \in Y - \mu(Y), y \in \mu(Y)\}$,

$\boldsymbol{D}(\alpha) := \{X \in \mathcal{Y} : x \in \mu(X), y \in X\}$,

$\Pi(\boldsymbol{O}, \alpha) := \Pi\{\mu(Y) : Y \in \boldsymbol{O}(\alpha)\}$,

$\Pi(\boldsymbol{D}, \alpha) := \Pi\{\mu(X) : X \in \boldsymbol{D}(\alpha)\}$.

The following lemma is probably the main technical result of the chapter.

Lemma 12.44. *Let U be the universe, $\mu : \mathcal{Y} \to \mathcal{P}(U)$. Let μ satisfy $(\mu \subseteq) + (\mu \subseteq\supseteq)$.*

Let $\mathcal{X}$ be a level 1 preferential structure, $\alpha : \langle y, j\rangle \to \langle x, i\rangle$, $\boldsymbol{O}(\alpha) \neq \emptyset$, $\boldsymbol{D}(\alpha) \neq \emptyset$.

We can modify $\mathcal{X}$ to a level 3 structure $\mathcal{X}'$ by introducing level 2 and level 3 arrows s.t. no copy of α is valid in any $X \in \boldsymbol{D}(\alpha)$, and in every $Y \in \boldsymbol{O}(\alpha)$ at least one copy of α is valid. (More precisely, we should write $\mathcal{X}' \restriction X$, etc.)

Thus, in $\mathcal{X}'$,

(1) $\langle x, i\rangle$ will not be minimal in any $Y \in \boldsymbol{O}(\alpha)$,

(2) if α is the only arrow minimizing $\langle x, i\rangle$ in $X \in \boldsymbol{D}(\alpha)$, $\langle x, i\rangle$ will now be minimal in X.

The construction is made independently for all such arrows $\alpha \in \mathcal{X}$.

Proof. (1) The construction

Make $\Pi(\boldsymbol{D}, \alpha)$ many copies of α : $\{\langle\alpha, f\rangle : f \in \Pi(\boldsymbol{D}, \alpha)\}$, all $\langle\alpha, f\rangle : \langle y, j\rangle \to \langle x, i\rangle$. Note that $\langle\alpha, f\rangle \in X$ for all $X \in \boldsymbol{D}(\alpha)$ and $\langle\alpha, f\rangle \in Y$ for all $Y \in \boldsymbol{O}(\alpha)$.

Add to the structure $\langle\beta, f, X_r, g\rangle : \langle f(X_r), i_r\rangle \to \langle\alpha, f\rangle$, for any $X_r \in \boldsymbol{D}(\alpha)$, and $g \in \Pi(\boldsymbol{O}, \alpha)$ (and some or all i_r - this does not matter).

For all $Y_s \in \boldsymbol{O}(\alpha)$:

if $\mu(Y_s) \not\subseteq X_r$ and $f(X_r) \in Y_s$, then add to the structure $\langle\gamma, f, X_r, g, Y_s\rangle$: $\langle g(Y_s), j_s\rangle \to \langle\beta, f, X_r, g\rangle$ (again for all or some j_s),

if $\mu(Y_s) \subseteq X_r$ or $f(X_r) \notin Y_s$, $\langle\gamma, f, X_r, g, Y_s\rangle$ is not added.

See Figure 12.13.

(2) Let $X_r \in \boldsymbol{D}(\alpha)$. We have to show that no $\langle\alpha, f\rangle$ is valid in X_r. Fix f.

$\langle\alpha, f\rangle$ is in X_r, so we have to show that for at least one $g \in \Pi(\boldsymbol{O}, \alpha)$ $\langle\beta, f, X_r, g\rangle$ is valid in X_r, i.e. that for this g, no $\langle\gamma, f, X_r, g, Y_s\rangle$: $\langle g(Y_s), j_s\rangle \to \langle\beta, f, X_r, g\rangle$, $Y_s \in \boldsymbol{O}(\alpha)$ attacks $\langle\beta, f, X_r, g\rangle$ in X_r.

We define g. Take $Y_s \in \boldsymbol{O}(\alpha)$.

Case 1: $\mu(Y_s) \subseteq X_r$ or $f(X_r) \notin Y_s$: choose arbitrary $g(Y_s) \in \mu(Y_s)$.

Case 2: $\mu(Y_s) \not\subseteq X_r$ and $f(X_r) \in Y_s$: Choose $g(Y_s) \in \mu(Y_s) - X_r$.

In Case 1, $\langle\gamma, f, X_r, g, Y_s\rangle$ does not exist, so it cannot attack $\langle\beta, f, X_r, g\rangle$.

In Case 2, $\langle\gamma, f, X_r, g, Y_s\rangle : \langle g(Y_s), j_s\rangle \to \langle\beta, f, X_r, g\rangle$ is not in X_r, as $g(Y_s) \notin X_r$.

Thus, no $\langle\gamma, f, X_r, g, Y_s\rangle : \langle g(Y_s), j_s\rangle \to \langle\beta, f, X_r, g\rangle$, $Y_s \in \boldsymbol{O}(\alpha)$ attacks $\langle\beta, f, X_r, g\rangle$ in X_r.

So $\forall\langle\alpha, f\rangle : \langle y, j\rangle \to \langle x, i\rangle$

$$
\begin{aligned}
y \in X_r \Rightarrow \exists\langle\beta, f, X_r, g\rangle &: \langle f(X_r), i_r\rangle \to \langle\alpha, f\rangle \\
&(f(X_r) \in \mu(X_r) \wedge \neg\exists\langle\gamma, f, X_r, g, Y_s\rangle : \\
&\quad \langle g(Y_s), j_s\rangle \to \langle\beta, f, X_r, g\rangle.g(Y_s) \in X_r)).
\end{aligned}
$$

But $\langle\beta, f, X_r, g\rangle$ was constructed only for $\langle\alpha, f\rangle$, so was $\langle\gamma, f, X_r, g, Y_s\rangle$, and there was no other $\langle\gamma, i\rangle$ attacking $\langle\beta, f, X_r, g\rangle$, so we are done.

(3) Let $Y_s \in \boldsymbol{O}(\alpha)$. We have to show that at least one $\langle\alpha, f\rangle$ is valid in Y_s.

We define $f \in \Pi(\boldsymbol{D}, \alpha)$. Take X_r.

If $\mu(X_r) \not\subseteq Y_s$, choose $f(X_r) \in \mu(X_r) - Y_s$. If $\mu(X_r) \subseteq Y_s$, choose arbitrary $f(X_r) \in \mu(X_r)$.

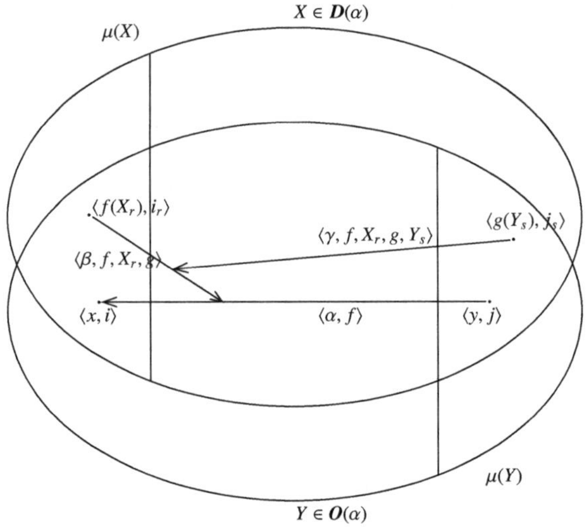

Fig. 12.13: The construction

All attacks on $\langle x, f\rangle$ have the form $\langle \beta, f, X_r, g\rangle : \langle f(X_r), i_r\rangle \to \langle \alpha, f\rangle$, $X_r \in \boldsymbol{D}(\alpha)$, $g \in \Pi(\boldsymbol{O}, \alpha)$. We have to show that they are either not in Y_s, or that they are themselves attacked in Y_s.

Case 1: $\mu(X_r) \not\subseteq Y_s$. Then $f(X_r) \notin Y_s$, so $\langle \beta, f, X_r, g\rangle : \langle f(X_r), i_r\rangle \to \langle \alpha, f\rangle$ is not in Y_s (for no g).

Case 2: $\mu(X_r) \subseteq Y_s$. Then $\mu(Y_s) \not\subseteq X_r$ by ($\mu \subseteq\supseteq$) and $f(X_r) \in Y_s$, so $\langle \beta, f, X_r, g\rangle : \langle f(X_r), i_r\rangle \to \langle \alpha, f\rangle$ is in Y_s (for all g). Take any $g \in \Pi(\boldsymbol{O}, \alpha)$. As $\mu(Y_s) \not\subseteq X_r$ and $f(X_r) \in Y_s$, $\langle \gamma, f, X_r, g, Y_s\rangle : \langle g(Y_s), j_s\rangle \to \langle \beta, f, X_r, g\rangle$ is defined, and $g(Y_s) \in \mu(Y_s)$, so it is in Y_s (for all g). Thus, $\langle \beta, f, X_r, g\rangle$ is attacked in Y_s.

Thus, for this f, all $\langle \beta, f, X_r, g\rangle$ are either not in Y_s, or attacked in Y_s, thus for this f, $\langle \alpha, f\rangle$ is valid in Y_s.

So for this $\langle x, i\rangle$

$\exists \langle \alpha, f\rangle : \langle y, j\rangle \to \langle x, i\rangle . y \in \mu(Y_s) \wedge$

(a) $\neg\exists \langle \beta, f, X_r, g\rangle : \langle f(X_r), i\rangle \to \langle \alpha, f\rangle . f(X_r) \in Y_s$

or

(b) $\forall \langle \beta, f, X_r, g\rangle : \langle f(X_r), i\rangle \to \langle \alpha, f\rangle$

$(f(X_r) \in Y_s \Rightarrow$

$\exists \langle \gamma, f, X_r, g, Y_s\rangle : \langle g(Y_s), j_s\rangle \to \langle \beta, f, X_r, g\rangle . g(Y_s) \in \mu(Y_s))$.

As we made copies of α only, introduced only βs attacking the α-copies, and γs attacking the βs, the construction is independent for different αs.

Proposition 12.45. *Let U be the universe, $\mu : \mathcal{Y} \to \mathcal{P}(U)$.*

Then any μ satisfying $(\mu \subseteq)$, $(\cap)$, (μCUM) (or, alternatively, $(\mu \subseteq)$ and $(\mu \subseteq\supseteq)$) can be represented by a level 3 essentially smooth structure.

Proof. In stage 1, consider as usual $\mathcal{U} := \langle \mathcal{X}, \{\alpha_i : i \in I\}\rangle$ where $\mathcal{X} := \{\langle x, f\rangle : x \in U, f \in \Pi\{\mu(X) : X \in \mathcal{Y}, x \in X - \mu(X)\}\}$, and set $\alpha : \langle x', f'\rangle \to \langle x, f\rangle :\Leftrightarrow x' \in ran(f)$.

For stage 2:

Any level 1 arrow $\alpha : \langle y, j\rangle \to \langle x, i\rangle$ was introduced in stage 1 by some $Y \in \mathcal{Y}$ s.t. $y \in \mu(Y)$, $x \in Y - \mu(Y)$. Do the construction of Lemma 12.44 for all level 1 arrows of $\mathcal{X}$ in parallel or successively.

We have to show that the resulting structure represents μ and is essentially smooth. (Level 3 is obvious.)

(1) Representation

Suppose $x \in Y - \mu(Y)$. Then there was in stage 1 for all $\langle x, i\rangle$ some $\alpha : \langle y, j\rangle \to \langle x, i\rangle$, $y \in \mu(Y)$. We examine the y.

If there is no X s.t. $x \in \mu(X)$, $y \in X$, then there were no βs and γs introduced for this $\alpha : \langle y, j\rangle \to \langle x, i\rangle$, so α is valid.

If there is X s.t. $x \in \mu(X)$, $y \in X$, consider $\alpha : \langle y, j\rangle \to \langle x, i\rangle$. So $X \subset \boldsymbol{D}(\alpha)$, $Y \in \boldsymbol{O}(\alpha)$, so we did the construction of Lemma 12.44, and by its result, $\langle x, i\rangle$ is not minimal in Y.

Thus, in both cases, $\langle x, i\rangle$ is successfully attacked in Y, and no $\langle x, i\rangle$ is a minimal element in Y.

Suppose $x \in \mu(X)$ (we change notation to conform to Lemma 12.44. Fix $\langle x, i\rangle$.

If there is no $\alpha : \langle y, j\rangle \to \langle x, i\rangle$, $y \in X$, then $\langle x, i\rangle$ is minimal in X, and we are done.

If there is α or $\langle \alpha, k\rangle : \langle y, j\rangle \to \langle x, i\rangle$, $y \in X$, then α originated from stage 1 through some Y s.t. $x \in Y - \mu(Y)$, and $y \in \mu(Y)$. (Note that stage 2 of the construction did not introduce any new level 1 arrows - only copies of existing level 1 arrows.) So $X \in \boldsymbol{D}(\alpha)$, $Y \in \boldsymbol{O}(\alpha)$, so we did the construction of Lemma 12.44, and by its result, $\langle x, i\rangle$ is minimal in X, and we are done again.

In both cases, all $\langle x, i\rangle$ are minimal elements in X.

(2) Essential smoothness. We have to show the conditions of Definition 12.24. We will, however, work with the reformulation given in Remark 12.25.

Case (1), $x \in \mu(X)$.

Case (1.1), there is $\langle x, i\rangle$ with no $\langle \alpha, f\rangle : \langle y, j\rangle \to \langle x, i\rangle$, $y \in X$. There is nothing to show.

Case (1.2), for all $\langle x, i\rangle$ there is $\langle \alpha, f\rangle : \langle y, j\rangle \to \langle x, i\rangle$, $y \in X$.

α was introduced in stage 1 by some Y s.t. $x \in Y - \mu(Y)$, $y \in X \cap \mu(Y)$, so $X \in \boldsymbol{D}(\alpha)$, $Y \in \boldsymbol{O}(\alpha)$. In the proof of Lemma 12.44, at the end of (2), it was shown that

$$\begin{aligned}
&\exists\langle \beta, f, X_r, g\rangle : \langle f(X_r), i_r\rangle \to \langle \alpha, f\rangle \\
&\quad (f(X_r) \in \mu(X_r) \wedge \\
&\quad \neg\exists\langle \gamma, f, X_r, g, Y_s\rangle : \langle g(Y_s), j_s\rangle \to \langle \beta, f, X_r, g\rangle . g(Y_s) \in X_r).
\end{aligned}$$

By $f(X_r) \in \mu(X_r)$, condition (1) of Remark 12.25 is true.

Case (2), $x \notin \mu(Y)$. Fix $\langle x, i\rangle$. (We change notation back to Y.)

In stage 1, we constructed $\alpha : \langle y, j\rangle \rightarrow \langle x, i\rangle$, $y \in \mu(Y)$, so $Y \in \boldsymbol{O}(\alpha)$.

If $\boldsymbol{D}(\alpha) = \emptyset$, then there is no attack on α, and the condition (2) of Remark 12.25 is trivially true.

If $\boldsymbol{D}(\alpha) \neq \emptyset$, we did the construction of Lemma 12.44, so

$\exists\langle\alpha, f\rangle : \langle y, j\rangle \rightarrow \langle x, i\rangle.y \in \mu(Y_s) \wedge$

(a) $\neg\exists\langle\beta, f, X_r, g\rangle : \langle f(X_r), i\rangle \rightarrow \langle\alpha, f\rangle.f(X_r) \in Y_s$

or

(b) $\forall\langle\beta, f, X_r, g\rangle : \langle f(X_r), i\rangle \rightarrow \langle\alpha, f\rangle$

$(f(X_r) \in Y_s \Rightarrow$

$\exists\langle\gamma, f, X_r, g, Y_s\rangle : \langle g(Y_s), j_s\rangle \rightarrow \langle\beta, f, X_r, g\rangle.g(Y_s) \in \mu(Y_s))$.

As the only attacks on $\langle\alpha, f\rangle$ had the form $\langle\beta, f, X_r, g\rangle$, and $g(Y_s) \in \mu(Y_s)$, condition (2) of Remark 12.25 is satisfied.

As we said after Example 12.39, we do not know if level 3 is necessary for representation. We also do not know whether the same can be achieved with level 3, or higher, totally smooth structures.

Translation to logics

We turn to the translation to logics.

Proposition 12.46. *Let $\mid\!\sim$ be a logic for $\mathcal{L}$. Set $T^{\mathcal{M}} := Th(\mu_{\mathcal{M}}(M(T)))$, where $\mathcal{M}$ is a generalized preferential structure, and $\mu_{\mathcal{M}}$ its choice function. Then*

(1) there is a level 2 preferential structure $\mathcal{M}$ s.t. $\overline{\overline{T}} = T^{\mathcal{M}}$ iff (LLE), (CCL), (SC) hold for all $T, T' \subseteq \mathcal{L}$.

(2) there is a level 3 essentially smooth preferential structure $\mathcal{M}$ s.t. $\overline{\overline{T}} = T^{\mathcal{M}}$ iff (LLE), (CCL), (SC), ($\subseteq\supseteq$) hold for all $T, T' \subseteq \mathcal{L}$.

Proof. The proof is an immediate consequence of Corollary 12.35 (2), Fact 12.26, and Propositions 12.45 and 12.11 (10) and (11).

(More precisely, for (2): Let $\mathcal{M}$ be an essentially smooth structure, then by Definition 12.30 for all X $\mu(X) \sqsubseteq X$. Consider (μCUM). So by Fact 12.26 (2) $\mu(X') \subseteq X'' \subseteq X' \Rightarrow \mu(X') \sqsubseteq X''$, so by Fact 12.26 (1) $\mu(X') = \mu(X'')$. ($\mu \subseteq\supseteq$) is analogous, using Fact 12.26 (3).

We leave aside the generalization of preferential structures to attacking structures relative to η, as this can cause problems, without giving real insight: It might well be that $\rho(X) \not\subseteq \eta(X)$, but, still, $\rho(X)$ and $\eta(X)$ might define the same theory - due to definability problems.

References

1. M. Abraham, D. Gabbay and U. Schild. Contrary to time conditionals in Talmudic logic. In *Artificial Intelligence and Law*, 20(2), 145–179, 2012. DOI: 10.1007.s10506-012-9123-xOnlineFirst.
2. M. Abraham, I. Belfer, D. Gabbay, and U. Schild. Delegation, count as and security in Talmudic logic, a preliminary study. In *Logic without Frontiers: Festschrift for Walter Alexandre Carnielli on the occasion of his 60th Birthday*. Jean-Yves Béziau and Marcelo Esteban Coniglio, eds., pp. 73–96. Volume 17 of Tribute Series, College Publications. London, 2011.
3. S. Abramsky. Semantics of interaction: an introduction to game semantics. In *Proceedings of the 1996 CLiCS Summer School, Isaac Newton Institute*, P. Dybjer and A. Pitts, eds, pp. 1–31. Cambridge University Press, 1997.
4. A. V. Aho. Algorithms for finding patterns in strings. In *Handbook of Theoretical Computer Science, Volume A: Algorithms and Complexity (A)*, J. van Leeuwen, ed., pages 255–300. MIT Press, Cambridge, 1990.
5. A.V. Aho. Indexed Grammars — An Extension of Context-Free Grammars. *Journal of the ACM*, 15(4): 647–671, 1968)
6. T. Ågotnes, W. Van Der Hoek and M. Wooldridge. Quantified coalition logic. In *Proceedings of IJCAI'07, Proceedings of the 20th International Joint Conference on Artificial Intelligence*, R. Sangal, H. Mehta and R. K. Bagga, eds., Morgan Kaufmann, San Francisco.
7. L. Aqvist. A New Approach to the Logical Theory of Interrogatives, Part I, Analysis, Philosophical Society and Department of Philosophy, University of Uppsala, sect. 6.2, 1969.
8. C. Areces. *Logic Engineering. The Case of Description and Hybrid Logics*. PhD thesis, Institute for Logic, Language and Computation, University of Amsterdam, Amsterdam, The Netherlands, October 2000.
9. C. Areces, P. Blackburn, and S. R. Delany. Bringing them all together. *Journal of Logic and Computation*, 11:657–669, 2001.
10. C. Areces, P. Blackburn, and M. Marx. A road-map on complexity for hybrid logics. In *Proceedings of the 13th International Workshop and 8th Annual Conference of the EACSL on Computer Science Logic, CSL '99*, J. Flum and M. Rodriguez-Artalejo, eds., pages 307–321. Springer-Verlag, 1999.

D.M. Gabbay, *Reactive Kripke Semantics*, Cognitive Technologies,
DOI 10.1007/978-3-642-41389-6,

11. C. Areces, F. Carreiro, S. Figueira, and S. Mera. Basic model theory for memory logics. In *Proceedings of the 18th Intl. Workshop, WoLLIC 2011*, L. D. Beklemishev and R. de Queiroz, eds., pages 20–34. Springer LNAI 6642, Berlin, 2011.
12. M. Abraham, D. Gabbay, I. Belfer and U. Schild. Future determination of entities in Talmudic Logic. *Journal of Applied logic*, 11(1), 63–90, 2013.
13. N. Asher and M. Morreau. Commonsense entailment: a modal theory of nonmonotonic reasoning, in: *Proc. JELIA'90*, J. van Eijck, ed., Lecture Notes in Artificial Intelligence 478, Springer, Berlin 1990.
14. H. Barringer, D. M. Gabbay, J. Woods, Temporal dynamics of support and attack networks: From argumentation to zoology, *Mechanizing Mathematical Reasoning, Volume dedicated to Joerg Siekmann*, D. Hutter and W. Stephan eds., Springer Lecture Notes in Computer Science 2605, pp 59–98. Springer, Berlin, 2005.
15. H. Barringer, D Gabbay and J. Woods. Temporal, Numerical and Metalevel Dynamics in Argumentation Networks. *Argumentation and Computation*, 3(2-3), 143–202, 2012.
16. H. Barringer and D. M. Gabbay. Modal and temporal argumentation networks. In *Time for Verification. Essays in Memory of Amir Pnueli*, D. Peled and Z. Manna, eds., pages 1–25. LNCS 6200, Springer, Berlin 2010.
17. H. Barringer, D. M. Gabbay, and D. Rydeheard. Reactive grammars. In *Language, Culture, Computation. Essays Dedicated to Yaacov Choueka*, N. Dershowitz and E. Nissan, eds., Springer-Verlag, in preparation.
18. H. Barringer, M. Fisher, D. Gabbay, R. Owens and M. Reynolds. *The Imperative Future: Principles of Executable Temporal Logic*. Research Studies Press. 1996.
19. H. Barringer, A. Goldberg, K. Havelund and K. Sen. Rule-Based Runtime Verification. Proceedings of the *VMCAI'04, 5th International Conference on Verification, Model Checking and Abstract Interpretation*, B. Steffen and G. Levi, eds. Volume 2937, Lecture Notes in Computer Science, Springer-Verlag, 2004.
20. H. Barringer, K. Havelund and D. Rydeheard. Rule systems for run-time monitoring: from EAGLE to RULER (extended version). *Journal of Logic and Computation*, Oxford University Press, 20(3): 675 – 706, 2010.
21. M.-P. Béal, M. Crochemore, F. Mignosi, A. Restivo, and M. Sciortino. Computing forbidden words of regular languages. *Fundamenta Informaticae*, 56(1,2):121–135, 2003.
22. E. W. Beth. Semantic Entailment and Formal Derivability. *Mededelingen der Koninklijke Nederlandse Akademie van Wetenschappen*, 18:3, 309–342, 1955.
23. P. Blackburn. Representation, reasoning, and relational structures: a hybrid logic manifesto. *Logic Journal of the IGPL*, 8(3):339–365, 2000.
24. P. Blackburn, J. F. A. K. v. Benthem, and F. Wolter. *Handbook of Modal Logic, Volume 3 (Studies in Logic and Practical Reasoning)*. Elsevier Science, New York, 2006.
25. P. Blackburn, M. de Rijke, and Y. Venema. *Modal Logic*. Cambridge University Press, 2001.
26. P. Blackburn and B. ten Cate. Pure extensions, proof rules, and Hybrid axiomatics. *Studia Logica*, 84(2):277–322, 2006.
27. P. Blackburn and M. Tzakova. Hybrid languages and temporal logic. *Logic Journal of the IGPL*, 7(1):27–54, 1999.
28. M. de Boer, D. M. Gabbay, X. Parent and M. Slavkovic. Two-dimensional standard deontic logic. *Synthese*, 1–38, 2011. Published online: DOI: 10.1007/s11229-010-9866-4.
29. S. Borgo. Modal operators with adaptable semantics for multi-agent systems. In *AI*IA'05*, S. Manzani, ed., LNAI 3673, pp. 186–197, Springer, Berlin, 2005.

30. S. Borgo. Coalitions in action logic. In *IJCAI'07. Proceedings of the 20th International Joint Conference on Artificial Inteligence*, R. Sangal, M. Mehta and R. K. Bagga, eds., Morgan Kaufmann, San Francisco, 2007.
31. J. Broersen, A. Herzig and N. Troquard. From coalition logic to STIT. *Electronic Notes in Theoretical Computer Science*, 157, 23–35, 2006.
32. J. Broersen, D. M. Gabbay and L. van der Torre. Norms as automata and norm change as reactive automata. In preparation.
33. G. Bossu, P. Siegel. Saturation, nonmonotonic reasoning and the closed-world assumption,*Artificial Intelligence* 25, 13–63, 1985.
34. R. A. Bull. An approach to tense logic. *Theoria*, 36(3):282–300, 1970.
35. J. Carmo and A. J. I. Jones. Deontic logic and contrary-to-duties. In *Handbook of Philosophical Logic, Vol 8*. D. M. Gabbay and F. Guenthner, eds., pp. 265-344, Springer, 2002.
36. B.F. Chellas. *Modal Logic - An Introduction*. Cambridge University Press, Cambridge, 1980.
37. M. Crochemore and D. M. Gabbay. Reactive Automata. *Information and Computation*, 209(4), 692–704. Published online: DOI: 10.1016/j.ic.2011.01.002
38. R. M. Chisholm. Contrary-to-duty imperatives and deontic logic. *Analysis*, 24, 33–36, 1969.
39. N. Chomsky. On Certain Formal Properties of Grammars. *Information and Control*, 2: 137–167, 1959.
40. J. Dassow and G. Paun. *Regulated Rewriting in Formal Language Theory*. EATCS Monographs on Theoretical Computer Science 18. Springer-Verlag, Berlin, 1989.
41. M. D'Agostino. Tableau Methods for Classical Propositional Logic. In *Handbook of Tableau Methods*, M. D'Agostino, D. M. Gabbay, R. Hähnle and J. Posegga, eds., pp. 45–123. Kluwer, Dordrecht, 1999.
42. M. D'Agostino, D. Gabbay, R. Hähnle and J. Posegga. *Handbook of Tableau Methods*, Kluwer Academic Publishers, Dordrecht, 1999.
43. J.P. Delgrande. An approach to default reasoning based on a first-order conditional logic, Revised Report, *Artificial Intelligence*, 36, 1988.
44. M. d'Inverno and M. Luck. *Understanding Agent Systems*, Springer, Berlin, 2001.
45. P. M. Dung. On the acceptability of arguments and its fundamental role in nonmonotonic reasoning, logic programming and *n*-person games, *Artificial Intelligence*, 77, 321–357, 1995.
46. P. E. Dunne, W. van der Hoek, S. Kraus and M. Wooldridge. Cooperative Boolean games. In *Proceedings of AAMAS 2008*, pp. 1015–1022.
47. H. Fernau. Regulated grammars with leftmost derivations. *Proceedings of SOFSEM'98*, B. Rovan, ed., Lecture Notes in Computer Science, Volume 1521, 322–331, Springer-Verlag, Berlin, 1998.
48. M. Fitting. *Proof Methods for Modal and Intuitionistic Logics*. Reidel, Dordrecht 1983.
49. J.W. Forrester. *Gentle Murder, or the Adverbial Samaritan*, The Journal of Philosophy, Vol. 81, no. 4, 1984, pp.193–197.
50. D. M. Gabbay. *Investigations in Modal and Tense Logics*, Reidel, Dordrecht, 1976.
51. D. M. Gabbay. Theoretical foundations for non-monotonic reasoning in expert systems. In *Logics and Models of Concurrent Systems*, K. R. Apt, ed., , pp. 439–457. Springer, Berlin, 1985.
52. D. M. Gabbay. Temporal logic and quantifier translations. In *How We Say WHEN it Happens*, H. Kamp and U. Reyle, eds., pp. 231–279. Max Niemeyer Verlag, Tuebingen, 2002.

53. D. M. Gabbay, I.Hodkinson and M. Reynolds. *Temporal Logic*, Volume 1, Oxford University Press, 1994.
54. D. M. Gabbay, M. Reynolds and M. Finger. *Temporal Logic*, volume 2, Oxford University Press, 2000.
55. D. M. Gabbay. *Labelled Deductive Systems*, Volume 1. Oxford University Press, 1996.
56. D. M. Gabbay. An irreflexivity lemma with applications to axiomatizations of conditions on tense frames. In *Aspects of Philosophical Logic*, U. Monnich, ed., pp. 67–89. D. Reidel, Dordrecht, 1981.
57. D. M. Gabbay. *Fibring Logics*, Oxford University Press, 1998.
58. D. M. Gabbay. More on reactive Kripke semantics and proof theory. In preparation.
59. D. M. Gabbay, C. J. Hogger and J. A. Robinson, eds. *Handbook of Logic in Artificial Intelligence and Logic Programming. Volume 3: Nonmonotonic Reasoning and Uncertain Reasoning*. Oxford University Press, 1994.
60. D. M. Gabbay, O. Rodrigues and J. Woods. Belief Contraction, Anti-formulas and Resource Overdraft: Part I. *Logic Journal of the IGPL*, 10, 601–652, 2002.
61. D. M. Gabbay, O. Rodrigues and J. Woods. Belief Contraction, Anti-formulas and Resource Overdraft: Part II. In *Logic, Epistemology and the Unity of Science*, D. M. Gabbay, S. Rahman, J. Symons and J.-P. van Bendegem, eds, pp. 291–326, Kluwer, Dordrecht, 2004.
62. D. Gabbay and S. Marcelino. Modal logics of reactive fames. *Studia Logica*, 93, 403–444, 2009.
63. D. M. Gabbay and V. Shehtman. Flow products of modal logics. Incorporated in [70].
64. D. M. Gabbay. Reactive Kripke Models and Contrary-to-duty Obligations. Part A, semantics. *Journal of Applied Logic*, 11(1), 103-136, 2013. Part B, proof theory submitted to *Journal of Applied Logic*.
65. D. M. Gabbay. Reactive Kripke Models and Contrary-to-duty Obligations. In *DEON - 2008: Deontic Logic in computer science*, Ron van der Meyden and Leendert van der Torre, eds. LNAI 5076, pp 155-173 , Springer, Berlin, 2008
66. D. M. Gabbay Reactive Kripke Semantics and Arc Accessibility In *Pillars of Computer Science: Essays Dedicated to Boris (Boaz) Trakhtenbrot on the Occasion of His 85th Birthday*, A. Avron, N. Dershowitz, and A. Rabinovich, eds., Lecture Notes in Computer Science, vol. 4800, Springer-Verlag, Berlin, 2008 pp 292-341.
67. D. Gabbay. Reactive Kripke semantics. In *Proceedings of CompLog 2004*, W. Carnielli, ed., pp. 7–20. Centre of Logic and Computation, University of Lisbon, 2004.
68. D. M. Gabbay. Reactive Intuitionistic Tableaux. *Synthese*, Special issue in honour of E. W. Beth, edited by J. van Benthem, T. Kuipers, and H. Visser, 179(2), 253–269, 2011.
69. D. M. Gabbay. A Theory of Hypermodal Logics: Mode Shifting in Modal Logic. *Journal of Philosophical Logic*, 31(3):211–243, June 2002.
70. D. M. Gabbay, A. Kurucz, F. Wolter, and M. Zakharyaschev. *Many-Dimensional Modal Logics: Theory and Applications*, volume 148 of *Studies in Logic and the Foundations of Mathematics*. Elsevier, 2003, 742pp.
71. D. M. Gabbay and K. Schlechta. Defeasible inheritance systems and reactive diagrams. *Logic Journal of the IGPL*, 17(1):1–54, 2009.
72. D. M. Gabbay and K. Schlechta. Reactive Preferential Structures and Nonmonotonic Consequence. *Review of Symbolic Logic*, 2(2):414–450, 2009.
73. D. M. Gabbay and K. Schlechta. *Logical Tools for Handling Change in Agent-Based Systems*, Cognitive Technologies, pages 251–293. Springer, Berlin, 2010.
74. D. Gabbay, K. Schlechta. Size and logic, *Review of Symbolic Logic*, 2(2), 396–413, 2009.
75. D. M. Gabbay and K. Schlechta. A theory of hierarchical consequence and conditionals. *J. of Logic, Lang. and Inf.*, 19:3–32, January 2010.

76. D. M. Gabbay and A. S. d'Avila Garcez. Logical modes of attack in argumentation networks. *Studia Logica*, 93(2–3) 199–230, 2009.
77. D. M. Gabbay. Reactive Kripke models and contrary-to-duty obligations. In *Deontic Logic in Computer Science*, R. van der Meyden and L. van der Torre, eds., volume 5076 of *LNAI*, pages 155–173. Springer, Berlin, 2008.
78. D. M. Gabbay and M. d'Agostino. Reactive conditionals. Draft.
79. D. M. Gabbay and C. Strasser. Reactive standard deontic logic. To appear in *Journal of Logic and Computation*, Corner on deontic logic. doi: 10.1093/logcom/exs043
80. D. Gabbay. Bipolar argumentation frames and contrary to duty obligations, preliminary report . In *Proceedings of CLIMA 2012*, M. Fisher, L. van der Torre, M. Dastani and G. Governatori, eds. pp. 1–24. LNAI 7486, Springer, Berlin, 2012.
81. D. Gabbay. Temporal deontic logic for the generalised Chisholm set of contrary to duty obligations. In *DEON 2012*, T. Agotnes, J. Broersen, and D. Elgesem, eds., LNAI 7393, pp. 91–107. Springer, Heidelberg, 2012.
82. D. Gabbay. Semantics for higher level attacks in extended argumentation frames Part 1 : Overview. *Studia Logica*, 93: 355–379, 2009.
83. D. Gabbay. Fibring argumentation frames. *Studia Logica*, 93(2-3): 231–295, 2009.
84. D. Gabbay, I. Shapirovsky, and V. Shehtman. Flow products. Draft, 2012.
85. D. Gabbay, K. Schlechta. Roadmap for preferential logics. *Journal of Applied Nonclassical Logic*, 19(1), 43–95, 2009, see also hal-00311941, arXiv 0808.3073.
86. D. M. Gabbay and K. Schlechta. Cumulativity Without Closure of the Domain under Finite Unions. *The Review of Symbolic Logic* , **1**, 3:372–392, 2008.
87. D. Gabbay. Overview on the connection between reactive Kripke models and argumentation networks. *Annals of Mathematics and Artificial Intelligence*, Special Issue: New Developments in Reactive Semantics. 66:1–4, 1–5, 2012.
88. D. Gabbay. Introducing reactive Kripke semantics and arc accessibility. *Annals of Mathematics and Artificial Intelligence*, Special Issue: New Developments in Reactive Semantics, 66:1–4, 7–53, 2012.
89. D. Gabbay. Introducing reactive modal tableaux. *Annals of Mathematics and Artificial Intelligence*, Special Issue: New Developments in Reactive Semantics, 66:1–4, 55–79, 2012.
90. D. Gabbay. Completeness theorems for reactive modal logics: *Annals of Mathematics and Artificial Intelligence*, Special Issue: New Developments in Reactive Semantics, 66:1–4, 81–129, 2012.
91. D. Gabbay and S. Marcelino. Global view on reactivity: switch graphs and their logics. *Annals of Mathematics and Artificial Intelligence*, Special Issue: New Developments in Reactive Semantics, 66:1–4, 131–162, 2012.
92. D. Gabbay, J. Horty, R. van der Meyden, X. Parent, and L. van der Torre, eds. *Handbook of Deontic Logic and Normative Systems*, College Publications, 2013.
93. D. Gabbay. Selective Filtration in Modal logics I, *Theoria* 36, 323-330, 1970.
94. D. Gabbay and K. Schlechta. Critical analysis of the Carmo–Jones model of contrary-to-duty obligations. Draft paper 358, 2009.`http://arxiv.org/PS_cache/arxiv/pdf/1002/1002.3021v1.pdf`
95. D. M. Gabbay. *Modal and Tense Logics*. D. Reidel, Dordrecht, 1976.
96. D. M. Gabbay. *Deontic Logic, Reactivity and Contrary to Duty Obligations*, Springer, to appear.
97. G. Gargov and V. Goranko. Modal logic with names. *Journal of Philosophical Logic*, 22:607–636, 1993. 10.1007/BF01054038.
98. S. Ginsburg and E. H. Spanier. Control Sets on Grammars. *Mathematical Systems Theory*, 2(2): 159–177, 1968.

99. R. Goldblatt. Logics of Time and Computation. Center for the Study of Language and Information, Stanford, 1987.
100. V. Goranko. Completeness and Incompleteness in the Bimodal Base L(R,-R). In *Mathematical Logic*, P. Petkov, ed., pp. 311-327. Plenum Press, New York, 1990
101. R. Gore. Tableau Methods for Modal and Temporal Logics. In *Handbook of Tableau Methods*, M. D'Agostino, D. M. Gabbay, R. Hähnle and J. Posegga, eds., pp. 297–369. Kluwer, Dordrecht, 1999.
102. G. Governatori. Ideality and subideality from a computational point of view, *European Journal of Law, Philosophy and Computer Science, Practical Reason, History of Deontics, Computer Law*, 1–2: 315–329, 1998.
103. S. O. Hansson. A note on the Deontic System DL of Jones and Pörn. *Synthese*, 80 , 427 –428, 1989.
104. D. Harel and A. Pnueli. On the development of reactive systems. In *Logics and Models of Cocurrent Systems*, K. R. Apt, ed., pp. 477–498. Springer-Verlag, New York, 1985.
105. W. Hoek and M. Wooldridge. On the logic of cooperation and propositional control. *Artificial Intelligence*, 164(1-2): 81–119, May 2005.
106. J. E. Hopcroft, R. Motwani, and J. D. Ullman. *Introduction to Automata Theory, Languages, and Computation.* Addison-Wesley, third edition, 2006.
107. J. F. Horty. Some direct theories of nonmonotonic inheritance. In [59], pp. 111–188.
108. M. Huth and M. Ryan. *Logic in Computer Science: Modelling and Reasoning about Systems.* Cambridge University Press, 1999.
109. O. H. Ibarra. Simple matrix languages. *Information and Control*, 17: 359–394, 1970.
110. A. I. J. Jones and I. Pörn. A rejoinder to Hansson. *Synthese*, 80, 429-432, 1989.
111. A. J. I. Jones. On the Logic of Deontic Conditionals, *Ratio Juris*. 4:3, 355–66, 1991.
112. A. J. I. Jones. Towards a formal theory of defeasible deontic conditionals, *Annals of Mathematics and Artificial Intelligence*, 9, 151-166, 1993.
113. A. J. I. Jones, Intentions and the logic of norms. In *MEDLAR (Mechanising Deduction in the Logics of Practical Reasoning)*, J.R. Cunningham, ed., Imperial College, London, 1990.
114. A. J. I. Jones and I. Pörn. On the logic of deontic conditionals, in: *DEON91 - Proc. 1st Int. Workshop on Deontic Logic in Computer Science*, ed. J.-J. Ch. Meyer and R.J. Wieringa, Amsterdam, 1991.
115. A. J. I. Jones and I. Pörn. Ideality, sub-ideality and deontic logic, *Synthese*, 65, 1985.
116. A.J. I. Jones and I. Pörn. "Ought" and "Must", *Synthese*, 66, 1986.
117. A. J. I. Jones and M. Sergot. Formal specification of security requirements using the theory of normative positions. In *Computer Security ESORICS 92*, Y. Deswarte, G. Eizenberg and J. J. Quisquater, eds., LNCS vol 648, pp. 103–121. Springer, Berlin, 1992.
118. T. Kasai. An Hierarchy between Context-Free and Context-Sensitive Languages. *Journal of Computer and System Sciences*, 4: 492–508, 1970.
119. S. Kraus, D. Lehmann, M. Magidor. Nonmonotonic reasoning, preferential models and cumulative logics, *Artificial Intelligence*, 44 (1–2), pp. 167–207, July 1990.
120. A. Kurz, Sequence Frames, Proc. Verif. in New Orientation, Univ. Maribor (1995) http://www.cs.le.ac.uk/people/akurz/LMU/maribor.ps.gz
121. V. Lifschitz. Circumscription. In *Handbook of Logic in Artificial Intelligence and Logic Programming. Volume 3: Nonmonotonic Reasoning and Uncertain Reasoning*, C. J. Hogger D. M. Gabbay and J. A. Robinson, eds. Oxford University Press, 1994.
122. A. Lindenmayer. Mathematical models for cellular interaction in development, I and II. *Journal of Theoretical Biology*, 18: 280–315, 1968.
123. B. Loewer and M Belzer, Dyadic deontic detachment, *Synthese*, 54, 1983.

124. M. Lothaire, editor. *Appplied Combinatorics on Words*. Cambridge University Press, 2005.
125. L. L. Maksimova. Craig's Theorem in superintuitionistic logic and amalgamable varieties of pseudo-Boolean algebras. *Algebra and Logic*, 16, 427–455, 1977.
126. S. Marcelino. *Modal Logic for Changing Systems*. PhD thesis, King's College London, 2011.
127. D. Makinson. Stenius' approach to disjunctive permission. *Theoria*, 50(2-3): 138-147, 1984.
128. R. Meersman and G. Rozenberg. Cooperating grammar systems. In *Mathematical Foundations of Computer Science 1978*, J. WInkowski, ed., LNCS 64, Springer, Berlin, 1978.
129. E. Moriya. Some remarks on State Grammars and Matrix Grammars. *Information and Control*, 23: 48–57, 1973.
130. S. Modgil. Reasoning about preferences in argumentation frameworks. *Artif. Intell.*, 173:901–934, June 2009.
131. S. H. Nielsen and S. Parsons. A generalization of Dung's abstract framework for argumentation: arguing with sets of attacking arguments. In *Proceedings of the Third International Workshop on Argumentation in Multi-agent Systems*, N. Maudet, S. Parsons nd I. Rahwan, eds., pp. 7–19, LNAI 4049, Spirnger, Berlin, 2007.
132. H. J. Ohlbach. Semantics-based translation methods for modal logic. *Journal of Logic and Computation*, 1(5), 1991.
133. M. Pauly. *Logic for Social Software*. PhD Thesis, University of Amsterdam, 2001.
134. G. Paun. *The Generative Mechanisms of Some Economic Processes*. Techn. Publ. House. Bucharest, 1980.
135. H. Prakken and M. Sergot. Dyadic deontic logic and contrary-to-duty obligations. In *Defeasible Deontic Logic*, D. Nute, ed. pp. 223–262. Synthese Library, Vol 263, Kluwer, Dordrecht, 1997.
136. H. Prakken and M. Sergot. Contrary-to-duty obligations, *Studia Logica*, 91–115, 1996.
137. A. N. Prior. *Past, Present and Future*. Oxford University Press, 1967.
138. A. N. Prior. Modality and quantification in S5. *The Journal of Symbolic Logic*, 21(1), 60–62, 1956.
139. A.N. Prior. Escapism: The Logical Basis of Ethics. In *Essays in Moral Philosophy*, A.I. Melden, ed., pp. 135–146. University of Washington Press, , 1958.
140. H. Reichenbach. *Elements of Symbolic Logic*, Macmillan, New York, 1947. Republished, Free Press, New York, 1966.
141. P. Rohde. Moving in a crumbling network: the balanced case. In *CSL 2004*, J. Marcinkowski and A. Tarlecki, eds., LNCS 3210, pp. 1–25. Springer, Berlin, 2004.
142. D.J. Rosenkrantz. Programmed grammars and classes of formal languages. *Journal of the ACM*, 16(1): 107–131, 1969.
143. A. Ross: 1941, Imperatives and Logic. *Theoria* 7, 53-71, 1941.
144. H. Sahlqvist. Completeness and Correspondence in the first and second order semantics for modal logic. In *Proceedings of the Third Scandinavian Logic Symposium*, S. Kanger, ed. pp. 110–143. North-Holland, 1975.
145. U. Sattler and M. Vardi. The hybrid μ-calculus. In *Proceedings of IJCAR'01*, R. Goré, A. Leitsch and T. Nipkow, eds., LNAI 2083, Springer, Berlin, 2001.
146. L. Sauro, J. Gerbrandy, W. van der Hoek and M. Wooldridge. Reasoning about action and cooperation. In *AAMAS 2006*, H. Nakashima, M. P. Wellmann, G. Weiss and P. Stone, eds., pp, 185–192. ACM, New York, 2006.
147. K. Schlechta. *Coherent Systems*, Elsevier, Amsterdam, 2004.

148. Yoav Shoham. A semantical approach to nonmonotonic logics. in *Proc. Logics in Computer Science*, J. P. McDermott, ed., pp. 275–279. IEEE Computer Society, Ithaca, NY, 1987; and *Proceedings IJCAI 87*, pp. 388–392. Morgan Kaufmann, San Francisco, 1987.
149. E. Stenius. Ross' paradox and well-formed codices. *Theoria*, 48(2): 49-77, 1982.
150. C. Strasser. A deontic logic framework allowing for factual detachment. *Journal of Applied Logic*, 9(1), 61–80, 2010.
151. L. van der Torre and Y.-H. Tan. Contrary-to-duty reasoning with preference-based dyadic obligations, *Annals of Mathematics and Artificial Intelligence*, 27, 49–78, 1999.
152. D. Ulrich. The shortest possible length of the longest implicational axiom. *Journal of Philosophical Logic*, 25: 101–108, 1996.
153. J. van Benthem, J. van Eijck and B. Kook. Logics of Communciation and Change. *Information and Computation*, 204(11), 1620–1662, 2006.
154. J. van Benthem. An essay on sabotage and obstruction. In *Mechanizing Mathematical Reasoning*, D. Hutter and W. Stephan, eds., pp. 268–176, LNCS 2605, Springer, Berlin, 2005.
155. J. van Benthem. Dynamic logic for belief revision. *Journal of Applied Non-classical Logics*, 17:129–155, 2007.
156. H. van Ditmarsch, W. van der Hoek and B. Kooi. *Dynamic Epistemic Logic*, Springer, Berlin, 2007.
157. F. Veltman. Defaults in update semantics, *ITLI prepublication Series LP-91-02*, Amsterdam, 1991.
158. M. Wooldridge and N. R. Jennings. Agent theories, architectures and languages: a survey. In *Proc. ECAI-94*, M. J. Wooldridge and N. R. Jennings, eds., pp. 1–39, LNCS vol 890, Springer, Berlin, 1995.
159. A. Zanardo. Branching-Time Logic with Quantification over Branches: The Point of View of Modal Logic. *The Journal of Symbolic Logic*, 61(1):pp. 1–39, 1996.

Index

D.M. Gabbay, *Reactive Kripke Semantics*, Cognitive Technologies,
DOI 10.1007/978-3-642-41389-6, © Springer-Verlag Berlin Heidelberg 2013

MIX
Papier aus verantwortungsvollen Quellen
Paper from responsible sources
FSC® C105338

If you have any concerns about our products,
you can contact us on
ProductSafety@springernature.com

In case Publisher is established outside the EU,
the EU authorized representative is:
Springer Nature Customer Service Center GmbH
Europaplatz 3, 69115 Heidelberg, Germany

Printed by Libri Plureos GmbH
in Hamburg, Germany